FUNDAMENTALS OF ELECTRONIC COMMUNICATIONS SYSTEMS

Wayne Tomasi
Mesa Community College

 PRENTICE HALL, Englewood Cliffs, New Jersey 07632

Library of Congress Cataloging-in-Publication Data

TOMASI, WAYNE.
  Fundamentals of electronic communications systems.

  Includes index.
    1. Telecommunication.   2. Information theory.   I. Title.
TK5101.T63 1988   621.38′0413   87–7228
ISBN   0–13–336579–4

Editorial/production supervision and
interior design: **Kathryn Pavelec**
Cover design: **Diane Saxe**
Cover photo: Courtesy of **GTE**
Manufacturing buyer: **Margaret Rizzi/Lorraine Fumoso**

**To my son, Aaron
and his lovely wife, Margo**

© 1988 by **Prentice-Hall, Inc.**
A Division of Simon & Schuster
Englewood Cliffs, New Jersey 07632

Printed in the United States of America

10   9   8   7   6   5   4   3   2   1

ISBN   0-13-336579-4

PRENTICE-HALL INTERNATIONAL (UK) LIMITED, *London*
PRENTICE-HALL OF AUSTRALIA PTY. LIMITED, *Sydney*
PRENTICE-HALL CANADA INC., *Toronto*
PRENTICE-HALL HISPANOAMERICANA, S.A., *Mexico*
PRENTICE-HALL OF INDIA PRIVATE LIMITED, *New Delhi*
PRENTICE-HALL OF JAPAN, INC., *Tokyo*
SIMON & SCHUSTER ASIA PTE. LTD., *Singapore*
EDITORA PRENTICE-HALL DO BRASIL, LTDA., *Rio de Janeiro*

CONTENTS

PREFACE   vii

Chapter 1   **INTRODUCTION TO ELECTRONIC COMMUNICATIONS**   1

Introduction   1
Signal Analysis   5
Electrical Noise   18
Multiplexing   29
Questions   32
Problems   33

Chapter 2   **FREQUENCY GENERATION**   35

Introduction   35
Oscillators   35
Frequency Multipliers   70
Mixing   70
Questions   78
Problems   78

Chapter 3   **AMPLITUDE MODULATION TRANSMISSION**   81

Introduction   81
Mathematical Analysis of AM Double-Sideband Full Carrier   89
Circuits for Generating AM DSBFC   98
AM Double-Sideband Full-Carrier Transmitters   113
Trapezoidal Patterns   117
Carrier Shift   120
Questions   122
Problems   122

Chapter 4   **AMPLITUDE MODULATION RECEPTION**   126

Introduction   126
AM Receivers   130
AM Receiver Circuits   145
Automatic Gain Control and Squelch   163
Double-Conversion AM Receivers   168
Questions   169
Problems   170

Chapter 5   **PHASE-LOCKED LOOPS AND FREQUENCY SYNTHESIZERS**   172

Introduction   172
Phase-Locked Loop   172
Frequency Synthesizers   187
Large-Scale Integration Programmable Timers   205
Questions   214
Problems   214

Chapter 6   **SINGLE-SIDEBAND COMMUNICATIONS SYSTEMS**   217

Introduction   217
Single-Sideband Systems   217
Circuits for Generating SSB   224
Single-Sideband Transmitters   232

Single-Sideband Receivers   254
Questions   261
Problems   262

Chapter 7   **ANGLE MODULATION
TRANSMISSION**   264

Introduction   264
Angle Modulation   264
Frequency Modulation Transmission   290
Questions   314
Problems   315

Chapter 8   **ANGLE MODULATION RECEIVERS
AND SYSTEMS**   318

Introduction   318
FM Receivers   319
FM Systems   340
Questions   358
Problems   358

Chapter 9   **TRANSMISSION LINES**   360

Introduction   360
Transverse Electromagnetic Waves   360
Types of Transmission Lines   363
Transmission-Line Equivalent Circuit   367
Transmission-Line Wave Propagation   373
Transmission-Line Losses   376
Incident and Reflected Waves   378
Standing Waves   380
Transmission-Line Input Impedance   387
Questions   395
Problems   396

Chapter 10   **WAVE PROPAGATION**   398

Introduction   398
Rays and Wavefronts   398

Electromagnetic Radiation   400
Spherical Wavefront and the Inverse Square Law   401
Wave Attenuation and Absorption   403
Optical Properties of Radio Waves   405
Propagation of Waves   412
Propagation Terms and Definitions   418
Questions   421
Problems   421

Chapter 11   ANTENNAS   423

Introduction   423
Terms and Definitions   424
Basic Antennas   433
Antenna Arrays   442
Special-Purpose Antennas   445
Questions   449
Problems   450

Chapter 12   BASIC TELEVISION PRINCIPLES   452

Introduction   452
History of Television   452
Monochrome Television Transmission   454
The Composite Video Signal   456
Monochrome Television Reception   467
Color Television Transmission and Reception   475
Questions   483
Problems   484

Appendix A   THE SMITH CHART   486

SOLUTIONS TO ODD-NUMBERED PROBLEMS   508

INDEX   514

PREFACE

During the past two decades, the electronic communications industry has undergone some remarkable technological changes, primarily in the form of miniaturization. In late 1959 and in the early 1960s, vacuum tubes were replaced by transistors. More recently, transistors are being replaced with multipurpose large-scale integrated circuits. Also, digital electronics principles are being implemented into electronic communications circuits and systems more and more each year. However, fundamental communications concepts are still an intricate part of understanding modern communications systems.

This book is intended for use as an introductory text in electronic communications fundamentals. It was written so that a reader with previous knowledge in basic electronic principles and an understanding of mathematics through trigonometry will have little trouble grasping the concepts presented. An understanding of basic calculus principles (i.e., differentiation and integration) would be helpful but is not a prerequisite. Within the text, there are numerous examples that emphasize the important concepts, and questions and problems are included at the end of each chapter. Also, answers to the odd-numbered problems are given at the end of the book.

Chapter 1 is an introduction to electronic communications. Fundamental communications terms and concepts such as modulation, demodulation, bandwidth, and information capacity are explained. Signal analysis using both frequency and time domain is discussed. Nonsinusoidal periodic waves are examined using fourier analysis, and the effects of bandlimiting on signals is discussed. Electrical noise, signal-to-noise ratio, and noise figure are explained and discussed. The basic principles of frequency and time division multiplexing are also explained in Chapter 1. Chapter 2 covers frequency generation. The basic requirements for oscillations to occur are outlined and discussed. Standard LC, crystal, and negative resistance oscillator configurations are explained.

The basic concepts of frequency multiplication are discussed. Linear and nonlinear mixing with single and multiple frequency input signals are analyzed. Chapter 3 defines and explains the basic concepts of amplitude modulation transmission. A detailed analysis is presented on the voltage, power, and bandwidth considerations of amplitude modulation in both the frequency and time domain. Circuits for generating amplitude modulation are explained. High and low level transmitters are discussed. Chapter 4 introduces the basic concepts of radio receivers including a detailed analysis of tuned radio frequency and superheterodyne receivers. The primary functions of each receiver stage are explained. Amplitude modulation receivers and the detection of amplitude modulated signals are explained. The basic concepts of automatic gain control and squelch are discussed and double conversion receivers are introduced. Chapter 5 is dedicated to the analysis of phase-locked loops and frequency synthesizers. Basic phase-locked loop concepts are introduced and a detailed explanation of the operation of a phase-locked loop is given. Both direct and indirect frequency synthesizers with single and multiple crystals are discussed. Chapter 6 extends the coverage of amplitude modulation given in Chapters 3 and 4 to single sideband transmission. The various types of sideband transmission are explained and contrasted. Circuits for generating single sideband waveforms are explained. Several types of single sideband receivers are discussed in Chapter 6. Chapter 7 introduces the basic concepts of angle modulation. Frequency and phase modulation are explained and contrasted. The amplitude, power, and frequency characteristic of an angle modulated wave are explained in detail. Both direct and indirect frequency and phase modulation transmitters are shown and discussed in detail. Chapter 8 extends the coverage of angle modulation to receivers. Several types of angle modulation demodulators are introduced and discussed. Frequency modulation stereo transmission, two-way frequency modulation communications, and mobile telephone communications including cellular radio are discussed. Chapter 9 explains the characteristics of an electromagnetic wave and wave propagation on a metallic transmission line. Several basic transmission line configurations are discussed and contrasted. Incident and reflected energy and the concept of standing waves are discussed in Chapter 9. Transmission line characteristic impedance and input impedance are also introduced. The concept of a matched line is explained and the consequences of a mismatched line are discussed. Chapter 10 extends the coverage of wave propagation to free space. Electromagnetic radiation concepts are explained. Spherical wavefronts are analyzed and the inverse square law is derived. Wave attenuation and absorption are covered. Chapter 10 gives a complete explanation of the optical properties of electromagnetic waves: refraction, reflection, diffraction, and interference. Ground, space, and sky wave propagation are discussed, and the fundamental limits for free space wave propagation are defined and explained. Chapter 11 introduces the antenna and describes basic antenna operation. Fundamental antenna terms are defined and explained. Radiation patterns are explained. The most basic antenna, the elementary doublet, is explained. The basic half- and quarter-wave antenna are explained along with the effects of the ground on the wave, and several antenna-loading techniques are described. Some of the more common antenna arrays and special-purpose configurations are explained including the following: folded dipole, log-periodic and loop antennas. Chapter 12 introduces the basic concepts of television broadcasting.

Both monochrome and color television transmission and reception are explained. Generation of the composite video signal is explained. Basic transmitter and receiver circuits are explained. The basic concepts of scanning, blanking, and synchronization are discussed. Appendix A describes the Smith chart and how it is used for transmission line calculations. Examples are given for calculating input impedance, quarter-wave transformer matching, and shorted stub matching.

WAYNE TOMASI

ACKNOWLEDGMENTS

I would like to acknowledge the following individuals for their contributions to this book: Gregory Burnell, Executive Editor and Assistant Vice President, Electronic Technology, for giving me the opportunity to write this book; Kathryn Pavelec, Production Editor, for deciphering my manuscript and providing a pleasant and professional working environment; and the two reviewers of my manuscript who corrected several of my mathematical errors and provided invaluable constructive criticism—Robert E. Greenwood, Ryerson Polytechnical Institute; and James W. Stewart, DeVry, Woodbridge. I would also like to thank my daughter, Belinda, for helping me type the original manuscript.

Chapter 1

INTRODUCTION TO ELECTRONIC COMMUNICATIONS

INTRODUCTION

In essence, *electronic communications* is the transmission, reception, and processing of information with the use of electronic circuits. The basic concepts involved in electronic communications have not changed much since their inception, although the methods by which these concepts are implemented have undergone dramatic changes.

Figure 1-1 shows a communications system in its simplest form, which comprises three primary sections: a *source* (transmitter), a *destination* (receiver), and a *transmission medium* (wire pair, coaxial cable, fiber link, or free space).

The *information* that is propagated through a communications system can be *analog* (proportional), such as the human voice, video picture information, or music; or it can be *digital pulses*, such as binary-coded numbers, alpha/numeric codes, graphic symbols, microprocessor op-codes, or data base information. However, very often the source information is unsuitable for transmission in its original form and therefore must be converted to a more suitable form prior to transmission. For example, with digital communications systems, analog information is converted to digital form prior to transmission, and with analog communications systems, digital data are converted to analog signals prior to transmission. This book is concerned primarily with the basic concepts of conventional analog radio communications.

Modulation and Demodulation

For reasons that are explained in Chapter 10, it is impractical to propagate low-frequency electromagnetic energy through the earth's atmosphere. Therefore, with *radio communi-*

FIGURE 1-1 Block diagram of a communications system in its simplest form.

cations, it is necessary to superimpose a relatively low-frequency intelligence signal onto a relatively high-frequency signal for transmission. In electronic communications systems, the source information (intelligence signal) acts upon or modulates a single-frequency sinusoidal signal. *Modulate* simply means to vary or change. Therefore, the source information is called the *modulating signal*, the signal that is acted upon (modulated) is called the *carrier*, and the resultant signal is called the *modulated wave*. In essence, the source information is transported through the system on the carrier.

With analog communications systems, *modulation* is the process of changing some property of an analog carrier in accordance with the original source information and then transmitting the modulated carrier. Conversely, *demodulation* is the process of converting the changes in the analog carrier back to the original source information. The total or *composite information signal* that modulates the main carrier is called *baseband*. The baseband is converted from its original frequency band to a band more suitable for transmission through the communications system. Baseband signals are *up-converted* at the transmitter and *down-converted* at the receiver. *Frequency translation* is the process of converting a frequency or band of frequencies to another location in the total frequency spectrum.

Equation 1-1 is the general expression for a time varying sine wave of voltage such as an analog carrier. There are three properties of a sine wave that can be varied: the *amplitude* (V), the *frequency* (F), or the *phase* (θ). If the amplitude of the carrier is varied proportional to the source information, *amplitude modulation* (AM) results. If the frequency of the carrier is varied proportional to the source information, *frequency modulation* (FM) results. If the phase of the carrier is varied proportional to the source information, *phase modulation* (PM) results.

$$v = V \sin (2\pi Ft + \theta) \qquad (1-1)$$

where

$\quad v$ = time-varying sine wave of voltage
$\quad V$ = peak amplitude (V)

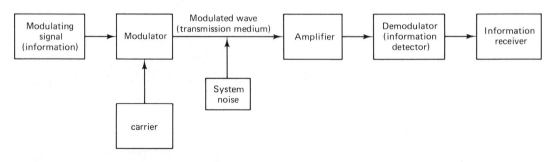

FIGURE 1-2 Communications system block diagram.

F = frequency (Hz)

θ = phase (deg)

Figure 1-2 is a simplified block diagram for a communications system showing the relationships among the modulating signal (information), the modulated signal (carrier), the modulated wave (resultant), and the system noise. Again, the frequency of the carrier is relatively high as compared to the frequency of the intelligence signal.

Transmission Frequencies

In the United States, frequency assignments for *free-space radio propagation* are assigned by the *Federal Communications Commission* (FCC). The exact frequencies assigned specific transmitters operating in the various classes of service are constantly being updated and altered to meet the nation's communications needs. However, the general division of the total usable frequency spectrum is decided at the *International Telecommunications Conventions*, which are held approximately once every 10 years.

The usable *radio-frequency* (RF) spectrum is divided into narrower frequency bands which are given descriptive names and band numbers. *The International Radio Consultative Committee's* (CCIR) designations are listed in Table 1-1. Several of these bands are further broken down into various services, which include shipboard search, microwave, satellite, mobile land-based search, shipboard navigation, aircraft approach, airport surface detection, airborne weather, mobile telephone, and many more.

TV CH. 14-83
470 - 890 MHz

TABLE 1-1   CCIR BAND DESIGNATIONS

Band number	Frequency range[a]	Designation
2	30–300 Hz	ELF (extremely low frequencies)
3	0.3–3 kHz	VF (voice frequencies)
4	3–30 kHz	VLF (very low frequencies)
5	30–300 kHz	LF (low frequencies)
6	0.3–3 MHz	MF (medium frequencies)
7	3–30 MHz	HF (high frequencies)
8	30–300 MHz	VHF (very high frequencies)
9	0.3–3 GHz	UHF (ultra high frequencies)
10	3–30 GHz	SHF (super high frequencies)
11	30–300 GHz	EHF (extremely high frequencies)
12	0.3–3 THz	Infrared light
13	3–30 THz	Unassigned
14	30–300 THz	Visible-light spectrum
15	0.3–3 PHz	Ultraviolet light
16	3–30 PHz	X-ray
17	30–300 PHz	Unassigned
18	0.3–3 EHz	Gamma rays
19	3–30 EHz	Cosmic rays

[a] 10^0, hertz (Hz); 10^3, kilohertz (kHz); 10^6, megahertz (MHz); 10^9, gigahertz (GHz); 10^{12}, terahertz (THz); 10^{15}, petahertz (PHz); 10^{18}, exahertz (EHz).

TABLE 1-2   EMISSION CLASSIFICATIONS

Type of modulation or emission		Type of information		Supplementary characters
A	Amplitude	0	Carrier on only	None	Double sideband, full carrier
F	Frequency	1	Carrier on–off		
P	Pulse	2	Carrier on, keyed tone—on–off	a	Single sideband
		3	Telephony, voice, or music	b	Two independent sidebands
		4	Facsimile, nonmoving or slow-scan TV	c	Vestigal sideband
		5	Vestigal sideband, commercial TV	d	Pulse amplitude modulation (PAM)
		6	Four-frequency diplex telephony	e	Pulse width modulation (PWM)
		7	Multiple sidebands	f	Pulse position modulation
		8	Unassigned	g	Digital video
		9	General, all others	h	Single sideband, full carrier
				i	Single sideband, no carrier

Classification of Transmitters

For licensing purposes in the United States, radio transmitters are classified according to their bandwidth, type of modulation, and type of intelligence information. The emission classification is identified by combinations of letters and numbers as shown in Table 1-2. Emission designations include an uppercase letter which identifies the type of modulation, a number which identifies the type of emission, and a lowercase subscript which further defines the emission characteristics. For example, the designation A3a describes a single-sideband, reduced-carrier, amplitude-modulated signal carrying voice or music information.

Bandwidth and Information Capacity

The two most significant limitations on system performance are *noise* and *bandwidth*. The significance of noise is discussed later in this chapter. The bandwidth of a communications system is the minimum passband required to propagate the source information through the system. The bandwidth of a communications system must be sufficiently large to pass all significant information frequencies.

The *information capacity* of a communications system is a measure of how much source information can be carried through the system in a given period of time. The amount of information that can be propagated through a transmission system is proportional to the product of the system bandwidth and the time of transmission. The relationship among bandwidth, transmission time, and information capacity was developed by R. Hartley of Bell Telephone Laboratories in 1928. Simply stated, Hartley's law is

$$C \propto B \times T \qquad (1\text{-}2)$$

where

C = information capacity
B = bandwidth (Hz)
T = transmission time (s)

Equation 1-2 shows that information capacity is a linear function and is directly proportional to both the system bandwidth and the transmission time. If either the bandwidth or the transmission time changes, the information capacity changes by the same proportion.

Approximately 3 kHz of bandwidth is required to transmit basic voice-quality telephone signals. Approximately 200 kHz of bandwidth is required for FM transmission of high-fidelity music, and almost 6 MHz of bandwidth is required for broadcast-quality television signals (i.e., the more information per unit time, the more bandwidth required).

SIGNAL ANALYSIS

When designing electronic communications' circuits, it is often necessary to analyze and predict the performance of the circuit based on the power distribution and frequency composition of the information signal. This is done with mathematical *signal analysis*. Although all signals in electronic communications are not single-frequency sine or cosine waves, many of them are, and the signals that are not can be represented by a series of sine and cosine functions.

Sinusoidal Signals

In essence, signal analysis is the mathematical analysis of the frequency, bandwidth, and voltage level of a signal. Electrical signals are voltage- or current-time variations that can be represented by a series of sine or cosine waves. Mathematically, a single-frequency voltage or current waveform is

$$v = V \sin (2\pi Ft + \theta) \qquad \text{or} \qquad v = V \cos (2\pi Ft + \theta)$$

$$i = I \sin (2\pi Ft + \theta) \qquad \text{or} \qquad i = I \cos (2\pi Ft + \theta)$$

where

v = time-varying voltage wave
V = peak voltage (V)
i = time-varying current wave
I = peak current (A)
F = frequency (Hz)
θ = phase (deg)

Whether a sine or a cosine function is used to represent a signal is purely arbitrary and depends on which is chosen as the reference (however, it should be noted that sin θ = cos θ − 90°). Therefore, the following relationships hold true:

$$v = V \sin (2\pi Ft + \theta) = V \cos (2\pi Ft + \theta - 90°)$$

$$v = V \cos (2\pi Ft + \theta) = V \sin (2\pi Ft + \theta + 90°)$$

The preceding formulas are for a single-frequency, repetitive waveform. Such a waveform is called a periodic wave because it repeats at a uniform rate (i.e., each successive cycle of the signal takes exactly the same length of time and has exactly the same amplitude variations as every other cycle—each cycle has exactly the same shape). A series of sine waves is an example of a periodic wave. Periodic waves can be analyzed in either the *time* or the *frequency domain*. In fact, it is often necessary when analyzing system performance to switch from the time domain to the frequency domain, and vice versa.

Time domain.   A standard oscilloscope is a time-domain instrument. The display on the CRT is an amplitude-versus-time representation of the input signal and is commonly called a signal waveform. Essentially, a signal waveform shows the shape and the instantaneous value of the signal but is not necessarily indicative of its frequency content. With an oscilloscope, the vertical deflection is proportional to the instantaneous amplitude of the total input signal, and the horizontal deflection is a function of time (sweep rate). Figure 1-3 shows the signal waveform for a single-frequency sinusoidal signal with a peak amplitude of V volts and a frequency of F hertz.

Frequency domain.   A spectrum analyzer is a frequency-domain instrument. Essentially, there is no waveform displayed on the CRT. Instead, an amplitude-versus-frequency plot is shown (this is called a frequency spectrum). With a spectrum analyzer,

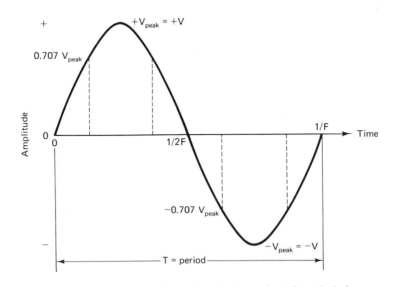

FIGURE 1-3   **Time-domain representation (signal waveform) for a single-frequency sinusoidal wave.**

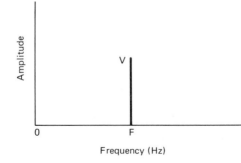

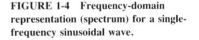

FIGURE 1-4 Frequency-domain representation (spectrum) for a single-frequency sinusoidal wave.

the horizontal axis represents frequency and the vertical axis amplitude. Therefore, there is a vertical deflection for each input frequency. Effectively, the input waveform is swept with a variable-frequency, high-Q bandpass filter whose center frequency is synchronized to the horizontal sweep rate of the CRT. Each frequency present in the input waveform produces a vertical line on the CRT (these are called spectral components). The height of each line is proportional to the amplitude of that frequency. A frequency-domain representation of a wave shows the frequency content but is not necessarily indicative of the shape of the waveform or the instantaneous amplitude. Figure 1-4 shows the spectrum for a single-frequency sinusoidal signal with a peak amplitude of V volts and a frequency of F hertz.

Nonsinusoidal Periodic Waves (Complex Waves)

Essentially, any repetitive waveform that comprises more than one sine or cosine wave is a *nonsinusoidal* or *complex periodic wave*. To analyze a complex periodic waveform, it is necessary to use a mathematical series developed in 1826 by the French physicist and mathematician Baron Jean Fourier. This series is appropriately called the *Fourier series*.

The Fourier series. The Fourier series is used in signal analysis to represent the sinusoidal components of a nonsinusoidal periodic waveform. In general, a Fourier series can be written for any series of terms that include trigonometric functions with the following mathematical expression:

$$f(t) = A_0 + A_1 \cos \alpha + A_2 \cos 2\alpha + A_3 \cos 3\alpha + \cdots + A_N \cos N\alpha$$
$$+ B_1 \sin \beta + B_2 \sin 2\beta + B_3 \sin 3\beta + \cdots + B_N \sin N\beta \tag{1-3}$$

Equation 1-3 states that the waveform $f(t)$ comprises an average value (A_0), a series of cosine functions in which each successive term has a frequency that is an integer multiple of the frequency of the first cosine term in the series, and a series of sine functions in which each successive term has a frequency that is an integer multiple of the frequency of the first sine term in the series. There are no restrictions on the values or relative values of the amplitudes for the sine or cosine terms. Equation 1-3 is stated in words as follows: Any *periodic waveform* comprises an average component and a series of harmonically related sine and cosine waves. A *harmonic* is an integral multiple of the fundamental frequency. The *fundamental frequency* is the first harmonic, the second multiple of the fundamental is called the second harmonic, the third multiple is called the third harmonic, and so on. The fundamental frequency is the minimum frequency necessary to represent a waveform and is also the frequency of the waveform (i.e., the repetition rate). Therefore, Equation 1-3 can be rewritten as

$$f(t) = \text{dc} + \text{fundamental} + \text{2nd harmonic}$$

$$+ \text{ 3rd harmonic} + \cdots + n\text{th harmonic}$$

Wave symmetry

Even symmetry. If a periodic voltage waveform is symmetric about the vertical (amplitude) axis, it is said to have *axes* or *mirror symmetry* and is called an *even function*. For all even functions, the B coefficients in Equation 1-3 are zero. Therefore, the signal simply contains a dc component and the cosine terms (note that a cosine wave is itself an even function). The sum of a series of even functions is an even function. Even functions satisfy the condition

$$f(t) = f(-t) \tag{1-4}$$

Equation 1-4 states that the magnitude of the function at $+t$ is equal to the magnitude at $-t$. A waveform that contains only the even functions is shown in Figure 1-5a.

Odd symmetry. If a periodic voltage waveform is symmetric about a line midway between the vertical and horizontal axes and passing through the coordinate origin, it is said to have *point* or *skew symmetry* and is called an *odd function*. For all odd functions, the A coefficients in Equation 1-3 are zero. Therefore, the signal simply contains a dc component and the sine terms (note that a sine wave is itself an odd function). The sum of a series of odd functions is an odd function. This form must be mirrored first in the Y axis, then in the X axis for superposition. Thus

$$f(t) = -f(-t) \tag{1-5}$$

Equation 1-5 states that the magnitude of the function at $+t$ is equal to the negative of the magnitude at $-t$. A periodic waveform that contains only the odd functions is shown in Figure 1-5b.

Half-wave symmetry. If a periodic voltage waveform is such that the waveform for the first half cycle ($t = 0$ to $T/2$) repeats itself except with the opposite sign for the second half cycle ($t = T/2$ to T), it is said to have *half-wave symmetry*. For all waveforms with half-wave symmetry, the even harmonics in the series for both the sine and cosine terms are zero. Therefore, half-wave functions satisfy the condition

$$f(t) = -f(T/2 + t) \tag{1-6}$$

A periodic waveform that exhibits half-wave symmetry is shown in Figure 1-5c.
The coefficients A_0, B_1 to B_N, and A_1 to A_N can be evaluated using the following integral formulas:

$$A_0 = \frac{1}{T} \int_0^T f(t)\, dt \qquad (1\text{-}7)$$

$$A_N = \frac{2}{T} \int_0^T f(t) \cos N\omega t\, dt \qquad (1\text{-}8)$$

$$B_N = \frac{2}{T} \int_0^T f(t) \sin N\omega t\, dt \qquad (1\text{-}9)$$

Solving Equations 1-7, 1-8, and 1-9 requires integral calculus, which is beyond the intent of this book. Therefore, in subsequent discussions, the appropriate solutions are given.

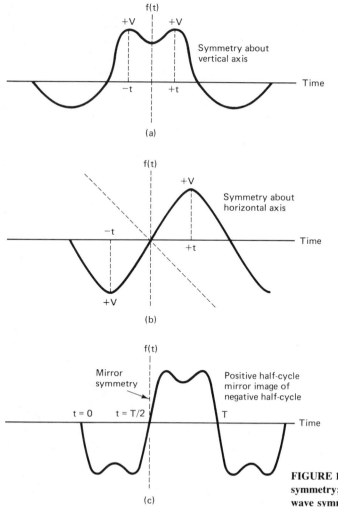

FIGURE 1-5 Wave symmetries: (a) even symmetry; (b) odd symmetry; (c) half-wave symmetry.

EXAMPLE 1-1

For the train of square waves shown in Figure 1-6:
> (a) Determine the coefficients for the first 9 harmonics.
> (b) Draw the frequency spectrum.
> (c) Sketch the time-domain signal for frequency components up to the ninth harmonic.

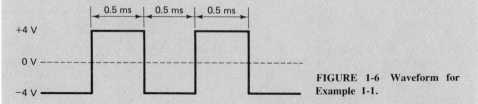

FIGURE 1-6 Waveform for Example 1-1.

Solution   (a) From inspection of the waveform, it can be seen that the average dc component is 0 V and the waveform has half-wave symmetry. Evaluating Equations 1-7, 1-8, and 1-9 yields the following Fourier series for a square wave:

$$v = \frac{4V}{N\pi} \left(\cos \omega t - \frac{1}{3} \cos 3\omega t + \frac{1}{5} \cos 5\omega t - \frac{1}{7} \cos 7\omega t + \cdots \right) \qquad (1\text{-}10)$$

From Equation 1-10 the following frequencies and coefficients are derived:

$$V_N = \frac{4V}{N\pi} \ (\text{odd})$$

where

> N = Nth harmonic (odd harmonics only)
> V = peak amplitude of the complex waveform

N	Harmonic	Frequency (kHz)	Voltage (V)
0	0	dc	0
1	1st	1	5.09
2	2nd	2	0
3	3rd	3	1.69
4	4th	4	0
5	5th	5	1.02
6	6th	6	0
7	7th	7	0.728
8	8th	8	0
9	9th	9	0.566

(b) The spectrum is shown in Figure 1-7. (Note that although both + and − components are included in the waveform, all magnitudes are shown in the + direction on a waveform spectrum.)

(c) The time-domain signal for the first nine harmonics is shown in Figure 1-8. Although the waveform shown is not an exact square wave, it does closely resemble one.

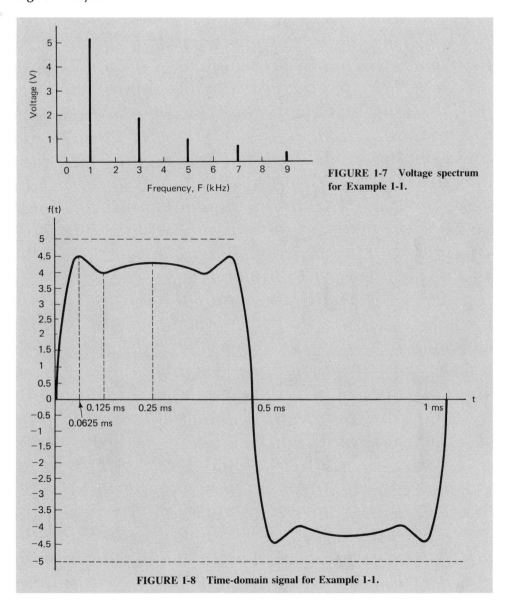

FIGURE 1-7 Voltage spectrum for Example 1-1.

FIGURE 1-8 Time-domain signal for Example 1-1.

Table 1-3 is a summary of the Fourier series for several of the more common nonsinusoidal periodic waveforms.

Fourier Series for a Rectangular Waveform

When analyzing electronic communications circuits, it is often necessary to use *rectangular pulses*. A waveform for a string of rectangular pulses is shown in Figure 1-9. The

TABLE 1-3   FOURIER SERIES SUMMARY

Waveform	Fourier series

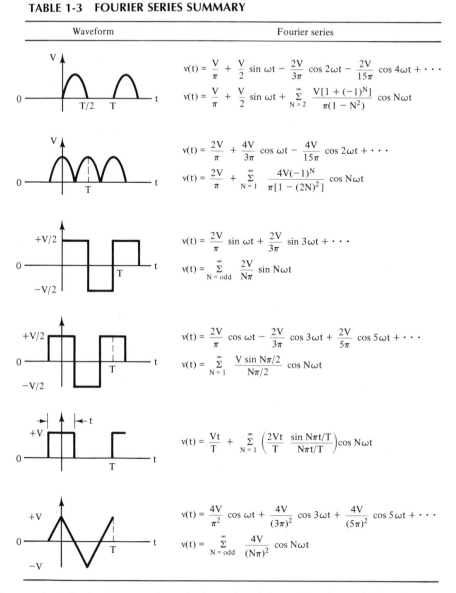

$$v(t) = \frac{V}{\pi} + \frac{V}{2} \sin \omega t - \frac{2V}{3\pi} \cos 2\omega t - \frac{2V}{15\pi} \cos 4\omega t + \cdots$$

$$v(t) = \frac{V}{\pi} + \frac{V}{2} \sin \omega t + \sum_{N=2}^{\infty} \frac{V[1 + (-1)^N]}{\pi(1 - N^2)} \cos N\omega t$$

$$v(t) = \frac{2V}{\pi} + \frac{4V}{3\pi} \cos \omega t - \frac{4V}{15\pi} \cos 2\omega t + \cdots$$

$$v(t) = \frac{2V}{\pi} + \sum_{N=1}^{\infty} \frac{4V(-1)^N}{\pi[1 - (2N)^2]} \cos N\omega t$$

$$v(t) = \frac{2V}{\pi} \sin \omega t + \frac{2V}{3\pi} \sin 3\omega t + \cdots$$

$$v(t) = \sum_{N=\text{odd}}^{\infty} \frac{2V}{N\pi} \sin N\omega t$$

$$v(t) = \frac{2V}{\pi} \cos \omega t - \frac{2V}{3\pi} \cos 3\omega t + \frac{2V}{5\pi} \cos 5\omega t + \cdots$$

$$v(t) = \sum_{N=1}^{\infty} \frac{V \sin N\pi/2}{N\pi/2} \cos N\omega t$$

$$v(t) = \frac{Vt}{T} + \sum_{N=1}^{\infty} \left(\frac{2Vt}{T} \frac{\sin N\pi t/T}{N\pi t/T} \right) \cos N\omega t$$

$$v(t) = \frac{4V}{\pi^2} \cos \omega t + \frac{4V}{(3\pi)^2} \cos 3\omega t + \frac{4V}{(5\pi)^2} \cos 5\omega t + \cdots$$

$$v(t) = \sum_{N=\text{odd}}^{\infty} \frac{4V}{(N\pi)^2} \cos N\omega t$$

duty cycle (DS) for the waveform is the ratio of the active time of the pulse (t) to the period of the waveform (T). Mathematically, the duty cycle is

$$DS = \frac{t}{T} \qquad (1\text{-}11a)$$

$$DS\,(\%) = \frac{t}{T} \times 100 \qquad (1\text{-}11b)$$

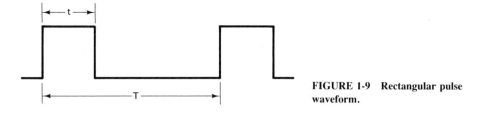

FIGURE 1-9   Rectangular pulse waveform.

Regardless of the duty cycle, a rectangular waveform is made up of a series of harmonically related sine waves. However, the amplitude of the spectrum components are dependent on the duty cycle. The Fourier series for a rectangular voltage waveform is

$$v = \frac{Vt}{T} + \frac{2Vt}{T}\left[\frac{\sin x}{x}(\cos \omega t) + \frac{\sin 2x}{2x}(\cos 2\omega t)\right.$$
$$\left. + \frac{\sin 3x}{3x}(\cos 3\omega t) + \cdot\ \cdot\ \cdot + \frac{\sin Nx}{Nx}(\cos N\omega t)\right] \tag{1-12}$$

where

$$x = \pi t/T$$
$$N = N\text{th harmonic and can be any whole integer}$$

From Equation 1-12 it can be seen that a rectangular waveform has a 0-Hz (dc) component equal to

$$V \times \frac{t}{T} \quad \text{or} \quad V \times \text{duty cycle} \tag{1-13}$$

The narrower the pulse width, the smaller the dc component. Also, from Equation 1-12, the amplitude of the Nth harmonic is

$$V_N = \frac{2Vt}{T}\frac{\sin Nx}{Nx} \tag{1-14}$$

The (sin x)/x function is used to describe repetitive pulse waveforms. Sin x is simply a sinusoidal waveform whose instantaneous amplitude depends on x. With only x in the denominator, the denominator increases with x. Therefore, a (sin x)/x function is simply a damped sine wave. A (sin x)/x function is shown in Figure 1-10.

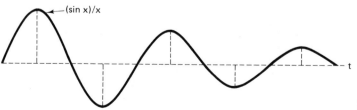

FIGURE 1-10   (sin x)/x function.

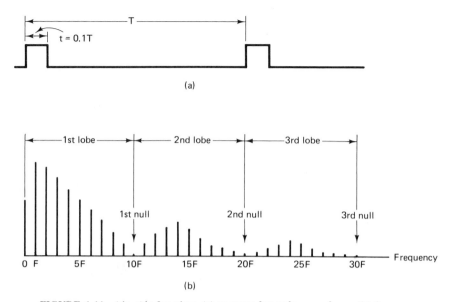

(a)

(b)

FIGURE 1-11 **(sin *x*)/*x* function: (a) rectangular pulse waveform; (b) frequency spectrum.**

Figure 1-11 shows the *frequency spectrum* for a rectangular pulse with a pulse width-to-period ratio of 0.1. It can be seen that the amplitudes of the harmonics follow a damped sinusoidal shape. The frequency whose period equals 1/*t* (i.e., at frequency 10*F* Hz), there is a 0-V component. A second null occurs at 20*F* Hz (period = 2/*t*), a third at 30*F* Hz (period = 3/*t*), and so on. All harmonics between 0 Hz and the first null frequency are considered in the first *lobe* of the frequency spectrum. All spectrum components between the first and second null frequencies are in the second lobe, frequencies between the second and third nulls are in the third lobe, and so on.

The following characteristics are true for all repetitive rectangular waveforms:

1. The dc component is equal to the pulse amplitude times the duty cycle.
2. There are 0-V components at frequency 1/*t* Hz and all integer multiples of that frequency.
3. The amplitude-versus-frequency time envelope of the spectrum components take on the shape of a damped sine wave.

EXAMPLE 1-2.

For the pulse waveform shown in Figure 1-12:
    (a) Determine the dc component.
    (b) Determine the peak amplitudes of the first 10 harmonics.
    (c) Plot the (sin *x*)/*x* function.
    (d) Sketch the frequency spectrum.

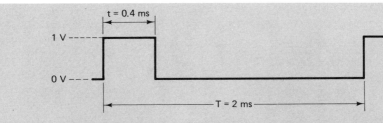

FIGURE 1-12   Pulse waveform for Example 1-2.

Solution   (a) From Equation 1-13 the dc component is

$$V(0 \text{ Hz}) = \frac{1 (0.4 \text{ ms})}{2 \text{ ms}} = 0.2 \text{ V}$$

(b) The amplitudes of the first 10 harmonics are determined from Equation 1-14:

$$V_N = 2(1) \left(\frac{0.4 \text{ ms}}{2 \text{ ms}}\right) \left[\frac{\sin N180(0.4 \text{ ms}/2 \text{ ms})}{N3.14 (0.4 \text{ ms}/2 \text{ ms})}\right]$$

N	Frequency (Hz)	Amplitude (V)
0	0	0.2
1	500	0.374
2	1000	0.303
3	1500	0.202
4	2000	0.094
5	2500	0
6	3000	−0.063
7	3500	−0.087
8	4000	−0.076
9	4500	−0.042
10	5000	0

(c) The $(\sin x)/x$ function is shown in Figure 1-13.

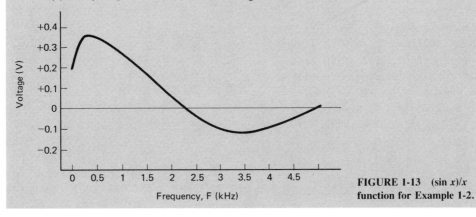

FIGURE 1-13 $(\sin x)/x$ function for Example 1-2.

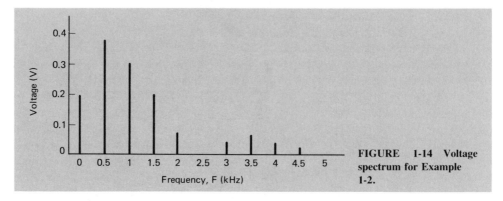

FIGURE 1-14 Voltage spectrum for Example 1-2.

(d) The frequency spectrum is shown in Figure 1-14.

Although the frequency components in the second lobe are negative, it is customary to plot all voltages in the positive direction on the frequency spectrum.

Figure 1-15 shows the effect that reducing the duty cycle (i.e., reducing the t/T ratio) has on the frequency spectrum for a nonsinusoidal waveform. It can be seen that narrowing the pulse width produces a frequency spectrum with a more uniform amplitude. In fact, for infinitely narrow pulses, the frequency spectrum comprises an infinite number of frequencies of equal amplitude. Increasing the period of a rectangular waveform while keeping the pulse width constant has the same effect on the frequency spectrum.

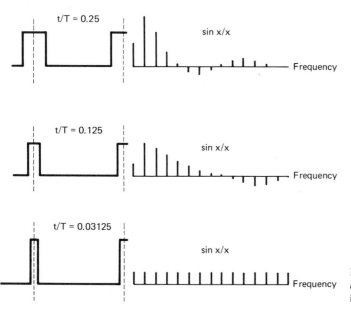

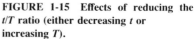

FIGURE 1-15 Effects of reducing the t/T ratio (either decreasing t or increasing T).

Effects of Bandlimiting on Signals

Every communications channel has a limited bandwidth and therefore has a limiting effect on signals that are propagated through them. We can consider a communications channel to be equivalent to an ideal linear phase filter with a finite bandwidth. If a

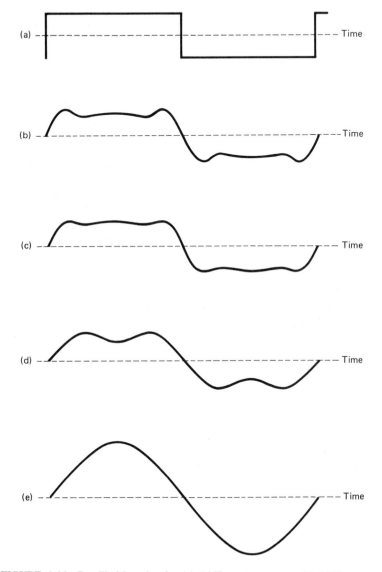

FIGURE 1-16   Bandlimiting signals: (a) 1-kHz square wave; (b) 1-kHz square wave bandlimited to 8 kHz; (c) 1-kHz square wave bandlimited to 6 kHz; (d) 1-kHz square wave bandlimited to 4 kHz; (e) 1-kHz square wave bandlimited to 2 kHz.

nonsinusoidal repetitive waveform passes through an ideal low-pass filter, the harmonic frequency components that are higher in frequency than the upper cutoff frequency for the filter are removed. Consequently, the shape of the waveform is changed. Figure 1-16a shows the time-domain waveform for the square wave used in Example 1-1. If this waveform is passed through a low-pass filter with an upper cutoff frequency of 8 kHz, frequencies above the eighth harmonic (9 kHz and above) are cut off and the waveform shown in Figure 1-16b results. Figures 1-16c, d, and e show the waveforms produced when low-pass filters with upper cutoff frequencies of 6, 4, and 2 kHz are used, respectively.

It can be seen from Figure 1-16 that *bandlimiting* a signal changes the frequency content and thus the shape of its waveform and, if sufficient bandlimiting is imposed, the waveform eventually comprises only the fundamental frequency. In a communications system, bandlimiting reduces the information capacity of the system and, if excessive bandlimiting is imposed, a portion of the information signal can be removed from the composite waveform.

ELECTRICAL NOISE

In general terms, *electrical noise* is defined as any unwanted electrical energy present in the usable passband of a communications circuit. For instance, in audio recording any undesired signals that fall into the band 0 to 15 kHz are audible and will interfere with the audio information. Consequently, for audio circuits, any unwanted electrical energy in the band 0 to 15 kHz is considered noise.

Essentially, noise can be divided into two general categories: correlated and uncorrelated. Correlation implies a relationship between the signal and the noise. Uncorrelated noise is noise that is present in the absence of any signal.

Correlated Noise

Correlated noise is unwanted electrical energy that is present as a direct result of a signal such as harmonic and intermodulation distortion. Harmonic and intermodulation distortion are both forms of nonlinear distortion; they are produced from nonlinear amplification. Correlated noise can not be present in a circuit unless there is an input signal. Simply stated, no signal, no noise! Both harmonic and intermodulation distortion change the shape of the wave in the time domain and the spectral content in the frequency domain.

Harmonic distortion.   *Harmonic distortion* is the generation of unwanted multiples of a single-frequency sine wave when the sine wave is amplified in a nonlinear device such as a large-signal amplifier. *Amplitude distortion* is another name for harmonic distortion. Generally, the term "amplitude distortion" is used for analyzing a waveform in the time domain, and the term "harmonic distortion" is used for analyzing a waveform in the frequency domain. The original input frequency is the first harmonic and, as stated previously, is called the fundamental frequency.

There are various degrees or orders of harmonic distortion. Second-order harmonic distortion is the ratio of the amplitude of the second harmonic to the amplitude of the fundamental frequency. Third-order harmonic distortion is the ratio of the amplitude of the third harmonic to the amplitude of the fundamental frequency, and so on. The ratio of the combined amplitudes of the higher harmonics to the amplitude of the fundamental frequency is called *total harmonic distortion* (THD). Mathematically, total harmonic distortion is

$$\% \text{ THD} = \frac{V_{\text{higher}}}{V_{\text{fundamental}}} \times 100 \tag{1-15}$$

where

$\% \text{ TDH}$ = percent total harmonic distortion

V_{higher} = quadratic sum of the root-mean-square (rms) voltages of the higher harmonics

$V_{\text{fundamental}}$ = rms voltage of the fundamental frequency

---

EXAMPLE 1-3

Determine the percent second-order, third-order, and total harmonic distortion for the output spectrum shown in Figure 1-17.

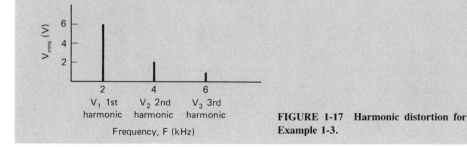

FIGURE 1-17 Harmonic distortion for Example 1-3.

Solution

$$\% \text{ 2nd-order harmonic distortion} = \frac{V_2}{V_1} \times 100 = \frac{2}{6} \times 100 = 33\%$$

$$\% \text{ 3rd-order harmonic distortion} = \frac{V_3}{V_1} \times 100 = \frac{1}{6} \times 100 = 16.7\%$$

$$\% \text{ TDH} = \frac{\sqrt{V_2^2 + V_3^2}}{V_1} \times 100 = \frac{\sqrt{2^2 + 1^2}}{6} \times 100 = 37.3\%$$

Intermodulation distortion.   *Intermodulation distortion* is the generation of unwanted *cross products* (sum and difference frequencies) created when two or more frequencies are amplified in a nonlinear device such as a large-signal amplifier. As with harmonic distortion, there are various degrees of intermodulation distortion. It would be impossible to measure all of the intermodulation components produced when two or more frequencies mix in a nonlinear device. Therefore, for comparison purposes, a common method used to measure intermodulation distortion is percent *second-order* intermodulation distortion. Second-order intermodulation distortion is the ratio of the total amplitude of the second-order cross products to the combined amplitude of the original input frequencies. Generally, to measure second-order intermodulation distortion, four test frequencies are used; two designated the A band (FA1 and FA2) and two designated the B band (FB1 and FB2). The second-order cross products (2A-B) are: 2FA1 − FB1, 2FA1 − FB2, 2FA2 − FB1, 2FA2 − FB2, (FA1 + FA2) − FB1, and (FA1 + FA2) − FB2. Mathematically, percent second-order intermodulation distortion (IMD) is

$$\% \text{ 2nd-order IMD} = \frac{V_{\text{2nd-order cross products}}}{V_{\text{original}}} \times 100 \qquad (1\text{-}16)$$

where

$V_{\text{2nd order}}$ = quadratic sum of the amplitudes of the 2nd-order cross products
V_{original} = quadratic sum of the amplitudes of the input frequencies

EXAMPLE 1-4

Determine the percent second-order intermodulation distortion for the A-band, B-band, and second-order intermodulation components shown in Figure 1-18.

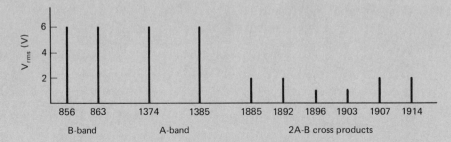

FIGURE 1-18   Intermodulation distortion for Example 1-4.

Solution

$$\% \text{ 2nd-order IMD} = \frac{V_{\text{2nd-order cross products}}}{V_{\text{original}}}$$

$$\% \text{ 2nd-order IMD} = \frac{\sqrt{2^2 + 2^2 + 2^2 + 2^2 + 1^2 + 1^2}}{\sqrt{6^2 + 6^2 + 6^2 + 6^2}} \times 100$$

$$\% \text{ IMD} = 35.4\%$$

Harmonic and intermodulation distortion are caused by the same thing, *nonlinear amplification*. Essentially, the only difference between the two is that harmonic distortion can occur when there is a single input frequency, and intermodulation distortion can occur only when there are two or more input frequencies. The generation of harmonic and intermodulation components is explained in Chapter 2.

Uncorrelated Noise

Uncorrelated noise is noise that is present regardless of whether or not there is a signal present. Uncorrelated noise is divided into two general categories: external and internal.

External noise. *External noise* is noise generated external to a circuit and allowed to enter into the circuit only if the frequency of the noise falls into the passband of the input filter. There are three primary types of external noise: atmospheric noise, extraterrestrial noise, and man-made noise.

Atmospheric noise. Atmospheric noise is naturally occurring electrical energy that originates within the earth's atmosphere. Atmospheric noise is commonly called static electricity. The source of most static electricity is natural electrical disturbances such as lightning. Static electricity generally comes in the form of impulses which spread its energy throughout a wide range of radio frequencies. The magnitude of the impulses measured from naturally occurring events has been observed to be inversely proportional to frequency. Consequently, at frequencies above 30 MHz, atmospheric noise is insignificant. Also, frequencies above 30 MHz are limited predominantly to line-of-sight propagation, which limits their interfering range to approximately 80 km.

Atmospheric noise is the summation of the energy from all sources both local and distant. Atmospheric noise propagates through the earth's atmosphere in the same manner as radio waves. Therefore, the magnitude of the static noise received depends on the propagation conditions at the time and is dependent, in part, on diurnal and seasonal variations. Atmospheric noise is the familiar sputtering, cracking, and so on, heard on a radio receiver predominantly in the absence of a received signal and is relatively insignificant compared to the other sources of noise.

Extraterrestrial noise. Extraterrestrial noise is noise that originates outside the earth's atmosphere and is, therefore, sometimes called *deep-space noise*. Extraterrestrial noise originates from the Milky Way, other galaxies, and the sun. Extraterrestrial noise is divided into two categories: solar and cosmic (galactic).

Solar noise is noise generated directly from the sun's heat. There are two components of solar noise: a "quiet" condition when a relatively constant radiation intensity exists and high-intensity sporadic disturbances caused by sun spot activity and solar flare-ups. The sporadic disturbances come from specific locations on the sun's surface. The magnitude of the disturbances caused from sun spot activity follows a cyclic pattern that repeats every 11 years. Also, these 11-year periods follow a supercycle pattern where approximately every 100 years a new maximum intensity is realized.

Cosmic noise sources are continuously distributed throughout our galaxy and other galaxies. Distant stars are also suns and have high temperatures associated with them. Consequently, they radiate noise in the same manner as our sun. Because the sources of galactic noise are located much farther away than our sun, their noise intensity is relatively small. Cosmic noise is often called *black body noise* and is distributed fairly evenly throughout the sky. Extraterrestrial noise contains frequencies from approximately 8 MHz to 1.5 GHz, although frequencies below 20 MHz seldom penetrate the earth's atmosphere and are, therefore, insignificant.

Man-made noise. *Man-made noise* is simply noise that can be attributed to man. The sources of man-made noise include spark-producing mechanisms such as commutators in electric motors, automobile ignition systems, power switching equipment, and fluorescent lights. Man-made noise is also impulsive in nature and therefore contains a wide range of frequencies that are propagated through space in the same manner as radio waves. Man-made noise is most intense in more populated metropolitan and industrial areas and is sometimes called *industrial noise*.

Internal noise.   *Internal noise* is electrical interference generated within a device. There are three primary kinds of internally generated noise: thermal, shot, and transit-time.

Thermal noise. Thermal noise is a phenomenon associated with *Brownian movement* of electrons within a conductor. In accordance with the kinetic theory of matter, electrons within a conductor are in thermal equilibrium with the molecules and in constant random motion. This random movement is accepted as being a confirmation of the kinetic theory of matter and was first noted by the English botanist, Robert Brown (hence the name ''Brownian noise''). Brown first observed evidence for the *kinetic* (moving-particle) nature of matter while observing pollen grains under a microscope. Brown noted an extraordinary agitation of the pollen grains that made them extremely difficult to examine. He later noted that this same phenomenon existed for smoke particles in the air. Brownian movement of electrons was first recognized in 1927 by J. B. Johnson of Bell Telephone Laboratories. In 1928, a quantitative theoretical treatment was furnished by H. Nyquist (also of Bell Telephone Laboratories). Electrons within a conductor carry a unit negative charge, and the mean-square velocity of an electron is proportional to the absolute temperature. Consequently, each flight of an electron between collisions with molecules constitutes a short pulse of current. Because the electron movement is totally random and in all directions, the average voltage produced by their movement is 0 V dc. However, such a random movement gives rise to an ac component. This ac component has several names, which include *thermal noise* (because it is temperature dependent), *Brownian noise* (after its discoverer), *Johnson noise* (after the person who related Brownian particle movement to electron movement), *random noise* (because the direction of electron movement is totally random), *resistance noise* (because the magnitude of its voltage is dependent upon resistance), and *white noise* (because it contains all frequencies). Hence thermal noise is the random motion of free electrons within a conductor caused by thermal agitation.

The *equipartition law* of Boltzmann and Maxwell combined with the works of

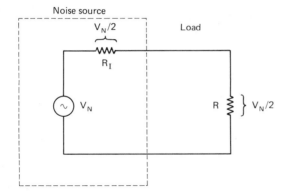

FIGURE 1-19 Noise source equivalent circuit.

Johnson and Nyquist states that the thermal noise power generated within a source for a 1-Hz bandwidth is

$$N_o = KT \qquad (1\text{-}17)$$

where

N_o = noise power density (W/Hz)
K = Boltzmann's constant
  $= 1.38 \times 10^{-23} \text{J/K}$
T = absolute temperature (K) (room temperature = 17°C or 290 K)[*]

Thus at room temperature, the available noise power density is

$$N_o = 1.38 \times 10^{-23} \text{ J/K} \times 290 \text{ K} = 4 \times 10^{-21} \text{ W/Hz}$$

The total noise power is equal to the product of the bandwidth and the noise density. Therefore, the total noise power present in bandwidth (B) is

$$N = KTB \qquad (1\text{-}18)$$

where

N = total noise power in bandwidth B (W)
$KT = N_o$ = noise power density (W/Hz)
B = bandwidth of the device or system (Hz)

Figure 1-19 shows the equivalent circuit for an electrical noise source. The internal resistance of the noise source (R_I) is in series with the rms noise voltage (V_N). For the worst case condition (maximum transfer of noise power), the load resistance (R) is made equal to R_I. Therefore, the noise voltage dropped across R is equal to $V_N/2$, and the noise power (N) developed across the load resistor is equal to KTB. Therefore, V_N is determined as follows:

[*] 0 K = −273°C.

$$N = KTB = \frac{(V_N/2)^2}{R} = \frac{V_N^2}{4R} \tag{1-19}$$

and

$$V_N^2 = 4RKTB$$

$$V_N = \sqrt{4RKTB}$$

Thermal noise is equally distributed throughout the frequency spectrum. Because of this property, a thermal noise source is called a "white noise source" (this is an analogy to white light, which contains all visible-light frequencies). Therefore, the noise power measured at any frequency is equal to the noise power measured at any other frequency. Similarly, the noise measured in any given bandwidth is equal to the noise measured in any other equal bandwidth regardless of the center frequency. In other words, the thermal noise power present in the band 1000 to 2000 Hz is equal to the thermal noise power present in the band 1,001,000 to 1,002,000 Hz.

EXAMPLE 1-5

For a device operating at a temperature of 17°C with a bandwidth of 10 kHz, determine:
(a) The noise power density (N_o).
(b) The total noise power (N).
(c) The rms noise voltage (V_N) for a 100-Ω internal resistance and a 100-Ω load resistor.

Solution   (a) By definition, the noise density is the noise measured in any 1-Hz bandwidth with constant temperature. Therefore, substituting into Equation 1-15, noise density equals

$$N_o = KT = 1.38 \times 10^{-23} \times 290 = 4 \times 10^{-21} \text{ W/Hz}$$

(b) Substituting into Equation 1-17, the total thermal noise is found to be

$$N = KTB$$

$$= 1.38 \times 10^{-23} \times 290 \times 10^4 = 4 \times 10^{-17} \text{ W}$$

(c) Substituting into Equation 1-19, we find that the rms noise voltage is

$$V_N = \sqrt{4RKTB}$$

$$V_n = \sqrt{4 \times 100 \times 1.38 \times 10^{-23} \times 290 \times 10^4} = 0.1265 \text{ } \mu\text{V}$$

For equal-value load and internal resistances, the noise voltage dropped across the load resistor is equal to one-half the noise voltage, or 0.06325 μV. Therefore, the total noise power is

$$N = \frac{E^2}{R} = \frac{(V_N/2)^2}{R} = \frac{(0.1265/2)^2}{100} = \frac{(0.06325)^2}{100} \text{ } \mu\text{V} = 4 \times 10^{-17} \text{ W}$$

Thermal noise is random, continuous, and occurs at all frequencies. Also, thermal noise is present in all devices, is predictable, and is additive. This is why thermal noise is generally the most significant of all the noise sources. Also, as previously stated, there are several names for thermal noise. In subsequent discussions in this book it will be referred to as simply thermal or random white noise.

Shot noise. Shot noise is caused by the random arrival of carriers (electrons and holes) at the output element of an active device such as a diode, field-effect transistor (FET), bipolar transistor (BJT) or tube. Shot noise is randomly varying and is superimposed on top of any signal present. Shot noise, when amplified, sounds like a shower of pellets falling on a tin roof.

Shot noise is proportional to the charge of an electron (1.6×10^{-19}), direct current, and system bandwidth. Also, shot noise is additive with thermal noise.

Transit-time noise. Any modification to a stream of carriers as they pass from the input to the output of a device (such as from the emitter to the collector of a transistor) produces an irregular random variation categorized as *transit noise.*

When the time it takes a carrier to propagate through a device is an appreciable part of the time of one cycle of the signal, the noise becomes noticeable. Transit-time noise in transistors is determined by ion mobility, the bias voltages, and the actual transistor construction. Carriers traveling from the emitter to the collector suffer from emitter delay times, base transit-time delays, and collector recombination and propagation delay times. At high frequencies and if transit delays are excessive, the device may add more noise than amplification to the signal.

Miscellaneous types of noise

Excess noise. Excess noise is a form of uncorrelated internal noise that is not totally understood. It is found in transistors and is directly proportional to emitter current and junction temperature and inversely proportional to frequency. Excess noise is also called *low-frequency* or *flicker noise, 1/F noise,* and *modulation noise.* It is believed to be caused by or at least associated with *carrier traps* in the emitter depletion layer. These traps capture and release holes and electrons at different rates but with energy levels that vary inversely with frequency. Excess noise is insignificant above approximately 1 kHz.

Resistance noise. Resistance noise is a form of thermal noise that is associated with the internal resistances of the base, emitter, and collector of a transistor. Resistance noise is fairly constant from about 500 Hz up and therefore may be *lumped* into an equivalent resistance with thermal and shot noise.

Precipitation noise. Precipitation noise is a type of static noise caused when an airplane passes through snow or rain. The airplane becomes electrically charged to a potential difference high enough in respect to the surrounding space that a *corona* discharge occurs at a sharp point on the airplane. The interference from precipitation static is most annoying at shortwave frequencies and lower.

Signal-to-Noise Ratio

Signal-to-noise ratio (*S/N*) is a simple mathematical relationship of the signal level in respect to the noise level at a given point in a circuit, amplifier, or system. Signal-to-noise can be expressed as a voltage ratio or as a power ratio. Mathematically, *S/N* is:

As a voltage ratio:

$$\frac{S}{N} = \frac{\text{signal voltage}}{\text{noise voltage}} = \frac{V_s}{V_n} \qquad (1\text{-}20)$$

As a power ratio:

$$\frac{S}{N} = \frac{\text{signal power}}{\text{noise power}} = \frac{P_s}{P_n} \qquad (1\text{-}21)$$

The signal-to-noise ratio is often expressed as a logarithmic function with the unit dB (decibel):

For power ratios:

$$\frac{S}{N} (\text{dB}) = 10 \log \frac{P_s}{P_n} \qquad (1\text{-}22)$$

For voltage ratios:

$$\frac{S}{N} (\text{dB}) = 20 \log \frac{V_s}{V_n} \qquad (1\text{-}23)$$

Signal-to-noise is probably the most important and most often used parameter for evaluating the performance of a radio communications system. The higher the signal-to-noise ratio, the better the system performance. From the signal-to-noise ratio, the general quality of a system can be determined.

Noise Figure

Noise figure (F) (sometimes called *noise factor*) is a figure of merit that indicates the degradation of the signal-to-noise ratio as a signal passes through an amplifier. Noise figure is the ratio of the input signal-to-noise ratio (S_i/N_i) to the output signal-to-noise ratio (S_o/N_o). Therefore, noise figure is a ratio of ratios. Mathematically, noise figure is

$$F = \frac{S_i/N_i}{S_o/N_o} \qquad (1\text{-}24)$$

$$F \text{ (db)} = 10 \log \left(\frac{S_i/N_i}{S_o/N_o} \right) \qquad (1\text{-}25)$$

An amplifier will amplify equally all signals and noise present at its input that fall within its passband. Therefore, if the amplifier is ideal and noiseless, the signal and noise are amplified by the same amount and the signal-to-noise ratio at the output is equal to the signal-to-noise ratio at the input. However, in reality, amplifiers are not ideal noiseless devices. Therefore, although the input signal and noise are amplified equally, the device adds internally generated noise to the waveform, which reduces the

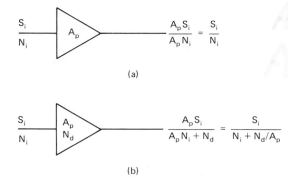

FIGURE 1-20 Noise figure: (a) ideal noiseless amplifier; (b) amplifier with internal noise.

overall signal-to-noise ratio. As stated previously, the most predominant form of electrical noise is thermal noise.

Figure 1-20a shows an ideal noiseless amplifier with a power gain (A_p), an input signal level (S_i), and an input noise level (N_i). It can be seen that the output S/N ratio is the same as the input S/N ratio. In Figure 1-20b, the same amplifier is shown except that instead of being ideal and noiseless, the amplifier adds an internally generated noise, N_d.

From Figure 1-20b, it can be seen that the output S/N ratio is less than the input S/N ratio by an amount proportional to N_d. Also, it can be seen that the noise figure of a perfect noiseless device is 1 or 0 dB.

EXAMPLE 1-6

For the amplifier shown in Figure 1-20b and the following parameters, determine:
(a) Input S/N.
(b) Output S/N.
(c) Noise figure.

$$\text{input signal voltage } (S_i) = 0.1 \times 10^{-3} \text{ V}$$

$$\text{input signal power } (P_i) = 2 \times 10^{-10} \text{ W}$$

$$\text{input noise voltage } (N_i) = 0.01 \times 10^{-6} \text{ V}$$

$$\text{input noise power } (P_n) = 2 \times 10^{-18} \text{ W}$$

$$\text{voltage gain } (A_v) = 1000$$

$$\text{power gain } (A_p) = 1{,}000{,}000$$

$$\text{amplifier noise voltage } (N_d) = 1 \times 10^{-6} \text{ V}$$

$$\text{amplifier noise power } (N_p) = 6 \times 10^{-12} \text{ W}$$

Solution (a) For the input signal and noise voltages given and substituting into Equation 1-20, we obtain

$$\frac{S_i}{N_i} = \frac{0.1 \text{ mV}}{0.01 \text{ } \mu\text{V}} = 10{,}000$$

Substituting into Equation 1-23 yields

$$\frac{S_i}{N_i} = 20 \log 10{,}000 = 80 \text{ dB}$$

For the input signal and noise powers given and substituting into Equation 1-21, we have

$$\frac{P_i}{P_n} = \frac{2 \times 10^{-10} \text{ W}}{2 \times 10^{-18} \text{ W}} = 100{,}000{,}000$$

Substituting into Equation 1-22 gives us

$$\frac{P_i}{P_n} = 10 \log 100{,}000{,}000 = 80 \text{ dB}$$

(b) For the calculated output signal and noise voltages and substituting into Equation 1-20, we obtain

$$\frac{S_o}{N_o} = \frac{A_v S_i}{A_v N_i + N_d} = \frac{(1000)\,(0.1 \text{ mV})}{(1000)\,(0.01 \text{ } \mu\text{V}) + 10 \mu\text{V}} = 5000$$

Substituting into Equation 1-23 yields

$$\frac{S_o}{N_o} = 20 \log 5000 = 74 \text{ dB}$$

For the calculated output signal and noise powers and substituting into Equation 1-21, we have

$$\frac{S_o}{N_o} = \frac{A_p P_i}{A_p P_n + N_p} = \frac{(1 \times 10^6)(2 \times 10^{-10} \text{ W})}{(1 \times 10^6)(2 \times 10^{-18} \text{ W}) + 6 \times 10^{-12} \text{ W}}$$

$$= 25 \times 10^6$$

Substituting into Equation 1-22 gives us

$$\frac{S_o}{N_o} = 10 \log 25 \times 10^6 = 74 \text{ dB}$$

(c) From the results of parts (a) and (b) and using Equations 1-24 and 1-25, we obtain

$$\text{voltage noise figure} = \frac{10{,}000}{5000} = 2$$

$$F \text{ (db)} = 20 \log 2 \quad = 6 \text{ dB}$$

$$\text{power noise figure} = \frac{100{,}000{,}000}{25{,}000{,}000} = 4$$

$$F \text{ (dB)} = 10 \log 4 \quad = 6 \text{ dB}$$

A noise figure of 6 dB indicates that the power signal-to-noise ratio decreased by a factor of 4 and the voltage signal-to-noise ratio decreased by a factor of 2 as the signal propagated from the input to the output of the amplifier.

When two or more amplifiers or devices are cascaded together, the total noise figure (NF) is the accumulation of the individual noise figures. Mathematically, the total noise figure is[*]

$$NF = F_1 + \frac{F_2 - 1}{A_1} + \frac{F_3 - 1}{A_1 A_2} + \frac{F_4 - 1}{A_1 A_2 A_3} + \cdots \qquad (1\text{-}26)$$

where

NF = total noise figure
F_1 = noise figure of amplifier 1
F_2 = noise figure of amplifier 2
F_3 = noise figure of amplifier 3
A_1 = gain of amplifier 1
A_2 = gain of amplifier 2
A_3 = gain of amplifier 3

It can be seen that the noise figure of the first amplifier (F_1) contributes the most toward the overall noise figure. The noise introduced in the first stage is amplified by each of the succeeding amplifiers. Therefore, when compared to the noise introduced in the first stage, the noise added by each succeeding amplifier is effectively reduced by a factor equal to the product of the gains of the preceding amplifiers.

EXAMPLE 1-7

For three cascaded amplifiers each with noise figures of 3 dB and gains of 10 dB, determine the total noise figure.

Solution   Substituting into Equation 1-26 gives us

$$NF = F_1 + \frac{F_2 - 1}{A_1} + \frac{F_3 - 1}{A_1 A_2}$$

$$= 2 + \frac{2 - 1}{10} + \frac{2 - 1}{100}$$

$$= 2.11$$

and

$$10 \log 2.11 = 3.24 \text{dB}$$

An overall noise figure of 3.24 dB indicates that the *S/N* ratio at the output of A_3 is 3.24 dB less than the *S/N* ratio at the input to A_1.

MULTIPLEXING

Multiplexing is the transmission of information (either voice or data) from more than one source to more than one destination on the same transmission medium. Transmissions occur on the same medium but not necessarily at the same time. The transmission

[*] Noise figures and gains are given in absolute ratios rather than in decibels.

medium may be a metallic wire pair, a coaxial cable, a microwave radio, a satellite radio, or a fiber optic link. There are several ways in which multiplexing can be achieved, although the two most common methods are frequency-division multiplexing and time-division multiplexing.

Frequency-Division Multiplexing

In *frequency-division multiplexing* (FDM), multiple sources that originally occupied the same frequency band are transmitted simultaneously over a single transmission medium. Thus many relatively narrowband channels can be transmitted over a single wideband transmission system.

FDM is an analog multiplexing scheme; the information entering the system is analog and it remains analog throughout transmission. An example of FDM is the AM commercial broadcast band, which occupies a frequency spectrum from 535 to 1605 kHz. Each station carries an audio intelligence signal with a bandwidth of 5 kHz. If the audio from each station were transmitted with the original frequency spectrum, it would be impossible to separate one station's transmissions from another's. Instead, each station amplitude modulates a different carrier frequency and produces a signal with a 10-kHz bandwidth. Because adjacent stations' carrier frequencies are separated by 10 kHz, the total commerical AM band is divided into one hundred and seven 10-kHz frequency slots stacked next to each other in the frequency domain. To receive a particular station, a receiver is simply tuned to the frequency band associated with that station's transmissions. Figure 1-21 shows how commercial AM broadcast station signals are frequency-division multiplexed and transmitted over a single transmission medium (free space). There are many other applications for FDM, such as commercial FM and television broadcasting and high-volume telecommunications systems. Within any of the commercial broadcast bands, each station's transmissions are independent of all the other station's transmissions.

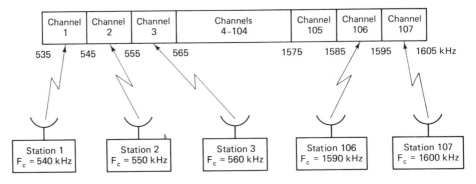

FIGURE 1-21   Frequency-division-multiplexing commercial AM broadcast-band stations.

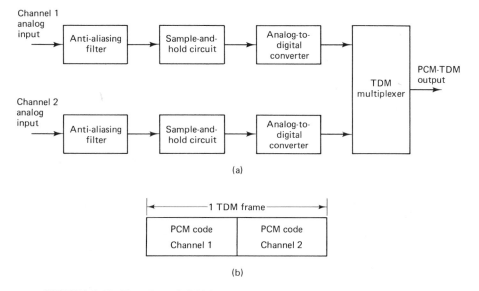

(a)

(b)

FIGURE 1-22 Two-channel PCM-TDM system: (a) block diagram; (b) TDM frame.

Time-Division Multiplexing

With *time-division multiplexing* (TDM), transmissions from multiple sources occur on the same facility but not at the same time. Transmissions from various sources are *interleaved* in the time domain. The most common type of modulation used with TDM systems is *pulse code modulation* (PCM). PCM is a type of digital transmission where analog signals are periodically *sampled* and converted to a series of binary codes, and the codes are transmitted as binary digital pulses. With a PCM-TDM system, several voice band channels are sampled, converted to PCM codes, then time-division multiplexed onto a single metallic cable pair.

Figure 1-22a shows a simplified block diagram of a two-channel PCM-TDM carrier system. Each channel is alternately sampled and converted to a PCM code. While PCM code for channel 1 is being transmitted, channel 2 is being sampled and converted to a PCM code. While the PCM code from channel 2 is being transmitted, the next sample is taken from channel 1 and converted to a PCM code. This process continues and samples are taken alternately from each channel, converted to PCM codes, and transmitted. The multiplexer is simply a switch with two inputs and one output. Channel 1 and channel 2 are alternately selected and connected to the multiplexer output. The time it takes to transmit one sample from each channel is called the *frame time*.

The PCM code for each channel occupies a fixed time slot (*epoch*) within the total TDM frame. With a two-channel system, the time allocated for each channel is equal to one-half the total frame time. A sample from each channel is taken once during each frame. Therefore, the total frame time is equal to the reciprocal of the sample rate. Figure 1-22b shows the TDM frame allocation for the two-channel system shown in Figure 1-22a.

QUESTIONS

1-1. Define *electronic communications.*

1-2. What three primary components make up a communications system?

1-3. Define *modulation.*

1-4. Define *demodulation.*

1-5. Define *carrier frequency.*

1-6. Explain the relationships among the source information, the carrier, and the modulated wave.

1-7. What are the three properties of an analog carrier that can be varied?

1-8. What organization assigns frequencies for free-space radio propagation in the United States?

1-9. Briefly describe the significance of Hartley's law and give the relationships between information capacity and bandwidth; information capacity and transmission time.

1-10. What are the two primary limitations on the performance of a communications system?

1-11. Describe signal analysis as it pertains to electronic communications.

1-12. Describe a time-domain display of a signal waveform; a frequency-domain display.

1-13. What is meant by the term *even symmetry*? What is another name for even symmetry?

1-14. What is meant by the term *odd symmetry*? What is another name for odd symmetry?

1-15. What is meant by the term *half-wave symmetry*?

1-16. Describe the term *duty cycle.*

1-17. Describe a $(\sin x)/x$ function.

1-18. Define *electrical noise.*

1-19. What is meant by the term *correlated noise*? List and describe two common forms of correlated noise.

1-20. What is meant by the term *uncorrelated noise*? List several types of uncorrelated noise and state their sources.

1-21. Briefly describe thermal noise.

1-22. What are four alternative names for thermal noise?

1-23. Describe the relationship between thermal noise and temperature; thermal noise and bandwidth.

1-24. Define *signal-to-noise ratio.* What does a signal-to-noise ratio of 100 indicate? 100 dB?

1-25. Define *noise figure.* An amplifier has a noise figure of 20 dB; what does this mean?

1-26. What is the noise figure for a totally noiseless device?

1-27. Define *multiplexing.*

1-28. Describe frequency-division multiplexing.

1-29. Describe time-division multiplexing.

PROBLEMS

1-1. For the train of square waves shown below:
    (a) Determine the coefficients for the first five harmonics.
    (b) Draw the frequency spectrum.
    (c) Sketch the time-domain signal for frequency components up to the first five harmonics.

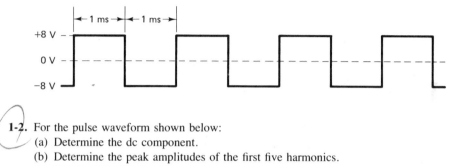

1-2. For the pulse waveform shown below:
    (a) Determine the dc component.
    (b) Determine the peak amplitudes of the first five harmonics.
    (c) Plot the $(\sin x)/x$ function.
    (d) Sketch the frequency spectrum.

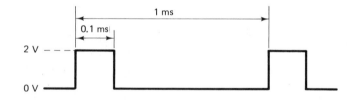

1-3. Determine the percent second-order, third-order, and total harmonic distortion for the output spectrum shown below.

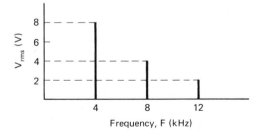

1-4. Determine the percent second-order intermodulation distortion for the A-band, B-band, and second-order intermodulation components shown below.

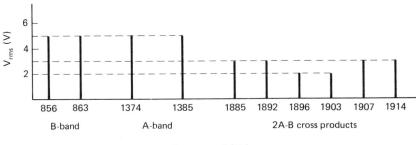

B-band   A-band   2A-B cross products

Frequency, F (Hz)

1-5. Determine the second-order cross-product frequencies for the following A- and B-band frequencies: B = 822 and 829 Hz, A = 1356 and 1365 Hz.

1-6. For an amplifier operating at a temperature of 27°C with a bandwidth of 20 kHz, determine:
   (a) The noise power density (N_o) in watts and dBm.
   (b) The total noise power (N) in watts and dBm.
   (c) The rms noise voltage (V_N) for a 50-Ω internal resistance and a 50-Ω load resistor.

1-7. (a) Determine the noise power (N) in watts and dBm for an amplifier operating at a temperature of 400°C with a bandwidth of 1 MHz.
   (b) Determine the decrease in noise power in decibels if the temperature decreased to 100°C.
   (c) Determine the increase in noise power in decibels if the bandwidth doubled.

1-8. Determine the overall noise figure for three cascaded amplifiers, each with individual noise figures of 3 dB and power gains of 20 dB.

1-9. Determine the duty cycle for the pulse waveform shown below.

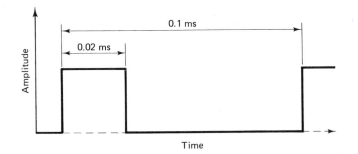

1-10. Determine the % THD for a fundamental frequency voltage $V_{fundamental}$ = 12 V rms and a total voltage for the higher harmonics V_{higher} = 1.2 V rms.

1-11. Determine the percent second-order IMD for a quadratic sum of the A- and B-band components $V_{original}$ = 2.6 V rms and a quadratic sum of the second-order cross products V_{2nd} order = 0.02 V rms.

1-12. If an amplifier with a bandwidth B = 20 kHz and a total noise power $N = 2 \times 10^{-17}$ W, determine:
   (a) Noise density.
   (b) Total noise if the bandwidth is increased to 40 kHz.
   (c) Noise density if the bandwidth is increased to 30 kHz.

Chapter 2

FREQUENCY GENERATION

INTRODUCTION

In electronic communications systems, there are many applications that require repetitive waveforms (both sinusoidal and rectangular). In most of these applications, more than one frequency is required, and very often these frequencies must be synchronized to each other. Therefore, *frequency generation* is an essential part of electronic communications.

OSCILLATORS

By definition, to *oscillate* is to fluctuate between two states or conditions. "To oscillate" is to vibrate, change, or to undergo oscillations. *Oscillating* is the act of fluctuating from one state to another. An *oscillator* is a device that produces *oscillations*. There are many applications for oscillators in electronic communications, such as high-frequency *carrier supplies*, *pilot supplies*, and *clock circuits*.

In electronic applications, an oscillator is a device or a circuit that produces electrical oscillations. An electrical oscillation is a repetitive change in a voltage or current waveform. If an oscillator is *self-sustaining*, the changes in the waveform are *continuous* and *repetitive*; they occur at a periodic rate. A self-sustaining oscillator is also called a *free-running oscillator*. Oscillators that are not self-sustaining require an external input signal or trigger to produce a change in the output waveform. Oscillators that are not self-sustaining are called *triggered* or *one-shot oscillators*. The remainder of this chapter is restricted to self-sustaining oscillators, which require no external input other than a dc supply voltage. Essentially, an oscillator converts a dc input voltage to an ac output voltage. The shape of the output waveform can be a sine wave, a square wave, a sawtooth, or any other shape as long as it repeats at periodic intervals.

Free-running oscillators, once started, generate an ac output signal of which a small portion is fed back to the input, where it is amplified. The amplified input signal appears at the output and the process repeats; a *regenerative* process occurs. The output is dependent on the input and the input is dependent on the output.

According to the *Barkhausen criterion*, for a feedback circuit to sustain oscillations, the net gain around the feedback loop must be unity or greater and the net phase shift around the loop must be a positive integer multiple of 360°.

Essentially, there are four requirements for a feedback oscillator to work: amplification, positive feedback, frequency dependency, and a source of power.

1. *Amplification.* The circuit must be capable of amplifying. In fact, at times it must be capable of infinite gain.
2. *Positive feedback.* There must be a complete path for the output signal to feed back to the input. The feedback signal must be regenerative, which means that it must have the correct phase and amplitude to sustain oscillations. If the phase is incorrect or if the amplitude is insufficient, oscillations will cease. If the amplitude is excessive, the device will saturate. Regenerative feedback is called *positive* feedback, where "positive" simply means that its phase aids the oscillation process and is not necessarily indicative of a positive (+) or negative (−) polarity.
3. *Frequency dependency.* There must be frequency-dependent components to allow the frequency of the oscillator to be set or changed.
4. *Power source.* There must be a source of electrical power, such as a dc power supply.

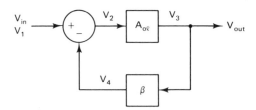

FIGURE 2-1   **Model of an amplifier with feedback.**

Figure 2-1 shows an electrical model for an amplifier with feedback. It includes an amplifier with an *open-loop gain* (A_{ol}) and a frequency-dependent regenerative feedback path with a *feedback ratio* of β.

From Figure 2-1, the following mathematical relationships are derived:

$$\text{transfer function} = \frac{V_{out}}{V_{in}} = \frac{V_3}{V_1}$$

$$V_2 = V_1 - V_4$$

$$V_3 = A_{ol}V_2$$

$$V_4 = \beta V_3$$

where

V_1 = external input voltage
V_2 = voltage input to amplifier
V_3 = output voltage
V_4 = feedback voltage

Substituting for V_4 gives us

$$V_2 = V_1 - \beta V_3$$

Thus

$$V_3 = (V_1 - \beta V_3)A_{ol}$$

And

$$V_3 = V_1 A_{ol} - V_3 \beta A_{ol}$$

Rearranging and factoring yield

$$V_3 + V_3 \beta A_{ol} = V_1 A_{ol} = V_3(1 + \beta A_{ol}) = V_1 A_{ol}$$

Then

$$\frac{V_{out}}{V_{in}} = \frac{V_3}{V_1} = \frac{A_{ol}}{1 + \beta A_{ol}} \qquad (2\text{-}1)$$

$A_{ol}/(1 + \beta A_{ol})$ is the standard formula used for an amplifier with feedback. If at some frequency βA_{ol} goes to -1, the denominator goes to 0 and V_{out}/V_{in} is infinity. When this happens, the circuit will oscillate and the external input may be removed.

For self-sustained oscillations to occur, a circuit must fulfill the four basic requirements for oscillation outlined previously, meet the criterion of Equation 2-1, and fit the basic feedback circuit model shown in Figure 2-1. Although oscillator action can be accomplished in many different ways, the most common configurations use *RC phase-shift networks*, *LC tank circuits*, *crystals*, or *negative resistance devices*.

Wien-Bridge Oscillator

The *Wien-bridge oscillator* is an *RC* phase-shift oscillator that uses both positive and negative feedback. It is a relatively stable low-frequency oscillator that is easily tuned and is commonly used in signal generators to produce frequencies between 5 Hz and 1 MHz. The Wien-bridge oscillator is the circuit that Mr. Hewlett and Mr. Packard used in their original signal generator design.

Figure 2-2a shows a simple *lead-lag network*. At the frequency of oscillation (F_o), $R = X_c$ and the signal undergoes a $-45°$ phase shift across Z_1 and a $+45°$ phase shift across Z_2. Consequently, at F_o, the total phase shift across the network is exactly

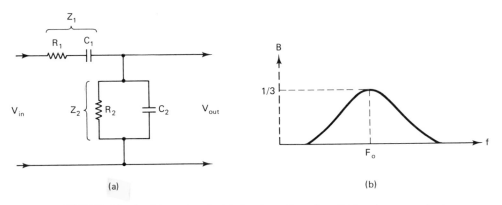

FIGURE 2-2   Lead-lag network: (a) circuit configuration; (b) input-versus-output transfer curve (β).

$0°$. At frequencies below resonance, the phase shift across the network leads and for frequencies above resonance it lags. At extreme low frequencies, C_1 looks open and there is no output. At extreme high frequencies C_2 looks like a short and there is no output.

A lead-lag network is a *reactive* voltage divider where the input voltage is divided between the series combination of R_1 and C_1 (Z_1) and the parallel combination of R_2 and C_2 (Z_2). Therefore, the lead-lag network is frequency selective and the output voltage is maximum at F_o. The transfer function for the feedback network (β) equals $Z_2/(Z_1 + Z_2)$ and is maximum and equal to $\frac{1}{3}$ at F_o. Figure 1-12b shows a plot of β versus frequency. If $R_1 = R_2$ and $C_1 = C_2$, F_o is determined from the following expression:

$$F_o = \frac{1}{2\pi RC}$$

where

$$R = R_1 = R_2$$
$$C = C_1 = C_2$$

Figure 2-3 shows a Wien-bridge oscillator. The lead-lag network and the resistive voltage divider make up a Wien bridge (hence the name "Wien-bridge oscillator"). When the bridge is balanced, the difference voltage equals zero. The voltage divider provides negative or degenerative feedback that offsets the positive or regenerative feedback from the lead-lag network. The ratio of the resistors in the voltage divider is 2:1, which sets the noninverting gain of amplifier A_1 to 3 ($R_f/R_i + 1$). Thus, at F_o, the signal at the output of A_1 is reduced by a factor of 3 as it passes through the lead-lag network, then amplified by 3 in amplifier A_1. Thus, at F_o, the closed-loop gain is equal to Aβ or 1 ($\frac{1}{3} \times 3$).

To compensate for imbalances in the bridge, *automatic gain control* (AGC) is added to the circuit. A simple way of providing automatic gain is to replace R_I in Figure 2-3 with a variable resistance device such as a FET. The resistance of the FET

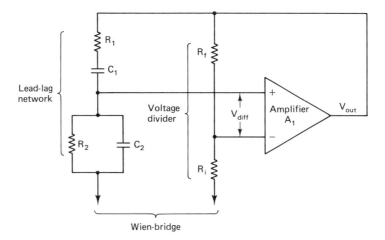

FIGURE 2-3 Wien-bridge oscillator.

is made directly proportional to V_{out}. If V_{out} goes up, the resistance of the FET is decreased and if V_{out} goes down, the resistance of the FET is increased. Therefore, the gain of the amplifier automatically adjusts to changes in the output signal amplitude.

The operation of the circuit shown in Figure 2-3 is as follows. On initial power up, noise (at all frequencies) appears at V_{out} and is fed back through the lead-lag network. Only noise at F_o passes through the lead-lag network with a 0° phase shift and a transfer ratio of $\frac{1}{3}$. Consequently, only a single frequency (F_o) is fed back in phase, undergoes a closed-loop gain of 1, and produces self-sustained oscillations.

LC Oscillators

LC oscillators are oscillators in which the frequency of oscillation is determined by an *LC tank circuit*. Tank circuit operation involves an exchange of energy between *kinetic* and *potential*.

Oscillator action. Figure 2-4 shows the basic oscillator action of an *LC* tank circuit. Initially, S_1 is open (t_0), there is no current flowing, and the charge across C is zero volts. V_{out} equals open-circuit voltage, V_{CC}. When S_1 is closed, current flows, charging C to V_{CC}. V_{out} equals $-V_C + V_{CC} = 0$ V (t_1). S_1 is opened and C discharges through L. The current flowing through L generates a magnetic field around its windings that produces a *counter* electromotive force (CEMF), which opposes the change in current. The inductor controls the rate at which C discharges. When C has completely discharged (t_2) $V_{out} = V_{CC}$ and current flow ceases. The magnetic field around L collapses, reversing the voltage polarity across it, which causes a current to flow that charges C to V_{CC} with the opposite polarity as before. The capacitor controls the rate at which L discharges. When the magnetic field has completely collapsed and C is fully charged, current once again ceases. $V_{out} = +V_C + V_{CC} = 2V_{CC}$ (t_3). C again discharges through

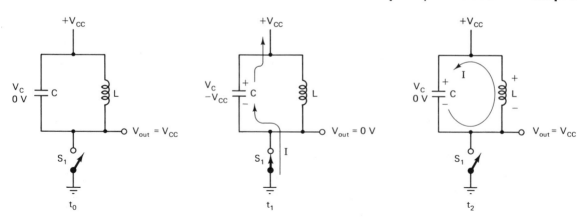

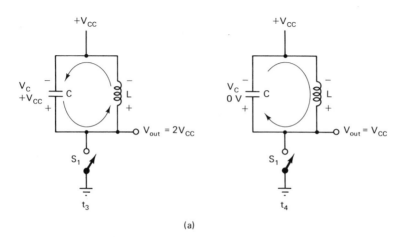

(a)

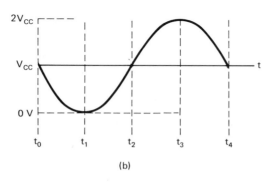

(b)

FIGURE 2-4   *LC* tank circuit: (a) oscillator action and flywheel effect; (b) output waveform.

L except in the opposite direction. When $V_C = 0$ V, current stops and the magnetic field around L again discharges, producing a current that charges C to V_{CC} with the original polarity. $V_{out} = -V_C + V_{CC} = 0$ V (t_4). The process just described repeats itself, producing a continuous sinusoidal waveform at V_{out}. The tank circuit is oscillating. Figure 2-4b shows a plot of V_{out} versus time.

The operation described above is called the *flywheel effect* and involves a continuous exchange of energy between the capacitor and inductor. The frequency of the waveform at V_{out} is determined by the rate at which the capacitor and inductor can take on and give off energy. By definition, resonance occurs when a capacitor and an inductor give off and take on energy at the same rate (i.e., $X_c = X_L$). Therefore, the frequency at V_{out} equals the resonant frequency of the tank circuit and is determined from the following mathematical expression:

$$F_o = \frac{1}{2\pi\sqrt{LC}} \tag{2-1}$$

Because the inductor and capacitor in Figure 2-4 are not ideal, they dissipate power whenever current is flowing in the circuit. Consequently, on each successive cycle of V_{out}, the amplitude of the output waveform decreases. This is called a *damped oscillation*. Eventually, all of the energy is dissipated as heat, V_{out} goes to zero volts, and oscillations cease. If oscillations are to continue, it is necessary that S_1 be closed at periodic intervals to supply energy back into the circuit and compensate for power dissipated in the imperfect components.

The two most common oscillator configurations using LC tank circuits are the Hartley and Colpitts oscillators.

Hartley oscillator.   Figure 2-5a shows the schematic diagram of a *Hartley oscillator*. The transistor amplifier (Q_1) provides the amplification necessary for unity gain at the resonant frequency. The coupling capacitor (C_c) provides the path for regenerative feedback, L_1 and C_1 are the frequency-determining components, and V_{CC} is the dc supply voltage.

Figure 2-5b shows the dc equivalent circuit for the Hartley oscillator. C_c is a blocking capacitor which isolates the dc base bias voltage and prevents it from being shorted to ground through L_{1b}. C_2 is also a blocking capacitor that prevents the collector supply voltage from being shorted to ground through L_{1a}. The radio frequency choke (RFC) is a dc short.

Figure 2-5c shows the ac equivalent circuit for the Hartley oscillator. C_c is a coupling capacitor for ac and provides a path for regenerative feedback from the tank circuit to the base of Q_1. C_2 couples ac signals from the collector of Q_1 to the tank circuit. The RFC looks open to ac and isolates the dc power supply from ac oscillations.

The Hartley oscillator operates as follows. On initial power-up, a multitude of frequencies appear at the collector of Q_1 and are coupled through C_2 into the tank circuit. The initial noise provides the energy necessary to charge C_1. Once C_1 is partially charged, oscillator action begins. The tank circuit will only oscillate efficiently at its resonant frequency. A portion of the oscillating tank circuit energy is dropped across

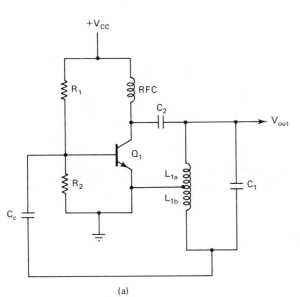

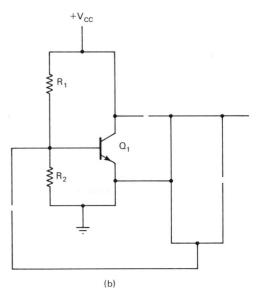

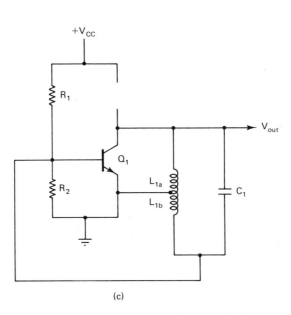

FIGURE 2-5 Hartley oscillator: (a) schematic diagram; (b) dc equivalent circuit; (c) ac equivalent circuit.

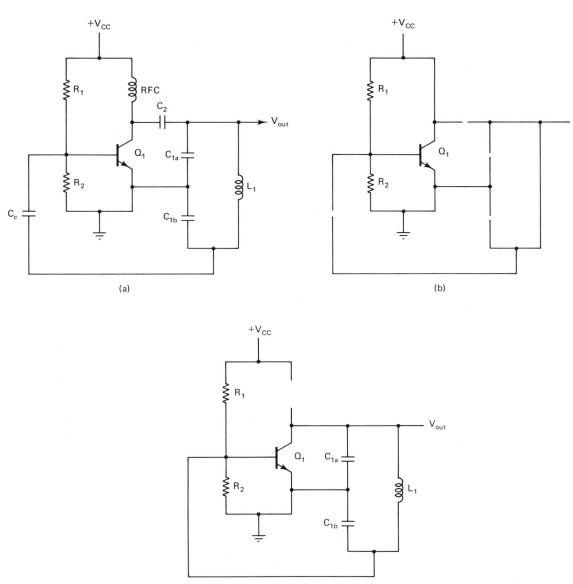

FIGURE 2-6 Colpitts oscillator: (a) schematic diagram; (b) dc equivalent circuit; (c) ac equivalent circuit.

L_{1b} and fed back to the base of Q_1, where it is amplified. The amplified signal appears at the collector 180° out of phase with the base signal. An additional 180° of phase shift is realized across L_1; consequently, the signal fed back to the base of Q_1 is amplified and phase shifted 360°. Thus the circuit is regenerative and will sustain oscillations with no external input.

The proportion of oscillating energy that is fed back to the base of Q_1 is determined by the ratio of L_{1b} to the total inductance $(L_{1a} + L_{1b})$. If insufficient energy is fed back, oscillations are damped. If excessive energy is fed back, the transistor saturates. Therefore, the position of the wiper on L_1 is adjusted until the amount of feedback energy is exactly what is required for self-sustained oscillations to continue.

The frequency of oscillation for the Hartley oscillator is closely approximated by the following formula:

$$F_o = \frac{1}{2\pi\sqrt{LC}}$$

where

$$L = L_{1a} + L_{1b}$$
$$C = C_1$$

Colpitts oscillator.   Figure 2-6a shows the schematic diagram of a *Colpitts oscillator*. The operation of the Colpitts oscillator is very similar to that of the Hartley oscillator. Q_1 provides the amplification, C_c provides the regenerative feedback path, L_1, C_{1a}, and C_{1b} are the frequency-determining components, and V_{CC} is the dc supply voltage.

Figure 2-6b shows the dc equivalent circuit for the Colpitts oscillator. C_2 is a blocking capacitor that prevents the collector supply voltage from appearing at the output. The RFC is a dc short.

Figure 2-6c shows the ac equivalent circuit for the Colpitts oscillator. C_c is a coupling capacitor for ac and provides the feedback path for regenerative feedback from the tank circuit to the base of Q_1. The RFC is open to ac and decouples oscillations from the dc power supply.

The operation of the Colpitts oscillator is almost identical to the Hartley oscillator. On initial power-up, noise appears at the collector of Q_1 and supplies energy to the tank circuit, causing it to begin oscillating. C_{1a} and C_{1b} make up an ac voltage divider. The voltage dropped across C_{1b} is fed back to the base of Q_1 through C_c. There is a 180° phase shift from the base to the collector of Q_1 and an additional 180° phase shift across C_1. Consequently, the total phase shift is 360° and the feedback signal is regenerative. The ratio of C_{1a} to C_{1b} determines the amplitude of the feedback signal.

The frequency of oscillation for the Colpitts oscillator is closely approximated by the following formula:

$$F_o = \frac{1}{2\pi\sqrt{LC}}$$

where

$$L = L_1$$
$$C = C_{1a}C_{1b}/(C_{1a} + C_{1b})$$

Crystal Oscillators

Frequency stability.   In communications systems, *frequency stability* is of primary importance. Frequency stability is the ability of an oscillator to remain at a fixed frequency. In the *LC* tank circuit and *RC* phase shift oscillators discussed previously, the frequency stability is inadequate for most radio communications applications. For example, commercial FM broadcast stations must maintain their carrier frequency to within ± 2 kHz of their assigned frequency. This is approximately a 0.002% tolerance. In commercial AM broadcasting, the maximum allowable carrier shift is only ± 20 Hz.

There are several factors that affect the stability of an oscillator. The most obvious are those that directly affect the value of the frequency-determining components. These include changes in the *L*, *C*, and *R* values due to environmental variations such as temperature and humidity and changes in the quiescent operating point of the transistor amplifier. Stability is also affected by ac ripple in the dc power supply. The frequency stability of an *RC* or *LC* oscillator can be greatly improved by regulating the dc power supply and minimizing the environmental variations. Also, special temperature-independent components can be used.

The FCC has established stringent regulations concerning the tolerances of radio-frequency carriers. Whenever the airway (free-space radio propagation) is used as the transmission medium, it is possible that transmissions from one source could interfere with transmissions from other sources if their transmit frequencies or transmission bandwidths overlap. Therefore, it is important that all sources maintain their frequency of operation within a specified tolerance.

Piezoelectric effect.   Simply stated, the *piezoelectric effect* occurs when mechanical vibrations in a crystal *lattice* structure generate electrical oscillations, and vice versa. When an alternating voltage is applied across a crystal at or near the natural resonant frequency of the crystal, the crystal will break into mechanical oscillations. This is called exciting a crystal into mechanical vibrations. The magnitude of the mechanical vibration is directly proportional to the magnitude of the applied voltage.

There are a number of natural crystal substances that exhibit piezoelectric properties: quartz, Rochelle salt, and tourmaline. The effect is most pronounced in Rochelle salt, which is why it is the substance commonly used in crystal microphones. Quartz, however, is used more often for frequency control in oscillators because of its *permanence*, low-*temperature coefficient*, and high *mechanical Q*. These properties are explained in subsequent discussions.

Crystal cuts.   In nature, complete quartz crystals have a hexagonal cross section with pointed ends, as shown in Figure 2-7a. There are three sets of *axes* associated

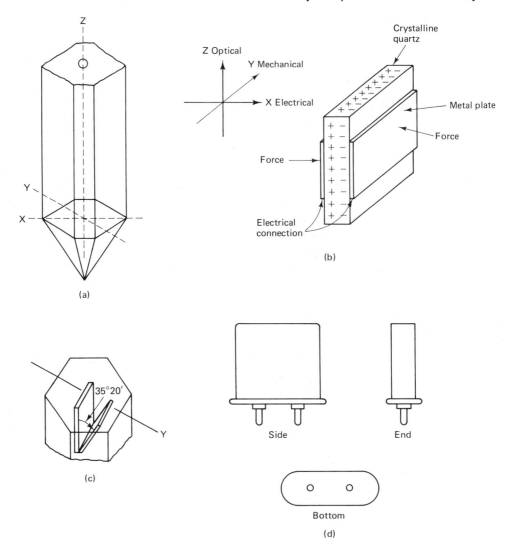

FIGURE 2-7   Quartz crystal: (a) basic crystal structure; (b) crystal axes; (c) crystal cuts; (d) crystal mountings.

with a crystal: *optical*, *electrical*, and *mechanical*. The longitudinal axis joining the points at the ends of the crystal is called the optical or *Z* axis. Electrical stresses applied to the optical axis do not produce the piezoelectric effect. The electrical or *X* axis passes diagonally through opposite corners of the hexagon. The axis that is perpendicular to the faces of the crystal is the *Y* or mechanical axis. Figure 2-7b shows the axes and the basic operation of a quartz crystal.

If a thin flat section is cut from a crystal such that the flat sides are perpendicular to an electrical axis, mechanical stresses along the *Y* axis will produce electrical charges on the flat sides. As the stress changes from compression to tension, and vice versa, the polarity of the charge is reversed. Conversely, if an alternating electrical charge is placed on the flat sides, a mechanical vibration is produced along the *Y* axis. This, of course, is the piezoelectric effect and is also exhibited when mechanical forces are applied across the faces of a crystal cut with its flat sides perpendicular to the *Y* axis. When a crystal wafer is cut parallel to the *Z* axis with its faces perpendicular to the *X* axis, it is called an *X*-cut crystal. When the faces are perpendicular to the *Y* axis, it is called a *Y*-cut crystal. A variety of cuts can be obtained by rotating the plane of the cut around one or more axes. If the *Y* cut is made at a 35°20′ angle from the vertical axis (Figure 2-7c), an AT cut is obtained. The type, length, and thickness of a cut and the mode of vibration determine the natural resonant frequency of the crystal. Resonant frequencies from a few kilohertz up to approximately 30 MHz are possible.

Crystal *wafers* are generally mounted in crystal *holders*, which include the mounting and housing assemblies. A crystal *unit* refers to the holder and the crystal itself. Figure 2-7d shows a common crystal mounting. Because a crystal's stability is somewhat temperature dependent, a crystal unit may be mounted in an *oven* to maintain a constant operating temperature.

Overtone crystal oscillator.   To increase the frequency of vibration of a quartz crystal, the quartz wafer is sliced thinner. This imposes an obvious physical limitation; the thinner the wafer, the more susceptible it is to damage and the less useful it becomes. The practical limit for fundamental-mode crystal oscillators is approximately 30 MHz. However, it is possible to operate the crystal in an *overtone* mode. That is, the oscillator is tuned to operate at the third, fifth, or even the seventh harmonic of the crystal's fundamental frequency. This increases the limits of crystal oscillators to approximately 200 MHz.

Temperature coefficient.   The natural resonant frequency of a crystal is influenced somewhat by its operating temperature. The magnitude of frequency change (ΔF) is expressed in hertz change per megahertz of crystal frequency per degree Celsius (Hz/MHz/°C). The fractional change in frequency is often given in parts per million (ppm) per °C. For example, a temperature coefficient of +20 Hz/MHz/°C is the same as +20 ppm/°C. If the direction of the frequency change is the same as the temperature change (i.e., an increase in temperature causes an increase in frequency and a decrease in temperature causes a decrease in frequency), it is called a *positive* temperature coefficient. If the change in frequency is in the opposite direction as the temperature (i.e., an increase in temperature causes a decrease in frequency and a decrease in temperature causes an increase in frequency), it is called a *negative* temperature coefficient. Mathematically, the change in frequency of a crystal is

$$\Delta F = k(F \times \Delta C) \qquad (2\text{-}3)$$

where

ΔF = change in frequency (Hz)
k = temperature coefficient (Hz/MHz/°C)

F = natural crystal frequency (MHz)

ΔC = change in temperature (°C)

The temperature coefficient (k) of a crystal varies depending on the type of crystal cut and its operating temperature. For a range of temperatures from approximately 20 to 50°C, both X- and Y-cut crystals have a temperature coefficient that is nearly constant. X-cut crystals are approximately 10 times more stable than Y-cut crystals. Typically, X-cut crystals have a temperature coefficient that ranges from -10 to -25 Hz/MHz/°C. Y-cut crystals have a temperature coefficient that ranges from approximately -25 to $+100$ Hz/MHz/°C.

Today there are so called *zero-coefficient* (GT-cut) crystals which have temperature coefficients as low as -1 to $+1$ Hz/MHz/°C. The GT-cut crystal is almost a perfect zero-coefficient crystal from freezing to boiling but is useful only at frequencies below a few hundred kilohertz.

EXAMPLE 2-1

For a 10-MHz crystal with a positive temperature coefficient of 10 Hz/MHz/°C, determine the frequency of operation if the temperature
(a) Increases 10°C.
(b) Decreases 5°C.

Solution　　(a) Substituting into Equation 2-3 yields

$$\Delta F = k(F \times \Delta C)$$

$$= 10(10 \times 10) = 1000 \text{ Hz}$$

$$F = 10 \text{ MHz} + \Delta F = 10 \text{ MHz} + 1 \text{ kHz} = 10.001 \text{ MHz}$$

(b) Again, substituting into Equation 2-3, we have

$$\Delta F = 10(10 \times -5) = -500 \text{ Hz}$$

$$F = 10 \text{ MHz} - 500 \text{ Hz} = 9.9995 \text{ MHz}$$

Crystal equivalent circuit.　　Figure 2-8a shows the equivalent electrical circuit for a crystal. Each of the electrical components is equivalent to a mechanical property of the crystal. C_2 is the actual capacitance formed between the electrodes of the crystal, with the crystal itself being the dielectric. C_1 is equivalent to the mechanical *compliance* of the crystal (also called *resilience* or *elasticity*). L_1 is equivalent to the mass of the crystal in vibration, and R is the mechanical *friction* loss. In a crystal, the mechanical *mass-to-friction ratio* (L/R) is quite high. Typical values of L range from 0.1 H to well over 100 H. Consequently, Q factors are quite high for crystals. Q factors in the range 10,000 to 100,000 are not uncommon (as compared to 100 to 1,000 for the discrete inductors used in LC tank circuits). This, of course, attributes to the high stability and accuracy of crystal oscillators as compared to LC tank circuit oscillators. Values for C_1 are typically less than 1 pF, while values for C_2 range between 4 and 40 pF.

Because there is a series and a parallel equivalent circuit for a crystal, there are also two resonant frequencies: a series and a parallel. The series impedance is the

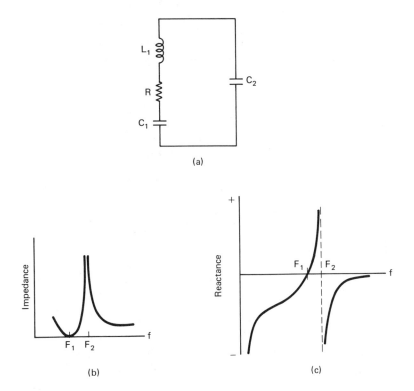

(a)

(b)

(c)

FIGURE 2-8 **Crystal equivalent circuit: (a) equivalent circuit; (b) impedance curve; (c) reactance curve.**

impedance of R, L, and C_1. The parallel impedance is the impedance of L and C_2. At extreme low frequencies, the series impedance of L, C_1, and R is very high and capacitive ($-$). This is shown in Figure 2-8c. As the frequency is increased, a point is reached where $X_L = X_{C1}$. At this frequency (F_1), the series impedance is minimum, resistive, and equal to R. As the frequency is increased even further (F_2), the series impedance becomes high and inductive ($+$). The parallel combination of L and C_2 causes the crystal to act like a parallel resonant circuit (maximum impedance at resonance). The difference between F_1 and F_2 is usually quite small (typically about 1% of the crystal operating frequency). A crystal can operate at either its series or parallel resonant frequency, depending on the circuit application. The relative steepness of the impedance curve (Fig. 2-8b) also attributes to the stability and accuracy of a crystal.

Crystal oscillator circuits.   Although there are many different crystal-based oscillator configurations, the most common are the discrete and IC Pierce and the RLC half-bridge. If you need very good frequency stability and reasonably simple circuitry, the discrete Pierce is a good choice. If low cost and simple digital interfacing capabilities are of primary concern, an IC-based Pierce oscillator will suffice. However, for the best frequency stability, the RLC half-bridge is the best choice.

Discrete Pierce oscillator. The Pierce crystal oscillator has many advantages. Its operating frequency spans the full fundamental crystal range (1 kHz to approximately 30 MHz). It uses relatively simple circuitry requiring few components (most medium-frequency versions require only one transistor). The Pierce oscillator design develops a high output signal power while dissipating very little power in the crystal. Finally, the short-term frequency stability of the Pierce crystal oscillator is excellent (this is because the in-circuit Q is almost as high as the crystal's internal Q). The only drawback to the Pierce oscillator is that it requires a high-gain amplifier (approximately 70). Consequently, you must use a single high-gain transistor of possibly even multiple amplifier stages.

Figure 2-9 shows a discrete 1-MHz Pierce crystal oscillator circuit. Q_1 provides all the gain necessary for self-sustained oscillations to occur. R_1 and C_1 provide a 65° phase lag to the feedback signal. The crystal impedance is basically resistive with a small inductive component. This impedance combined with C_2 provides an additional 115° of *RLC* phase lag. The transistor inverts the signal giving the circuit the necessary 360° phase shift. Because the crystal's load is primarily nonresistive (mostly the series combination of C_1 and C_2), this type of oscillator provides very good short-term frequency stability. Unfortunately, C_1 and C_2 introduce substantial losses, and consequently, the transistor must have a high current gain; this is obviously a drawback.

IC Pierce crystal oscillator. Figure 2-10 shows an IC-based Pierce crystal oscillator. Although it provides less frequency stability, it can be implemented using simple digital IC design and reduce costs substantially over conventional discrete designs.

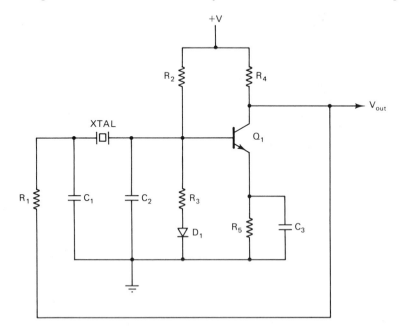

FIGURE 2-9   Discrete Pierce crystal oscillator.

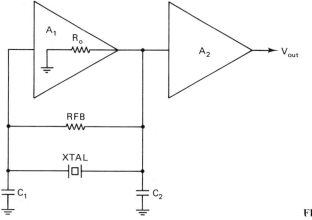

FIGURE 2-10 IC Pierce crystal oscillator.

To ensure that oscillations begin, RFB dc biases inverter A_1's input and output for class A operation. A_2 converts the output of A_1 to a full rail-to-rail swing reducing the rise and fall times and buffering A_1's output. The output resistance of A_1 combines with C_1 to provide the RC phase lag needed. CMOS versions operate up to approximately 2 MHz, and ECL versions operate as high as 20 MHz.

RLC half-bridge crystal oscillator. Figure 2-11 shows the Meacham version of the *RLC* half-bridge crystal oscillator. The original Meacham oscillator was developed in the 1940s and used a full four-arm bridge and a negative-temperature-coefficient tungsten lamp. The circuit configuration shown in Figure 2-10 uses only a two-arm bridge and employs a negative-temperature-coefficient thermister. Q_1 serves as a phase splitter and provides two 180° out-of-phase signals. The crystal must operate at its

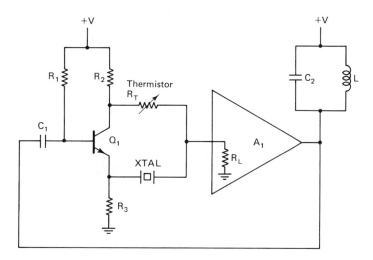

FIGURE 2-11 *RLC* half-bridge crystal oscillator.

series resonant frequency, so that its internal impedance is resistive and quite small. When oscillations begin, the signal amplitude increases gradually, decreasing the thermistor resistance until the bridge almost nulls. The amplitude of the oscillations stabilize and determine the final thermister resistance. The *LC* tank circuit at the output is tuned to the crystal's resonant frequency.

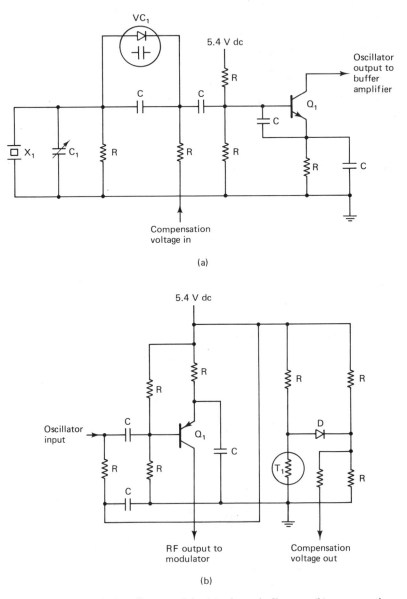

FIGURE 2-12   Crystal oscillator module: (a) schematic diagram; (b) compensation circuit.

Crystal oscillator module. A crystal oscillator *module* consists of a crystal-controlled oscillator and a voltage-variable component such as a varactor diode. The entire oscillator circuit is contained in a single metal *can*. A schematic diagram for a Colpitts crystal oscillator module is shown in Figure 2-12a. X_1 is the crystal itself and Q_1 is the active component for the oscillator. C_1 is a shunt capacitor that allows the crystal oscillator frequency to be varied over a narrow range of operating frequencies. VC_1 is a voltage-variable capacitor (*varicap* or *varactor diode*). A varactor diode is a specially constructed diode that exhibits capacitance when reverse biased, and by varying the reverse-bias voltage, the capacitance of the diode is changed. A varactor diode has a special depletion layer between the p- and n-type materials that is constructed with various degrees and types of doping material (the term "graded junction" is often used when describing varactor diode fabrication). Figure 2-13 shows the capacitance versus reverse-bias characteristics of a typical varactor diode. The capacitance of a varactor diode is approximated as

$$C_d = \frac{C}{\sqrt{1 + 2/V_r}}$$

(2-4)

where

C = diode capacitance with zero reverse bias
V_r = diode reverse bias
C_d = reverse-biased capacitance

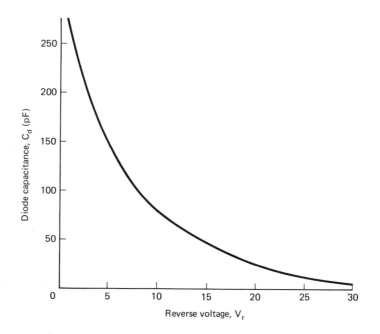

FIGURE 2-13 Varactor diode characteristics.

The frequency at which the crystal oscillates can be adjusted slightly by changing the capacitance of VC_1 (i.e., changing the value of the reverse-bias voltage). The varactor diode, in conjunction with a temperature-compensating module, provide instant frequency compensation for variations caused by changes in the temperature. A temperature-compensating module is shown in Figure 2-12b. The compensator module includes a buffer-amplifier (Q_1) and a temperature-compensating network (T_1). T_1 is a negative-temperature-coefficient thermister. When the temperature falls below the threshold value for the thermister, the compensation voltage increases to maintain the proper voltage on the varactor diode in the oscillator module. Compensation modules are available that can compensate for a frequency stability of $+0.0005\%$ from -30 to $+80°C$.

Negative resistance oscillator: tunnel diode.   Although there are several devices currently available that exhibit negative resistance properties, the *tunnel diode* (invented in the 1950s) is one of the most common and thus will be used as an example.

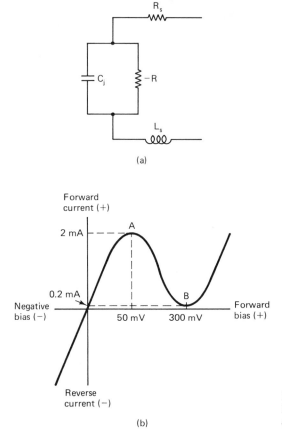

(a)

(b)

FIGURE 2-14   **Tunnel diode: (a) equivalent circuit; (b) voltage-versus-current characteristics.**

The tunnel diode (or *Esaki diode*, as it is sometimes called) is a thin-junction diode which, when slightly forward biased, exhibits negative resistance. Because of its thin junction it has an extremely short transit time and is therefore ideally suited for high-frequency applications.

Figure 2-14a shows the equivalent circuit for a forward-biased tunnel diode. Except for extremely high frequencies (above approximately 2 GHz), the series resistance (R_s) and inductance (L_s) can be ignored. Consequently, for practical applications, the equivalent circuit is reduced to the parallel combination of junction capacitance (C_j) and the negative resistance ($-R$). The junction capacitance of the tunnel diode is highly dependent on the dc bias voltage and the temperature. As a result, using a tunnel diode with a tuned circuit yields a very low Q and relatively unstable oscillator. However, if a high-Q mechanical cavity is loosely coupled to the diode, a very stable oscillator is obtained which is relatively independent of changes in temperature or dc bias.

A tunnel diode is similar to the conventional *p-n* junction semiconductor diode except that the semiconductor materials are much more heavily doped (perhaps as much as 1000 times more). Because of the heavy doping, the depletion layer is extremely thin (typically about 0.01 μm) and therefore has a very small junction capacitance and has an extremely short transit time. Figure 2-14b shows the voltage-versus-current characteristics of a typical tunnel diode. For forward bias voltages between 0 and 50 mV and greater than 300 mV, the characteristics resemble those of a standard *p-n* junction diode. In the portion of the curve between point *A* (the peak) and point *B* (the valley), an increase in forward bias causes a corresponding reduction in forward current (i.e., negative resistance). If the tunnel diode is operated between points *A* and *B*, self-sustained oscillations can be obtained.

Large-Scale Integration Oscillators

In recent years, the use of large-scale integration integrated circuits for frequency generation has increased at a tremendous rate because they have excellent frequency stability and a wide tuning range. Figure 2-15 shows the specification sheet for the XR-2207 monolithic voltage controlled oscillator. The XR-2207 is a precision oscillator that can provide simultaneous sine and square wave outputs over a frequency range of 0.01 Hz to 1 MHz. It is ideally suited for a wide range of applications in electronic communications including frequency modulation, frequency shift keying, sweep or tone generation, as well as for phase locked loop applications.

Figure 2-16 on page 66 shows the specification sheet for the XR-2209 monolithic precision oscillator. The XR-2209 is a variable frequency oscillator circuit featuring excellent temperature stability and a wide linear sweep range. The circuit provides simultaneous triangle and square-wave outputs over a frequency range of 0.01 Hz to 1 MHz. The XR-2209 is ideally suited for frequency modulation, voltage-to-frequency or current-to-frequency conversion, sweep or tone generation, as well as phase locked loop applications.

Voltage-Controlled Oscillator

GENERAL DESCRIPTION

The XR-2207 is a monolithic voltage-controlled oscillator (VCO) integrated circuit featuring excellent frequency stability and a wide tuning range. The circuit provides simultaneous triangle and squarewave outputs over a frequency range of 0.01 Hz to 1 MHz. It is ideally suited for FM, FSK, and sweep or tone generation, as well as for phase-locked loop applications.

The XR-2207 has a typical drift specification of 20 ppm/°C. The oscillator frequency can be linearly swept over a 1000:1 range with an external control voltage; and the duty cycle of both the triangle and the squarewave outputs can be varied from 0.1% to 99.9% to generate stable pulse and sawtooth waveforms.

FEATURES

Excellent Temperature Stability (20 ppm/°C)
Linear Frequency Sweep
Adjustable Duty Cycle (0.1% to 99.9%)
Two or Four Level FSK Capability
Wide Sweep Range (1000:1 Min)
Logic Compatible Input and Output Levels
Wide Supply Voltage Range (±4V to ±13V)
Low Supply Sensitivity (0.1%/V)
Wide Frequency Range (0.01 Hz to 1 MHz)
Simultaneous Triangle and Squarewave Outputs

APPLICATIONS

FSK Generation
Voltage and Current-to-Frequency Conversion
Stable Phase-Locked Loop
Waveform Generation
  Triangle, Sawtooth, Pulse, Squarewave
FM and Sweep Generation

ABSOLUTE MAXIMUM RATINGS

Power Supply	26 V
Power Dissipation (package limitation)	
Ceramic package	750 mW
Derate above +25°C	6.0 mW/°C
Plastic package	625 mW
Derate above +25°C	5 mW/°C
Storage Temperature Range	-65°C to +150°C

FUNCTIONAL BLOCK DIAGRAM

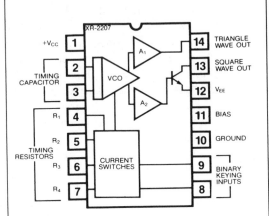

ORDERING INFORMATION

Part Number	Package	Operating Temperature
XR-2207M	Ceramic	-55°C to +125°C
XR-2207N	Ceramic	0°C to +70°C
XR-2207P	Plastic	0°C to +70°C
XR-2207CN	Ceramic	0°C to +70°C
XR-2207CP	Plastic	0°C to +70°C

SYSTEM DESCRIPTION

The XR-2207 utilizes four main functional blocks for frequency generation. These are a voltage controlled oscillator (VCO), four current switches which are activated by binary keying inputs, and two buffer amplifiers for triangle and squarewave outputs. The VCO is actually a current controlled oscillator which gets its input from the current switches. As the output frequency is proportional to the input current, the VCO produces four discrete output frequencies. Two binary input pins determine which timing currents are channelled to the VCO. These currents are set by resistors to ground from each of the four timing terminals.

The triangle ouput buffer provides a low impedance output (typically 10 Ω) while the squarewave is a open-collector type. A programmable reference point allows using the XR-2207 in either single or split supply configurations.

EXAR Corporation, 750 Palomar Avenue, Sunnyvale, CA 94086 • (408) 732-7970 • TWX 910-339-9233

FIGURE 2-15   XR-2207 precision oscillator. (Courtesy of EXAR Corporation.)

ELECTRICAL CHARACTERISTICS

Test Conditions: Test Circuit of Figure 1, $V^+ = V^- = 6V$, $T_A = +25°C$, C = 5000 pF, $R_1 = R_2 = R_3 = R_4 = 20$ KΩ, $R_L = 4.7$ KΩ, Binary Inputs grounded, S_1 and S_2 closed unless otherwise specified.

PARAMETERS	XR-2207/XR-2207M			XR-2207C			UNITS	CONDITIONS
	MIN.	TYP.	MAX.	MIN.	TYP.	MAX.		
GENERAL CHARACTERISTICS								
Supply Voltage								
Single Supply	8		26	8		26	V	See Figure 3
Split Supplies	±4		±13	±4		±13	V	
Supply Current								
Single Supply		5	7		5	8	mA	Measured at pin 1, S_1 and S_2 open; See Figure 2
Split Supplies								
Positive		5	7		5	8	mA	Measured at pin 1, S_1, S_2 open
Negative		4	6		4	7	mA	Measured at pin 12, S_1, S_2 open
OSCILLATOR SECTION — FREQUENCY CHARACTERISTICS								
Upper Frequency Limit	0.5	1.0		0.5	1.0		MHz	C = 500 pF, R_3 = 2 KΩ
Lowest Practical Frequency		0.01			0.01		Hz	C = 50 μF, R_3 = 2 MΩ
Frequency Accuracy		±1	±3		±1	±5	% of f_o	
Frequency Matching		0.5			0.5		% of f_o	
Frequency Stability								
Temperature		20	50		30		ppm/°C	$0° < T_A < 75°C$
Power Supply		0.15			0.15		%/V	
Sweep Range	1000:1	3000:1		1000:1			f_H/f_L	R_3 = 1.5 KΩ for f_{H1}; R_3 = 2 MΩ for f_L
Sweep Linearity							%	C = 5000 pF
10:1 Sweep		1	2		1.5			f_H = 10 kHz, f_L = 1 kHz
1000:1 Sweep		5			5			f_H = 100 kHz, f_L = 100 Hz
FM Distortion		0.1			0.1		%	±10% FM Deviation
Recommended Range of Timing Resistors	1.5		2000	1.5		2000	KΩ	See Characteristic Curves
Impedance at Timing Pins		75			75		Ω	Measured at pins 4, 5, 6, or 7
DC Level at Timing Terminals		10			10		mV	
BINARY KEYING INPUTS								
Switching Threshold	1.4	2.2	2.8	1.4	2.2	2.8	V	Measured at pins 8 and 9, Referenced to pin 10
Input Impedance		5			5		KΩ	
OUTPUT CHARACTERISTICS								
Triangle Output								Measured at pin 13
Amplitude	4	6		4	6		V_{pp}	
Impedance		10			10		Ω	
DC Level		+100			+100		mV	Referenced to pin 10
Linearity		0.1			0.1		%	From 10% to 90% to swing
Squarewave Output								Measured at pin 13, S_2 closed
Amplitude	11	12		11	12		V_{pp}	
Saturation Voltage		0.2	0.4		0.2	0.4	V	Referenced to pin 12
Rise Time		200			200		nsec	$C_L \leq 10$ pF
Fall Time		20			20		nsec	$C_L \leq 10$ pF

PRECAUTIONS

The following precautions should be observed when operating the XR-2207 family of integrated circuits:

1. Pulling excessive current from the timing terminals will adversely effect the temperature stability of the circuit. To minimize this disturbance, it is recommended that the *total current drawn* from pins 4, 5, 6, and 7 be limited to ≤6 mA. In addition, permanent damage to the device may occur if the total timing current exceeds 10 mA.
2. Terminals 2, 3, 4, 5, 6, and 7 have very low internal impedance and should, therefore, be protected from accidental shorting to ground or the supply voltages.
3. The keying logic pulse amplitude should not exceed the supply voltage.

DEFINITIONS OF TERMS

Frequency Accuracy

The difference between the actual operating frequency and the theoretical frequency determined from the design equations in Table 1, expressed as a percent of the calculated value.

Frequency Matching

The change in operating frequency as different timing terminals are activated for fixed timing resistor and timing capacitor values, expressed as a percent of the original operating frequency.

Binary Input Switching Threshold

The logic level at Pins 8 and 9 activate the binary current switches. The voltage level is referenced to Pin 10.

Frequency Sweep Range

The ratio of the highest and lowest operating frequencies (f_H/f_L) obtainable with a given value of timing capacitor.

Sweep Linearity

The maximum deviation of the sweep characteristics from a best-fit straight line extending over the frequency range.

Triangle Nonlinearity

The maximum deviation from a best-fit straight line extending along the rising and falling edges of the waveform, measured between 10% and 90% of each excursion.

PRINCIPLES OF OPERATION

The XR-2207 oscillator is a modified emitter-couple multivibrator type. As shown in the Functional Block Diagram, the oscillator also contains four current switches which activate the timing terminals, Pins 4, 5, 6, and 7. The oscillator frequency is inversely proportional to the value of timing capacitance, C, between Pins 2 and 3, and directly proportional to the total current, I_T, pulled out of the activated timing terminals.

Figure 3 provides greater detail of the oscillator control mechanism. Timing Pins 4, 5, 6, and 7 correspond to the emitters of switching transistors T_1, T_2, T_3 and T_4 respectively, which are internal to the integrated circuit. The current switches (and corresponding timing terminals) are activated by external logic signals applied to the keying terminals, Pins 8 and 9. The logic table for keying is given in Table 1.

As an example, logic inputs of 0, 0 at Pins 8 and 9 (i.e., both inputs "low") will result in turning on transistor pairs T_3, and only the timing terminal 6 will be activated. Under this condition, the total timing current, I_T, is equal to current I_3 pulled from Pin 6. This current is determined by external resistor R_3, resulting in a frequency $f_0 = f_1 = 1/R_3C$ Hz.

It is important to observe that timing Pins 4, 5, 6, and 7 are low impedance points in the circuit. Care must be taken to avoid shorting these pins to the supply voltages or to ground.

EQUIVALENT SCHEMATIC DIAGRAM

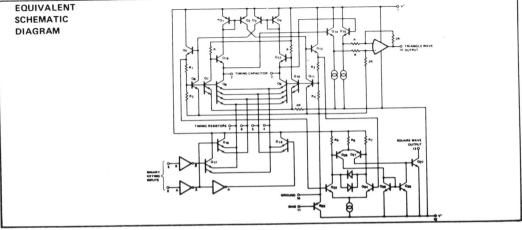

XR-2207

OPERATING INSTRUCTIONS

Split Supply Operation

Figure 4 is the recommended circuit connections for split supply operation. With the generalized connections of Figure 4A, the frequency of operation is determined by timing capacitor, C, and the activated timing resistors (R_1 through R_4). The timing resistors are activated by the logic signals at the binary keying inputs (Pins 8 and 9), as shown in the logic table, Table 1. If a fixed frequency of oscillation is required, the circuit connections can be simplified as shown in Figure 4B. In this connection, the input logic is set at (0,0) and the operating frequency is equal to $(1/R_3C)$ Hz.

The squarewave output is obtained at Pin 13 and has a peak-to-peak voltage swing equal to the supply voltages. This output is an "open-collector" type and requires an external pull-up load resistor (nominally 5 KΩ) to the positive supply. The triangle waveform obtained at Pin 14 is centered about ground and has a peak amplitude of $V^+/2$.

The circuit operates with supply voltages ranging from ±4 V to ±13 V. Minimum drift occurs with ±6 volt supplies. For operation with unequal supply voltages, see Figure 7.

The logic levels at the keying inputs (Pins 8 and 9) are referenced to ground. A logic "0" corresponds to a keying voltage $V_k < 1.4$ V, and a logic "1" corresponds to $V_k > 3$ V. An open circuit at the keying inputs also corresponds to a "0" level.

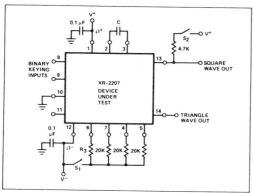

Figure 1. Test Circuit for Split Supply Operation

LOGIC LEVEL		SELECTED TIMING PINS	FREQUENCY	DEFINITIONS
A	B			
0	0	6	f_1	$f_1=1/R_3C, \Delta f_1=1/R_4C$
0	1	6 and 7	$f_1 + \Delta f_1$	$f_2=1/R_2C, \Delta f_2=1/R_1C$
1	0	5	f_2	Logic Levels: 0=ground
1	1	4 and 5	$f_2 + \Delta f_2$	1=>3V

Table 1. Logic Table for Binary Keying Controls

Note: For Single Supply Operation, Logic Levels are Referenced to Voltage at Pin 10

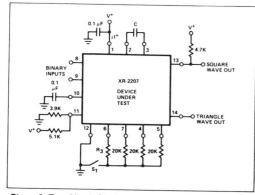

Figure 2. Test Circuit for Single Supply Operation

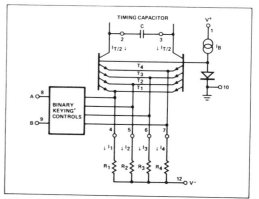

Figure 3. Simplified Schematic of Frequency Control Mechanism

Single Supply Operation

The circuit should be interconnected as shown in Figure 12 for single supply operation. Pin 12 should be grounded, and Pin 11 biased from V$^+$ through a resistive divider to a value of bias voltage between V$^+$/3 and V$^+$/2. Pin 10 is bypassed to ground through a 1 μF capacitor.

For single supply operation, the dc voltage at Pin 10 and the timing terminals (Pins 4 through 7) are equal and approximately 0.6 V below V$_B$, the bias voltage at Pin 11. The logic levels at the binary keying terminals are referenced to the voltage at Pin 10.

For a fixed frequency of f$_3$ = 1/R$_3$C, the external circuit connections can be simplified as shown in Figure 12B.

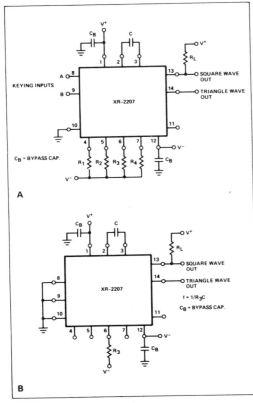

Figure 4. Split Supply Operation
A: General   B: Fixed Frequency

SELECTION OF EXTERNAL COMPONENTS

Timing Capacitor (Pins 2 and 3)

The oscillator frequency is inversely proportional to the timing capacitor, C, as indicated in Table 1. The minimum capacitance value is limited by stray capacitances and the maximum value by physical size and leakage current considerations. Recommended values range from 100 pF to 100 μF. The capacitor should be non-polar.

Timing Resistors (Pins 4, 5, 6, and 7)

The timing resistors determine the total timing current, I$_T$, available to charge the timing capacitor. Values for timing resistors can range from 2 KΩ to 2MΩ; however, for optimum temperature and power supply stability, recommended values are 4 KΩ to 200 KΩ (see Figures 8, 10 and 11). To avoid parasitic pick up, timing resistor leads should be kept as short as possible. For noisy environments, unused or deactivated timing terminals should be bypassed to ground through 0.1 μF capacitors.

Supply Voltage (Pins 1 and 12)

The XR-2207 is designed to operate over a power supply range of ±4 V to ±13 V for split supplies, or 8 V to 26 V for single supplies. At high supply voltages, the frequency sweep range is reduced (see Figures 7 and 8). Performance is optimum for ±6 V, or 12 V single supply operation.

Binary Keying Inputs (Pins 8 and 9)

The internal impedance at these pins is approximately 5 KΩ. Keying levels are < 1.4 V for "zero" and >3 V for "one" logic levels referenced to the dc voltage at Pin 10 (see Table 1).

Bias for Single Supply (Pin 11)

For single supply operation, Pin 11 should be externally biased to a potential between V$^+$/3 and V$^+$/2 volts (see Figure 12). The bias current at Pin 11 is nominally 5% of the total oscillation timing current, I$_T$.

Ground (Pin 10)

For split supply operation, this pin serves as circuit ground. For single supply operation, Pin 10 should be ac grounded through a 1 μF bypass capacitor. During split supply operation, a ground current of 2I$_T$ flows out of this terminal, where I$_T$ is the total timing current.

XR-2207

TYPICAL CHARACTERISTICS

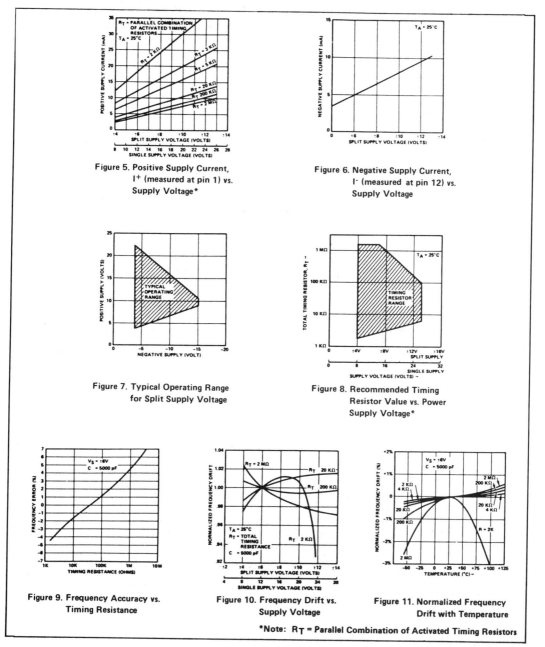

Figure 5. Positive Supply Current, I⁺ (measured at pin 1) vs. Supply Voltage*

Figure 6. Negative Supply Current, I⁻ (measured at pin 12) vs. Supply Voltage

Figure 7. Typical Operating Range for Split Supply Voltage

Figure 8. Recommended Timing Resistor Value vs. Power Supply Voltage*

Figure 9. Frequency Accuracy vs. Timing Resistance

Figure 10. Frequency Drift vs. Supply Voltage

Figure 11. Normalized Frequency Drift with Temperature

*Note: R_T = Parallel Combination of Activated Timing Resistors

6

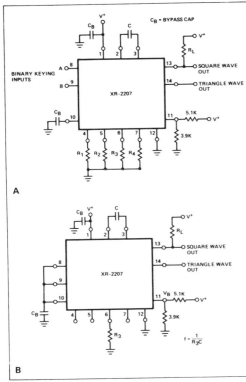

Figure 12. Single Supply Operation

A: General   B: Fixed Frequency

Squarewave Output (Pin 13)

The squarewave output at Pin 13 is an open collector stage capable of sinking up to 20 mA of load current. R_L serves as a pull-up load resistor for this output. Recommended values for R_L range from 1 $K\Omega$ to 100 $K\Omega$.

Triangle Output (Pin 14)

The output at Pin 14 is a triangle wave with a peak swing of approximately one-half of the total supply voltage. Pin 14 has a very low output impedance of 10 Ω and is internally protected against short circuits.

Bypass Capacitors

The recommended value for bypass capacitors is 1 μF, although, larger values are required for very low frequency operation.

Frequency Control (Sweep and FM)

The frequency of operation is controlled by varying the total timing current, I_T, drawn from the activated timing Pins 4, 5, 6, or 7. The timing current can be modulated by applying a control voltage, V_C, to the activated timing pin through a series resistor R_C, as shown in Figures 13 and 14.

For split supply operation, a *negative* control voltage, V_C, applied to the circuits of Figures 13 and 14 causes the total timing current, I_T, and the frequency, to increase.

As an example, in the circuit of Figure 13, the binary keying inputs are grounded. Therefore, only timing Pin 6 is activated. The frequency of operation, normally $f = \dfrac{1}{R_3 C}$

is now proportional to the control voltage, V_C, and determined as:

$$f = \frac{1}{R_3 C}\left[1 - \frac{V_C R_3}{R_C V^-}\right] Hz$$

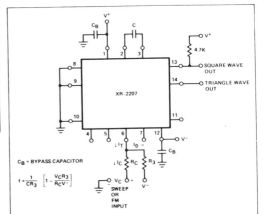

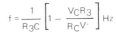

Figure 13. Frequency Sweep Operation

The frequency, f, will increase as the control voltage is made more negative. If R_3 = 2 $M\Omega$, R_C = 2 $K\Omega$, C = 5000 pF, then a 1000:1 frequency sweep would result for a negative sweep voltage $V_C \cong V^-$. The voltage to frequency conversion gain, K, is controlled by the series resistance R_C and can be expressed as:

$$K = \frac{\Delta f}{\Delta V_C} = -\frac{1}{R_C C V^-} Hz/volt$$

7

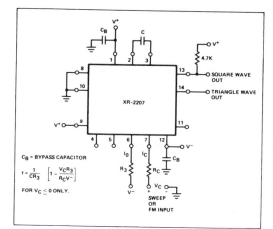

Figure 14. Alternate Frequency Sweep Operation

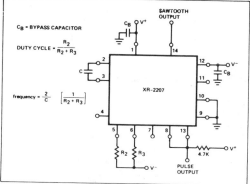

Figure 15. Sawtooth and Pulse Outputs

The circuit of Figure 13 can operate both with positive and negative values of control voltage. However, for positive values of V_C with small (R_C/R_3) ratio, the direction of the timing current I_T is reversed and the oscillations will stop.

Figure 14 shows an alternate circuit for frequency control where two timing pins, 6 and 7, are activated. The frequency and the conversion gain expressions are the same as before, except that the circuit will operate only with negative values of V_C. For $V_C > 0$, Pin 7 becomes deactivated and the frequency is fixed at

$$f = \frac{1}{R_3 C}$$

CAUTION

For operation of the circuit, total timing current I_T must be less than 6 mA over the frequency control range.

Duty Cycle Control

The duty cycle of the output waveforms can be controlled by frequency shift keying at the end of every half cycle of oscillator output. This is accomplished by connecting one or both of the binary keying inputs (Pins 8 or 9) to the squarewave output at Pin 13. The output waveforms can then be converted to positive or negative pulses and sawtooth waveforms.

Figure 15 is the recommended circuit connection for duty cycle control. Pin 8 is shorted to Pin 13 so that the circuit switches between the "0, 0" and the "1, 0" logic states given in Table 1. Timing Pin 5 is activated when the output is high, and the timing pin is activated when the squarewave output goes to a low state.

The duty cycle of the output waveform is given as:

$$\text{Duty Cycle} = \frac{R_2}{R_2 + R_3}$$

and can be varied from 0.1% to 99.9% by proper choice of timing resistors. The frequency of oscillation, f, is given as:

$$f = \frac{2}{C}\left[\frac{1}{R_2 + R_3}\right]$$

The frequency can be modulated or swept without changing the duty cycle by connecting R_2 and R_3 to a common control voltage V_C, instead of to V^- (see Figure 13). The sawtooth and the pulse output waveforms are shown in Figure 16.

On — Off Keying

The XR-2207 can be keyed on and off by simply activating an open circuited timing pin. Under certain conditions, the circuit may exhibit very low frequency (<1 Hz) residual oscillations in the off state due to internal bias currents. If this effect is undesirable, it can be eliminated by connecting a 10 MΩ resistor from Pin 3 to V^+.

APPLICATIONS

Two-Channel FSK Generator (Modem Transmitter)

The multi-level frequency shift-keying capability of XR-2207 makes it ideally suited for two-channel FSK generation. A recommended circuit connection for this application is shown in Figure 17.

For two-channel FSK generation, the "mark" and "space" frequencies of the respective channels are determined by the timing resistor pairs (R_1, R_2) and (R_3, R_4). Pin 8 is the "channel-select" control in accord with Table 1. For a "high" logic level at Pin 8, the timing resistors R_1 and R_2 are activated. Similarly, for a "low" logic level, timing resistors R_3 and R_4 are enabled.

The "high" and "low" logic levels at Pin 9 determine the respective high and low frequencies within the selected FSK channel.

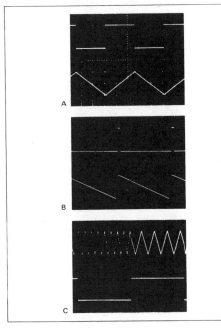

Figure 16. Output Waveforms
A: Squarewave and Triangle Outputs
B: Pulse and Sawtooth Outputs
C: Frequency-Shift Keyed Output
Top: FSK Output with $f_2 = 2f_1$
Bottom: Keying Logic Input

Recommended component values for various commonly used FSK frequencies are given in Table 2. When only a single FSK channel is used, the remaining channel can be deactivated by connecting Pin 8 to either V^+ or ground. In this case, the unused timing resistors can also be omitted from the circuit.

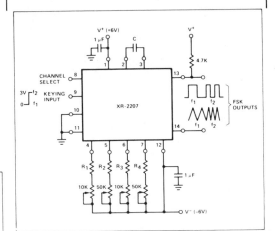

Figure 17. Multi-Channel FSK Generation

The low and high frequencies, f_1 and f_2, for a given FSK channel can be fine tuned using potentiometers connected in series with respective timing resistors. In fine tuning the frequencies, f_1 should be set first with the logic level at Pin 9 in a "low" level.

Typical frequency drift of the circuit for $0°C$ to $+70°C$ operation is ±0.2%. Since the frequency stability is directly related to the external timing components, care must be taken to use timing components with low temperature coefficients.

FSK Transceiver (Full-Duplex Modem)

The XR-2207 can be used in conjunction with the XR-210 FSK Demodulator to form a full-duplex FSK transceiver, or modem. A recommended circuit connection for this application is shown in Figure 18. Table 2 shown the recommended component values for 300 Baud (103-type) and 1200 Baud (202-type) modem applications.

XR-2207

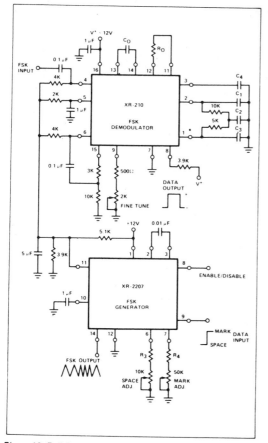

Figure 18. Full Duplex FSK Modem using XR-210 and XR-2207
(See Table 2 for Component Values)

OPERATING CONDITIONS	TYPICAL COMPONENT	VALUE
300 Baud	XR-210	XR-2207
Low Band: f_1 = 1070 Hz f_2 = 1270 Hz	R_0 = 5.1 kΩ, C_0 = 0.22 μF C_1 = C_2 = 0.047 μF C_3 = 0.033 μF	R_3 = 91 k R_4 = 470 k
High Band: f_1 = 2025 Hz f_2 = 2225 Hz	R_0 = 8.2 kΩ, C_0 = 0.1 μF C_1 = C_2 = C_3 = 0.033 μF	R_3 = 47 k R_4 = 470 k
1200 Baud f_1 = 1200 Hz f_2 = 2200 Hz	R_0 = 2 kΩ, C_0 = 0.14 μF C_1 = 0.033 μF C_3 = 0.02 μF C_2 = 0.01 μF	R_3 = 75 k R_4 = 91 k

Table 2. Recommended Component Values for Full Duplex
FSK Modem of Figure 18

XR-2209

Precision Oscillator

GENERAL DESCRIPTION

The XR-2209 is a monolithic variable frequency oscillator circuit featuring excellent temperature stability and a wide linear sweep range. The circuit provides simultaneous triangle and squarewave outputs over a frequency range of 0.01 Hz to 1 MHz. The frequency is set by an external RC product. It is ideally suited for frequency modulation, voltage to frequency or current to frequency conversion, sweep or tone generation as well as for phase-locked loop applications when used in conjunction with a phase comparator such as the XR-2208.

The circuit is comprised of three functional blocks: a variable frequency oscillator which generates the basic periodic waveforms and two buffer amplifiers for the triangle and the squarewave outputs.

The oscillator frequency is set by an external capacitor, C, and the timing resistor R. With no sweep signal applied, the frequency of oscillation is equal to 1/RC. The XR-2209 has a typical drift specification of 20 ppm/°C. Its frequency can be linearly swept over a 1000:1 range with an external control signal.

FEATURES

Excellent Temperature Stability (20 ppm/°C)
Linear Frequency Sweep
Wide Sweep Range (1000:1 Min)
Wide Supply Voltage Range (±4V to ±13V)
Low Supply Sensitivity (0.15%/V)
Wide Frequency Range (0.01 Hz to 1 MHz)
Simultaneous Triangle and Squarewave Outputs

APPLICATIONS

Voltage and Current-to-Frequency Conversion
Stable Phase-Locked Loop
Waveform Generation
FM and Sweep Generation

ABSOLUTE MAXIMUM RATINGS

Power Supply	26 volts
Power Dissipation (package limitation)	
Ceramic package	385 mW
Plastic Package	300 mW
Derate above +25°C	2.5 mW/°C
Temperature Range	
Operating	
XR-2209M	−55°C to +125°C
XR-2209C	0°C to +75°C
Storage	−65°C to +150°C

AVAILABLE TYPES

Part Number	Package	Operating Temperature
XR-2209M	Ceramic	−55°C to +125°C
XR-2209CN	Ceramic	0°C to +75°C
XR-2209CP	Plastic	0°C to +75°C

EQUIVALENT SCHEMATIC DIAGRAM

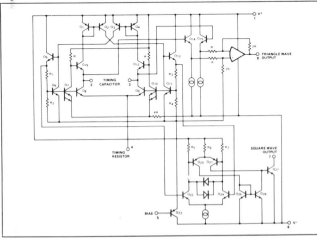

FUNCTIONAL BLOCK DIAGRAM

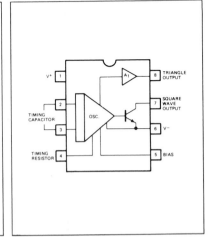

EXAR INTEGRATED SYSTEMS, INC.
750 Palomar Ave., P.O. Box 62229, Sunnyvale, CA 94088
(408) 732-7970   TWX 910-339-9233

FIGURE 2-16   XR-2209 precision oscillator. (Courtesy of EXAR Corporation.)

ELECTRICAL CHARACTERISTICS – PRELIMINARY

Test Conditions: Test Circuit of Figure 1, $V^+ = V^- = 6V$, $T_A = +25°C$, $C = 5000$ pF, $R = 20$ KΩ, $R_L = 4.7$ KΩ. S_1 and S_2 closed unless otherwise specified.

PARAMETERS	XR-2209M			XR-2209C			UNITS	CONDITIONS
	MIN.	TYP.	MAX.	MIN.	TYP.	MAX.		
GENERAL CHARACTERISTICS								
Supply Voltage								
Single Supply	8		26	8		26	V	See Figure 2
Split Supplies	±4		±13	±4		±13	V	See Figure 1
Supply Current								
Single Supply		5	7		5	8	mA	Measured at pin 1, S_1, S_2 open
								See Figure 2
Split Supplies								
Positive		5	7		5	8	mA	Measured at pin 1, S_1, S_2 open
Negative		4	6		4	7	mA	Measured at pin 4, S_1, S_2 open
OSCILLATOR SECTION – FREQUENCY CHARACTERISTICS								
Upper Frequency Limit	0.5	1.0		0.5	1.0		MHz	$C = 500$ pF, $R = 2$ KΩ
Lowest Practical Frequency		0.01			0.01		Hz	$C = 50$ μF, $R = 2$ MΩ
Frequency Accuracy		±1	±3		±1	±5	% of f_o	
Frequency Stability								
Temperature		20	50		30		ppm/°C	$0° < T_A < 75°C$
Power Supply		0.15			0.15		%/V	
Sweep Range	1000:1	3000:1		1000:1			f_H/f_L	$R = 1.5$ KΩ for f_{H1}
								$R = 2$ MΩ for f_L
Sweep Linearity							%	$C = 5000$ pF
10:1 Sweep		1	2		1.5			$f_H = 10$ kHz, $f_L = 1$ kHz
1000:1 Sweep		5			5			$f_H = 100$ kHz, $f_L = 100$Hz
FM Distortion		0.1			0.1		%	±10% FM Deviation
Recommended Range of Timing Resistors	1.5		2000	1.5		2000	KΩ	See Characteristic Curves
Impedance at Timing Pin		75			75		Ω	Measured at pin 4
OUTPUT CHARACTERISTICS								
Triangle Output								Measured at pin 8
Amplitude	4	6		4	6		Vpp	
Impedance		10			10		Ω	
Linearity		0.1			0.1		%	10% to 90% of swing
Squarewave Output								Measured at pin 7, S_2 closed
Amplitude	11	12		11	12		Vpp	
Saturation Voltage		0.2	0.4		0.2	0.4	V	Referenced to pin 6
Rise Time		200			200		nsec	$C_L \leq 10$ pF, $R_L = 4.7$ KΩ
Fall Time		20			20		nsec	$C_L \leq 10$ pF

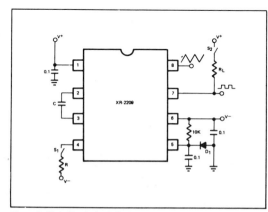

Figure 1. Test Circuit for Split Supply Operation (D_1 = 1N 4148 or Equivalent)

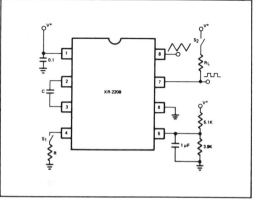

Figure 2. Test Circuit for Single Supply Operation

CHARACTERISTIC CURVES

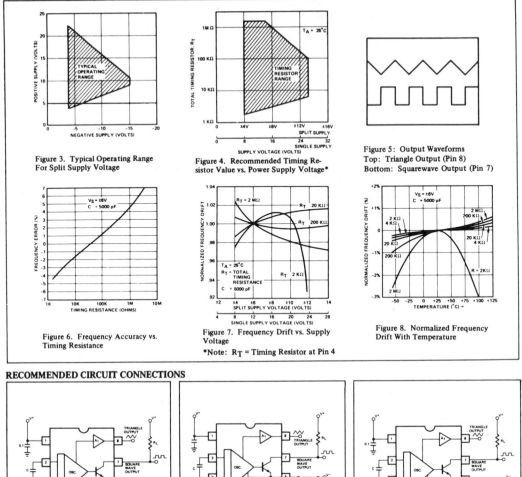

Figure 3. Typical Operating Range For Split Supply Voltage

Figure 4. Recommended Timing Resistor Value vs. Power Supply Voltage*

Figure 5: Output Waveforms
Top: Triangle Output (Pin 8)
Bottom: Squarewave Output (Pin 7)

Figure 6. Frequency Accuracy vs. Timing Resistance

Figure 7. Frequency Drift vs. Supply Voltage

*Note: R_T = Timing Resistor at Pin 4

Figure 8. Normalized Frequency Drift With Temperature

RECOMMENDED CIRCUIT CONNECTIONS

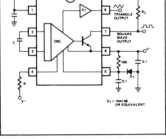

Figure 9. Circuit Connection for Single Supply Operation

Figure 10. Generalized Circuit Connection for Split Supply Operation

Figure 11. Simplified Circuit Connection for Split Supply Operation with $V_{CC} = V_{EE} > \pm7V$ *(Note: Triangle wave output has +0.6V offset with respect to ground.)*

PRECAUTIONS

The following precautions should be observed when operating the XR-2209 family of integrated circuits:

1. Pulling excessive current from the timing terminal will adversely effect the temperature stability of the circuit. To minimize this disturbance, it is recommended that the *total current* drawn from pin 4 be limited to ≤6 mA.

2. Terminals 2, 3, and 4 have very low internal impedance and should, therefore, be protected from accidental shorting to ground or the supply voltages.

3. Triangle waveform linearity is sensitive to parasitic coupling between the square and the triangle-wave outputs (pins 7 and 8). In board layout or circuit wiring care should be taken to minimize stray wiring capacitances between these pins.

DESCRIPTION OF CIRCUIT CONTROLS

TIMING CAPACITOR (PINS 2 and 3)

The oscillator frequency is inversely proportional to the timing capacitor, C. The minimum capacitance value is limited by stray capacitances and the maximum value by physical size and leakage current considerations. Recommended values range from 100 pF to 100 μF. The capacitor should be non-polar.

TIMING RESISTOR (PIN 4)

The timing resistor determines the total timing current, I_T, available to charge the timing capacitor. Values for the timing resistor can range from 1.5 KΩ to 2 MΩ; however, for optimum temperature and power supply stability, recommended values are 4 KΩ to 200 KΩ (see Figures 4, 7, and 8). To avoid parasitic pick up, timing resistor leads should be kept as short as possible.

SUPPLY VOLTAGE (PINS 1 AND 6)

The XR-2209 is designed to operate over a power supply range of $\pm$4V to $\pm$13V for split supplies, or 8V to 26V for single supplies. At high supply voltages, the frequency sweep range is reduced (see Figures 3 and 4). Performance is optimum for $\pm$6V, or 12V single supply operation.

BIAS FOR SINGLE SUPPLY (PIN 5)

For single supply operation, pin 5 should be externally biased to a potential between $V^+/3$ and $V^+/2$ volts (see Figure 9). The bias current at pin 5 is nominally 5% of the total oscillation timing current, I_T, at pin 4. This pin should be bypassed to ground with 0.1 μF capacitor.

SQUAREWAVE OUTPUT (PIN 7)

The squarewave output at pin 7 is a "open-collector" stage capable of sinking up to 20 mA of load current. R_L serves as a pull-up load resistor for this output. Recommended values for R_L range from 1 KΩ to 100 KΩ.

TRIANGLE OUTPUT (PIN 8)

The output at pin 8 is a triangle wave with a peak swing of approximately one-half of the total supply voltage. Pin 8 has a very low output impedance of 10Ω and is internally protected against short circuits.

OPERATING INSTRUCTIONS

SPLIT SUPPLY OPERATION

The recommended circuit for split supply operation is shown in Figure 10. Diode D_1 in the figure assures that the triangle output swing at pin 8 is symmetrical about ground. This circuit operates with supply voltages ranging from $\pm$4V to $\pm$13V. Minimum drift occurs at $\pm$6V supplies. See Figure 3 for operation with unequal supplies.

Simplified Connection

For operation with split supplies in excess of $\pm$7 volts, the simplified circuit connection of Figure 11 can be used. This circuit eliminates the diode D_1 used in Figure 10; however the triangle wave output at pin 8 now has a +0.6 volt DC offset with respect to ground.

SINGLE SUPPLY OPERATION

The recommended circuit connection for single-supply operation is shown in Figure 9. Pin 6 is grounded; and pin 5 is biased from V^+ through a resistive divider, as shown in the figure, and is bypassed to ground with a 1 μF capacitor.

For single supply operation, the DC voltage at the timing terminal, pin 4, is approximately 0.6 volts above V_B, the bias voltage at pin 5.

The frequency of operation is determined by the timing capacitor C and the timing resistor R, and is equal to 1/RC. The squarewave output is obtained at pin 7 and has a peak-to-peak voltage swing equal to the supply voltage. This output is an "open-collector" type and requires an external pull-up load resistor (nominally 5 KΩ) to V^+. The triangle waveform obtained at pin 8 is centered about a voltage level V_O where:

$$V_O = V_B + 0.6V$$

where V_B is the bias voltage at pin 5. The peak-to-peak output swing of triangle wave is approximately equal to $V^+/2$.

FREQUENCY CONTROL (SWEEP AND FM)

The frequency of operation is proportional to the *total* timing current I_T drawn from the timing pin, pin 4. This timing current, and the frequency of operation can be modulated by applying a control voltage, V_C, to the timing pin, through a series resistor, R_S, as shown in Figure 12. If V_C is negative with respect to V_A, the voltage level at pin 4, then an additional current I_O is drawn from the timing pin causing I_T to increase, thus increasing the frequency. Conversely, making V_C higher than V_A causes the frequency to decrease by decreasing I_T.

The frequency of operation, is determined by:

$$f = f_O \left[1 + \frac{R}{R_S} - \frac{V_C}{V_A} \frac{R}{R_S} \right]$$

where $f_O = 1/RC$.

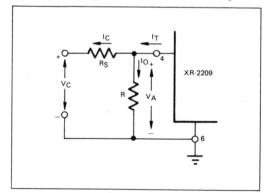

Figure 12. Frequency Sweep Operation

FREQUENCY MULTIPLIERS

In electronic communications systems, it is often necessary or desirable to produce exact multiples of a frequency. A *frequency multiplier* is just such a circuit; it multiplies the frequency of the input signal and generates harmonics of that frequency. A frequency multiplier takes advantage of the nonlinear characteristics of a circuit. In Chapter 1, it was shown that a nonsinusoidal repetitive waveform is made up of a series of harmonically related sine waves. If a pure sine wave is distorted in such a way that it produces a repetitive nonsinusoidal waveform, higher-order harmonics are produced and frequency multiplication is accomplished. The more the signal is distorted, the higher the amplitude of the harmonics. A frequency multiplier consists of an oscillator, a nonlinear amplifier, and a frequency-selective circuit such as an *LC* tank circuit. The schematic diagram of a frequency multiplier is shown in Figure 2-17a. Q_1 is a class C amplifier, and its input voltage and collector current waveforms are shown in Figure 2-17b and c, respectively. It can be seen that the fundamental frequency of the input voltage and output current waveforms are the same. However, the collector current waveform is not a pure sine wave, although it is a repetitive waveform. Consequently, it is made up of a series of harmonically related sine waves. If the resonant frequency of the tank circuit is tuned to a multiple of the input frequency, frequency multiplication is accomplished. The oscillating voltage waveform for the second harmonic is shown in Figure 2-17d. It can be seen that energy is injected into the tank circuit during every other cycle of the output waveform. During each intermediate cycle, the signal is somewhat damped due to the power dissipated in the resistive component of the inductor. However, if the Q factor of the coil is made sufficiently high, the damping factor is negligible and the output is a continuous sine wave at twice the frequency of the input signal.

In Chapter 1 it was shown that the higher the harmonic, the lower its amplitude. Consequently, frequency multipliers, like the one shown in Figure 2-17, are extremely inefficient and are impractical for multiplication factors above 3. A device that is used to generate several different output frequencies is called a *frequency synthesizer*. Frequency synthesizers are explained in Chapter 5.

MIXING

Mixing is the process of combining two or more signals and is an essential process in electronic communications. In essence, there are two ways in which signals are mixed: linearly and nonlinearly.

Linear Summing

Linear summing is when two or more signals combine in a linear device such as a resistive or inductive network or a small-signal amplifier. In the audio recording industry,

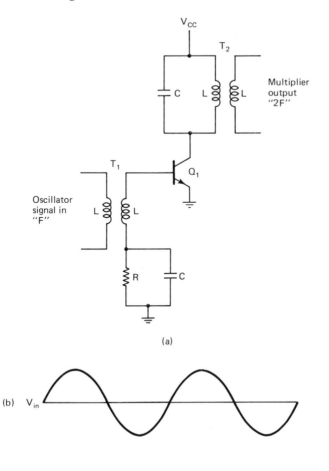

(a)

(b) V_{in}

(c) I_c

(d) V_{out}

FIGURE 2-17 Frequency multiplication: (a) schematic drawing; (b) input voltage waveform; (c) collector current waveform; (d) tank circuit oscillating voltage waveform, 2nd harmonic.

linear summing is sometimes called *linear mixing*. However, in communications work mixing almost always implies a nonlinear process.

Single-input frequency. Figure 2-18a shows the amplification of a single-frequency input signal by a linear amplifier. The output is simply the original input frequency

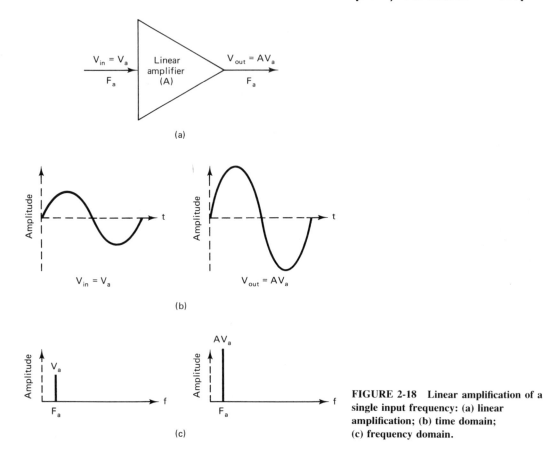

(a)

(b)

(c)

FIGURE 2-18   Linear amplification of a single input frequency: (a) linear amplification; (b) time domain; (c) frequency domain.

amplified by the gain (A). Figure 2-18b shows the output signal in the time domain, and Figure 2-18c shows the frequency domain. Mathematically, the output is

$$V_{out} = AV_{in} = AV_a \sin 2\pi F_a t \qquad (2\text{-}5)$$

where $V_{in} = V_a \sin 2\pi F_a t$.

Multiple-input frequencies.   Figure 2-19a shows two input frequencies combining in a small-signal amplifier. Each input signal is amplified by the gain (A). Therefore, the output is expressed mathematically as

$$V_{out} = AV_{in}$$

where

$$V_{in} = V_a \sin 2\pi F_a t + V_b \sin 2\pi F_b t \qquad (2\text{-}6)$$

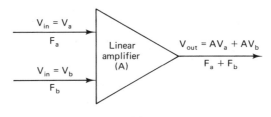

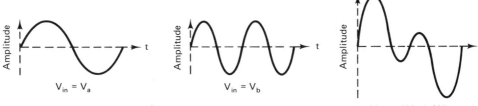

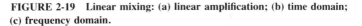

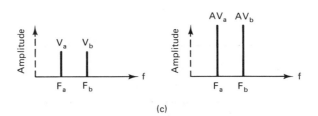

FIGURE 2-19   Linear mixing: (a) linear amplification; (b) time domain;
(c) frequency domain.

Therefore,

$$V_{out} = A(V_a \sin 2\pi F_a t + V_b \sin 2\pi F_b t)$$

or

$$V_{out} = AV_a \sin 2\pi F_a t + AV_b \sin 2\pi F_b t$$

V_{out} is simply an *additive* waveform and is equal to the algebraic sum of V_a and
V_b. Figure 2-19b shows the *linear summation* of V_a and V_b in the time domain, and
Figure 2-19c shows the linear summation in the frequency domain. If additional input
frequencies are applied to the circuit, they are linearly summed with V_a and V_b. In
high-fidelity audio systems, it is important that the output spectrum contain only the
original input frequencies, and therefore linear operation is desired. However, in electronic communications systems where modulation is essential, nonlinear mixing is often
required.

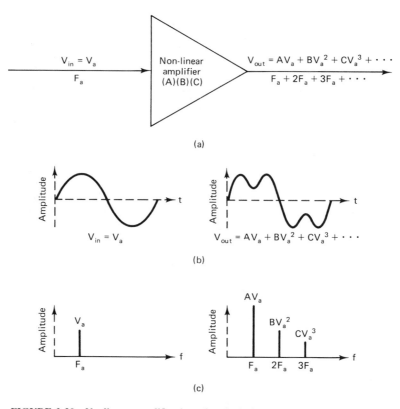

(a)

(b)

(c)

FIGURE 2-20   Nonlinear amplification of a single input frequency: (a) nonlinear amplification; (b) time domain; (c) frequency domain.

Nonlinear Mixing

Nonlinear mixing is when two or more signals are combined in a nonlinear device such as a diode or a large-scale amplifier.

Single-input frequency.   Figure 2-20a shows the amplification of a single-frequency input signal by a nonlinear amplifier. The output from a nonlinear amplifier with a single-frequency input signal is not a single sine wave. Mathematically, the output is in the infinite power series

$$V_{\text{out}} = AV_{\text{in}} + BV_{\text{in}}^2 + CV_{\text{in}}^3 + \cdots \tag{2-7a}$$

where $V_{\text{in}} = V_a \sin 2\pi F_a t$. Therefore

$$V_{\text{out}} = A(V_a \sin 2\pi F_a t) + B(V_a \sin 2\pi F_a t)^2 + C(V_a \sin 2\pi F_a t)^3 + \cdots \tag{2-7b}$$

where

AV_{in} = linear term or simply the input signal (F_a) amplified by the gain (A)

BV_{in}^2 = quadratic term that generates the second harmonic frequency $(2F_a)$

CV_{in}^3 = cubic term that generates the third harmonic frequency $(3F_a)$

V_{in}^n produces a frequency equal to n times F. BV_{in}^2 generates a frequency equal to $2F_a$, CV_{in}^3 generates a frequency equal to $3F_a$, and so on. Multiples of a *base* frequency are called *harmonics*. As stated in Chapter 1, the original input frequency (F_a) is the first harmonic or the *fundamental* frequency, $2F_a$ is the second harmonic, $3F_a$ the third harmonic, and so on. Figure 2-20b shows the output waveform in the time domain for a nonlinear amplifier with a single input frequency. It can be seen that the output waveform is simply the summation of the input frequency and its higher harmonics (multiples of the fundamental frequency). Figure 2-20c shows the output spectrum in the frequency domain. Note that adjacent harmonics are separated in frequency by a value equal to the fundamental frequency (F_a).

Nonlinear amplification of a single frequency results in the generation of multiples or harmonics of that frequency. If the harmonics are undesired it is called *harmonic distortion*.

A JFET is a special-case nonlinear device that has *square-law* properties. The output from a square-law device is simply

$$V_{out} = AV_{in} + BV_{in}^2 \qquad (2\text{-}8)$$

The output from a square-law device with a single input frequency is the fundamental frequency (i.e., the original input frequency) and its second harmonic. There are no additional harmonics generated. Therefore, there is less harmonic distortion produced with a JFET than with a comparable BJT.

Multiple-input frequencies. Figure 2-21a shows the nonlinear amplification of two input frequencies by a large-signal (nonlinear) amplifier.

Mathematically, the output of a large-signal amplifier with two input frequencies is

$$V_{out} = AV_{in} + BV_{in}^2 + CV_{in}^3 + \cdots \qquad (2\text{-}9a)$$

where $V_{in} = V_a \sin 2\pi F_a t + V_b \sin 2\pi F_b t$. Therefore,

$$V_{out} = A(V_a \sin 2\pi F_a t + V_b \sin 2\pi F_b t) + B(V_a \sin 2\pi F_a t + V_b \sin 2\pi F_b t)^2$$
$$+ C(V_a \sin 2\pi F_a t + V_b \sin 2\pi F_b t)^3 + \cdots \qquad (2\text{-}9b)$$

The preceding formula is an infinite series and there is no limit to the number of terms it can have. If the binomial theorem is applied to each higher-power term, the formula can be rearranged and written as

$$V_{out} = (AV_a + BV_a^2 + CV_a^3 + \cdots) + (AV_b + BV_b^2 + CV_b^3 + \cdots)$$
$$+ (2BV_aV_b + 3CV_a^2V_b + 3CV_aV_b^2 + \cdots) \qquad (2\text{-}9c)$$

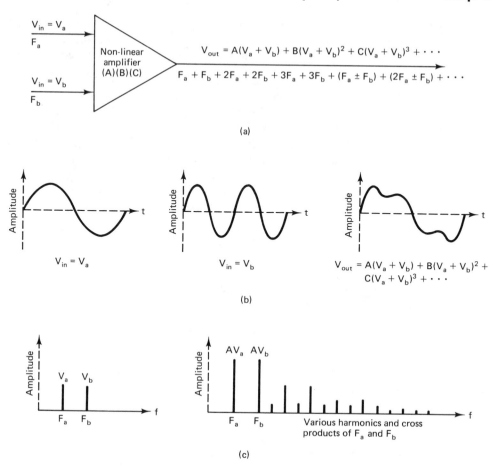

FIGURE 2-21   Nonlinear amplification of two sine waves: (a) nonlinear amplification; (b) time domain; (c) frequency domain.

The terms in the first set of parentheses generate harmonics of F_a ($2F_a$, $3F_a$, etc.). The terms in the second set of parentheses generate harmonics of F_b ($2F_b$, $3F_b$, etc.). The terms in the third set of parentheses generate the cross products ($F_a + F_b$, $F_a - F_b$, $2F_a + F_b$, $2F_a - F_b$, etc.). The cross products are produced from intermodulation among the two original frequencies and their harmonics. The *cross products* are the *sum* and *difference frequencies*; they are the sum and difference of the two original frequencies and the sums and differences of their harmonics. There is an infinite number of harmonic and cross-product frequencies produced when two or more frequencies mix in a nonlinear device. If the cross products are undesired, it is called *intermodulation distortion*. Mathematically, the sum and difference frequencies are

$$\text{cross products} = mF_a \pm nF_b \qquad (2\text{-}10)$$

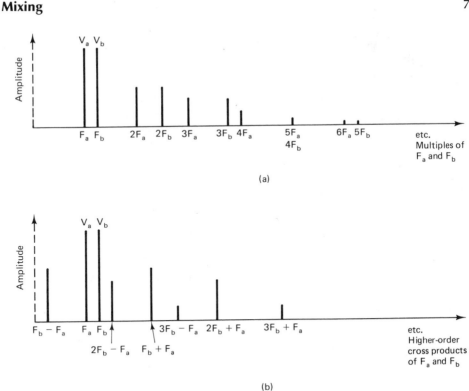

FIGURE 2-22   Output spectrum from a nonlinear amplifier with two input frequencies: (a) harmonic distortion; (b) intermodulation distortion.

where m and n are all positive integers between 1 and infinity. Figure 2-22 shows the output spectrum from a nonlinear amplifier with two input frequencies.

Intermodulation distortion is any unwanted cross product that is generated when two or more frequencies are mixed in a nonlinear device. Consequently, whenever two or more frequencies are amplified by a nonlinear device, harmonic and intermodulation distortion are present in the output.

EXAMPLE 2-2

For a nonlinear amplifier with two input frequencies, 5 kHz and 7 kHz:
(a) Determine the first three harmonics present in the output for each input frequency.
(b) Determine the cross products produced in the output for values of m and n of 1 and 2.
(c) Draw the output spectrum for the harmonics and cross products determined in parts (a) and (b).

Solution (a) The first three harmonics include the two original frequencies, 5 and 7 kHz; two times each of the original frequencies, 10 and 14 kHz; and three times each of the original frequencies, 15 and 21 kHz.

(b) The cross products for values of m and n of 1 and 2 are determined from Equation 2-10 and are summarized below.

m	n	Cross products
1	1	7 kHz ± 5 kHz = 2 and 12 kHz
1	2	7 kHz ± 10 kHz = 3 and 17 kHz
2	1	14 kHz ± 5 kHz = 9 and 19 kHz
2	2	14 kHz ± 10 kHz = 4 and 24 kHz

(c) The output spectrum is shown in Figure 2-23.

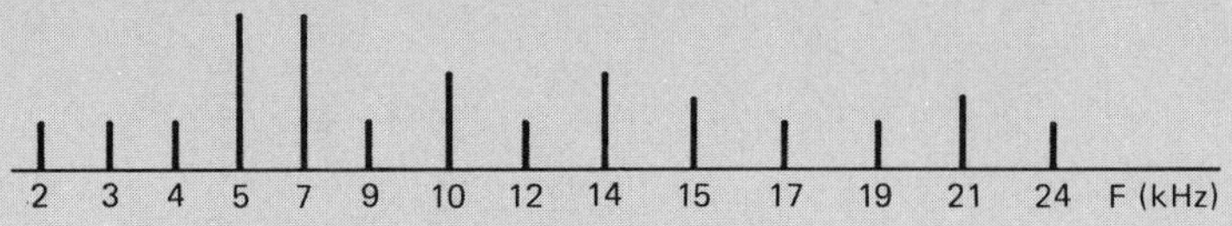

FIGURE 2-23   Output spectrum for Example 2-2.

QUESTIONS

2-1. Define *oscillate*; *oscillator*.

2-2. Describe the following terms: *self-sustaining*; *repetitive*; *free-running*; *one-shot*.

2-3. Describe the regenerative process necessary for self-sustained oscillations to occur.

2-4. List and describe the four requirements for a feedback oscillator to work.

2-5. What is meant by the terms *positive feedback* and *negative feedback*?

2-6. Define *open-loop gain*; *closed-loop gain*.

2-7. List the four most common oscillator configurations.

2-8. Describe the operation of a Wien-bridge oscillator.

2-9. Describe oscillator action for an *LC* tank circuit.

2-10. What is the flywheel effect?

2-11. What is meant by *damped oscillations*? What causes them?

2-12. Describe the operation of a Hartley oscillator; a Colpitts oscillator.

2-13. Define *frequency stability*.

2-14. List several factors that affect the frequency stability of an oscillator.

2-15. Describe the piezoelectric effect.

2-16. What is meant by *crystal cut*? List and describe several crystal cuts and contrast their stabilities.

2-17. Describe how an overtone crystal oscillator works.

2-18. What is the advantage of an overtone crystal oscillator over a conventional crystal oscillator?

2-19. What is meant by *positive temperature coefficient*; *negative temperature coefficient*?

2-20. What is meant by *zero coefficient crystal*?

2-21. Sketch the electrical equivalent circuit for a crystal and describe the various components.

2-22. Which crystal oscillator configuration has the best frequency stability?

2-23. Which crystal oscillator configuration is the least expensive and most adaptable to digital interfacing?

2-24. Describe a crystal oscillator module.

2-25. What is the predominant advantage of crystal oscillators over *LC* tank circuit oscillators?

2-26. Describe the operation of a varactor diode.

2-27. Describe the operation of a negative resistance oscillator.

2-28. Define *frequency multiplier*.

2-29. Define *linear summing*.

2-30. Contrast the input and output frequency spectra for a linear amplifier.

2-31. Define *nonlinear mixing*.

2-32. Contrast the input and output frequency spectra for a nonlinear amplifier.

2-33. When will harmonic distortion and intermodulation distortion occur?

PROBLEMS

2-1. For a 20-MHz crystal with a negative temperature coefficient $k = 8$ Hz/MHz/°C, determine the frequency of operation for the following temperature changes.
  (a) Increase of 10°C.
  (b) Increase of 20°C.
  (c) Decrease of 20°C.

2-2. For the Wien-bridge oscillator shown in Figure 2-3 and the following component values, determine the frequency of oscillation: $R_1 = R_2 = 1$ kΩ; $C_1 = C_2 = 100$ pF.

2-3. For the Hartley oscillator shown in Figure 2-5a and the following component values, determine the frequency of oscillation: $L_{1a} = L_{1b} = 50$ µH; $C_1 = 0.01$ µF.

2-4. For the Colpitts oscillator shown in Figure 2-6 and the following component values, determine the frequency of oscillation: $C_{1a} = C_{1b} = 0.01$ µF; $L_1 = 100$ µH.

2-5. Determine the capacitance for a variactor diode with the following values: $C = 0.005$ µF; $v_r = -2$ V dc.

2-6. Describe the spectrum shown below. Determine the type of amplifier and the frequency content of the input signal.

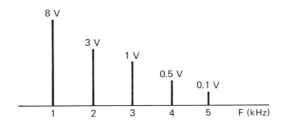

2-7. Repeat Problem 2-1 for the spectrum shown below.

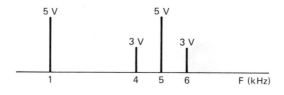

2-8. For a nonlinear amplifier with two input frequencies, 7 kHz and 4 kHz:

(a) Determine the first three harmonics present in the output for each frequency.

(b) Determine the cross products produced in the output for values of m and n of 1 and 2.

(c) Draw the output spectrum for the harmonics and cross products determined in parts (a) and (b).

Chapter 3

AMPLITUDE MODULATION TRANSMISSION

INTRODUCTION

Information signals must be carried between a transmitter and a receiver over some form of transmission medium. However, the information signals are seldom in a form suitable for transmission. *Modulation* is defined as the process of transforming information from its original form to a form that is more suitable for transmission between a transmitter and a receiver. *Demodulation* is the reverse process (i.e., the modulated signal is converted back to its original form). Modulation takes place in a circuit called a *modulator*, and demodulation takes place in a circuit called a *demodulator*.

Amplitude modulation (AM) is the process of changing the amplitude of a high-frequency carrier in accordance with a modulating signal (information). Frequencies that are high enough to be efficiently radiated by an antenna and propagated through free space are commonly called *radio frequencies* or simply RF. In AM, the information is impressed onto the carrier in the form of amplitude changes. Amplitude modulation is a relatively inexpensive, low-quality form of modulation that is used for commercial broadcasting of both audio and video signals. The commercial AM broadcast band extends from 535 to 1605 kHz. Commercial television broadcasting is divided into three bands (two VHF and one UHF). The low-band VHF channels are 2 to 6 (54 to 88 MHz), the high-band VHF channels are 7 to 13 (174 to 216 MHz), and the UHF channels are 14 to 83 (470 to 890 MHz). Amplitude modulation is also used for two-way mobile radio communications such as CB radio (citizens' band, 26.965 to 27.405 MHz).

An AM modulator is a nonlinear device with two inputs: a single-frequency, constant-amplitude carrier and the information. The information acts on or modulates the carrier and may be a single frequency or a complex waveform made up of many frequencies. Because the information acts on the carrier, it is called the *modulating signal*. The carrier is acted upon and is therefore called the *modulated signal*. The resultant is called the *modulated wave*.

AM Envelope

Although mathematically it is not the simplest form of AM, double-sideband full carrier (AM DSBFC) will be discussed first because it is probably the most often used form of AM.

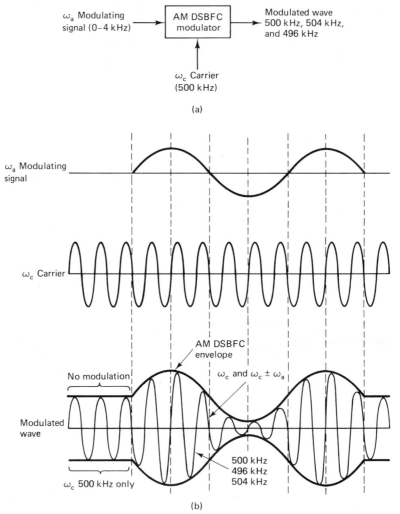

FIGURE 3-1   AM generation: (a) AM DSBFC modulator; (b) producing an AM DSBFC envelope—time domain.

Figure 3-1a shows a AM DSBFC modulator and the relationship among the carrier (ω_c), the modulating signal (ω_a), and the modulated wave (v). Figure 3-1b shows in the time domain how an AM wave is produced from a single-frequency modulating signal. Because the modulated wave contains all of the frequencies that make up the AM signal and is used to carry the information through the system, it is called an AM *envelope*. With no modulating signal, the output wave is simply the amplified carrier. When a modulating signal is applied, the amplitude of the output wave is varied in accordance with the modulating signal. Note the shape of the AM envelope; it is identical to the shape of the modulating signal. Also, the time of one cycle of the envelope is the same as the period of the modulating signal. Consequently, the repetition rate of the envelope is equal to the frequency of the modulating signal.

AM Spectrum and Bandwidth

As stated previously, an AM modulator is a nonlinear device. Therefore, nonlinear mixing occurs in an AM modulator and the output envelope is a complex wave made up of the carrier and the sum ($F_c + F_a$) and difference ($F_c - F_a$) frequencies (i.e., the cross products). The difference between the sum or difference frequency and the carrier frequency is equal to the modulating signal frequency. Therefore, an AM envelope contains both the amplitude and frequency components of the original information signal.

Figure 3-2 shows the frequency spectrum for the AM envelope produced in Figure 3-1. The AM spectrum extends from $F_c - F_a$ to $F_c + F_a$, where F_a is the highest modulating signal frequency. The band of frequencies between $F_c - F_a$ and F_c is called the *lower sideband* (LSB), and any frequency within this band is called a *lower side frequency* (LSF). The band of frequencies between F_c and $F_c + F_a$ is called the *upper sideband* (USB), and any frequency within this band is called an *upper side frequency* (USF). Therefore, the bandwidth (B) of an AM DSBFC wave is equal to the difference between the highest USF and the lowest LSF (i.e., $B = 2F_{a(\max)}$). The carrier and all of the frequencies within the upper and lower sidebands must be RF to propagate through the earth's atmosphere.

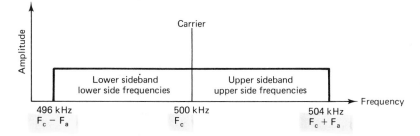

FIGURE 3-2 Frequency spectrum of the AM DSBFC wave produced in Figure 2-1.

EXAMPLE 3-1

For an AM modulator with a carrier frequency of 100 kHz and a maximum modulating signal frequency of 5 kHz:
(a) Determine the frequency limits for the upper and lower sidebands.
(b) Determine the upper and lower side frequencies produced when the modulating signal is a single-frequency 3-kHz tone.
(c) Determine the bandwidth.
(d) Draw the output spectrum.

Solution   (a) The lower sideband extends from the lowest possible side frequency to the carrier frequency.

$$\text{LSB} = F_c - F_{a(\max)} \text{ to } F_c$$

$$= (100 - 5) \text{ kHz to } 100 \text{ kHz} = 95 \text{ to } 100 \text{ kHz}$$

The upper sideband extends from the highest possible side frequency to the carrier frequency.

$$\text{USB} = F_c \text{ to } F_c + F_{a(\max)}$$

$$= 100 \text{ kHz to } (100 + 5) \text{ kHz} = 100 \text{ to } 105 \text{ kHz}$$

(b) The USF is the sum of the carrier frequency and the modulating signal frequency, or

$$\text{USF} = F_c + F_a = 100 \text{ kHz} + 3 \text{ kHz} = 103 \text{ kHz}$$

The LSF is the difference between the carrier frequency and the modulating signal frequency, or

$$\text{LSF} = F_c - F_a = 100 \text{ kHz} - 3 \text{ kHz} = 97 \text{ kHz}$$

(c) $B = 2F_a = 2(5 \text{ kHz}) = 10 \text{ kHz}$
(d) The output spectrum is shown in Figure 3-3.

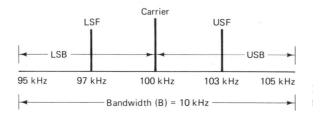

FIGURE 3-3   Output spectrum for Example 3-1.

Phasor Representation of an AM Wave

For a single-frequency modulating signal, an AM envelope is produced from the vector addition of the carrier and the upper and lower side frequencies. Figure 3-4a shows this phasor addition. On a time basis, the carrier and the upper and lower side frequencies all rotate in a counterclockwise direction. However, the USF rotates faster than the carrier ($\omega USF > \omega_c$) and the LSF rotates slower ($\omega LSF < \omega_c$). Consequently, if the

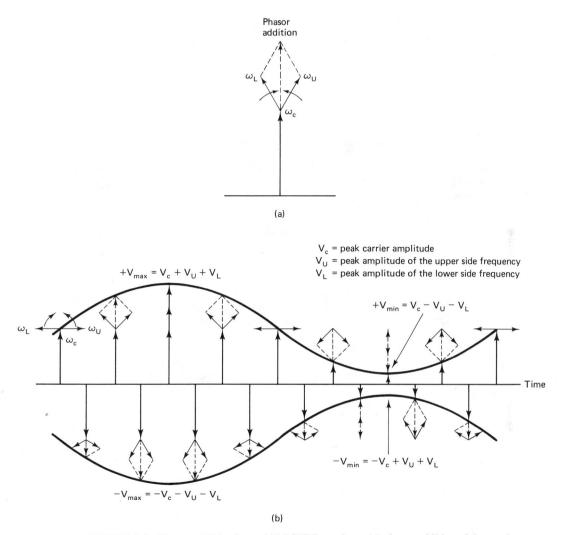

FIGURE 3-4   Phasor addition in an AM DSBFC envelope: (a) phasor addition of the carrier and the upper and lower side frequencies; (b) phasor addition producing an AM envelope.

carrier vector is held stationary, the vector for the USF will continue in a counterclockwise direction and the vector for the LSF will fall back in a clockwise direction. The carrier, the USF, and the LSF combine, sometimes in phase (adding) and sometimes out of phase (subtracting). For the waveform shown in Figure 3-4b, the maximum positive amplitude of the envelope ($+V_{max}$) occurs when the carrier and the two side frequencies are at their maximum positive values at the same time. The minimum positive amplitude ($+V_{min}$) occurs when the carrier is at its maximum positive value at the same time that the USF and LSF are at their maximum negative values. The maximum negative amplitude

$(-V_{max})$ occurs when the carrier, the USF, and the LSF are at their maximum negative values at the same time. The minimum negative amplitude $(-V_{min})$ occurs when the carrier is at its maximum negative value at the same time that the USF and LSF are at their maximum positive values $(-V_{min})$.

Coefficient of Modulation and Percent Modulation

Coefficient of modulation is a term that describes the amount of amplitude change (modulation) in an AM envelope. *Percent modulation* is simply the coefficient of modulation stated as a percentage. More specifically, percent modulation gives the percentage change in the amplitude of the output wave when the carrier is acted on by a modulating signal. Mathematically, modulation coefficient is

$$m = \frac{E_m}{E_c} \qquad (3\text{-}1a)$$

where

m = modulation coefficient
E_m = peak change in the amplitude of the output wave
E_c = peak amplitude of the unmodulated carrier

Equation 3-1a can be rearranged to solve for E_m or E_c:

$$E_m = mE_c \qquad \text{and} \qquad E_c = \frac{E_m}{m} \qquad (3\text{-}1b)$$

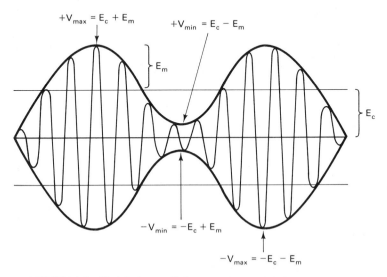

FIGURE 3-5   Modulation coefficient, percent modulation, E_m, and E_c.

and percent modulation (M) is

$$M = \frac{E_m}{E_c} \times 100 \qquad \text{or simply} \qquad m \times 100 \qquad (3\text{-}2)$$

The relationship among m, M, E_m, and E_c is shown in Figure 3-5.

If the modulating signal is a pure sine wave and the modulation process is linear (i.e., the positive and negative swings of the envelope's amplitude are equal), percent modulation can be derived as follows:

$$E_m = \tfrac{1}{2}(V_{\max} - V_{\min}) \qquad (3\text{-}3\text{a})$$

and

$$E_c = \tfrac{1}{2}(V_{\max} + V_{\min}) \qquad (3\text{-}3\text{b})$$

Therefore

$$M = \frac{\tfrac{1}{2}(V_{\max} - V_{\min})}{\tfrac{1}{2}(V_{\max} + V_{\min})} \times 100$$

$$= \frac{V_{\max} - V_{\min}}{V_{\max} + V_{\min}} \times 100 \qquad (3\text{-}3\text{c})$$

where

$$V_{\max} = E_c + E_m \qquad (3\text{-}3\text{d})$$
$$V_{\min} = E_c - E_m \qquad (3\text{-}3\text{e})$$

The peak change in the amplitude of the output wave (E_m) is the sum of the voltages from the upper and lower side frequencies. Therefore,

$$E_U = E_L = \frac{E_m}{2} = \frac{\tfrac{1}{2}(V_{\max} - V_{\min})}{2} = \tfrac{1}{4}(V_{\max} - V_{\min}) \qquad (3\text{-}4)$$

where

E_U = peak voltage of the USF
E_L = peak voltage of the LSF

From Equation 3-2, it can be seen that the percent modulation goes to 100% when $E_m = E_c$. This condition is shown in Figure 3-6d. It can also be seen that at 100% modulation, the minimum amplitude of the AM envelope $V_{\min} = 0$V. Figure 3-6c shows a 50% modulated AM envelope, the peak change in the amplitude of the envelope is equal to one-half the amplitude of the unmodulated wave. The maximum percent modulation that can be imposed without causing excessive distortion to the modulated wave is 100%. Sometimes percent modulation is expressed as the peak change in output voltage in respect to the unmodulated carrier voltage (i.e., percent change = $\Delta V_c / V_c \times 100$).

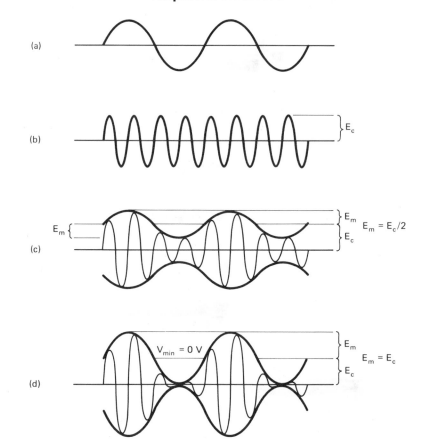

FIGURE 3-6   Percent modulation of an AM DSBFC envelope: (a) modulating signal; (b) unmodulated carrier; (c) 50% modulated wave; (d) 100% modulated wave.

EXAMPLE 3-2

For the AM envelope shown in Figure 3-7, determine:
(a) The peak amplitude of the upper and lower side frequencies.
(b) The peak amplitude of the unmodulated carrier.
(c) The peak change in the amplitude of the envelope.
(d) The modulation coefficient.
(e) The percent modulation.

Solution   (a) From Equation 3-4,

$$E_U = E_L = \tfrac{1}{4}(V_{max} - V_{min}) = \tfrac{1}{4}(18 - 2) = 4 \text{ V}$$

(b) From Equation 3-3b,

$$E_c = \tfrac{1}{2}(V_{max} + V_{min}) = \tfrac{1}{2}(18 + 2) = 10 \text{ V}$$

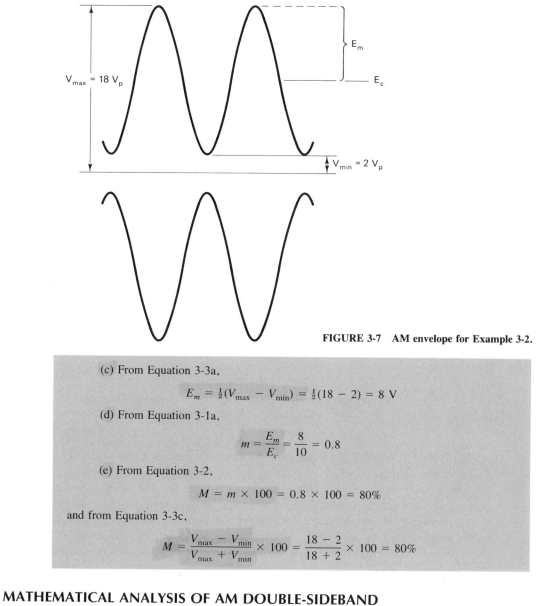

FIGURE 3-7 AM envelope for Example 3-2.

(c) From Equation 3-3a,

$$E_m = \tfrac{1}{2}(V_{max} - V_{min}) = \tfrac{1}{2}(18 - 2) = 8 \text{ V}$$

(d) From Equation 3-1a,

$$m = \frac{E_m}{E_c} = \frac{8}{10} = 0.8$$

(e) From Equation 3-2,

$$M = m \times 100 = 0.8 \times 100 = 80\%$$

and from Equation 3-3c,

$$M = \frac{V_{max} - V_{min}}{V_{max} + V_{min}} \times 100 = \frac{18 - 2}{18 + 2} \times 100 = 80\%$$

MATHEMATICAL ANALYSIS OF AM DOUBLE-SIDEBAND FULL CARRIER

Voltage Distribution in an AM DSBFC Wave

An unmodulated carrier can be described mathematically as

$$v_c = E_c \sin \omega_c t$$

In a previous discussion it was pointed out that the repetition rate of an AM envelope is equal to the frequency of the modulating signal, the amplitude of the AM wave is varied proportional to the amplitude of the modulating signal, and the amplitude of the modulated wave is equal to $E_c + E_m$. Therefore, the instantaneous amplitude of the modulated wave (v_c) can be expressed as

$$v_c = \underbrace{(E_c + E_m \sin \omega_a t)}_{\substack{\text{modulated wave} \\ \text{amplitude}}} \sin \omega_c t \qquad \text{(3-5a)}$$

From Equation 3-1b, mE_c can be substituted for E_m:

$$v_c = \underbrace{(E_c + mE_c \sin \omega_a t)}_{\substack{\text{modulated wave} \\ \text{amplitude}}} \sin \omega_c t \qquad \text{(3-5b)}$$

Multiplying yields

$$v_c = E_c \sin \omega_c t + (mE_c \sin \omega_a t)(\sin \omega_c t)$$

The trigonometric identity for the product of two sines is

$$(\sin A)(\sin B) = -\tfrac{1}{2}\cos (A + B) + \tfrac{1}{2}\cos (A - B)$$

Therefore,

$$v_c = \underbrace{E_c \sin \omega_c t}_{\text{carrier}} - \underbrace{\frac{mE_c}{2} \cos (\omega_c + \omega_a)\, t}_{\text{USF}} + \underbrace{\frac{mE_c}{2} \cos (\omega_c - \omega_a)t}_{\text{LSF}} \qquad \text{(3-5c)}$$

The first component in Equation 3-5c is the carrier (ω_c), the second component is the upper side frequency ($\omega_c + \omega_a$), and the third component is the lower side frequency ($\omega_c - \omega_a$). Several interesting properties of an AM DSBFC wave can be pointed out from Equation 3-5c. First, note that the amplitude of the carrier after modulation is equal to the amplitude of the carrier before modulation (E_c). Therefore, the carrier amplitude is unaffected by the modulation process. Second, the amplitude of the USF and LSF are dependent on both the carrier amplitude and the coefficient of modulation. For 100% modulation, $m = 1$ and the amplitude of the USF and the LSF are each equal to one-half the carrier amplitude ($E_c/2$). Therefore, at 100% modulation,

$$v_{c(\max)} = E_c + \frac{E_c}{2} + \frac{E_c}{2} = 2E_c$$

and

$$v_{c(\min)} = E_c - \frac{E_c}{2} - \frac{E_c}{2} = 0 \text{ V}$$

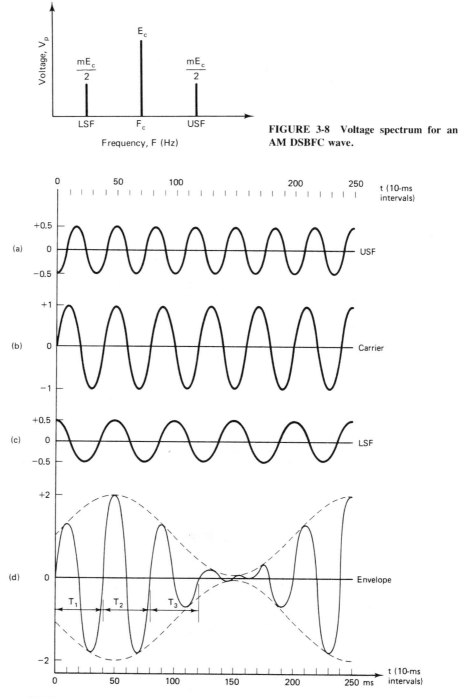

FIGURE 3-8 Voltage spectrum for an AM DSBFC wave.

FIGURE 3-9 Generation of an AM DSBFC envelope shown in the time domain: (a) $-\frac{1}{2} \cos 2\pi 30t$; (b) $\sin 2\pi 25t$; (c) $+\frac{1}{2} \cos 2\pi 20t$; (d) summation of (a), (b), and (c).

From the relationships above and using Equation 3-5c, it is evident that as long as we do not exceed 100% modulation, the maximum positive or negative amplitude of an AM envelope $V_{max} = 2E_c$, and the minimum positive or negative amplitude of an AM envelope $V_{min} = 0$. This relationship was shown in Figure 3-6d. Figure 3-8 shows the voltage spectrum for an AM DSBFC wave (note that all of the voltages are given in peak values).

Also, from Equation 3-5c, the relative phase relationship between the carrier and the upper and lower side frequencies is evident. The carrier component of the envelope is a +sine function, the USF a −cosine function, and the LSF a +cosine function. Therefore, the envelope is a repetitive wave. At the beginning of each cycle of the envelope, the carrier is 90° out of phase with both the USF and LSF, and the USF and LSF are 180° out of phase with each other. This relationship can be seen in Figure 3-9.

EXAMPLE 3-3

One input to an AM modulator is a 500-kHz carrier with a peak amplitude of 20 V_p. The second input is a 10-kHz modulating signal that is of sufficient amplitude to cause a peak change in the output wave of ±7.5 V.
(a) Determine the USF and LSF.
(b) Determine the modulation coefficient and the percent modulation.
(c) Determine the peak amplitude of the modulated carrier, the USF, and the LSF.
(d) Determine the maximum and minimum amplitudes of the AM envelope.
(e) Determine the expression for the modulated AM wave.
(f) Draw the output spectrum.
(g) Sketch the modulated AM DSBFC envelope.

Solution   (a) The USF and LSF are the sum and difference frequencies, respectively:

$$USF = F_c + F_a = 500 \text{ kHz} + 10 \text{ kHz} = 510 \text{ kHz}$$

$$LSF = F_c - F_a = 500 \text{ kHz} - 10 \text{ kHz} = 490 \text{ kHz}$$

(b) The modulation coefficient is determined from Equation 3-1a:

$$m = \frac{E_m}{E_c} = \frac{7.5}{20} = 0.375$$

Percent modulation is determined from Equation 3-2:

$$\% M = m \times 100 = 0.375 \times 100 = 37.5\%$$

(c) The peak amplitude of the carrier, USF, and LSF are determined as follows:

$$E_{c(\text{modulated})} = E_{c(\text{unmodulated})} = 20 \text{ V}_p$$

The peak amplitude of the USF and LSF are determined from Equation 3-5:

$$E_L = E_U = \frac{mE_c}{2} = \frac{(0.375)\,20}{2} = 3.75 \text{ V}_p$$

(d) The maximum and minimum amplitudes of the modulated wave are determined from Equations 3-3d and 3-3e, respectively:

$$V_{max} = E_c + E_m = 20 + 7.5 = 27.5 \text{ V}_p$$

$$V_{min} = E_c - E_m = 20 - 7.5 = 12.5 \text{ V}_p$$

(e) The expression for the modulated wave follows the format of Equation 3-5:

$$v_c = 20 \sin (2\pi 500kt) - 3.75 \cos (2\pi 510kt) + 3.75 \cos (2\pi 490kt)$$

(f) The output spectrum is shown in Figure 3-10.

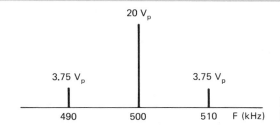

FIGURE 3-10   Output spectrum for Example 3-3.

(g) The modulated envelope is shown in Figure 3-11.

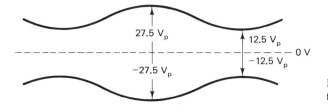

FIGURE 3-11   AM envelope for Example 3.3.

Generating an AM DSBFC Envelope in the Time Domain

Figure 3-9 showed how an AM DSBFC envelope is produced from the algebraic addition of the carrier and the upper and lower side frequency waveforms. For simplicity, the following waveforms are used for the modulating and carrier input signals:

$$\text{carrier} = v_c = E_c \sin (2\pi 25t) \tag{3-6a}$$

$$\text{modulating signal} = v_a = E_a \sin (2\pi 5t) \tag{3-6b}$$

substituting Equations 3-6a and 3-6b into Equation 3-5c, the expression for the modulated wave $M(t)$ is

$$M(t) = \underbrace{E_c \sin (2\pi 25t)}_{\text{carrier}} - \underbrace{\frac{mE_c}{2} \cos (2\pi 30t)}_{\text{USF}} + \underbrace{\frac{mE_c}{2} \cos (2\pi 20t)}_{\text{LSF}}$$

The waveforms for 100% modulation and an unmodulated carrier voltage $E_c = 1V_p$ are shown in Figure 3-9. Table 3-1 lists the instantaneous voltages for the carrier, the upper and lower side frequencies, and the total envelope at 10-ms intervals.

TABLE 3-1 INSTANTANEOUS VOLTAGES

USF, $-\frac{1}{2}\cos 2\pi 30t$	Carrier, $\sin 2\pi 25t$	LSF, $+\frac{1}{2}\cos 2\pi 25t$	Envelope, $M(t)$	Time, t (ms)
−0.5	0	+0.5	0	0
+0.155	+1	+0.155	+1.31	10
+0.405	0	−0.405	0	20
−0.405	−1	−0.405	−1.81	30
−0.155	0	+0.155	0	40
+0.5	+1	+0.5	2	50
−0.155	0	+0.155	0	60
−0.405	−1	−0.405	−1.81	70
+0.405	0	−0.405	0	80
+0.155	+1	+0.155	+1.31	90
−0.5	0	+0.5	0	100
+0.155	−1	+0.155	−0.69	110
+0.405	0	−0.405	0	120
−0.405	+1	−0.405	+0.19	130
−0.155	0	+0.155	0	140
+0.5	−1	+0.5	0	150
−0.155	0	+0.155	0	160
−0.405	+1	−0.405	+0.19	170
+0.405	0	−0.405	0	180
+0.155	−1	+0.155	−0.69	190
−0.5	0	+0.5	0	200
+0.155	+1	+0.155	+1.31	210
+0.405	0	−0.405	0	220
−0.405	−1	−0.405	−1.81	230
+0.405	0	−0.405	0	240
+0.155	+1	+0.155	+1.31	250

In Figure 3-9, note that the time between zero crossings is constant for the envelope (i.e., $T_1 = T_2 = T_3$, etc). Also note that the peak amplitudes of successive peaks are not equal. This indicates that a cycle within the envelope is not a pure sine wave; as mentioned previously, the modulated wave is the summation of the carrier and the upper and lower side frequencies. Thus, with AM DSBFC, the amplitude of the carrier is not varied, but rather, the amplitude of the envelope is varied in accordance with the modulating signal.

Power Distribution In An AM DSBFC Wave

In any electrical circuit, the power dissipated is equal to the rms voltage squared, divided by the resistance (i.e., $P = E^2/R$). Thus, the power developed across a load by an unmodulated carrier is equal to the carrier voltage squared, divided by the load resistance. Therefore, in an AM DSBFC wave, the unmodulated carrier power is

$$P_c = \frac{(E_c)^2}{R}$$

where

P_c = unmodulated carrier power
E_c = unmodulated carrier voltage
R = load resistance

The total power in an AM DSBFC wave is equal to the sum of the powers of the carrier, the upper side frequency, and the lower side frequency. (For a complex modulating signal, the total power equals the sum of the carrier and the upper and lower sideband powers.) Mathematically, the total power in an AM envelope is

$$P_t = \underbrace{\frac{(E_c)^2}{R}}_{\substack{\text{modulated} \\ \text{carrier} \\ \text{power}}} + \underbrace{\frac{\left(\frac{mE_c}{2}\right)^2}{R}}_{\substack{\text{USF} \\ \text{power}}} + \underbrace{\frac{\left(\frac{mE_c}{2}\right)^2}{R}}_{\substack{\text{LSF} \\ \text{power}}} \qquad (3\text{-}7)$$

where

P_t = total power of an AM envelope

$\dfrac{(E_c)^2}{R}$ = modulated carrier power

$\dfrac{\left(\dfrac{mE_c}{2}\right)^2}{R} = \dfrac{m^2 E_c{}^2}{4R}$ = USF and LSF power

From Equation 3-7, it can be seen that the term for the modulated carrier power is the same as the expression for the unmodulated carrier power. Thus it is evident that the carrier power is unaffected by the modulation process. Also, because the total power in the AM wave is the sum of the carrier power and the power in the side frequencies (or sidebands), the total power in an AM envelope increases with modulation. Equation 3-7 can be expanded and rewritten as

$$P_t = P_c + \frac{m^2}{4}\frac{E_c^2}{R} + \frac{m^2}{4}\frac{E_c^2}{R} \qquad (3\text{-}8)$$

and because $E_c^2/R = P_c$,

$$P_t = \underbrace{P_c}_{\substack{\text{carrier} \\ \text{power}}} + \underbrace{\frac{m^2 P_c}{4}}_{\substack{\text{USF} \\ \text{power}}} + \underbrace{\frac{m^2 P_c}{4}}_{\substack{\text{LSF} \\ \text{power}}} \tag{3-9}$$

Combining the powers from the upper and lower side frequencies gives us

$$P_t = P_c + \frac{m^2 P_c}{2} \tag{3-10}$$

Factoring P_c yields

$$P_t = P_c \left(1 + \frac{m^2}{2} \right) \tag{3-11}$$

Equation 3-11 is the general expression for the total power in an AM envelope. From Equation 3-9 it can be seen that the power in the USF is equal to the power in the LSF. Therefore,

$$P(\text{LSF}) = P(\text{USF}) = \frac{m^2 P_c}{4} \tag{3-12}$$

Thus the total power of both sidebands is

$$P_{sb} = \frac{2\, m^2 P_c}{4} = \frac{m^2 P_c}{2} \tag{3-13}$$

Equations 3-8 through 3-13 use the peak voltages of the carrier and the side frequencies and, consequently, solve for peak powers. To convert to rms, simply multiply each voltage by 0.707 or divide the total peak power by 2.

Figure 3-12 shows the power spectrum for an AM DSBFC wave. Note that with 100% modulation, the maximum power in either the USF or the LSF is equal to only one-fourth the power in the carrier. Thus the maximum total sideband power P(USB) + P(LSB) is equal to one-half of the carrier power. This is one significant disadvantage

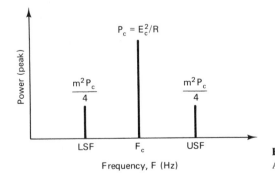

FIGURE 3-12 Power spectrum for an AM DSBFC wave.

of AM DSBFC transmission; the information is contained in the side frequencies, although most of the envelope power is wasted in the carrier. Actually, the power in the carrier is not totally wasted; it allows for the use of simple and inexpensive demodulators in the receiver using only a single diode, which is the predominant advantage of AM DSBFC.

EXAMPLE 3-4

For an AM DSBFC wave with a peak unmodulated carrier voltage of 10 V and a load resistance of 10 Ω:
(a) Determine the peak and rms powers of the carrier, the USF, and the LSF.
(b) Determine the total power of the modulated wave for a modulation coefficient of 0.5.
(c) Draw the power spectrum.

Solution (a) The peak modulated carrier power is:

$$P_c = \frac{E_c^2}{R} = \frac{10^2}{10} = \frac{100}{10} = 10 \text{ W}_p$$

The rms carrier power is

$$P_c = \frac{(0.707E_c)^2}{R} = \frac{7.07^2}{10} = \frac{50}{10} = 5 \text{ W rms}$$

The peak power of the USF and LSF are equal and determined from Equation 3-12:

$$P(USF) = P(LSF) = \frac{m^2 P_c}{4} = \frac{(0.5)^2(10)}{4} = 0.625 \text{ W}_p$$

The rms power is

$$P(USF) = P(LSF) = \frac{(0.5)^2(5)}{4} = 0.3125 \text{ W rms}$$

The total sideband power is then

$$P_{sb} = P(USF) + P(LSF) = 2(0.625) = 1.25 \text{ W}_p$$
$$P_{sb} = P(USF) + P(LSF) = 2(0.3125) = 0.625 \text{ W rms}$$

or from Equation 3-13:

$$P_{sb} = \frac{m^2 P_c}{2} = \frac{(0.5)^2(5)}{2} = 0.625 \text{ W rms}$$

(b) The total peak power of the modulated wave is the sum of the carrier and side frequency powers:

$$P_t = P_c + P(USF) + P(LSF)$$
$$= 10 + 0.625 + 0.625 = 11.25 \text{ W}_p$$

and the total rms power is

$$P_t = 5 + 0.3125 + 0.3125 = 5.625 \text{ W rms}$$

or from Equation 3-11,

$$P_t = P_c \left(1 + \frac{m^2}{2}\right) = 10\left(1 + \frac{0.5^2}{2}\right) = 11.25 \text{ W}_p$$

(c) The peak power spectrum is shown in Figure 3-13.

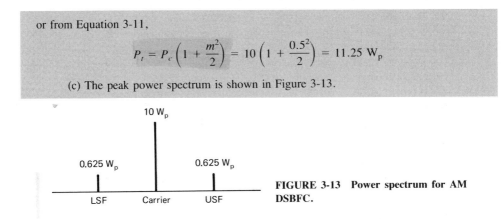

FIGURE 3-13   Power spectrum for AM DSBFC.

CIRCUITS FOR GENERATING AM DSBFC

Low-Power AM DSBFC Transistor Modulator

Figure 3-14 shows a simple low-power transistor AM modulator circuit which has only a single active component (Q_1). The carrier (ω_c) is applied to the base, and the modulating signal (ω_a) is applied to the emitter. Therefore, this is called *emitter modulation*. With

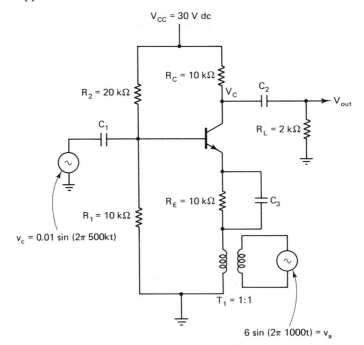

FIGURE 3-14   Single transistor, emitter modulator.

emitter modulation, it is important that the transistor be biased *class A* with a centered Q point.

Circuit operation. Figure 3-14 shows that the amplitude of the carrier (10 mV$_p$) is appreciably smaller than the amplitude of the modulating signal (6 V$_p$). If the modulating signal is removed or held constant at 0 V, Q_1 is a linear class A amplifier. The base input signal is simply amplified and inverted $180°$ at the collector. The amplification by Q_1 is determined by the ratio of the ac collector resistance (r_c) to the ac emitter resistance (r'_e) (i.e., $A_v = rc/r'_e$). For the circuit shown in Figure 3-14, r_c and r'_e are determined as follows:

$$rc = \text{parallel combination of } R_C \text{ and } R_L$$

$$= R_L \| R_C = 10 \text{ k}\Omega \| 2 \text{ k}\Omega$$

$$= \frac{(10{,}000)(2000)}{10{,}000 + 2000} = 1667$$

$$r'_e = \frac{25 \text{ mV}}{I_e}$$

$$I_E = \frac{V_{th} - V_{be}}{(R_{th}/B) + R_e}$$

$$V_{th} = \frac{V_{cc}(R_1)}{R_1 + R_2} = \frac{30(10 \text{ k}\Omega)}{30 \text{ k}\Omega} = 10 \text{ V}$$

$$R_{th} = R_1 \| R_2 = \frac{(20{,}000)(10{,}000)}{20{,}000 + 10{,}000} = 6667$$

Thus for $\beta = 100$,

$$I_E = \frac{10 - 0.7}{(6667/\beta) + 10{,}000} = 0.924 \text{ mA}$$

Therefore,

$$r'_e = \frac{25 \text{ mV}}{I_E} = \frac{0.025}{0.000924} = 27$$

and

$$A_q = A_v = \frac{rc}{r'_e} = \frac{1667}{27} = 61.7$$

where Aq = quiescent gain.

With no input modulating signal, Q_1 is a linear small-signal amplifier with a *quiescent* voltage gain (A_q) of 61.7. With the carrier input voltage shown,

$$V_{out} = A_q V_{in} = 61.7(0.01 \text{ V}) = 0.617 \text{ V}_p$$

When the modulating signal (V_a) is applied to the circuit, its voltage combines with the dc Thévenin voltage. The result is a bias voltage that has a constant term and a term that varies at a low-frequency sinusoidal rate. Thus

$$V_{bias} = \underbrace{V_{th}}_{\substack{\text{constant} \\ \text{term}}} + \underbrace{V_a \sin \omega_a t}_{\substack{\text{sinusoidal} \\ \text{term}}}$$

and

$$V_{bias} = 10 + 6 \sin (2\pi 1000t)$$

To analyze the operation of this circuit, it is not necessary to consider every possible value of V_{bias}. Instead, several key values are calculated and the others interpolated from them. The three most significant values for V_{bias} are when the sinusoidal voltage term is 0, maximum positive, and maximum negative. When $V_a = 0$ V, the circuit bias is equal to its quiescent value (V_{th}). For this example, the voltage gain at quiescence (A_q) is equal to 61.7.

When the sinusoidal term is maximum and negative:

$$\sin (2\pi 1000t) = -1$$

$$V_{bias} = V_{th} - 6(-1) = 10 + 6 = 16 \text{ V}$$

$$I_E = \frac{16 - 0.7}{(6667/100) + 10,000} = 1.52 \text{ mA}$$

$$r'_e = \frac{0.025}{0.00152} = 16.45$$

$$A_v = A_{max} = \frac{1667}{16.45} = 101.3$$

$$V_{out} = V_{in}A_{max} = (0.01)(101.3) = 1.013 \text{ V}_p$$

When the sinusoidal term is maximum and positive:

$$\sin 2\pi 1000t = +1$$

$$V_{bias} = V_{th} - 6(1) = 10 - 6 = 4 \text{ V}$$

$$I_E = \frac{4 - 0.7}{(6667/100) + 10,000} = 0.328 \text{ mA}$$

$$r'_e = \frac{0.025}{0.000328} = 76.3$$

$$A_v = A_{min} = \frac{1667}{76.3} = 21.9$$

$$V_{out} = V_{in}A_{min} = 0.01(21.9) = 0.219 \text{ V}_p$$

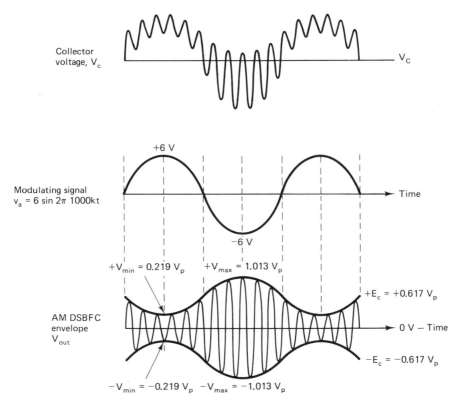

Collector voltage, V_c

Modulating signal $v_a = 6 \sin 2\pi 1000kt$

AM DSBFC envelope V_{out}

FIGURE 3-15 Output waveform for the circuit shown in Figure 3-14.

In this example, A_v varies at a sinusoidal rate equal to ω_a from a quiescent value $A_q = 61.7$ to a maximum value $A_{\max} = 101.3$ and a minimum value $A_{\min} = 21.9$. The collector voltage (V_c) and output voltage (V_{out}) are shown in Figure 3-15. The collector waveform is an AM DSBFC envelope riding on top of the low-frequency modulating signal. The average voltage is equal to the quiescent collector voltage V_C. The dc collector voltage and the low-frequency modulating signal component are removed from the waveform by coupling capacitor C_2. Consequently, the output waveform is an AM DSBFC envelope with an average voltage equal to 0 V, an unmodulated carrier amplitude of 0.617 V_p, a maximum positive or negative amplitude of 1.013 V_p, and a minimum positive or negative amplitude of 0.219 V_p. It is interesting to note that with emitter modulation the maximum amplitude of the envelope occurs when the modulating signal is maximum negative, and the minimum amplitude occurs when the modulating signal is maximum positive.

From the preceding example it can be seen that A_v varies at a sinusoidal rate equal to ω_a. Therefore, the voltage gain can be expressed mathematically as

$$A_v = A_q(1 + m \sin \omega_a t) \tag{3-14a}$$

sin $\omega_a t$ goes from a maximum value of $+1$ to a minimum value of -1. Thus

$$A_v = A_q(1 \pm m) \tag{3-14b}$$

where

$\quad m =$ modulation coefficient

$\quad A_q =$ quiescent gain

Therefore, at 100% modulation ($m = 1$)

$$A_{max} = A_q(1 + 1) = 2A_q$$

$$A_{min} = A_q(1 - 1) = 0$$

and the maximum and minimum amplitudes of V_{out} are

$$V_{out(max)} = A_{max}(V_{in}) = 2A_q(V_{in})$$

$$V_{out(min)} = A_{min}(V_{in}) = 0(V_{in}) = 0$$

For the preceding example, the modulation coefficient is

$$m = \frac{V_{max} - V_{min}}{V_{max} + V_{min}} = \frac{1.013 - 0.2195}{1.013 + 0.2195} = 0.645$$

and

$$\% \text{ modulation} = 0.645(100) = 64.5\%$$

Substituting into Equation 3-14b yields

$$A_{max} = A_q(1 + m) = 61.7(1 + 0.645) = 101.5$$

$$A_{min} = A_q(1 - m) = 61.7(1 - 0.645) = 21.9$$

Also, in the preceding example, the voltage gain changed symmetrically with modulation. In other words, the amount of increase in gain is equal to the amount of decrease in gain. The gain is approximately $A_q \pm 40$. This is called linear modulation, and is desired. Essentially, a conventional AM DSBFC receiver reproduces the original modulating signal from the shape of the envelope. If modulation is not linear, the envelope does not accurately represent the shape of the modulating signal and the demodulator will produce a distorted output waveform.

EXAMPLE 3-6

For a low-power transistor AM modulator similar to the one shown in Figure 3-10 with a modulation coefficient of 0.8, a quiescent gain of 100, and an input carrier amplitude of 0.005 V_p:
(a) Determine the maximum and minimum gains for the transistor.
(b) Determine the maximum and minimum amplitude of V_{out}.
(c) Sketch the output AM envelope.

Solution (a) substituting into Equation 3-14b

$$A_{max} = A_q(1 + m) = 100(1 + 0.8) = 180$$

$$A_{min} = A_q(1 - m) = 100(1 - 0.8) = 20$$

(b) $V_{out(max)} = A_{max}(V_{in})$,
$V_{out(max)} = 180(0.005) = 0.9 \ V_p$
$V_{out(min)} = A_{min}(V_{in})$
$V_{out(min)} = 20(0.005) = 0.1 \ V_p$

(c) The AM DSBFC output waveform is shown in Figure 3-16.

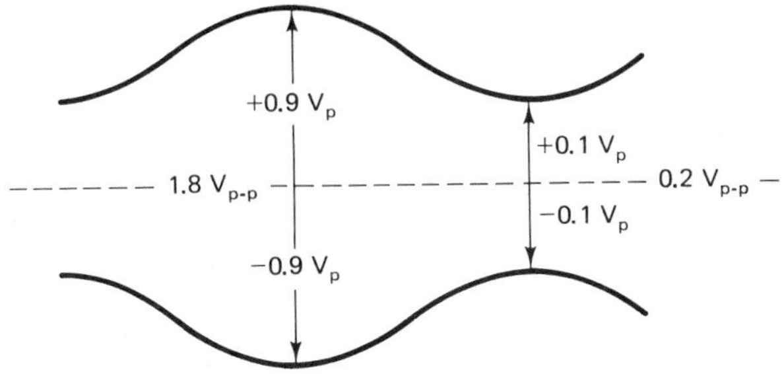

FIGURE 3-16 AM envelope for Example 3-6.

With no modulating signal, the transistor modulator shown in Figure 3-14 is a linear small-signal amplifier. When a modulating signal is applied, the *Q-point* of the amplifier is driven first toward *saturation,* then toward *cutoff* (i.e., the transistor is forced to operate over a nonlinear portion of its operating curve). Therefore, a linear amplifier operates nonlinearly when a modulating signal is applied.

The transistor modulator shown in Figure 3-14 is adequate for low-power applications but is not a practical circuit for high-power applications. This is because the transistor is operating class A, which is extremely inefficient. With AM, any amplifiers that follow the modulator circuit must be linear. If they are not linear, intermodulation between the upper and lower side frequencies and the carrier will generate additional cross-product frequencies that could interfere with signals from other transmitters.

Low-Level and High-Level Modulators

The terms *low-level* and *high-level* modulation simply refer to the point in a transmitter where the modulation is performed. With low-level modulation, the modulation takes place prior to the output element of the *final* stage. In other words, prior to the collector of the output transistor in a transistorized transmitter, prior to the drain of the output FET in a FET transmitter, or prior to the plate of the final stage in a tube transmitter.

The modulator shown in Figure 3-14 is a low-level modulator regardless of whether it is the final stage or not because modulation takes place in the emitter, which is not the output element. If the modulation took place in the collector and the modulator were the final stage, it would be a high-level modulator.

An advantage of low-level modulation is that less modulating signal power is required. In high-level modulators, the modulation takes place in the final element of the final stage, where the carrier is at its maximum amplitude and requires a much higher modulating signal amplitude to achieve a reasonable percent modulation. With high-level modulation, the final audio amplifier must supply all of the sideband power, which could be as much as 50% of the power in the carrier. An obvious disadvantage of low-level modulation is in high-power applications when all of the amplifiers that follow the modulator stage must operate class A, which is extremely inefficient.

Medium-Power AM DSBFC Transistor Modulator

Early medium- and high-power AM transmitters were limited to those that used tubes for the active devices. However, since the mid-1970s, solid-state AM transmitters have been available with output powers as high as several thousand watts. This is usually not accomplished with a single output stage, but rather, by placing several output stages in parallel such that their output signals combine in phase and are, thus, additive.

Figure 3-17a shows a simplified single-transistor medium-power AM DSBFC modulator. The modulation takes place in the collector circuit, which is the output element of the transistor. Therefore, if this is the final stage of the transmitter (i.e., there are no amplifier stages following it), it is a high-level modulator. If the output of the modulator goes through additional amplifiers prior to transmission, it is a low-level modulator. High- and low-level transmitter configurations are discussed in a later section of this chapter.

To achieve high efficiency, medium- and high-power AM modulators generally operate *class C*. Therefore, a theoretical efficiency of as high as 80% is possible. The circuit shown in Figure 3-17a is a class C transistor amplifier with two inputs: an unmodulated carrier (v_c) and a single-frequency modulating signal (v_a). Because the transistor is biased class C, it is operating nonlinear and therefore is capable of nonlinear mixing (modulation). This circuit is a *collector modulator*; the modulating signal is applied to the collector. The RFC is a radio-frequency choke. At low frequencies the RFC looks like a short and at high frequencies it looks like an open. Consequently, it allows the low-frequency intelligence signal to modulate the collector of Q_1 and, at the same time, prevents the high-frequency carrier from entering the dc power supply.

Circuit operation.   For the following explanation, refer to the circuit shown in Figure 3-17a and the waveforms shown in Figure 3-17b. When the amplitude of the carrier exceeds the barrier potential of the base–emitter junction (approximately 0.7 V for a silicon transistor), Q_1 turns on and collector current saturates. When the carrier voltage drops below 0.7 V, the transistor turns off and collector current ceases. Thus Q_1 is switched between saturation and cutoff and collector current flows for less than 180° of each input cycle. Thus Q_1 operates class C. Each successive positive half-cycle of the input signal turns Q_1 on and allows collector current to flow, which produces a negative going voltage waveform at the collector of Q_1. The collector current and

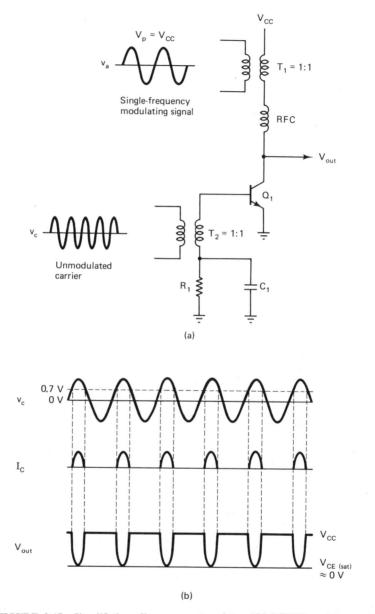

FIGURE 3-17 Simplified medium-power transistor AM DSBFC modulator: (a) schematic diagram; (b) collector waveforms with no modulating signal; (c) collector waveforms with modulating signal.

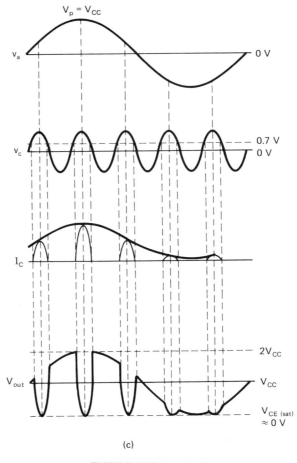

(c)

FIGURE 3-17 (*continued*)

voltage waveforms are shown in Figure 3-17b. The output voltage waveform resembles a repetitive half-wave rectified signal with a fundamental frequency equal to ω_c.

When a modulating signal is applied to the collector in series with V_{CC}, it adds to and subtracts from V_{CC}. The waveforms shown in Figure 3-17c are produced when the maximum peak modulating signal amplitude equals V_{CC}. It can be seen that the output waveform swings from a maximum value of $2V_{CC}$ to approximately 0 V ($V_{CE(\text{sat})}$). The change in collector voltage is equal to V_{CC}. Again, the waveform resembles a half-wave rectified signal superimposed onto a low-frequency ac component.

Because Q_1 is operating nonlinear, the collector waveform contains the two original input frequencies (ω_a and ω_c) and their sum and difference frequencies ($\omega_c \pm \omega_a$). The output waveform also contains all of the higher-order harmonics and intermodulation products (i.e., $2\omega_c$, $2\omega_a$, $3\omega_c$, $3\omega_a$, $2\omega_c + \omega_a$, $3\omega_c + \omega_a$, etc). Consequently, before the output signal is transmitted, it must be bandlimited to $\omega_c \pm \omega_a$.

A more practical circuit for medium-power AM DSBFC generation is shown in Figure 3-18a with the corresponding waveforms shown in Figure 3-18b. This circuit is also a collector modulator with a maximum peak modulating signal amplitude equal to V_{CC}. Operation of this circuit is almost identical to the circuit shown in Figure 3-17a except for the addition of the *tuned* circuit (C_1 and L_1) in the collector of Q_1. Because the transistor is operating between saturation and cutoff, collector current is not dependent on base drive voltage. The voltage developed across the *tank* circuit is determined by the ac component of the collector current and the impedance of the tank circuit at *resonance* (which depends on the Q of the coil). The waveforms for the modulating signal, the carrier, and the collector current are identical to those of the previous example. The output voltage across R_L is a symmetrical, AM DSBFC envelope with an average voltage of 0 V, a maximum positive peak amplitude equal to $2V_{CC}$, and a maximum negative peak amplitude equal to $-2V_{CC}$. The positive half-cycle of the output waveform

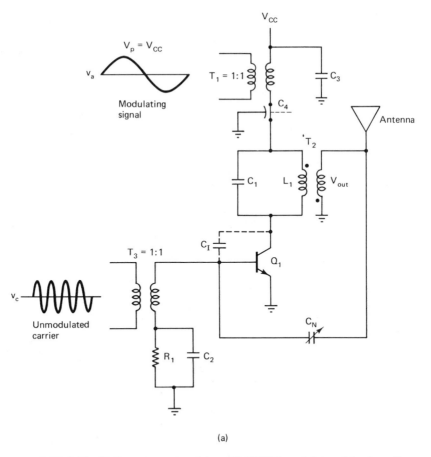

(a)

FIGURE 3-18 Medium-power transistor AM DSBFC modulator: (a) schematic diagram; (b) collector and output waveforms.

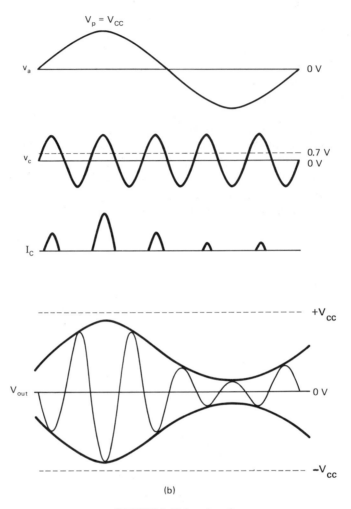

(b)

FIGURE 3-18 (*continued*)

is produced by the *flywheel effect* in the tank circuit. When Q_1 is conducting, C_1 charges to $V_{CC} + v_a$ (a maximum value of $2V_{CC}$) and, when Q_1 is off, C_1 discharges through L_1. When L_1 discharges, C_1 charges to $-2V_{CC}$. This produces the positive half cycle of the AM envelope. The resonant frequency of the tank circuit is equal to the carrier frequency and the bandwidth extends from $F_c - F_a$ to $F_c + F_a$. Consequently, the modulating signal, the harmonics, and all of the higher-order cross products are removed from the waveform, leaving a symetrical AM DSBFC envelope. One hundred percent modulation occurs when the peak amplitude of the modulated wave equals $2V_{CC}$.

There are several components shown in Figure 3-18a that have not been explained. R_1 is the bias resistor for Q_1. R_1, in conjunction with C_2 and the barrier potential of the transistor, determine the turn-on voltage for Q_1. Consequently, Q_1 can be biased to turn on only during the most positive peaks of the input carrier signal. This produces a narrower collector current waveform and enhances class C efficiency.

C_3 is a bypass capacitor. It looks like a short circuit to audio frequencies and thus prevents the information from entering the dc power supply. C_I is the base-to-collector *junction* capacitance of Q_1. At radio frequencies, the relatively small junction capacitances within a transistor are significant. If the capacitive reactance of C_I is sufficiently small, the collector signal may be returned to the base with sufficient amplitude to cause Q_1 to oscillate. Therefore, a signal of equal amplitude and frequency and 180° out of phase must be fed back to the base to cancel or *neutralize* the *intercapacitive feedback*. C_N is a *neutralizing capacitor*. Its purpose is to provide a feedback path for a signal that is equal in amplitude but 180° out of phase with the signal fed back through C_I. C_4 is an RF bypass capacitor. Its purpose is to isolate the dc power supply from RF frequencies. Its operation is quite simple; at the carrier frequency, C_4 looks like a short circuit preventing the carrier from leaking into the power supply or the audio circuitry and distributing throughout the transmitter.

High-Power AM DSBFC Transistor Modulators

Medium-power collector modulators produce a more linear (symmetrical) envelope than low-power base modulators. Also, collector modulators are more power efficient. However, collector modulators require a higher modulating signal drive power, and collector modulators cannot achieve 100% modulation because the collector current saturates and does not allow a full output voltage swing of $\pm V_{CC}$. Therefore, to achieve linear modulation, operate at maximum efficiency, develop a high output power, and require as little modulating signal drive power as possible, base and collector modulation are sometimes used simultaneously.

Circuit operations. Figure 3-19 shows a modulator that uses a combination of base and collector modulation. The modulating signal is simultaneously fed into the collectors of the push-pull modulators (Q_2 and Q_3) and the collector of the driver amplifier (Q_1). Therefore, collector modulation occurs in Q_1, Q_2, and Q_3. The carrier signal supplied to the base circuits of Q_2 and Q_3 has already been partially modulated. Thus the modulating signal power can be reduced, and the modulators are not required to operate over their entire operating curve to achieve 100% modulation.

Vacuum-Tube AM DSBFC Modulators

With the advent of medium- and high-power solid-state AM transmitters, *vacuum-tube* modulators have somewhat gone by the wayside. Vacuum-tube circuits have many disadvantages compared to solid-state circuits. Vacuum tubes require higher dc supply voltages (often, both positive and negative polarities for a single circuit), a separate *filament* (heater) supply voltage, very high modulating signal amplitudes, and vacuum tubes take up considerably more space and weigh much more than their solid-state counterparts.

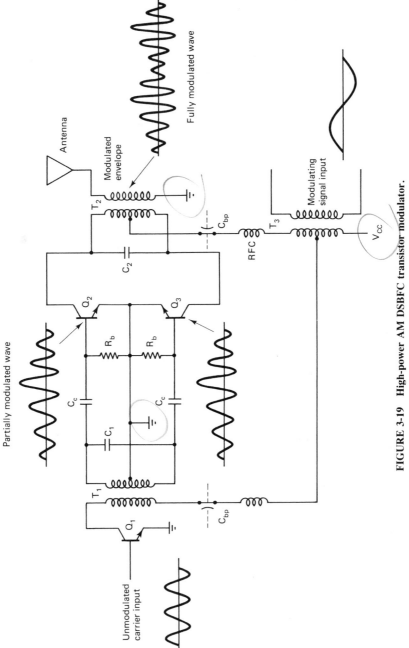

FIGURE 3-19 High-power AM DSBFC transistor modulator.

110

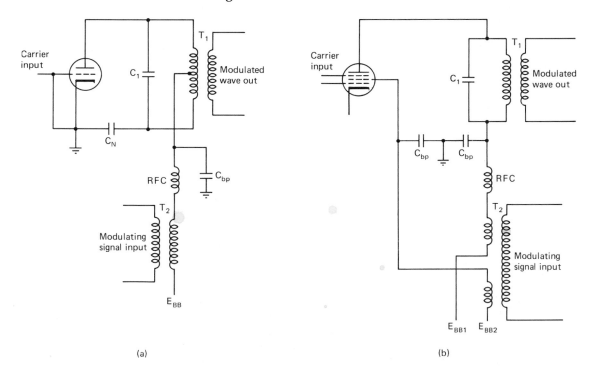

(a)   (b)

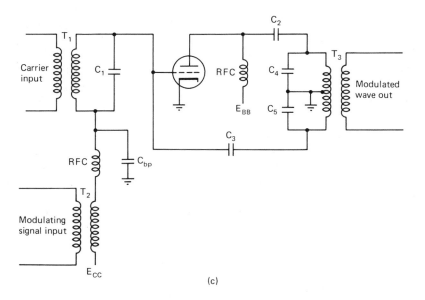

(c)

FIGURE 3-20   Vacuum-tube AM DSBFC modulator circuits: (a) triode plate modulator; (b) multi-grid plate modulator; (c) grid-bias modulator; (d) suppressor-grid modulator; (e) screen-grid modulator.

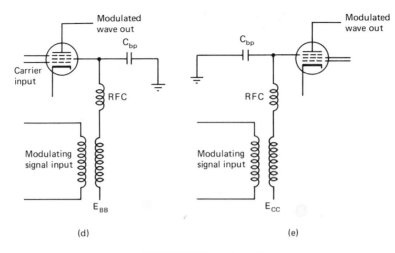

FIGURE 3-20 (*continued*)

Consequently, a vacuum-tube modulator (or, for that matter, any other vacuum-tube circuit) is seldom used today. Essentially, vacuum-tube modulators are used in applications where extreme high powers are required, such as in commercial AM and TV broadcast-band transmitters. For the more common low- and medium-power applications, such as two-way radio communications, solid-state circuitry dominates.

As a historical perspective (i.e., a look back into the "cavetronic" era), several vacuum-tube AM DSBFC modulators are shown in Figure 3-20. Assuming that the reader is familiar with vacuum-tube terminology (i.e., *plate*, *grid*, *cathode*, etc.), the operation of a vacuum-tube modulator is quite similar to a transistor modulator. A plate modulator is analogous to a collector modulator and a grid-bias modulator to a base modulator. Because there are more elements in a vacuum tube, there are more modulator configurations possible with vacuum tubes than with transistors (i.e., *suppressor* and *screen-grid modulators*).

High-level vacuum-tube triode modulators are class C biased and use plate modulation (Figure 3-20a). A high-power modulating signal is transformer coupled into the plate circuit in series with the plate supply voltage. Consequently, the modulating signal adds to and subtracts from E_{BB}. A relatively low-power carrier signal is coupled into the grid circuit, where it biases the tube into conduction only during its most positive peaks. Therefore, as with a transistor collector modulator, narrow pulses of plate current flow and supply the energy necessary to sustain oscillations in the plate tank circuit.

Figure 3-20b shows a *multigrid* vacuum-tube plate modulator. The vacuum tube is a pentode. It has five elements: plate, cathode, control grid, suppressor grid, and screen grid. *Pentode* vacuum-tube modulators are capable of high output powers and high efficiencies. However, to achieve 100% modulation, it is necessary to modulate the screen grid as well as the plate.

Figure 3-20c shows a vacuum-tube grid-bias modulator. The triode can be replaced with a *tetrode* or a pentode. The grid-bias modulator requires much less modulating power than the triode or pentode plate modulators, but it has poorer modulation linearity, more distortion, lower plate efficiency, and produces a lower output power.

   Figures 3-20d and e show suppressor-grid and screen-grid modulators, respectively. These two modulator configurations fall somewhere between the grid-bias and plate modulators discussed previously.

AM DOUBLE-SIDEBAND FULL-CARRIER TRANSMITTERS

Low-Level Transmitters

Figure 3-21 shows a block diagram for a low-level AM DSBFC transmitter. For voice or music transmission, the modulating signal source can be a microphone, a magnetic tape or disk, or a phonograph. The *preamplifier* is typically a sensitive, class A linear voltage amplifier with a high input impedance. The preamplifier's function is to raise the signal voltage from the source to a usable level and, at the same time, produce a minimum amount of nonlinear distortion. The modulating driver is also a linear class A amplifier. The modulating driver simply amplifies the signal from the preamplifier to an adequate level to modulate the carrier sufficiently. More than one driver amplifier may be required.

   The RF oscillator can be any of the oscillator configurations discussed. Cost and stability generally dictate which oscillator configuration is chosen. However, the FCC has stringent requirements on transmitter stability. The *buffer amplifier* is a low-gain, high-input impedance linear amplifier. Its function is to isolate the oscillator from the high-power amplifiers. The buffer provides a relatively constant load to the oscillator, which helps to reduce short-term frequency variations. An *emitter follower* or an *integrated-circuit op-amp* is commonly used for the buffer amplifier. The modulator can use either emitter or collector modulation. The intermediate and final power amplifiers are either linear class A or class B push-pull amplifiers. This is required with low-level transmitters to maintain the amplitude linearity of the AM DSBFC envelope. The

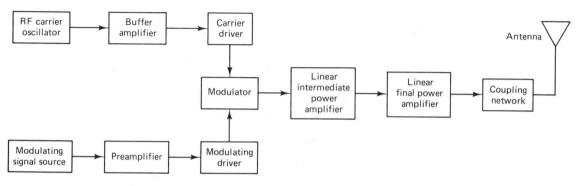

FIGURE 3-21   Block diagram of a low-level AM DSBFC transmitter.

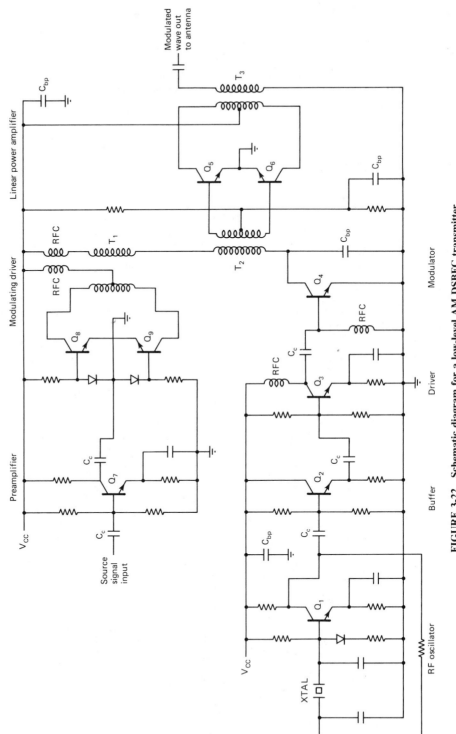

FIGURE 3-22 Schematic diagram for a low-level AM DSBFC transmitter.

antenna coupling network matches the output impedance of the final amplifier to the transmission line and antenna.

Low-level transmitters like the one shown in Figure 3-21 are used predominantly for low-power, low-capacity AM systems such as wireless intercoms, remote control units, and short-range walkie-talkies.

Figure 3-22 shows a schematic diagram for a simple low-level AM transmitter. The RF oscillator is the Pierce crystal oscillator that was discussed in Chapter 2. The buffer amplifier is an emitter follower which provides a constant high-impedance load to the oscillator and supplies the unmodulated carrier to the class A driver amplifier. The low-level source signal is amplified first by the small-signal class A preamplifier and then by the push-pull class B modulating driver. Class C collector modulation is used with this transmitter. Finally, the modulated wave is amplified to the desired transmit signal power by the class B power amplifier. A *class B* final power amplifier is chosen because it can develop more output power than its class A counterpart, it has a higher power efficiency, and it provides linear amplification of the modulated envelope. Note the presence of several radio-frequency (RF) bypass capacitors and radio-frequency chokes (RFCs). Also, note the biasing components used in the two class B amplifier configurations. The diodes in the base circuits of the modulating driver compensate for variations in the barrier potential of the transistors due to changes in the environmental temperature.

High-Level Transmitters

Figure 3-23 shows a block diagram for a high-level AM DSBFC transmitter. The modulating signal is processed the same as it is in the low-level transmitter except with the addition of a modulating signal power amplifier. With high-level modulation, the modulating signal power must be considerably higher than with low-level modulation. This is because the carrier is at full power when modulation occurs and, consequently, requires a high-amplitude modulating signal to produce 100% AM modulation.

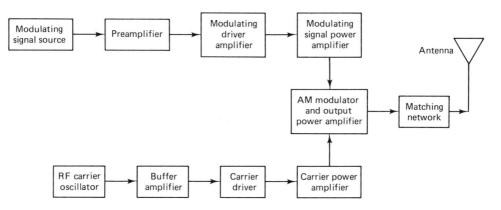

FIGURE 3-23 **Block diagram of a high-level AM DSBFC transmitter.**

The RF oscillator, its associated buffer, and the carrier driver are also the same as in a low-level transmitter. However, with high-level transmitters, the RF carrier undergoes additional power amplification prior to the modulator stage, and the final power amplifier is the modulator. Consequently, the modulator is generally a drain-, plate-, or collector-modulated class C power amplifier.

Figure 3-24 shows a partial schematic diagram for a high-level AM DSBFC transmitter. Q_1 is the active component in a standard crystal oscillator configuration. L_1 and the input impedance to Q_2 present a constant-impedance load to the oscillator. Q_2 is also the driver for the carrier. Class C amplifier, Q_3, is the modulator and also the final RF power amplifier. The modulating signal is coupled into the collector of Q_3 superimposed onto 13.5 V dc. The 13.5 V dc is the collector supply voltage for Q_3 and is present only when there is an input modulating signal. The 13.5 V dc is *keyed* on and off with the modulating signal. Consequently, the modulator has an output only when modulation is present. Although high-level modulation is accomplished in Q_3, in order to obtain 100% modulation, the collector of Q_2 is also collector modulated. The modulating signal is applied directly to the collector of Q_3 on top of the keyed 13.5 V dc. At the same time, dual diodes D_1 and D_2 allow Q_1 to be modulated. When the modulating signal voltage drops below 13.5 V, D_2 is reverse biased and shuts off, thus

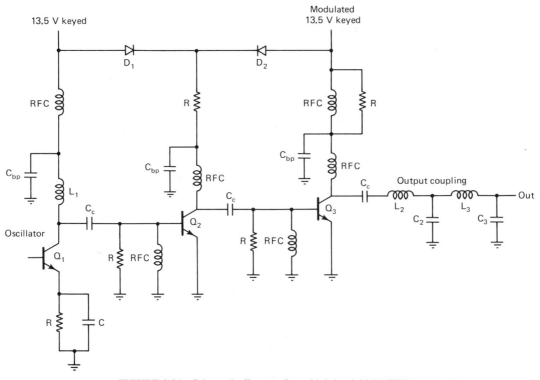

FIGURE 3-24   Schematic diagram for a high-level AM DSBFC transmitter.

removing modulation from Q_2. D_1 turns on and supplies the collector voltage for Q_2. When the modulating signal voltage rises above 13.5 V, D_2 turns on and applies modulation to the collector of Q_2 and shuts off D_1. This configuration allows the collector supply voltage for Q_3 to fluctuate above and below 13.5 V with the modulating signal and, at the same time, Q_2 can supply a partially modulated signal to the base of Q_3.

The output coupling circuit (L_2, L_3, C_2, and C_3) is an RF *trap*. An RF trap is simply a high-Q bandstop filter with a center frequency equal to twice the RF carrier frequency. Remember, the modulator is a nonlinear amplifier and, consequently, produces harmonic distortion. The second harmonic is the most significant harmonic produced and must be suppressed so that it does not interfere with transmissions from other transmitters an *octave* above in the radio spectrum.

In a high-level transmitter, the modulator circuit has three primary functions. It provides the circuitry necessary for AM modulation (nonlinearity), it is the final RF power amplifier (class C), and it is a frequency up-converter. An up-converter simply translates the low-frequency modulating signals to RF frequencies that can be more efficiently transmitted.

TRAPEZOIDAL PATTERNS

Trapezoidal patterns are used for testing AM transmitters and observing their modulation characteristics (i.e., percent modulation and modulation linearity). Although the modulation characteristics can be seen with an oscilloscope by observing the AM envelope, a

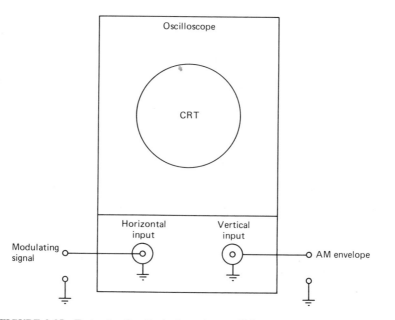

FIGURE 3-25 Test setup for displaying a trapezoidal pattern on an oscilloscope.

trapezoidal pattern is more easily interpreted, and is interpreted faster and more accurately. Figure 3-25 shows the basic test setup for producing a trapezoidal pattern on the CRT of a standard oscilloscope. The AM envelope is applied to the vertical input, and the modulating signal is applied to the external horizontal input (the internal horizontal sweep is disabled). Therefore, the horizontal sweep rate is determined by the modulating signal frequency, and the magnitude of the horizontal deflection is proportional to the amplitude of the modulating signal. The vertical deflection is totally dependent on the amplitude and rate of change of the AM envelope. In essence, the electron beam that is emitted from the cathode of the CRT is acted on in both the horizontal and vertical planes simultaneously.

Figure 3-26 shows how the AM envelope and the modulating signal produce a

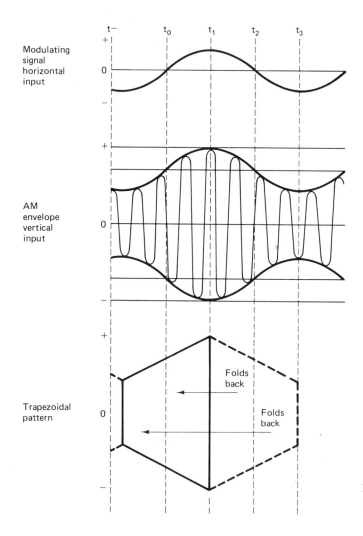

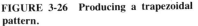

FIGURE 3-26 Producing a trapezoidal pattern.

trapezoidal pattern. With an oscilloscope, 0 V applied to the vertical input will center the electron beam vertically on the CRT, and 0 V applied to the external horizontal input will center the electron beam horizontally. Positive and negative voltages will deflect the beam up and down or right and left, respectively, for the vertical and horizontal inputs. If we begin when both the AM envelope and the modulating signal are 0 V (t_0), the electron beam is located in the exact center of the CRT. As the modulating signal goes positive, the beam is deflected to the right. At the same time the AM envelope is going positive, deflecting the beam upward. The beam deflects to the right until the modulating signal reaches its maximum positive value (t_1). While the beam is moving toward the right, the beam is deflected up and down as the AM envelope

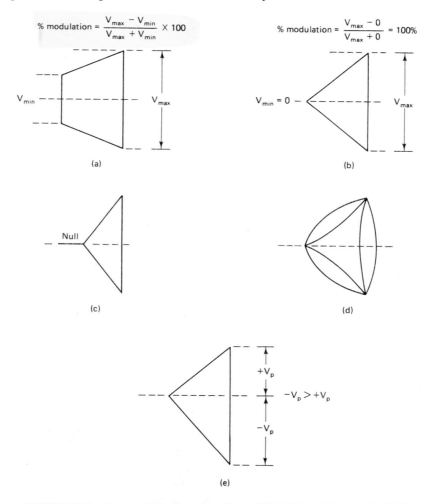

FIGURE 3-27 Trapezoidal patterns: (a) linear 50% AM modulation; (b) 100% AM modulation; (c) more than 100% AM modulation; (d) improper phase relationship; (e) nonlinear AM envelope.

alternately swings positive and negative. Notice that on each successive alternation, the AM envelope reaches a higher magnitude than the previous alternation. The AM envelope reaches its maximum value at the same time as the modulating signal. Therefore, as the CRT beam is deflected to the right, its peak-to-peak vertical deflection increases. As the modulating signal becomes less positive, the beam is deflected to the left (toward the center of the CRT). At the same time, the AM envelope alternately swings positive and negative, deflecting the beam up and down except now each successive alternation is lower in amplitude than the previous alternation. Consequently, as the beam moves horizontally back toward the center of the CRT, the vertical deflection decreases. The modulating signal and the AM envelope pass through 0 V at the same time and the beam is again in the center of the CRT (t_2). As the modulating signal goes negative, the beam is deflected to the left side of the CRT. At the same time, the AM envelope is decreasing in magnitude on each successive alternation. The modulating signal reaches its maximum negative value at the same time as the AM envelope (t_3). The trapezoidal pattern shown between times t_1 and t_3 folds back on top of the pattern displayed during times $t-$ and t_0. Thus a complete trapezoidal pattern is displayed on the screen during both the left-to-right and the right-to-left horizontal sweeps.

From Figure 3-26 it can be seen that the time of one cycle of the modulating signal equals the time of one cycle of the AM envelope and also the time for one complete horizontal alternation of the trapezoidal pattern. If the AM envelope is linear (the top half of the envelope is a mirror image of the bottom half), a linear trapezoidal pattern like the ones shown in Figure 3-27 are produced. Also, at 100% modulation, the peak-to-peak amplitude of the envelope goes to zero and the trapezoidal pattern comes to a point at one end (Figure 3-27b). If the percent modulation exceeds 100%, the pattern shown in Figure 3-27c is produced. The pattern shown in Figure 3-27a is for a 50% modulated wave. If the modulating signal and the AM envelope are out of phase, the pattern shown in Figure 3-27d is produced. If the magnitude of the positive and negative alternations of the AM envelope are not equal, the pattern shown in Figure 3-27e is produced. If the phase of the modulating signal is shifted 180° (inverted), the trapezoidal patterns would simply point in the opposite direction. As you can see, percent modulation and modulation linearity are more easily observed with a trapezoidal pattern than with a display of the AM envelope.

CARRIER SHIFT

Carrier shift is a term that is often misunderstood or misinterpreted. Carrier shift (sometimes called *upward* or *downward modulation*) has absolutely nothing to do with the frequency of the carrier. Carrier shift is a form of amplitude distortion introduced when the positive and negative alternations in the AM envelope are not equal (i.e., nonlinear modulation). Carrier shift may be either positive or negative. If the positive alternation of the envelope has a larger amplitude than the negative alternation, positive carrier shift results. If the negative alternation has a larger amplitude than the positive alternation, negative carrier shift results. Carrier shift is an indication of the average value of an

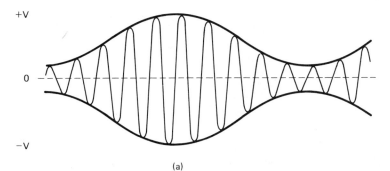

(a)

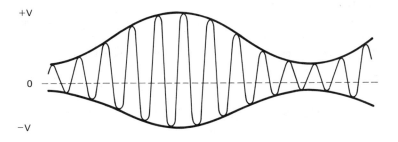

(b)

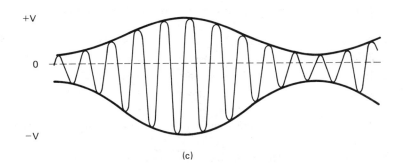

(c)

FIGURE 3-28   Carrier shift: (a) linear modulation; (b) positive carrier shift; (c) negative carrier shift.

AM envelope. If the positive and negative halves of the envelope are equal, the average voltage is 0. If the positive half is larger than the negative half, the average voltage is positive, and if the negative half is larger than the positive half, the average voltage is negative. Figure 3-28a shows a symmetrical AM envelope (no carrier shift); the average value for the envelope is 0 V. Figure 3-28b and c show positive and negative carrier shift, respectively.

QUESTIONS

3-1. Define *amplitude modulation.*

3-2. Define *RF*.

3-3. How many inputs are there to an amplitude modulator? What are they?

3-4. In an AM system, what is meant by the *modulating signal*; the *modulated signal*; the *modulated wave*?

3-5. Describe an AM envelope. Why is it called an envelope?

3-6. Describe upper and lower sidebands and upper and lower side frequencies.

3-7. Define *modulation coefficient.*

3-8. Define *percent modulation.*

3-9. What is the highest coefficient of modulation and percent modulation that can occur without causing excessive distortion?

3-10. Describe the meaning of each term in the following equation:

$$e_c = E_c \sin \omega_c t - \frac{mE_c}{2} \cos (\omega_c + \omega_a)t + \frac{mE_c}{2} \cos (\omega_c - \omega_a)t$$

3-11. What effect does modulation have on the amplitude of the carrier?

3-12. Describe the significance of the following formula:

$$P_t = P_c + \frac{m^2}{2} P_c$$

3-13. What does AM DSBFC stand for?

3-14. Describe the relationship between the carrier and sideband powers in an AM DSBFC modulator.

3-15. What is the predominant disadvantage of AM double-sideband transmission?

3-16. What is the predominant advantage of AM double-sideband transmission?

3-17. What is the maximum efficiency that can be achieved with AM DSBFC?

3-18. Why do any amplifiers that follow the modulator in an AM DSBFC system have to be linear?

3-19. What is the primary disadvantage of a low-power emitter-biased transistor modulator such as the one shown in Figure 3-14?

3-20. Describe the difference between a low- and a high-level modulator.

3-21. List the advantages of low-level modulation; high-level modulation.

3-22. What is the advantage of using a trapezoidal pattern to evaluate an AM envelope?

PROBLEMS

3-1. If a 20-V_p carrier changes in amplitude ± 5 V_p, determine the modulation coefficient and percent modulation.

3-2. For a maximum positive envelope voltage of 12 V_p and a minimum positive amplitude of 4 V_p, determine the modulation coefficient and percent modulation.

3-3. For an envelope with $+V_{max} = 40$ V_p and $+V_{min} = 10$ V_p, determine:
  (a) The unmodulated carrier amplitude.
  (b) The peak change in amplitude of the modulated wave.
  (c) The modulation coefficient.
  (d) The percent modulation.

3-4. Describe the following expression for an amplitude-modulated wave in terms of frequency content and voltage amplitude:

$$e_c = 10 \sin 2\pi 500kt - 5 \cos 2\pi 515kt + 5 \cos 2\pi 485kt$$

3-5. For an unmodulated carrier amplitude $V_c = 16$ V_p and a modulation coefficient $m = 0.4$, determine the amplitudes of the carrier and side frequencies.

3-6. Sketch the envelope for Problem 3-5 (label all pertinent voltages).

3-7. For the AM envelope shown below, determine:
  (a) The peak amplitude of the upper and lower side frequencies.
  (b) The peak amplitude of the carrier.
  (c) The peak change in the amplitude of the AM envelope.
  (d) The modulation coefficient.
  (e) The percent modulation.

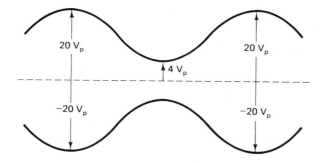

3-8. One input to an AM DSBFC modulator is a 800-kHz carrier with a peak amplitude $V_c = 40$ V. The second input is a 25-kHz modulating signal whose amplitude is sufficient to produce a ± 10 V_p change in the envelope.
  (a) Determine the USF and LSF frequencies.
  (b) Determine the modulation coefficient.
  (c) Determine the percent modulation.
  (d) Determine the maximum and minimum positive peak amplitudes of the envelope.
  (e) Draw the output spectrum.
  (f) Sketch the envelope (label all pertinent voltages).

3-9. For a modulation coefficient $m = 0.2$ and a peak carrier power $P_c = 1000$ W, determine:
  (a) The sideband power.
  (b) The total transmit power.

3-10. For an AM DSBFC wave with a peak unmodulated carrier voltage $V_c = 25$ V and a load resistance $R_1 = 50$ Ω:
  (a) Determine the peak and rms powers of the unmodulated carrier.
  (b) Determine the peak and rms powers of the modulated carrier, the USF, and the LSF for a modulation coefficient $m = 0.6$.

(c) Determine the total power in the modulated wave.

(d) Draw the output power spectrum.

3-11. Determine the quiescent, maximum, and minimum voltage gains for the emitter modulator shown below with the indicated carrier and modulating signal amplitudes.

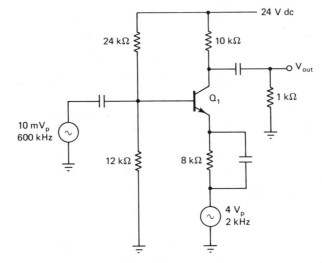

3-12. Sketch the output envelope and draw the output frequency spectrum for the circuit of Problem 3-11.

3-13. For a low-power transistor AM modulator with a modulation coefficient $m = 0.4$, a quiescent gain $A_q = 80$, and an input carrier amplitude of 0.002 V_p:

(a) Determine the maximum and minimum gains for the amplifier.

(b) Determine the maximum and minimum amplitudes for V_{out}.

(c) Sketch the AM envelope.

3-14. For the trapezoidal pattern shown below, determine:

(a) The modulation coefficient.

(b) The percent modulation.

(c) The carrier amplitude.

(d) The upper and lower side frequency amplitudes.

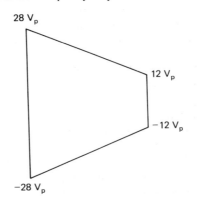

3-15. For an AM modulator with a carrier frequency of 200 kHz and a maximum modulating signal frequency of 10 kHz:
   (a) Determine the frequency limits for the upper and lower sidebands.
   (b) Determine the upper and lower side frequencies produced when the modulating signal is a 7-kHz tone.
   (c) Determine the bandwidth.
   (d) Draw the output spectrum.

Chapter 4

AMPLITUDE MODULATION RECEPTION

INTRODUCTION

In essence, AM reception is the reverse process of AM transmission. An AM receiver simply converts an amplitude-modulated wave back to the original source information (i.e., it demodulates the AM wave). However, the demodulation process can be quite different from the modulation process. To understand the demodulation process, it is necessary to have a basic understanding of the terminology associated with receivers and receiver circuits.

Figure 4-1 shows a simplified block diagram of an AM receiver. The RF section is the first stage of the receiver and is therefore often called the receiver *front end*. The primary functions of the RF section are bandlimiting and amplification of the received RF signals. The RF section comprises one or more of the following circuits: *antenna*, *antenna coupling network*, *receiver input filter* (*preselector*), and one or more *RF amplifiers*. The *mixer/converter* section down-converts the received RF frequencies to *intermediate frequencies* (IF). The *IF section* includes several *cascaded* amplifiers and bandpass filters. The primary functions of the IF section are amplification and selectivity. The *AM detector* demodulates the AM wave and recovers the original source information from the envelope. The audio section simply amplifies the recovered information to a usable level.

Receiver Parameters

Selectivity. The *selectivity* of a receiver is a measure of the ability of the receiver to accept a given band of frequencies and reject all others. For example, with the AM

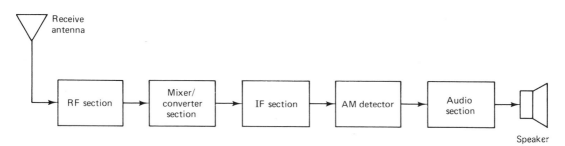

FIGURE 4-1 Simplified block diagram of an AM receiver.

commercial broadcast band, each station's transmitter is allocated a 10-kHz bandwidth (the carrier ± 5 kHz). Therefore, for a receiver to select a single channel, the input to the receiver must be bandlimited to the desired 10-kHz passband. If the passband of the receiver is greater than 10 kHz, more than one channel can be received and demodulated simultaneously. If the passband of the receiver is less than 10 kHz, a portion of the source information is rejected and information is lost.

Selectivity is often defined as the bandwidth of a receiver at some predetermined *attenuation factor* (commonly −60 dB) to the bandwidth at the −3 dB (half-power) points. This ratio is often called the shape factor (SF) and is determined by the number of *poles* and the *Q-factors* of the receiver's input filter. The *shape factor* defines the shape of the gain-versus-frequency plot for a filter and is expressed mathematically as

$$SF = \frac{B(-60 \text{ dB})}{B(-3 \text{ dB})} \tag{4-1a}$$

For a perfect filter, the attenuation factor is infinite and the bandwidth at the −3-dB frequencies is equal to the bandwidth at the −60-dB frequencies. Therefore, the ideal shape factor is 1. Selectivity is often given as a percentage and expressed mathematically as

$$\begin{aligned}
\% \text{ selectivity} &= SF \times 100 \\
&= \frac{B(-60 \text{ dB})}{B(-3 \text{ dB})} \times 100
\end{aligned} \tag{4-1b}$$

EXAMPLE 4-1

Determine the shape factor and the percent selectivity for the gain-versus-frequency plot shown in Figure 4-2.

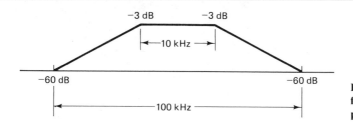

FIGURE 4-2 Gain-vs.-frequency plot for Example 4-1.

Solution   The shape factor is determined from Equation 4-1a:

$$SF = \frac{100 \text{ kHz}}{10 \text{ kHz}} = 10$$

The percent selectivity is determined from Equation 4-1b:

$$\% \text{ selectivity} = 10 \times 100 = 1000\%$$

Bandwidth improvement.   As stated in Chapter 1 and given in Equation 1-17, noise is directly proportional to bandwidth. Therefore, if the bandwidth is reduced, the noise is also reduced by the same proportion. Essentially, *bandwidth improvement* is the noise reduction ratio achieved by reducing the bandwidth. As a signal propagates from the antenna through the RF section, the mixer/converter section, and the IF section; the bandwidth is reduced. Therefore, the noise is also reduced. Effectively, this is the same as reducing (improving) the noise figure of the receiver. The bandwidth improvement factor is the ratio of the RF bandwidth to the IF bandwidth. Mathematically, bandwidth improvement is

$$BI = \frac{B(RF)}{B(IF)} \tag{4-2}$$

where

$$BI = \text{bandwidth improvement factor}$$
$$B(RF) = \text{RF bandwidth}$$
$$B(IF) = \text{IF bandwidth}$$

and the corresponding reduction in the noise figure is

$$NF \text{ (improvement)} = 10 \log BI$$

EXAMPLE 4-2

Determine the improvement in the noise figure for a receiver with an RF bandwidth equal to 200 kHz and an IF bandwidth equal to 10 kHz.

Solution

$$BI = \frac{B(RF)}{B(IF)} = \frac{200 \text{ kHz}}{10 \text{ kHz}} = 20$$

The noise-figure improvement is

$$10 \log BI = 10 \log 20 = 13 \text{ dB}$$

Consequently, a reduction in the bandwidth by a factor of 20 also reduces the noise by a factor of 20.

Sensitivity.   The *sensitivity* of a receiver is the minimum RF signal level that can be received and still produce a usable demodulated signal. Generally, the signal-to-noise ratio and the power of the signal at the output of the audio section determine whether the demodulated signal is usable or not. For commercial AM broadcast band receivers, 10 db *S/N* with $\frac{1}{2}$ W of audio power is generally considered usable. However, for broadband microwave receivers, 40 dB *S/N* and approximately 5 mW of audio power is necessary. The sensitivity of a receiver is usually stated in microvolts of received signal. For example, a common sensitivity level for a commercial AM receiver is approximately 50 μV, and a two-way mobile radio receiver has a sensitivity between 0.1 and 10 μV. Receiver sensitivity is also called *receiver threshold*. The sensitivity of an AM receiver depends on the noise power present at the input to the receiver, the receiver's noise figure, the sensitivity of the AM detector, and the bandwidth improvement factor. The best ways to improve the sensitivity of a receiver is to reduce the noise level (i.e., artificially cool the receiver's front end, reduce the receiver's bandwidth, or improve the receiver's noise figure).

Fidelity.   *Fidelity* is a measure of the ability of a communications system to produce, at the output of the receiver, an exact replica of the original source information. Any frequency, phase, or amplitude variation that is present in the demodulated waveform that was not present in the original source information is distortion.

Essentially, there are three forms of distortion that reduce the fidelity of a system: *amplitude*, *frequency*, and *phase*. Phase distortion is not particularly important for voice transmission because the human ear is relatively insensitive to phase variations. However, phase distortion can be devastating to data transmissions. The predominant cause of phase distortion is filtering. Frequencies at or near the break frequency of a filter undergo severe phase shift. Actually, absolute phase shift can be tolerated as long as all frequencies undergo the same amount of phase shift. Differential phase shift occurs when different frequencies undergo different amounts of phase shift. Phase shift is analogous to propagation delay. If all frequencies are not delayed by the same amount of time, the frequency/phase relationship of the received waveform is not consistent with the original source information.

Amplitude distortion occurs when the frequency-versus-amplitude characteristics of a signal at the output of a receiver differ from those of the original source information. Amplitude distortion is the result of *nonuniform* gain in amplifiers and filters.

Frequency distortion occurs when there are frequencies present in the received signal that were not present in the original source information. Frequency distortion is a result of harmonic and intermodulation distortion and is caused by nonlinear amplification. *Third-order intercept distortion* is the predominant form of frequency distortion. Third-order intercept distortion is a special case of intermodulation distortion. Third-order intermodulation components are the cross-product frequencies produced when the second harmonic of one signal is multiplied by the fundamental frequency of another signal (i.e., $2F_1 \pm F_2$, $2F_2 \pm F_1$, etc.). Frequency distortion can be reduced by using *square-law devices*, such as MOSFETs, in the front end of a receiver. Square-law devices have the unique advantage over BJTs in that they produce only second-order harmonic and intermodulation components.

Insertion loss.   *Insertion loss* is a parameter that is associated with the frequencies that fall within the passband of a filter and is generally defined as the ratio of the power transferred to a load with a filter in the circuit to the power transferred to the load without a filter. Because filters are generally constructed from lossy components such as resistors and imperfect capacitors, even signals within the passband of the filter are attenuated (reduced in value). Typical filter insertion losses are between a few tenths of a decibel to several decibels.

Noise temperature and equivalent noise temperatures.   Because thermal noise is directly proportional to temperature, it stands to reason that noise can be expressed in degrees as well as watts or volts. Rearranging Equation 1-17 yields

$$T = \frac{N}{KB} \tag{4-3}$$

where

T = Environmental temperature (K)
N = noise power (W)
K = Boltzmann constant (1.38×10^{-23} J/K)
B = bandwidth (Hz)

Equivalent noise temperature (T_e) is a hypothetical value that cannot be directly measured. T_e is a parameter that is often used in low-noise, sophisticated receivers rather than noise figure. T_e is an indication of the reduction in the *S/N* ratio as a signal propagates through a receiver. The lower the equivalent noise temperature, the better the quality of the receiver. Typical values for T_e range from 20° for "cool" receivers to 1000° for noisy receivers. The overall equivalent noise temperature for a receiver is simply the sum of the T_e values for each receiver stage. Mathematically, T_e is

$$T_e = T(F - 1) \tag{4-4}$$

where

T_e = equivalent noise temperature (K)
T = environmental temperature (K)
F = noise figure (unitless)

AM RECEIVERS

Essentially, there are two types of radio receivers: coherent and noncoherent. With a coherent or *synchronous receiver*, the carrier frequency used for demodulation in the receiver is synchronized to the carrier frequency used in the transmitter (the receiver has some means of recovering and synchronizing to the transmitter's carrier). With *noncoherent* or *asynchronous receivers*, either there is no carrier frequency generated in the receiver or the carrier frequency used for demodulation is totally independent

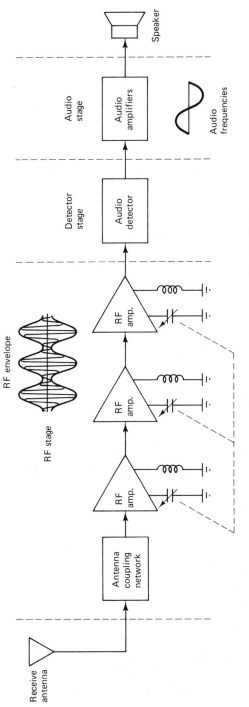

FIGURE 4-3 Noncoherent tuned radio-frequency receiver block diagram.

131

from the transmitter's carrier frequency. *Noncoherent detection* is often called *envelope detection* because the information is recovered by detecting the shape of the modulated envelope.

Tuned Radio-Frequency AM Receiver

The *tuned radio-frequency AM receiver* (TRF) was one of the earliest types of AM receivers and it is still probably the simplest design available. A block diagram of a noncoherent TRF is shown in Figure 4-3 (it is noncoherent because there is no synchronous carrier regenerated in the receiver). A TRF is essentially a three-stage receiver: an RF stage, a detector stage, and an audio stage. Generally, two or three RF amplifiers are required to develop sufficient signal amplitude to drive the detector stage. The detector converts the RF signals directly to audio signals, and the audio stage amplifies the audio signals to a usable level.

Tuning a TRF introduces three disadvantages that limit its usefulness to single-station applications. The primary disadvantage of a TRF is that its selectivity (bandwidth) varies when it is tuned over a wide range of input frequencies. The bandwidth of the RF input filter varies with the center frequency of the tuned circuit. This is caused by a phenomenon called the *skin effect*. At RF, current flow is limited to the outermost area of a conductor, and therefore the resistance of the conductor increases with frequency. (The skin effect is explained in detail in Chapter 9.) Consequently, the Q of the tank circuit (X_L/R) remains relatively constant over a wide range of frequencies and the bandwidth (F/Q) increases with frequency. As a result, the selectivity of the filter changes over any appreciable range of RF. If the bandwidth of the input filter is set to the desired value for low-band RF signals, it will be excessive for the high-band RF signals, and if the bandwidth is set to the desired value for high-band RF signals, it will be insufficient for the low-band RF signals. The second disadvantage of the TRF is that its gain is not uniform over a very wide frequency range. This is due to the nonuniform *L/C* ratio of the transformer-coupled tank circuits in the RF amplifiers (i.e., the ratio of the inductance to capacitance in one tuned amplifier is not the same as that of the other tuned amplifiers). The third disadvantage of the TRF is that it requires *multistage tuning*. To change stations, each RF filter must be tuned simultaneously to the new frequency band with a single adjustment, which requires exactly the same characteristics for each tuned circuit. With the development of the *superheterodyne receiver*, the TRF concept has essentially gone by the wayside except for single-station receivers and therefore does not warrant any further discussion.

EXAMPLE 4-3

For an AM commercial broadcast-band receiver (535 to 1605 kHz) with an input filter Q-factor of 54, determine the bandwidth at the low and high ends of the RF spectrum.

Solution   The bandwidth at the low end is centered around a carrier frequency of 540 kHz and is

$$B = \frac{F_c}{Q} = \frac{540 \text{ kHz}}{54} = 10 \text{ kHz}$$

The bandwidth at the high end is centered around a carrier frequency of 1600 kHz and is

$$B = \frac{1600 \text{ kHz}}{54} = 29{,}630 \text{ Hz}$$

The -3-dB bandwidth at the low end of the frequency spectrum is exactly 10 kHz, which is the desired value. However, the bandwidth at the high end is almost 30 kHz, which is three times the desired value. Consequently, when tuning for stations at the high end of the frequency spectrum, three stations would be received simultaneously.

To achieve a bandwidth of 10 kHz at the high end of the spectrum, a Q of 160 is required (1600 kHz/10 kHz). With a Q of 160 the bandwidth at the low end is

$$B = \frac{540 \text{ kHz}}{160} = 3375 \text{ Hz}$$

3375 Hz is too selective because approximately 66.25% of the information bandwidth is rejected.

Superheterodyne AM Receiver

The nonuniform selectivity of the TRF led to the development of the superheterodyne receiver at the end of World War I. The quality of the superheterodyne receiver has improved greatly since its original design and the basic receiver configuration is still used extensively in radio communications receivers today. The superheterodyne receiver has remained in use because its gain, selectivity, and sensitivity characteristics are superior to those of other receiver configurations.

Heterodyne means to mix two frequencies together in a nonlinear device or to translate one frequency to another using nonlinear mixing. A block diagram of a noncoherent superheterodyne receiver is shown in Figure 4-4 (it is noncoherent because the *local oscillator* is totally independent of the transmitter's carrier oscillator). Essentially, there are five primary sections that make up a superheterodyne receiver: the RF section, the mixer/converter section, the IF section, the detector section, and the audio amplifier section.

RF section. The RF section usually has a *preselector* stage and an RF amplifier stage. They can be two separate circuits or a single combined circuit. The preselector is a broad-tuned bandpass filter with an adjustable center frequency that is tuned to the desired RF carrier. The primary purpose of the preselector is to provide enough initial bandlimiting to prevent a specific unwanted radio frequency called the *image frequency* from entering the receiver. (Image frequency is explained later in this chapter.) The preselector also reduces the noise bandwidth of the receiver and provides the initial step toward reducing the overall receiver bandwidth to the minimum bandwidth required to pass the information signal. The RF amplifier determines the sensitivity of the receiver

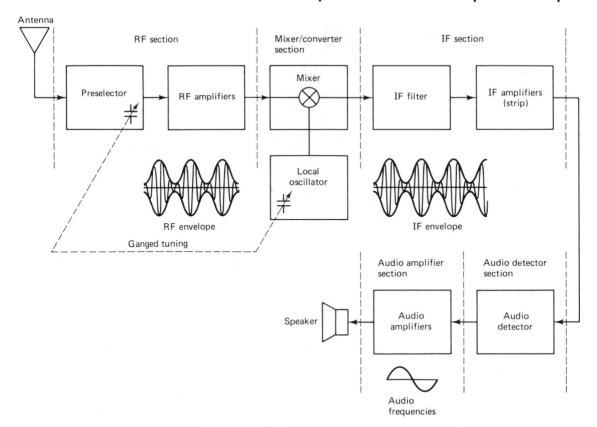

FIGURE 4-4 **AM superheterodyne receiver block diagram.**

(i.e., sets the signal threshold value). Also, because the RF amplifier is the first active device encountered by a received signal, it is the primary contributor of noise and therefore a predominant factor in determining the noise figure for the receiver. A receiver can have zero or only one RF amplifier, or there could be several, depending on the sensitivity desired.

Mixer/converter section. The mixer/converter section includes an RF oscillator stage (commonly called a local oscillator) and a mixer/converter stage (commonly called the *first detector*). The local oscillator can be any of the oscillator circuits discussed in Chapter 2, depending on the stability and accuracy desired. The mixer stage is a nonlinear device and its purpose is to convert RF to IF (i.e., frequency translation). Heterodyning takes place in the mixer stage and RF is down-converted to IF. Although the carrier and sideband frequencies are translated from RF to IF values, the shape of the AM envelope remains the same, and therefore the original information contained in the envelope remains unchanged. It is important to note that although the carrier and the upper and lower side frequencies are changed, the bandwidth of the spectrum is unchanged by the heterodyning process. The most common intermediate frequency used in AM broadcast-band receivers is 455 kHz.

IF section. The IF section comprises a series of IF amplifiers (commonly called an *IF strip*). Most of the receiver gain and selectivity is achieved in the IF section. The IF is constant for all stations and is chosen so that its frequency is less than any of the RF signals to be received. The IF is always lower because it is easier and less expensive to construct high-gain amplifiers for IF than for RF. Also, low-frequency IF amplifiers are less likely to oscillate than their RF counterparts. Therefore, it is not uncommon to see a receiver with five or six IF amplifiers and a single RF amplifier or possibly no RF amplification.

Detector section. The purpose of the detector section is to convert the IF envelope back to the original source information. The audio detector can be as simple as a diode or as complex as a phase-locked loop or balanced demodulator.

Audio section. The *audio section* comprises several cascaded audio amplifiers. The number of amplifiers depends on the audio signal power desired.

Operation of the superheterodyne receiver. During the demodulation process in a superheterodyne receiver, the received signals undergo two frequency *translations*: first the RF is converted to IF, then the IF is converted to audio.

Frequency conversion. *Frequency conversion* in the mixer/converter stage is identical to frequency conversion in the modulator stage of a transmitter except that in the receiver, the frequencies are *down-converted* rather than *up-converted*. In the mixer/converter, RF signals are mixed with the local oscillator frequency in a nonlinear device. The output of the mixer contains an infinite number of harmonic and cross-product frequencies which include the sum and difference frequencies between the desired RF and the local oscillator frequency. The IF filters are tuned to the difference frequencies. The local oscillator is designed such that its frequency of oscillation is always the IF above or below the desired RF. Therefore, the difference between the RF and the local oscillator frequency is the IF. The adjustment for the center frequency of the preselector and adjustment for the local oscillator frequency are *gang tuned*. Gang tuning means that the two adjustments are mechanically tied together so that a single adjustment will change the center frequency of the preselector and, at the same time, change the local oscillator frequency. When the local oscillator frequency is tuned above the RF, it is called *high-side* or *high-beat* injection. When the local oscillator frequency is tuned below the RF, it is called *low-side* or *low-beat* injection. In AM broadcast-band receivers, high-side injection is always used (the reason for this is explained later). Mathematically, the local oscillator frequency is:

For high-side injection:

$$F_{lo} = F_{rf} + F_{if} \qquad (4\text{-}5a)$$

For low-side injection:

$$F_{lo} = F_{rf} - F_{if} \qquad (4\text{-}5b)$$

where

F_{lo} = local oscillator frequency
F_{rf} = radio frequency
F_{if} = intermediate frequency

EXAMPLE 4-4

For an AM superheterodyne receiver that uses high-side injection and has a local oscillator frequency of 1355 kHz, determine the IF carrier, upper side frequency, and lower side frequency for an RF envelope which is made up of a carrier and upper and lower side frequencies of 900, 905, and 895 kHz, respectively.

Solution (Refer to Figure 4-5) Because high-side injection is used, the IF is the difference between the RF and the local oscillator frequency. Rearranging Equation 4-5a yields

$$F_{if} = F_{lo} - F_{rf}$$

$$= 1355 \text{ kHz} - 900 \text{ kHz} = 455 \text{ kHz}$$

The IF upper and lower side frequencies are

$$IF(usf) = F_{lo} - F_{rf}(lsf)$$

$$= 1355 \text{ kHz} - 895 \text{ kHz} = 460 \text{ kHz}$$

$$IF(lsf) = F_{lo} - F_{rf}(usf)$$

$$= 1355 \text{ kHz} - 905 \text{ kHz} = 550 \text{ kHz}$$

Note that the side frequencies undergo a reversal during the heterodyning process (i.e., the RF upper side frequency is translated to the IF lower side frequency, and the RF lower side frequency is translated to the IF upper side frequency—this is commonly called *sideband inversion*).

Local oscillator tracking. Tracking is the ability of the local oscillator in a receiver to remain the IF above or below the preselector center frequency. With high-side injection, the local oscillator should track above the incoming RF by a fixed frequency which is equal to the IF.

Figure 4-6a shows the schematic diagram for the tuned circuits in a preselector and local oscillator. The tuned circuit in the preselector is tunable from 540 to 1600 kHz (a ratio of 2.96:1), and the tuned circuit in the local oscillator is tunable from 995 to 2055 kHz (a ratio of 2.15:1). Because the resonant frequency of a tuned circuit is inversely proportional to the square root of the capacitance, the capacitor in the preselector must change by a factor of 8.8, while at the same time, the capacitor in the local oscillator must change by a factor of only 4.6. The local oscillator should track 455 kHz above the preselector over the entire RF spectrum and there should be a single tuning control. Fabricating such a unit is difficult, if not impossible. Therefore,

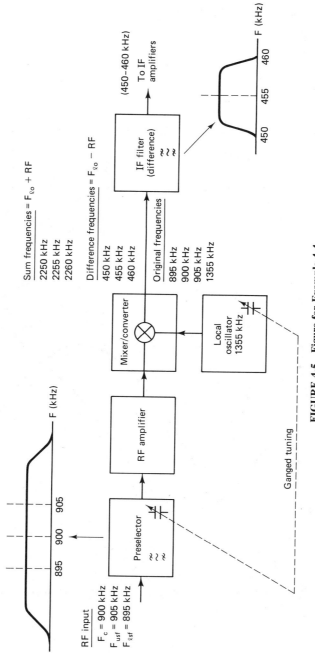

FIGURE 4-5 Figure for Example 4-4.

137

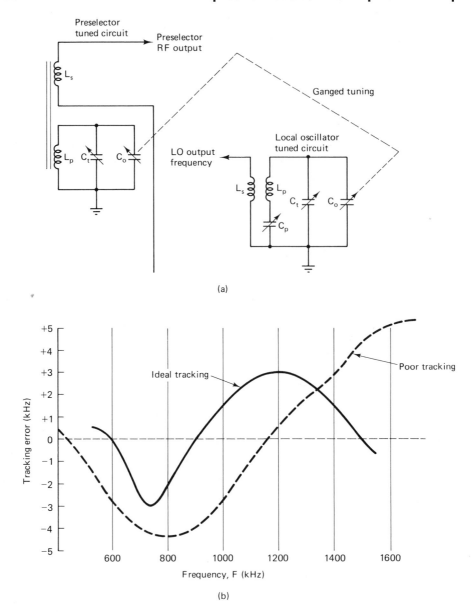

(a)

(b)

FIGURE 4-6   Receiver tracking: (a) preselector and local oscillator schematic; (b) tracking curve.

perfect tracking over the entire RF spectrum is unlikely. The difference between the actual tracking frequency and the IF is the tracking error. The *tracking error* is not uniform over the entire RF spectrum. A maximum tracking error of ±3 kHz is about the best that can be expected from a domestic AM broadcast-band receiver with a 455-kHz IF. Figure 4-6b shows a typical tracking curve. A tracking error of +3 kHz corresponds to a tracking frequency of 458 kHz, and a tracking error of −3 kHz corresponds to a tracking frequency of 452 kHz.

The tracking error is reduced by a technique called *three-point tracking*. The preselector and the local oscillator each have a *trimmer* capacitor (C_t) in parallel with the primary tuning capacitor (C_o) that compensates for minor tracking errors at the high end of the spectrum, and the local oscillator has an additional *padder* capacitor (C_p) placed in series with the tuning coil that compensates for minor tracking errors at the low end of the spectrum. With three-point tracking, the tracking error is adjusted to 0 Hz at approximately 600, 950, and 1500 kHz.

With low-side injection, the local oscillator is tunable from 85 to 1145 kHz (a ratio of 13.5:1). Consequently, the capacitance must change by a factor of 182. Standard variable capacitors seldom tune over more than a 10:1 ratio. This is why low-side injection is impractical for standard AM broadcast-band receivers.

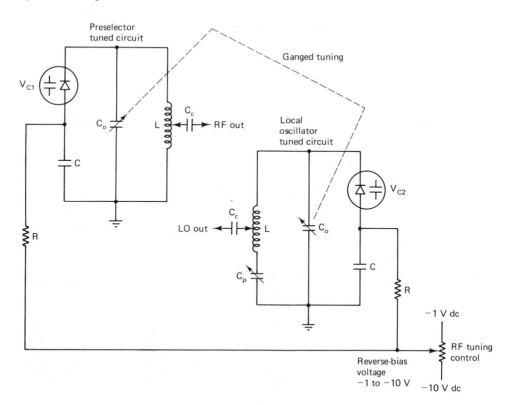

FIGURE 4-7 Electronic tuning.

Ganged capacitors are relatively large and expensive and they are somewhat difficult to compensate. Consequently, they are gradually being replaced with solid-state electronically tuned circuits. Electronically tuned circuits are smaller, less expensive, more accurate, more immune to environmental changes, more easily compensated, and more easily adapted to digital remote control and pushbutton tuning than are their mechanical counterparts. As with the crystal oscillator modules explained in Chapter 2, electronic tuned circuits use solid-state variable capacitance diodes (varactor diodes). Figure 4-7 shows the schematic diagrams for an electronically tuned preselector and local oscillator. The -1- to -10-V reverse bias comes from a single tuning control. By changing the position of the wiper on a precision variable resistor, the dc reverse bias for the two tuning diodes (VC_1 and VC_2) is changed. The diode capacitance and, consequently, the resonant frequency of the tuned circuit vary with the reverse bias. Three-point compensation with electronic tuning is accomplished the same as with mechanical tuning.

In a superheterodyne receiver, most of the receiver's selectivity is accomplished in the IF stage. For maximum noise reduction, the bandwidth of the IF filters is equal to the minimum bandwidth required to pass the information signal, which with double-sideband transmission is equal to two times the highest modulating frequency. For a maximum modulating signal frequency of 5 kHz, the minimum IF bandwidth is 10 kHz. For a 455-kHz IF and perfect tracking, a 450- to 460-kHz bandwidth is required. In reality, some RF stations are tracked 3 kHz above and some 3 kHz below 455 kHz. Therefore, the IF bandwidth must be expanded to allow the IF from the off-track stations to pass through the IF filters.

EXAMPLE 4-5

For the tracking curve shown in Figure 4-8a, a 455-kHz IF, and a maximum modulating signal frequency of 5 kHz; determine the minimum IF bandwidth.

Solution The maximum IF occurs for the RF carrier with the most positive tracking error (1400 kHz) and a 5-kHz modulating signal:

$$IF(maximum) = IF + tracking\ error + modulating\ signal$$

$$= 455\ kHz + 3\ kHz + 5\ kHz = 463\ kHz$$

The minimum IF occurs for the RF carrier with the most negative tracking error (800 kHz) and a 5-kHz modulating signal:

$$IF(minimum = IF + tracking\ error - modulating\ signal$$

$$= 455\ kHz + (-3\ kHz) - 5\ kHz = 447\ kHz$$

The minimum IF bandwidth necessary to pass the two sidebands is the difference between the maximum and minimum IFs:

$$minimum\ bandwidth = 463\ kHz - 447\ kHz = 16\ kHz$$

Figure 4-8b shows the IF bandpass characteristics for Example 4-5.

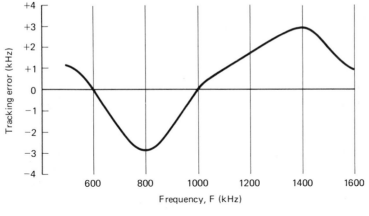

(a)

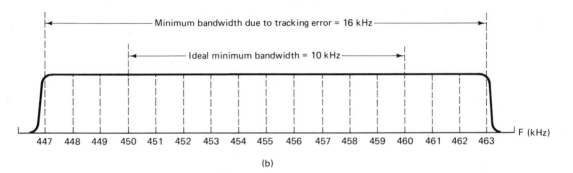

(b)

FIGURE 4-8 **Tracking error for Example 4-5: (a) tracking curve; (b) bandpass characteristics.**

Image frequency. An *image frequency* is any frequency other than the selected RF which, if allowed to enter a receiver and mix with the local oscillator will produce a cross-product frequency that is equal to the IF. Each RF carrier has an image frequency. Once an image frequency has been mixed down to IF, it cannot be filtered out or suppressed. If the selected RF carrier and its image frequency enter a receiver at the same time, they both mix with the local oscillator frequency in the mixer/converter and produce a difference frequency equal to the IF. Consequently, two different stations are received and demodulated simultaneously producing two audio signals. For a radio frequency to produce a cross product equal to the IF, it must be displaced from the local oscillator frequency by a value equal to the IF. With high-side injection, the selected RF is the IF below the local oscillator frequency. Therefore, the image frequency is the radio frequency that is the IF above the local oscillator frequency. Mathematically, for high-side injection, the image frequency is

$$F_{\text{image}} = F_{\text{lo}} + F_{\text{if}} \qquad (4\text{-}6a)$$

and since the desired RF equals the local oscillator frequency minus the IF:

$$F_{\text{image}} = F_{\text{rf}} + 2F_{\text{if}} \qquad (4\text{-}6b)$$

Figure 4-9 shows the relative frequency spectrum for the RF, the IF, the local oscillator frequency, and the image frequency in a superheterodyne receiver using high-side injection.

From Figure 4-9 it can be seen that the higher the IF, the farther away in the frequency spectrum the image frequency is from the selected RF. Therefore, for better *image frequency rejection*, a high IF is preferred. However, the higher the IF, the more difficult it is to build stable amplifiers with high gain. Therefore, there is a trade-off when selecting the IF for a radio receiver between image frequency rejection and IF gain.

Image frequency rejection ratio. The *image frequency rejection ratio* (IFRR) is a numerical measure of the ability of a preselector to reject the image frequency. For a single-tuned circuit, the ratio of its gain at the selected RF to the gain at the image frequency is the IFRR. Mathematically, IFRR is

$$\text{IFRR} = \sqrt{1 + Q^2\rho^2} \qquad (4\text{-}7a)$$

where

$$\rho = \frac{F(\text{image})}{F(\text{RF})} - \frac{F(\text{RF})}{F(\text{image})}$$

Q = quality factor of the tuned circuit   (4-7b)

IFRR (dB) = 20 log IFRR

If there is more than one tuned circuit in the front end of the receiver (perhaps a preselector filter and a separately tuned RF amplifier), the IFRR is simply the product of the two ratios.

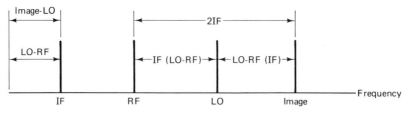

FIGURE 4-9   Image frequency.

EXAMPLE 4-6

For an AM broadcast-band superheterodyne receiver with an IF, RF, and local oscillator frequency of 455, 600, and 1055 kHz, respectively:
(a) Determine the image frequency.
(b) Calculate the IFRR for a preselector Q of 100.

Solution (a) From Equation 4-6a,

$$F_{image} = F_{lo} + F_{if}$$

$$= 1055 \text{ kHz} + 455 \text{ kHz} = 1510 \text{ kHz}$$

or from Equation 4-6b,

$$F_{image} = F_{rf} + 2F_{if}$$

$$= 600 \text{ kHz} + 2(455 \text{ kHz}) = 1510 \text{ kHz}$$

(b) From Equations 4-7a and 4-7b,

$$\rho = \frac{1510 \text{ kHz}}{600 \text{ kHz}} - \frac{600 \text{ kHz}}{1510 \text{ kHz}}$$

$$= 2.51 - 0.397 = 2.113$$

$$\text{IFRR} = \sqrt{1 + (100^2)(2.113^2)}$$

$$= 211.3 \text{ or } 46.5 \text{ dB}$$

See Figure 4-10.

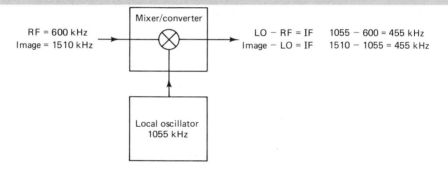

FIGURE 4-10 **Frequency conversion for Example 4-6.**

Once an image frequency has been down-converted to IF, it cannot be removed. Therefore, to reject the image frequency, it has to be removed prior to the mixer/converter stage. Image frequency rejection is the primary purpose of the RF preselector. If the bandwidth of the preselector is sufficiently narrow, the image frequency is prevented from entering the receiver. Figure 4-11 illustrates how proper RF and IF filtering can prevent an image frequency from interfering with the selected RF carrier.

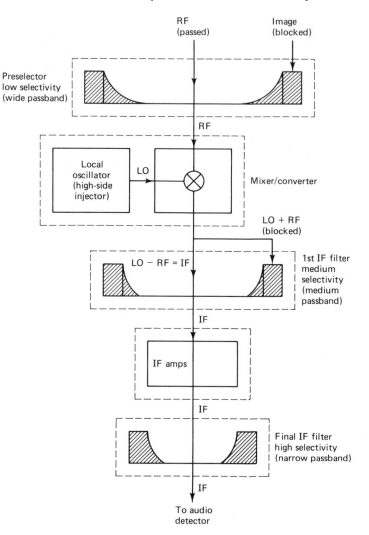

FIGURE 4-11   Image frequency rejection.

The ratio of the RF to the IF is also an important consideration for image frequency rejection. The closer the RF is to the IF, the closer the RF is to the image frequency.

EXAMPLE 4-7

For a citizens' band receiver using high-side injection with an RF of 27 MHz and an IF of 455 kHz determine:
(a) The local oscillator frequency.
(b) The image frequency.

(c) The IFRR for a preselector Q of 100.

(d) The preselector Q required to achieve the same IFRR as that achieved for an RF of 600 kHz.

Solution   (a) From Equation 4-5a,

$$F_{\text{lo}} = 27 \text{ MHz} + 455 \text{ kHz} - 27.455 \text{ MHz}$$

(b) From Equation 4-6a,

$$F_{\text{image}} = 27.455 \text{ MHz} + 455 \text{ kHz} = 27.91 \text{ MHz}$$

(c) From Equations 4-7a and 4-7b,

$$\text{IFRR} = 6.7 \text{ or } 16.5 \text{ dB}$$

(d) Rearranging Equation 4-7a yields

$$Q = \sqrt{\frac{\text{IFRR}^2 - 1}{\rho^2}} = \sqrt{\frac{211.3^2 - 1}{0.0663^2}} = 3187$$

From Examples 4-6 and 4-7 it can be seen that the higher the RF carrier, the more difficult it is to prevent the image frequency from entering the receiver. For the same IFRR, the higher RF carrier requires a much higher-quality filter in the preselector. This is illustrated in Figure 4-12.

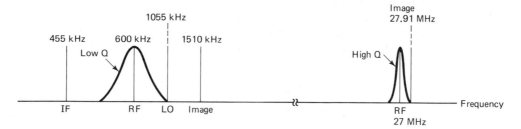

FIGURE 4-12   Frequency spectrum for Example 4-7.

AM RECEIVER CIRCUITS

RF Amplifier Circuits

An *RF amplifier* is a high-gain, low-noise, tuned amplifier that when used, is the first active stage encountered by a received signal. The primary purposes of an RF stage are selectivity, amplification, and sensitivity. Therefore, characteristics that are desirable in an RF amplifier are:

1. Low thermal noise
2. Low noise figure

3. Moderate to high gain
4. Low intermodulation and harmonic distortion
5. Moderate selectivity
6. High image frequency rejection ratio

Two of the most important parameters for a receiver are amplification and noise figure, of which both are dependent on the performance of the RF stage. An AM demodulator detects amplitude variations in its input signal and converts them to changes in its output signal. Consequently, amplitude variations caused by noise are converted to erroneous fluctuations in the detector output and the quality of the receive signal is degraded. The more gain that a signal experiences as it passes through a receiver, the more pronounced are its amplitude variations at the demodulator input, and the less noticeable are the variations caused by noise. The narrower the bandwidth, the less noise propagated through the receiver and, consequently, the less noise demodulated by the detector. From Equation 1-19 ($V_N = \sqrt{4RKTB}$), noise voltage is directly proportional to the square root of the temperature, the bandwidth, and the resistance. Therefore, if these three parameters are minimized, the thermal noise is reduced. The temperature of an RF stage can be reduced by artificially cooling the front end of a receiver. The bandwidth of an RF amplifier is reduced by using tuned amplifiers, and the resistance is reduced by using specially constructed solid-state components for the active device.

Noise figure is essentially a measure of the gain of the amplifier to the noise added by the amplifier. Therefore, the noise figure is improved (reduced) either by reducing the internal noise or by increasing the amplifier's gain.

Intermodulation and harmonic distortion are both forms of nonlinear distortion that reduce the noise figure by adding correlated noise to the total noise spectrum. The more linear an amplifier's operation, the less nonlinear distortion produced, and the better the receiver's noise figure. The IFRR ratio of an RF amplifier combines with the IFRR of the preselector to reduce the receiver input bandwidth sufficiently and prevent the image frequency from entering the mixer/converter stage. Consequently, moderate selectivity is all that is required from the RF stage.

Figure 4-13 shows several commonly used RF amplifier circuits. Keep in mind that RF is a relative term. RF simply means that the frequency is high enough to be efficiently radiated by an antenna and propagated through free space as an electromagnetic wave. RF for the AM broadcast band is between 535 and 1605 kHz, whereas RF for microwave radio is in excess of 1 GHz (1000 MHz). A common IF frequency used for FM broadcast band receivers is 10.7 MHz, which is considerably higher than the RF frequencies associated with the AM broadcast band. RF is simply the radiated or received frequency, and IF is an intermediate frequency within a transmitter or a receiver. Therefore, many of the considerations for RF amplifiers also apply to IF amplifiers such as neutralization, filtering, and coupling.

Figure 4-13a shows a schematic diagram for a bipolar RF amplifier. C_a, C_b, C_c, and L_1 form the coupling circuit from the antenna. Q_1 is class A biased to reduce nonlinear distortion. The collector circuit is transformer coupled to the mixer/converter through T_1, which is double tuned for more selectivity. C_x and C_y are RF bypass capaci-

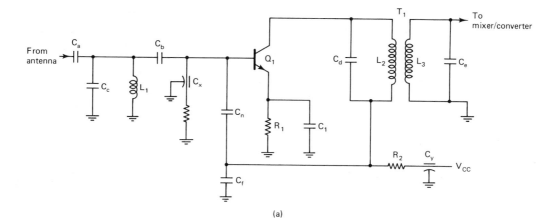

(a)

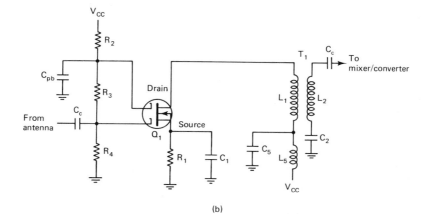

(b)

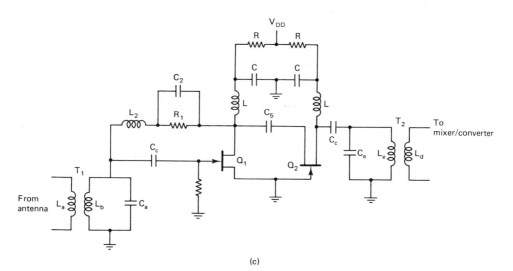

(c)

FIGURE 4-13   RF amplifier configurations: (a) bipolar transistor RF amplifier; (b) DEMOS-FET RF amplifier; (c) cascaded RF amplifier.

tors. Their symbols indicate that they are specially constructed *feedthrough* capacitors. Feedthrough capacitors offer less inductance, which prevents radiation from their leads. C_n is a neutralization capacitor. A portion of the collector signal is fed back to the base circuit to offset (or neutralize) the signal fed back through the collector-to-base lead capacitance of the transistor to prevent oscillations. C_f, in conjunction with C_n, form an ac voltage divider for the feedback signal. This neutralization configuration is called *off-ground* neutralization.

Figure 4-13b shows an RF amplifier using dual-gate field-effect transistors. This configuration uses DEMOS (depletion–enhancement metal-oxide semiconductor) FETs. The FETs feature high input impedance and low noise. A FET is a square-law device that generates only second-order harmonic and intermodulation distortion components, therefore producing less nonlinear distortion than a bipolar transistor. Q_1 is again biased class A for linear operation. T_1 is single tuned to the desired RF carrier to enhance the receiver's selectivity and improve the IFRR. L_5 is a RFC choke, and in conjunction with C_5, decouples the RF signals from the dc power supply.

Figure 4-13c shows the schematic diagram for a special RF amplifier configuration called a *cascoded* amplifier. A cascoded amplifier offers higher gain and less noise than conventional amplifiers. The active devices can be either bipolar transistors or FETs. Q_1 is a common-gate amplifier whose output is impedance-coupled to the source of Q_2. Because of the low input impedance of Q_2, Q_1 does not need to be neutralized; however, neutralization reduces the noise even further. Therefore, L_2, R_1, and C_2 provide the feedback path for neutralization. Q_2 is also a common-gate amplifier and because of its low input impedance requires no neutralization.

Mixer/Converter Circuits

As stated previously, the purpose of the mixer/converter stage is to down-convert the incoming RF to IF. This is done by mixing the RF with a local oscillator frequency in

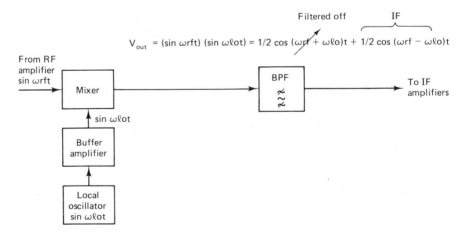

FIGURE 4-14   Mixer/converter block diagram.

a nonlinear device. In essence, this is heterodyning. A mixer is a nonlinear amplifier similar to a modulator except that the output is tuned to the difference between the RF and the local oscillator frequency. Figure 4-14 shows a block diagram for a mixer/converter stage. The output of a mixer is the product of the RF and the LO frequency and is expressed mathematically as

$$v_{out} = (\sin \omega_{rf}t)(\sin \omega_{lo}t)$$

where

ω_{rf} = RF input signal
ω_{lo} = LO input signal

Therefore, using the trigonometric identity for the product of two sines, the output of the mixer is

$$v_{out} = - \tfrac{1}{2} \cos (\omega_{rf} + \omega_{lo})t \quad + \quad \tfrac{1}{2} \cos (\omega_{rf} - \omega_{lo})t$$

sum frequency   difference frequency

The difference frequency ($\omega_{rf} - \omega_{lo}$) is the IF.

Although any nonlinear device can be used for a mixer, a transistor or a FET is preferred over a simple diode because they are also capable of amplification. However, because the actual output signal from a mixer is a cross-product frequency, a mixer has a net loss. This loss is called *conversion* loss (or sometimes, conversion gain) because a frequency conversion has occurred and, at the same time, the output signal (IF) is lower in amplitude than the input signal (RF). The conversion loss is generally about 6 dB (which corresponds to a conversion gain of -6 dB). Essentially, the conversion gain is the difference between the IF output level with an RF input to the IF output level with an IF input.

Figure 4-15 shows the schematic diagrams for several common mixer/converter circuits. Figure 4-15a shows what is probably the simplest mixer circuit available (other than diode mixers). The mixer shown in Figure 4-15a is used exclusively for inexpensive AM broadcast band receivers. RF from the antenna is filtered by the preselector tuned circuit (L_1 and C_1), then transformer coupled to the base of Q_1. The active device for the mixer (Q_1) is also the gain device for the local oscillator. This configuration is commonly called a *self-excited* mixer because the mixer excites itself by feeding energy back to the local oscillator tank circuit (C_2 and L_2) to sustain oscillations. When power is initially applied, Q_1 amplifies both the incoming RF and any noise present and supplies the oscillator tank circuit with enough energy to begin oscillator action. The LO frequency is the resonant frequency of the tank circuit. A portion of the resonant tank circuit energy is coupled through L3 and L5 to the emitter of Q_1. This signal drives Q_1 into its nonlinear region and, consequently, produces sum and difference frequencies at the collector. The difference frequency is the IF. The output tank circuit (L_3 and C_3) is tuned to the IF. Therefore, the IF signal is transformer coupled to the first IF amplifier. The process is regenerative as long as there is an incoming RF signal. The tuning

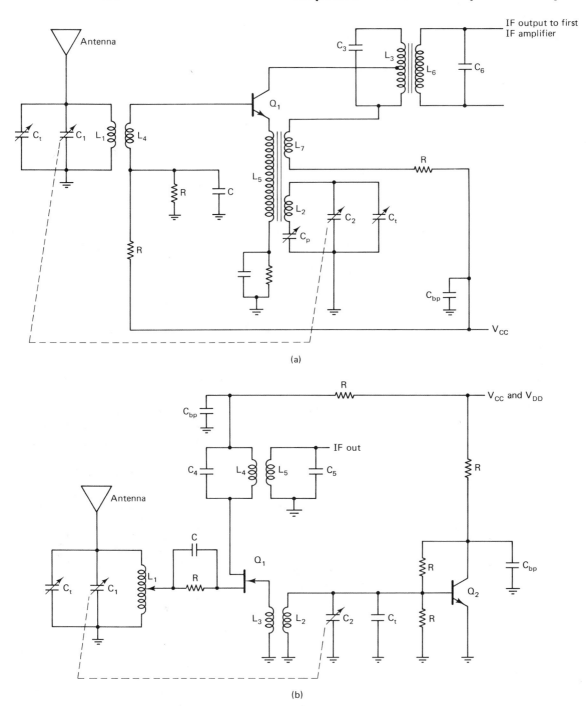

(a)

(b)

FIGURE 4-15

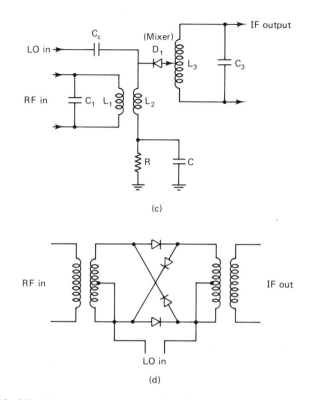

FIGURE 4-15 Mixer/converter circuits: (a) self-excited mixer; (b) separately excited mixer; (c) diode mixer; (d) balanced diode mixer.

capacitors in the RF and LO tank circuits are ganged together into a single tuning control. C_p and C_t are used for three-point tracking. This configuration has poor selectivity and poor image frequency rejection because there is no amplifier tuned to the RF signal and, consequently, the only RF selectivity is in the preselector. In addition there is essentially no RF gain and the transistor nonlinearities produce harmonic and intermodulation components that fall within the IF passband.

The mixer/converter circuit shown in Figure 4-15b is a separately excited mixer. Its operation is essentially the same as the self-excited mixer except that the local oscillator and the mixer each have their own gain device. The mixer itself is a FET which has nonlinear characteristics that are better suited for IF conversion than those of a bipolar transistor. This circuit is commonly used for high-frequency (HF) and very high-frequency (VHF) receivers.

The mixer/converter circuit shown in Figure 4-15c is a single-diode mixer. The concept is quite simple: the RF and LO signals are coupled into the diode, which is a nonlinear device. Therefore, nonlinear mixing occurs and the sum and difference frequencies are produced. The output tank circuit (C_3 and L_3) is tuned to the difference (IF) frequency. A single-diode mixer is inefficient because it has no gain. However, a diode mixer is commonly used for the audio detector in an AM receiver and to produce the audio subcarrier in a television receiver.

Figure 4-15d shows the schematic diagram for a *balanced diode mixer*. Balanced mixers are one of the most important circuits used in communications systems today. Balanced mixers are also called *balanced modulators*, *product modulators*, and *product detectors*. Balanced mixers are used extensively in both transmitters and receivers for AM, FM, and many of the digital modulation schemes, such as PSK and QAM. There are two inherent advantages that balanced mixers have over other types of mixers: noise reduction and carrier suppression. A detailed explanation of both descrete and integrated circuit balanced mixers is given in Chapter 5.

IF Amplifier Circuits

IF (intermediate-frequency) amplifiers are relatively high-gain tuned amplifiers that are very similar to RF amplifiers except IF amplifiers operate at a fixed frequency with a fixed passband. Consequently, it is easy to design and build IF amplifiers that are stable, do not radiate, and are easily neutralized. Because IF amplifiers operate at a fixed frequency, successive amplifiers can be inductively coupled with *double-tuned* circuits (with double-tuned circuits, both the primary and secondary sides of the transformer are tuned tank circuits). Therefore, it is easier to achieve an optimum (low) shape factor and good selectivity. Most of a receiver's gain and selectivity is achieved in the IF amplifier section. An IF stage generally has between two and five IF amplifiers. Figure 4-16 shows a schematic diagram for a three-stage IF section. T_1 and T_2 are double-tuned transformers; and L_1, L_2, and L_3 are tapped to reduce the effects of loading. The base of Q_3 is fed from the tapped capacitor pair, C_9 and C_{10}, for the same reason. C_1 and C_6 are neutralization capacitors.

Inductive coupling. *Inductive* or *transformer coupling* is the most common technique used for coupling RF and IF amplifiers. With inductive coupling, voltage that is induced in the primary windings of a transformer is transferred to the secondary windings. The proportion of the primary voltage that is coupled across to the secondary depends on several factors, including the number of turns in the primary and secondary windings (i.e., the turns ratio), the amount of *magnetic flux* in the primary winding, the *coefficient of coupling*, and the speed at which the flux is changing. Mathematically, the voltage induced in the secondary is

$$E_s = \omega M I_p \qquad (4\text{-}8)$$

where

E_s = voltage induced in the secondary
ω = angular velocity $(2\pi F)$
M = mutual inductance
I_p = primary current

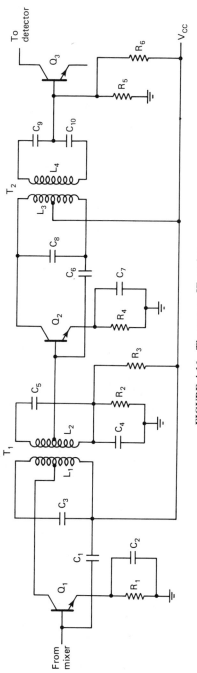

FIGURE 4-16 Three-stage IF section.

The ability of a coil to induce a voltage within its own windings is called *self-inductance* or simply *inductance* (L). When one coil induces a voltage in another coil, the two coils are said to be *coupled together*. The ability of one coil to induce a voltage in another coil is called *mutual inductance* (M). Mutual inductance in a transformer is caused by the magnetic lines of force (flux) that are produced in the primary windings cutting through the secondary windings and is directly proportional to the coefficient of coupling. Coefficient of coupling is the ratio of the secondary flux to the primary flux and is expressed mathematically as

$$k = \frac{\phi_s}{\phi_p} \qquad\qquad (4\text{-}9)$$

where

k = coefficient of coupling
ϕ_s = secondary flux
ϕ_p = primary flux

If all of the flux produced in the primary windings cuts through the secondary windings, the coefficient of coupling is 1. If none of the primary flux cuts through the secondary windings, the coefficient of coupling is 0. A coefficient of coupling of 1 is nearly impossible to achieve unless the two coils are wound around a common high-permeable iron core. Typically, the coefficient of coupling for standard RF and IF transformers is much less than 1. The transfer of flux from the primary windings to the secondary windings is called *flux linkage* and is directly proportional to the coefficient of coupling. The mutual inductance of a transformer is directly proportional to the coefficient of coupling and the square root of the product of the primary and secondary inductances. Mathematically, mutual inductance is

$$M = k\sqrt{L_s L_p} \qquad\qquad (4\text{-}10)$$

where

M = mutual inductance
L_s = inductance of the secondary winding
L_p = inductance of the primary winding
k = coefficient of coupling

Transformer-coupled amplifiers are divided into two general categories: single and double tuned.

Single-tuned transformers. Figure 4-17a shows a schematic diagram for a *single-tuned inductively* coupled amplifier. This configuration is called *untuned primary-tuned secondary*. The primary side of T_1 is simply the inductance of the primary windings, whereas a capacitor is in parallel with the secondary windings creating a tuned secondary. The transformer windings are not tapped because the loading effect of the FET is insignificant. Figure 4-17b shows the response curve for an untuned primary–tuned secondary transformer. E_s increases until the resonant frequency (F_o) of the secondary is reached,

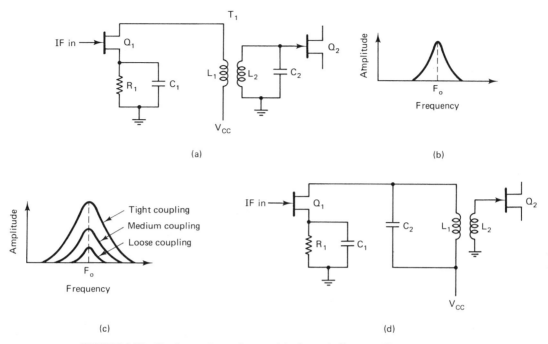

FIGURE 4-17 **Single-tuned transformer: (a) schematic diagram; (b) response curve; (c) effects of coupling; (d) tuned primary, untuned secondary.**

then E_s begins to decrease. The peaking of the response curve at F_o is caused by the reflected impedance. The impedance of the secondary is reflected back into the primary due to the mutual inductance between the two windings. For frequencies below resonance, the increase in ωM is greater than the decrease in I_p; therefore, E_s increases. For frequencies above resonance, the increase in ωM is less than the decrease in I_p; therefore, E_s decreases.

Figure 4-17c shows the effect of coupling on the response curve of an untuned primary–tuned secondary transformer. With *loose* coupling (low coefficient of coupling), the secondary voltage is relatively low and the bandwidth is narrow. As the degree of coupling is increased (coefficient of coupling increases), the secondary induced voltage increases and the bandwidth widens. Therefore, for a high degree of selectivity, loose coupling is desired, however, signal amplitude must be sacrificed. For high gain and a broad bandwidth, *tight* coupling is desired. Another single-tuned amplifier configuration is the *tuned primary–untuned secondary*, which is shown in Figure 4-17d.

Double-tuned transformers. Figure 4-18a shows a schematic diagram for a *double-tuned* inductively coupled amplifier. This configuration is called a *tuned primary–tuned secondary*; there is a capacitor in parallel with both the primary and secondary windings of T_1. Figure 4-18b shows the effect of coupling on the response curve of a double-tuned inductively coupled transformer. The response curve closely resembles that of a single-tuned circuit up to a coefficient of coupling called *critical coupling*. Critical

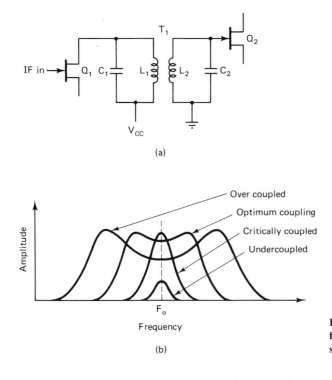

(a)

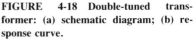

(b)

FIGURE 4-18 Double-tuned transformer: (a) schematic diagram; (b) response curve.

coupling is the point where the reflected resistance is equal to the primary resistance. At critical coupling, the primary's Q is halved, and consequently, the bandwidth is doubled. If the coefficient of coupling is increased beyond critical coupling, the response at the resonant frequency decreases and two new peaks occur on either side of the resonant frequency. This double peaking is caused from the reactive element of the reflected impedance being significant enough to change the resonant frequency of the primary tuned circuit. If the coefficient of coupling is increased further, the dip at resonance becomes more pronounced and the two peaks are spread even farther away from the resonant frequency. Increasing coupling beyond the critical value broadens the bandwidth but at the same time, produces a ripple in the response curve. An ideal response curve has a rectangular shape (i.e., a flat top with steep skirts). From Figure 1-18b it can be seen that a coefficient of coupling approximately 50% greater than the critical value yields a good compromise between flat response and steep *skirts.* This value of coupling is called *optimum coupling* and is expressed mathematically as

$$K_{\text{opt}} = 1.5k_c \qquad (4\text{-}11\text{a})$$

where

K_{opt} = optimum coupling

k_c = critical coupling = $\dfrac{1}{\sqrt{Q_p Q_s}}$

The bandwidth of a single double-tuned amplifier is

$$BW_{dt} = kF_o \qquad (4\text{-}11b)$$

Bandwidth reduction.   When several tuned amplifiers are cascaded, the total response is the product of all of the amplifier's individual responses. Figure 4-19a shows a response curve for a tuned amplifier. The gain at F_1 and F_2 is 0.707 of the

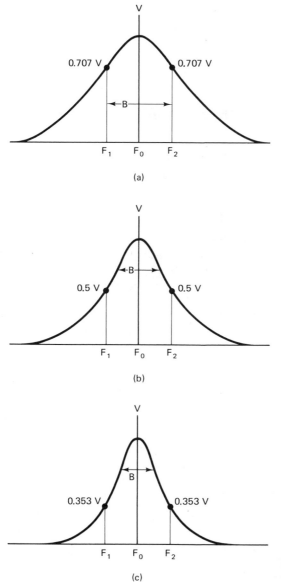

(a)

(b)

(c)

FIGURE 4-19   Bandwidth reduction: (a) single-tuned stage; (b) two cascaded stages; (c) three cascaded stages.

gain at F_o. If two identical tuned amplifiers are cascaded, the gain at F_1 and F_2 will be 0.5 of the gain at F_o ($0.707 \times 0.707 = 0.5$), and if three identical tuned amplifiers are cascaded, the gain at F_1 and F_2 is reduced to 0.353 of the gain at F_o. Consequently, as additional tuned amplifiers are added, the overall bandwidth of the stage is reduced. This bandwidth reduction is shown in Figure 4-19b and c. Mathematically, the overall bandwidth of n single-tuned stages is given as

$$BW_n = BW_1 \sqrt{2^{1/n} - 1} \qquad (4\text{-}12a)$$

where

BW_n = bandwidth of n stages
BW_1 = bandwidth of one single-tuned stage
n = number of stages

The bandwidth for n double-tuned stages is

$$BW_{ndt} = BW_{1dt} \sqrt[4]{2^{1/n} - 1} \qquad (4\text{-}12b)$$

where

BW_{ndt} = overall bandwidth of n double-tuned amplifiers
BW_{1dt} = bandwidth of a single double-tuned amplifier
n = number of stages

EXAMPLE 4-8

Determine the overall bandwidth for (a) two single-tuned amplifiers each with a bandwidth of 10 kHz, (b) three single-tuned amplifiers each with a bandwidth of 10 kHz, and (c) four single-tuned amplifiers each with a bandwidth of 10 kHz.

(d) Determine the bandwidth for a double-tuned amplifier with optimum coupling, a k_c of 0.02 and a resonant frequency of 1 MHz.

(e) Repeat parts (a), (b), and (c) for the double-tuned amplifier of part (d).

Solution   (a) From Equation 4-12a,

$$BW_2 = 10 \text{ kHz} \sqrt{2^{1/2} - 1}$$

$$= 6436 \text{ Hz}$$

(b) Again, from Equation 4-12a,

$$BW_3 = 10 \text{ kHz} \sqrt{2^{1/3} - 1}$$

$$= 5098 \text{ Hz}$$

(c) Again, from Equation 4-12a,

$$BW_4 = 10 \text{ kHz} \sqrt{2^{1/4} - 1}$$

$$= 4350 \text{ Hz}$$

(d) From Equation 4-11a,

$$K_{\text{opt}} = 1.5(0.02) = 0.03$$

From Equation 4-12b,

$$BW_{dt} = 0.03(1 \text{ MHz}) = 30 \text{ kHz}$$

(e) From Equation 4-12b,

n	BW (Hz)
2	24,067
3	21,420
4	19,786

AM Detector Circuits

The function of an AM detector is to demodulate the IF AM envelope and recover or reproduce the original source information. The recovered signal should contain the same frequencies as the original modulating signal and have the same relative amplitude characteristics.

Peak detector. Figure 4-20a shows a schematic diagram for a simple noncoherent AM demodulator which is commonly called a *peak detector*. Because a diode is a nonlinear device, nonlinear mixing occurs in D_1 when two or more frequencies are applied to its input. Therefore, the output contains the original input frequencies, their harmonics, and their cross products (the sum and difference frequencies). If a 300-kHz carrier is amplitude modulated by a 2-kHz sine wave, the modulated wave is made up of a LSF, carrier, and USF of 298, 300, and 302 kHz, respectively. If the resultant envelope is the input to the AM detector shown in Figure 4-20a, the output will comprise the three input frequencies, the harmonics of all three frequencies, and the cross products of all possible combinations of the three frequencies and their harmonics. Mathematically, the output is

$$V_{out} = \text{input frequencies} + \text{harmonics} + \text{sums and differences}$$

Because the *RC* network is a lowpass filter, only the difference frequencies are passed on to the audio section. Therefore, the output is simply

$$V_{out} = 300 - 298 = 2 \text{ kHz}$$
$$302 - 300 = 2 \text{ kHz}$$
$$302 - 298 = 4 \text{ kHz}$$

Therefore,

$$V_{out} = 2 \text{ kHz and } 4 \text{ kHz}$$

Because of the relative amplitude characteristics of the LSF, carrier, and USF, the difference between the carrier and either the upper or lower side frequency is the

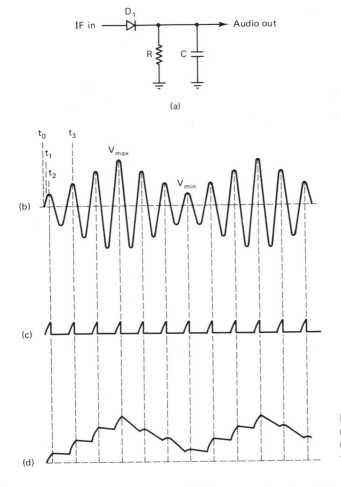

FIGURE 4-20 Peak detector: (a) schematic diagram; (b) AM input waveform; (c) diode current waveform; (d) output voltage waveform.

predominant output signal. Consequently, the original modulating signal (a 2-kHz sine wave) is recovered by the peak detector.

In the preceding analysis, the diode detector was analyzed as a simple mixer, which it is. Essentially, the difference between an AM modulator and an AM demodulator is that the output of a modulator is tuned to the sum frequency (up-converter), whereas the output of a demodulator is tuned to the difference frequency (down-converter). The demodulator shown in Figure 4-20a is commonly called a diode detector because the nonlinear device is a single diode, or a peak detector because it detects the peaks of the input envelope, or a *shape* or *envelope detector* because it detects the shape of the input envelope. Essentially, the carrier "*captures*" the diode and forces it to turn on and off (rectify) synchronously (both frequency and phase). Thus the sidebands can mix with the carrier and the original baseband signals are recovered.

Figure 4-20b, c, and d show a detector input voltage waveform, the corresponding diode current waveform, and the detector output voltage waveform. At time $t = 0$, the diode is reverse biased and off ($i_d = 0$ A), the capacitor is completely discharged ($V_C = 0$ V), and thus the output is 0 V. The diode remains off until the input voltage exceeds the barrier potential of D_1 (approximately 0.3 V). When V_{in} reaches 0.3 V ($t = 1$), the diode turns on and diode current begins to flow charging the capacitor. The capacitor voltage remains 0.3 V below the input voltage until V_{in} reaches its peak. When the input voltage begins to decrease, the diode turns off and i_d goes to 0 A ($t = 2$). The capacitor begins to discharge through the resistor but the RC time constant is made sufficiently long so that the capacitor cannot discharge as rapidly as V_{in} is decreasing. The diode remains off until the next input cycle when V_{in} goes 0.3 V more positive than V_C ($t = 3$). At this time the diode turns on, current flows, and the capacitor begins to charge again. It is relatively easy for the capacitor to charge to the new value because the R_C charging time constant is $R_d C$, where R_d is the "on" resistance of the diode, which is quite small. This sequence repeats itself on each successive positive peak of V_{in} and the capacitor voltage follows the positive peaks of V_{in} (hence the name "peak detector"). The output waveform resembles the shape of the input envelope (hence the name "shape detector"). The output waveform has a high-frequency ripple that is equal to the carrier frequency. This is due to the diode turning on during the positive peaks of the envelope. The ripple is easily removed by the audio amplifiers because the carrier frequency is much higher than the highest audio frequency. The circuit shown in Figure 4-20 responds only to the positive peaks of V_{in} and is, therefore, called a positive peak detector. By simply turning the diode around, the circuit becomes a negative peak detector. The output voltage reaches its peak positive amplitude at the same time that the input envelope reaches its maximum positive value (V_{max}), and the output voltage goes to its minimum peak amplitude at the same time that the input voltage goes to its minimum value (V_{min}). For 100% modulation, V_{out} swings from 0 V to $V_{max} - 0.3$ V.

Figure 4-21 shows the input and output waveforms for a peak detector with various

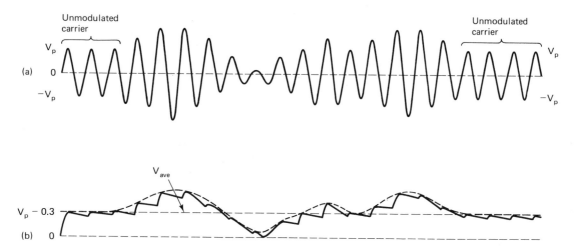

FIGURE 4-21   Positive peak detector: (a) input waveform; (b) output waveform.

percentages of modulation. With no modulation, a peak detector is simply a half-wave rectifier and the output voltage is approximately equal to the peak input voltage. As the percent modulation changes, the variations in the output voltage increase and decrease proportionally; the output waveform follows the shape of the AM envelope. However, regardless of whether there is modulation present or not, the average value of the output voltage is approximately equal to the peak value of the unmodulated carrier. Therefore, the output voltage variations reflect only the changes in the envelope (i.e., the detector is removing the audio information from the envelope). With no modulating signal, the output of the detector is a constant dc voltage approximately equal to the peak amplitude of the unmodulated carrier.

Detector distortion.   When the positive peaks of the input waveform are increasing, it is important that the capacitor hold its charge between successive peaks (i.e., a relatively long *RC* time constant is necessary). However, when the positive peaks are decreasing in amplitude, it is important that the capacitor discharge between peaks to a value less than the next peak (i.e., a short *RC* time constant is necessary). Obviously, a trade-off between a long and a short time constant is in order. If the *RC* time constant

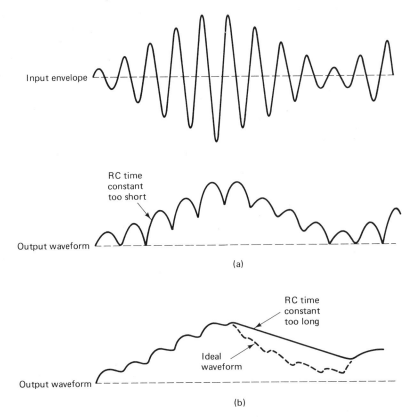

FIGURE 4-22   Detector distortion: (a) rectifier distortion; (b) diagonal clipping.

is too short, the output waveform resembles a half-wave rectified signal. This is sometimes called *rectifier distortion* and is shown in Figure 4-22a. If the *RC* time constant is too long, the slope of the output waveform cannot follow the trailing slope of the envelope. This type of distortion is called *diagonal clipping* and is shown in Figure 4-22b.

The *RC* network following the diode in a peak detector is a low-pass filter. The slope of the envelope is dependent on both the modulating signal frequency and the modulation coefficient (*m*). Therefore, the maximum slope (fastest rate of change) occurs when the envelope is crossing its zero axis in the negative direction (point *X* in Figure 4-22). The highest modulating signal frequency that can be demodulated without attenuation is given as

$$F_{a(\max)} = \frac{\sqrt{(1/m^2) - 1}}{2\pi RC} \qquad (4\text{-}13\text{a})$$

where

$$F_{a(\max)} = \text{maximum modulating frequency}$$
$$m = \text{modulation coefficient}$$
$$RC = RC \text{ time constant}$$

For 100% modulation, the numerator in Equation 4-13a goes to 0, which essentially means that all modulating signal frequencies are attenuated as they are demodulated. Typically, the modulating signal amplitude in a transmitter is limited or compressed such that approximately 90% modulation is the maximum that can be achieved. For 70.7% modulation, Equation 4-13a reduces to

$$F_{a(\max)} = \frac{1}{2\pi RC} \qquad (4\text{-}13\text{b})$$

Equation 4-13b is commonly used when designing peak detectors to determine an approximate maximum modulating signal frequency.

AUTOMATIC GAIN CONTROL AND SQUELCH

Automatic Gain Control Circuits

An *automatic gain control circuit* (AGC) compensates for minor variations in the received signal level. The AGC circuit automatically increases the receiver gain for weak RF input signals, and automatically decreases the receiver gain for strong RF signals. Weak signals can be buried in the receiver noise and, consequently, masked from the audio detector. Excessively strong signals can overdrive the RF and/or IF amplifiers and produce excessive nonlinear distortion. There are several types of AGC, including direct or simple AGC, delayed AGC, and forward AGC.

Simple AGC.   Figure 4-23 shows a block diagram for an AM superheterodyne receiver with *simple AGC*. The automatic gain control circuit monitors the received

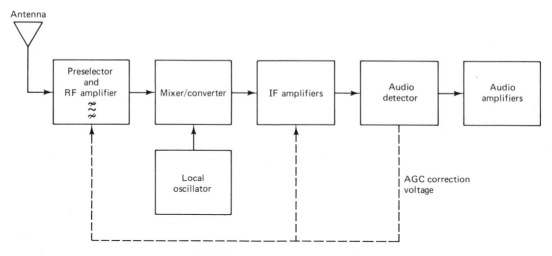

FIGURE 4-23 AM receiver with simple AGC.

signal level and sends a signal back to the RF and/or IF amplifiers to adjust their gain automatically. AGC is a form of *degenerative* or negative feedback. The purpose of AGC is to allow a receiver to detect and demodulate, equally well, signals that are transmitted from different stations whose output power and distance from the receiver varies. For example, an AM radio in a vehicle does not receive the same signal level from all of the transmitting stations or, for that matter, from a single station when the automobile is moving. The AGC circuit sends a voltage back to the RF and/or IF amplifiers to adjust the receiver gain and keep the IF carrier power at the input to the AM detector at a constant level. The AGC circuit is not a form of *automatic volume*

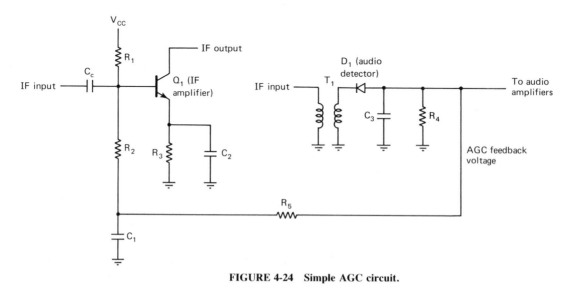

FIGURE 4-24 Simple AGC circuit.

control circuit; AGC is independent of modulation and totally unaffected by normal changes in the audio modulating signal amplitude.

Figure 4-24 shows a schematic diagram for a simple AGC circuit. As you can see, an AGC circuit is essentially a peak detector. In fact, very often the AGC correction voltage is taken from the output of the audio detector. In Figure 4-21 it was shown that the average dc voltage at the output of a peak detector is approximately equal to the peak unmodulated carrier amplitude and is totally independent of modulation. If the carrier amplitude increases, the AGC voltage increases and if the carrier amplitude decreases, the AGC voltage decreases. The circuit shown in Figure 4-24 is a negative peak detector, and therefore the output is a negative voltage. The higher the amplitude of the input carrier, the more negative the output voltage. The negative voltage from the AGC detector is fed back to the IF stage, where it controls the bias voltage on the base of Q_1. When the carrier amplitude increases, the voltage on the base of Q_1 goes more negative, causing the emitter current to decrease. As a result, r_e' increases and the amplifier gain (r_c/r_e') decreases, causing the carrier amplitude to decrease. If the carrier amplitude decreases, the AGC voltage goes less negative, the emitter current increases, r_e' decreases, and the amplifier gain increases. Capacitor C_1 is an audio bypass capacitor that prevents changes in the AGC voltage due to modulation from affecting the bias or gain of Q_1.

Delayed AGC.   Simple AGC is used in most inexpensive broadcast-band receivers. However, with simple AGC, the AGC bias begins to increase as soon as the received signal level exceeds the thermal noise of the receiver. Consequently, the receiver becomes less sensitive (this is sometimes called *automatic desensing*). *Delayed AGC* prevents the AGC voltage from reaching the RF and/or IF amplifiers until the RF level exceeds a predetermined level. Once the carrier signal has exceeded the threshold level, the delayed AGC voltage is proportional to the signal strength. Figure 4-25a shows the response characteristics for both simple and delayed AGC. It can be seen that with delayed AGC, the RF signal is unaffected until the AGC threshold level is exceeded, whereas with simple AGC, the RF signal is immediately affected. Delayed AGC is used with more sophisticated communications receivers. Figure 4-25b shows IF gain versus RF input signal level for both simple and delayed AGC.

Forward AGC.   An inherent problem with both simple and delayed AGC is the fact that they are both forms of *post-AGC* (after-the-fact compensation). With post-AGC, the circuit that monitors the carrier level and provides the AGC correction voltage is located after the IF amplifiers, and therefore the simple fact that the AGC voltage changed indicates that it may be too late (the carrier level has already changed). Therefore, simple and delayed AGC cannot compensate for rapid changes in the carrier amplitude. *Forward AGC* is similar to conventional AGC except that the carrier is monitored closer to the front end of the receiver and the correction voltage is fed to IF and/or RF amplifiers further back in the receiver. Consequently, when a signal change is detected, the change can be compensated for in succeeding stages. Figure 4-26 shows an AM superheterodyne

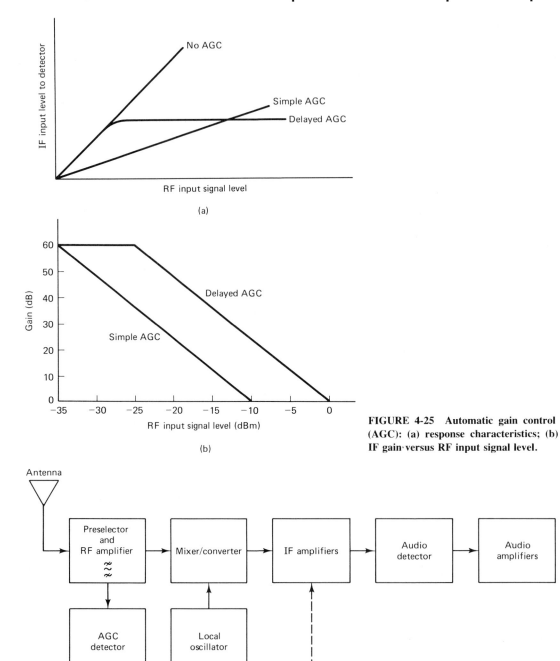

FIGURE 4-25 Automatic gain control (AGC): (a) response characteristics; (b) IF gain·versus RF input signal level.

FIGURE 4-26 Forward AGC.

receiver with forward AGC. For a more sophisticated method of accomplishing AGC, see Chapter 8 under the heading "Two-Way FM Receivers."

Squelch Circuits

The purpose of a *squelch circuit* is to "quiet" a receiver in the absence of a received RF signal. If an AM receiver is off-tuned to a location in the RF spectrum where there is no RF signal, the AGC circuit adjusts the receiver for maximum gain. Consequently, the receiver amplifies and demodulates its own internal noise. This is the familiar crackling and sputtering heard on the speaker. In domestic AM systems, each station is continuously transmitting a carrier regardless of whether there is any modulation or not. Therefore, the only time the idle receiver noise is heard is when tuning between stations. However, in two-way radio communications systems, the carrier in the transmitter is generally turned off unless a modulating signal is present. Therefore, during idle times, a receiver is simply amplifying and demodulating noise. A squelch circuit keeps the audio section turned off or muted in the absence of a received signal; the receiver is squelched. A disadvantage of a squelch circuit is that the receiver is *desensitized* and will not pick up weak RF signals.

Figure 4-27 shows a schematic diagram for a squelch circuit. This squelch circuit

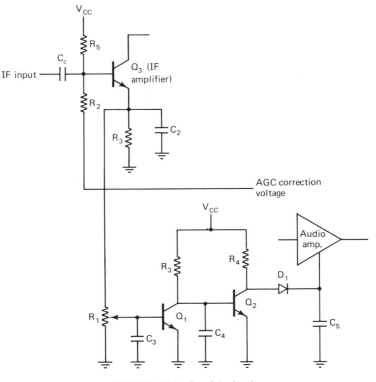

FIGURE 4-27   Squelch circuit.

uses the AGC voltage to determine how much RF signal is being received. The more AGC, the stronger the RF signal. When the AGC voltage drops below a preset level, the squelch circuit disables the audio section and mutes the receiver. In Figure 4-27 it can be seen that the squelch detector simply uses a resistive voltage divider to monitor the AGC voltage. When the RF signal drops below the squelch level, Q_1 turns on, Q_2 turns off, and D_1 turns on and shuts off the audio amplifier. When the RF level increases above squelch level, the AGC voltage goes more negative, turning off Q_1 and turning on Q_2. D_1 turns off, which, in turn, enables the audio amplifiers. The squelch threshold voltage is adjusted with R_1.

For a more sophisticated method of squelching a receiver, see Chapter 8 under the heading ''Two-Way FM Receivers.''

DOUBLE-CONVERSION AM RECEIVERS

In a previous section it was pointed out that for good image frequency rejection, a relatively high IF is selected. However, for high-gain, selective IF amplifiers that are easily neutralized, a low IF is more desirable. The solution is to use two intermediate frequencies. The first IF is relatively high for good image frequency rejection, and the second IF is relatively low for easy amplification. Figure 4-28 shows a block diagram for a *double-conversion* AM receiver. The first IF is 10.625 MHz, which pushes the image frequency 21.25 MHz away from the desired RF. The first IF is immediately down-converted to 455 kHz and fed to a series of high-gain IF amplifiers. Figure 4-29 illustrates the filtering requirements for a double-conversion AM receiver.

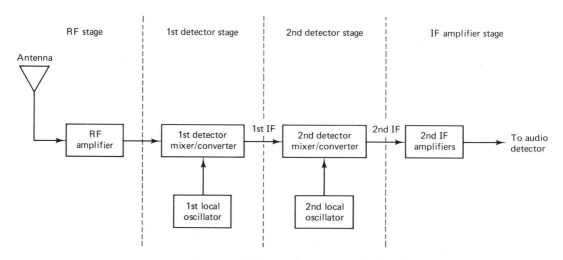

FIGURE 4-28 Double conversion AM receiver.

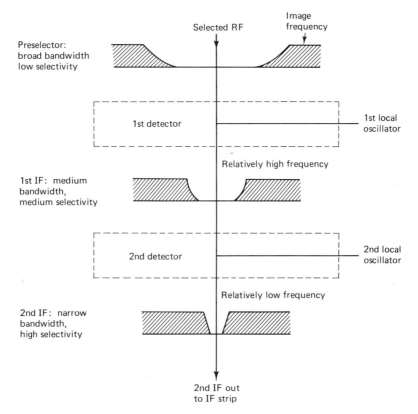

FIGURE 4-29 Filtering requirements for the double-conversion AM receiver shown in Figure 4–28.

QUESTIONS

4-1. What is meant by the *front end* of a receiver?

4-2. What are the primary functions of the front end?

4-3. Define *selectivity*; *shape factor*. What is the relationship between receiver noise and selectivity?

4-4. Describe *bandwidth improvement*. What is the relationship between bandwidth improvement and receiver noise?

4-5. Define *sensitivity*.

4-6. What is the relationship among receiver noise, bandwidth, and temperature?

4-7. Define *fidelity*.

4-8. List and describe the three types of distortion that reduce the fidelity of a receiver.

4-9. Define *insertion loss*.

4-10. Define *noise temperature*; *equivalent noise temperature*.

4-11. Describe the difference between a coherent and a noncoherent receiver.

4-12. Draw the block diagram for a TRF receiver and briefly describe its operation.

4-13. What are the three predominant disadvantages of a TRF receiver?

4-14. Draw the block diagram for an AM superheterodyne receiver and describe its operation and the primary functions of each stage.

4-15. Define *heterodyning*.

4-16. What are meant by the terms *high-* and *low-side injection*?

4-17. Define *local oscillator tracking*; *tracking error*.

4-18. Describe three-point tracking.

4-19. What is meant by *gang tuning*?

4-20. Define *image frequency*.

4-21. Describe image frequency rejection ratio.

4-22. List six characteristics that are desirable in an RF amplifier.

4-23. What advantage do FET RF amplifiers have over BJT amplifiers?

4-24. Define *neutralization*. Describe the neutralization process.

4-25. What is a cascoded amplifier?

4-26. Define *conversion gain*. What is another name for conversion gain?

4-27. What is the advantage of having a high IF; a low IF?

4-28. Define the following terms: *inductive coupling*; *self-inductance*; *mutual inductance*; *coefficient of coupling*; *critical coupling*; *optimum coupling*.

4-29. Describe loose coupling; tight coupling.

4-30. Describe the operation of a peak detector.

4-31. Describe rectifier distortion and what causes it. Describe diagonal clipping.

4-32. Describe automatic gain control.

4-33. Describe the following: simple AGC; delayed AGC; forward AGC.

4-34. Describe the purpose of a squelch circuit.

4-35. Explain the operation of a double-conversion superheterodyne receiver.

PROBLEMS

4-1. Determine the shape factor and percent selectivity for the gain-versus-frequency plot shown below.

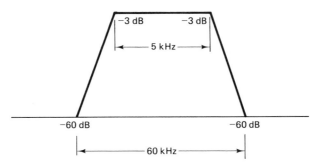

4-2. Determine the improvement in the noise figure for a receiver with an RF bandwidth equal to 40 kHz and an IF bandwidth equal to 16 kHz.

4-3. Determine the equivalent noise temperature for an amplifier with a noise figure F = 6 dB and an environmental temperature $T = 27°C$.

4-4. For an AM commercial broadcast band receiver with an input filter Q-factor of 85, determine the bandwidth at the low and high ends of the RF spectrum.

4-5. For an AM superheterodyne receiver using high-side injection with a local oscillator frequency of 1200 kHz, determine the IF carrier and upper and lower side frequencies for an RF envelope which is made up of a carrier and upper and lower side frequencies of 600, 604, and 596 kHz, respectively.

4-6. For a receiver with a $±2.5$-kHz tracking error and a maximum modulating signal frequency $F_a = 6$ kHz, determine the minimum IF bandwidth.

4-7. For a receiver with IF, RF, and local oscillator frequencies of 455, 900, and 1355 kHz, respectively, determine:
   (a) The image frequency.
   (b) The IFRR for a preselector $Q = 80$.

4-8. For a citizens' band receiver using high-side injection with an RF carrier of 27.04 MHz and a 10.645-MHz first IF, determine:
   (a) The local oscillator frequency.
   (b) The image frequency.

4-9. For a three-stage double-tuned RF amplifier with an RF equal to 800 kHz and a coefficient of coupling $k_{opt} = 0.025$, determine:
   (a) The bandwidth for each individual stage.
   (b) The overall bandwidth of the three stages.

4-10. Determine the maximum modulating signal frequency for a peak detector with the following parameters: $C = 1000$ pF, $R = 10$ kΩ, and $m = 0.5$. Repeat the problem for $m = 0.707$.

Chapter 5

PHASE-LOCKED LOOPS AND FREQUENCY SYNTHESIZERS

INTRODUCTION

With sophisticated electronic communications systems, such as single-sideband and frequency modulation systems, there are several circuits that are universally used. They include the *phase-locked loop* and the *frequency synthesizer*. Therefore, it is necessary that these circuits be discussed and understood before examining the more complicated communications systems.

PHASE-LOCKED LOOP

The phase-locked loop (PLL) is used extensively in electronic communications for modulation, demodulation, and frequency generation. PLLs are used in both transmitters and receivers with both analog and digital modulation and with the transmission of digital pulses. Phase-locked loops were first used in 1932 in receivers for synchronous detection of radio signals. However, for many years, PLLs were avoided because of their complexity and expense. With the advent of *large-scale integration* (LSI), phase-locked loops take up little space, are easy to use, and are more reliable. Therefore, they have become a universal building block with numerous applications.

Essentially, a PLL is a *closed-loop* feedback control system where the feedback signal is a frequency rather than simply a voltage. The basic phase-locked loop consists of a *phase comparator* (*frequency multiplier*), a *voltage-controlled oscillator*, and a *low-gain amplifier* (op-amp). The block diagram for a PLL is shown in Figure 5-1.

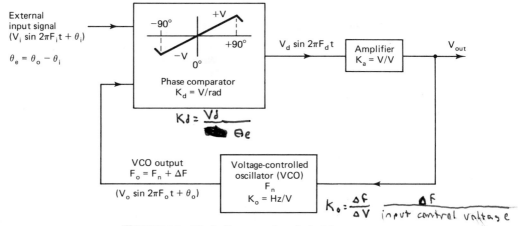

$$K_d = \frac{V_d}{\theta_e}$$

$$K_o = \frac{\Delta f}{\Delta V} \quad \frac{\Delta F}{input \; control \; voltage}$$

FIGURE 5-1   Block diagram: phase-locked loop.

Voltage-Controlled Oscillator

A *voltage-controlled oscillator* (VCO) is an oscillator (more specifically, a *free-running multivibrator*) with a stable frequency of oscillation that is dependent on an external bias voltage. The output from a VCO is a frequency and its input is a dc bias or *control* voltage. When a dc or a slowly changing ac voltage is applied to the VCO, its output frequency changes or deviates accordingly. Figure 5-2 shows the transfer curve (output frequency-versus-input bias voltage characteristics) for a typical VCO. The output frequency (F_o) with 0 V input bias is the VCO's *natural* frequency (F_n), and a change in the output frequency caused by a change in the input voltage is called frequency deviation (ΔF). Consequently, $F_o = F_n + \Delta F$. For a symmetrical ΔF, the natural fre-

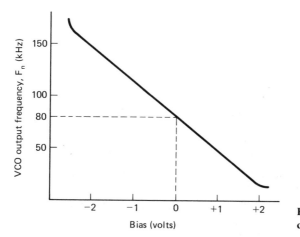

FIGURE 5-2   VCO input versus output characteristics.

quency of the VCO should be centered within the linear portion of the input/output curve. The transfer function for a VCO is

$$K_o = \frac{\Delta F}{\Delta V} \tag{5-1}$$

where

K_o = input/output transfer function (Hz/V)
ΔV = input control voltage (V)
ΔF = frequency deviation (Hz)

Phase Comparator

A *phase comparator* is a nonlinear mixer with two inputs: an externally generated frequency (F_i) and the VCO output signal (F_o). The phase comparator output is the product of F_o and F_i and therefore contains their sum and difference frequencies. Figure 5-3a shows the schematic diagram for a simple phase comparator. V_o is applied simultaneously to the two halves of input transformer T_1. D_1, R_1, and C_1 make up a half-wave rectifier, as do D_2, R_2, and C_2 (note that $C_1 = C_2$ and $R_1 = R_2$). During the positive alternation of V_o, D_1 and D_2 are forward biased and "on," charging C_1 and C_2 to equal values but with opposite polarities. Therefore, the average output voltage is V_d $= V_{C1} + (-V_{C2}) = 0$ V. This is shown in Figure 5-3b. During the negative half-cycle of V_o, D_1 and D_2 are reverse biased and "off." Therefore, C_1 and C_2 discharge equally through R_1 and R_2, respectively, keeping the output voltage at 0 V. This is shown in Figure 5-3c. The two half-wave rectifiers produce equal-magnitude output voltages with opposite polarities. Therefore, the output voltage due to V_o is constant and equal to 0 V. The corresponding input and output waveforms for a square-wave VCO signal are shown in Figure 5-3d.

   Circuit operation.   When an external signal $(V_i \sin 2\pi F_i t)$ is applied to the phase comparator, its voltage adds to V_o, causing C_1 and C_2 to charge and discharge, producing a proportional change in the output voltage. Figure 5-4a shows the unfiltered output waveform when $F_o = F_i$ and V_o leads V_i by 90°. For the phase comparator to operate properly, V_o is made much larger than V_i. Therefore, D_1 and D_2 are switched "on" only during the positive alternation of V_o and "off" during the negative alternation. During the first half of the "on" time, the voltage applied to $D_1 = V_o - V_i$ and $D_2 = V_o + V_i$. Therefore, C_1 is discharging and C_2 is charging. During the second half of the "on" time, the voltage applied to $D_1 = V_o + V_i$ and $D_2 = V_o - V_i$ and C_1 is charging while C_2 is discharging. During the "off" time, C_1 and C_2 are neither charging nor discharging. For each cycle of V_o, C_1 and C_2 charge and discharge equally and the average output voltage V_d remains at 0 V. Thus the average value of V_d is unaffected by V_i.
   Figure 5-4b shows the unfiltered output voltage waveform when V_o leads V_i by 45°. V_i is positive for 75% of the "on" time and negative for the remaining 25%. As

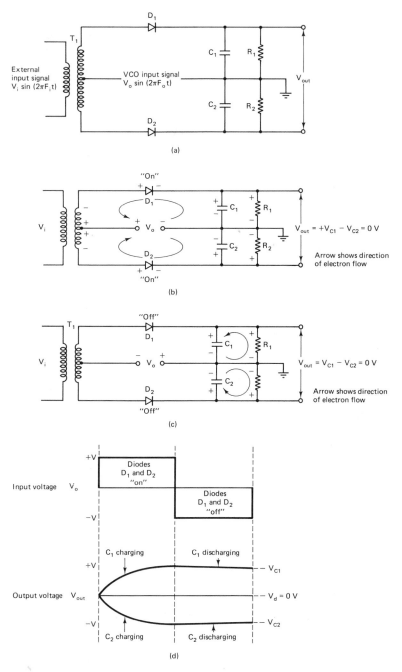

FIGURE 5-3 Phase comparator: (a) schematic diagram; (b) output voltage due to positive half-cycle of V_o; (c) output voltage due to negative half-cycle of V_o; (d) input and output voltage waveforms.

a result, the average output voltage for one cycle of V_o is positive and approximately equal to 0.3 V. Figure 5-4c shows the unfiltered output waveform when V_o and V_i are in phase. During the entire "on" time, V_i is positive. Consequently, the output voltage is positive and approximately equal to 0.636 V. Figures 5-4d and e show the unfiltered output waveform when V_o leads V_i by 135° and 180°, respectively. It can be seen that the output voltage goes negative when V_o leads V_i by more than 90° and reaches its maximum value when V_o leads V_i by 180°. In essence, a phase comparator rectifies the difference voltage between V_o and V_i and integrates it to produce an output voltage that is proportional to the difference between their phases. Simply stated, the magnitude

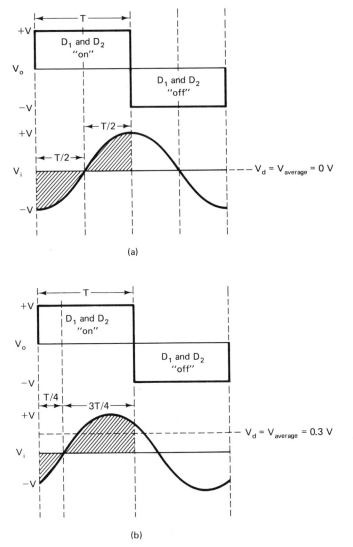

FIGURE 5-4 Phase comparator output voltage waveforms: (a) V_o leads V_i by 90°; (b) V_o leads V_i by 45°; (c) V_o and V_i in phase; (d) V_o leads V_i by 135°; (e) V_b leads V_i by 180°.

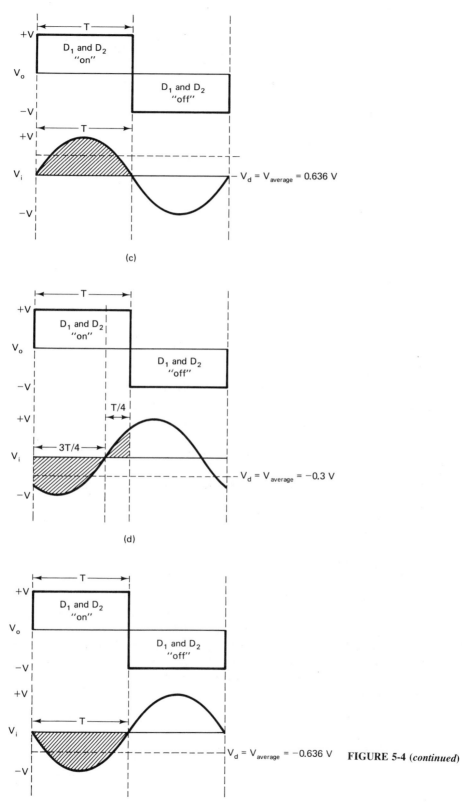

(c)

(d)

(e)

FIGURE 5-4 (*continued*)

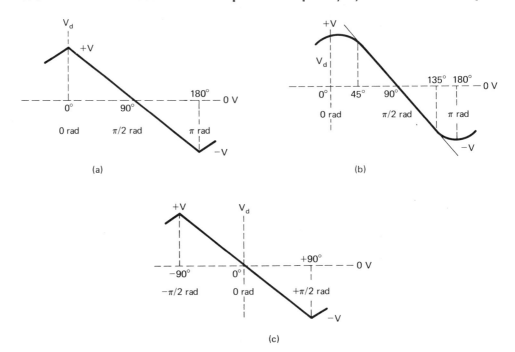

FIGURE 5-5 **Phase comparator output voltage (V_d) versus phase difference (θ_e) characteristics:**
(a) square-wave inputs; (b) sinusoidal inputs; (c) square-wave inputs, phase bias reference.

and polarity of the phase comparator output voltage is proportional to the difference in phase between V_o and V_i.

Figure 5-5 shows the output voltage (V_d)-versus-input phase difference (θ_e) characteristics for the phase comparator shown in Figure 5-3. Figure 5-5a shows the curve for a square-wave phase comparator. The curve has a triangular shape with a negative slope from 0 to 180°. V_d is maximum positive when V_o and V_i are in phase, 0 V when V_o leads V_i by 90°, and maximum negative when V_o leads V_i by 180°. If V_o advances more than 180°, the output voltage becomes less negative and, if V_o lags behind V_i, the output voltage becomes less positive. Therefore, the maximum phase difference that the phase comparator can track is 90° ± 90° or from 0 to 180°. The input to the phase comparator is the difference between the phase of F_o (θ_o) and the phase of the external input frequency (θ_i). The phase difference is called the *phase error* (θ_e), where $\theta_e = \theta_o - \theta_i$. The output of the phase comparator is a dc voltage which is linear between 0 and 180° (0 to πr). Therefore, the gain for a square-wave comparator for $\theta_e = 0$ to πr is given as

$$K_d = \frac{V_d}{\theta_e} = 2 \text{ V}/\pi \text{ (volts/radian)} \qquad (5\text{-}2)$$

where

K_d = gain (V/rad)
V_d = dc output voltage (V)
θ_e = phase error, $\theta_o - \theta_i$ (rad)
    = π radians
V = peak input voltage

Figure 5-5b shows the curve for an analog phase comparator with sinusoidal characteristics. The phase error-versus-output voltage is linear only from 45° to 135°. Therefore, the gain is given as

$$K_d = \frac{V_d}{\theta_e} = V \text{ (volts/radian)} \tag{5-3}$$

From Figures 5-5a and b, it can be seen that the phase comparator output voltage $V_d = 0$ V when $F_o = F_i$ and V_o and V_i are 90° out of phase. Therefore, if F_i is initially equal to F_n, a 90° phase difference is required to hold $V_d = 0$ V. This is understandable because the VCO does not require any correction bias. The 90° phase difference is always there and is equivalent to a bias or offset phase. Generally, the phase bias is considered the reference phase, which can be deviated $\pm\pi/2$ rad. Therefore, V_d goes from its maximum positive value at $-90°$ ($-\pi/2$ rad) and to its maximum negative value at $+90°$ ($+\pi/2$ rad). Figure 5-5c shows the V_d-versus-0_e characteristics for square-wave inputs with the phase bias as the reference.

Figure 5-6a shows the unfiltered output waveform when V_i leads V_o by 90°. Note that the average value of $V_d = 0$ V (the same as when V_o lead V_1 by 90°). When frequency *lock* occurs, it is uncertain whether V_o will lock onto V_i with a + or − 90° phase bias. Therefore, there is a 180° phase ambiguity in the phase of V_o. Figure 5-6b shows the output voltage-versus-phase difference characteristics for square-wave inputs when $F_o = F_n$ and V_o has locked onto V_i, lagging by 90°. Note that the opposite conditions to those shown in Figure 5-5 prevail; the maximum positive and negative voltages occur for the opposite direction phase error and the slope is positive rather than negative from $-\pi/2$ to $+\pi/2$ rad. When frequency lock occurs, the PLL produces a coherent frequency ($F_o = F_i$), but the phase of the recovered signal is uncertain (either F_o leads F_i by 90° $\pm\theta_e$, or vice versa).

Loop Operation

For the following explanations, refer to Figure 5-7a.

Loop acquisition. An external input signal ($V_i \sin 2\pi F_i t$) enters the phase comparator and mixes with the VCO output signal (a square wave with peak amplitude, V_o, and a fundamental frequency, F_o). Initially, $F_o \neq F_i$ and the loop is *unlocked*. Because the phase comparator is a nonlinear device, F_i and F_o mix and generate cross-product frequencies ($F_o + F_i$ and $F_o - F_i$). Therefore, the primary output frequencies from the phase comparator are F_i, F_o, $F_i + F_o$, and $F_i - F_o$. The LPF blocks F_o, F_i,

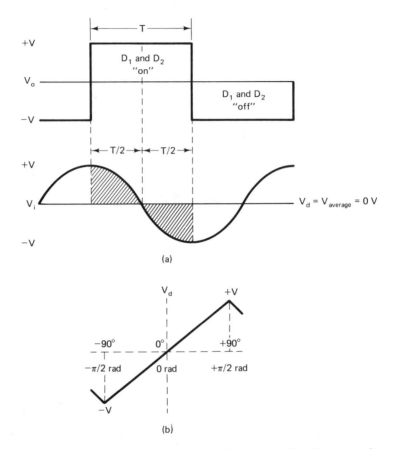

FIGURE 5-6 Phase comparator output voltage: (a) unfiltered output voltage waveform when V_i leads V_o by 90°; (b) output voltage versus phase difference characteristics.

and $F_o + F_i$; thus the input signal to the op-amp is the difference frequency $F_d = F_o - F_i$ (sometimes called the *beat* frequency). F_d is amplified by the op-amp, then applied to the input of the VCO, where it deviates F_o by an amount proportional to the polarity and frequency of F_d. The change in frequency $= \Delta F$ and the VCO output frequency $F_o = F_n + \Delta F$. As F_o changes frequency, the amplitude and frequency of F_d change proportionately until $F_o = F_i$. At this time the output from the phase comparator $V_d = F_o - F_i = 0$ Hz (dc) and the loop is said to be *locked* ($\Delta F = F_n - F_i$). Figure 5-7b shows the beat frequency produced when F_o is swept by F_d. It can be seen that once lock has occurred, F_d is a dc voltage which is necessary to bias the VCO and keep $F_o = F_i$. In essence, the phase comparator is a frequency comparator until frequency *acquisition* (*zero beat*) is achieved; then it becomes a phase comparator. Once the loop is locked, the difference in phase between F_o and F_i is converted to a dc bias voltage and fed back to the VCO to hold lock. Therefore, it is necessary that a phase error ($\theta_o - \theta_i$) is maintained. The change in F_o required to achieve lock $\Delta F = F_n - F_i$, and the

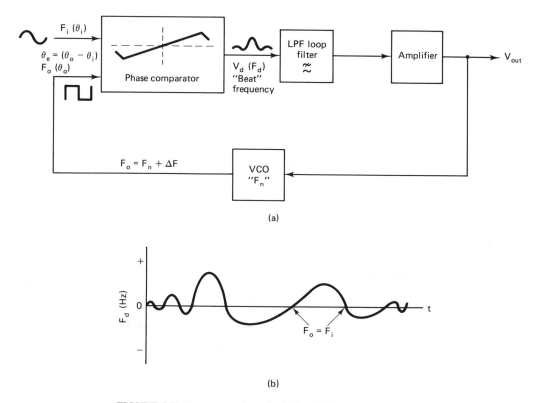

(a)

(b)

FIGURE 5-7   Loop operation of a PLL: (a) block diagram; (b) beat frequency.

time required to achieve lock (*acquisition* or *pull-in* time) for a PLL with no *loop* filter (loop filters are explained later in this chapter) is approximately equal to $1/K_v$ seconds, where K_v is the *open-loop gain* of the PLL. Once the loop is locked, any change in the frequency of F_i is seen as a phase error and the comparator produces a corresponding change in V_d. The change in V_d is amplified and fed back to the VCO to reestablish lock. Thus the loop dynamically adjusts itself to follow input frequency changes.

The input frequency range within which the PLL will lock onto the input signal is called the *capture range*. For a typical PLL, the capture range is between 1.1 and 1.7 times the natural frequency of the VCO. Once the loop is locked, V_d is proportional to the difference in phase between V_o and V_i ($\theta_o - \theta_i$). Mathematically, the output from the phase comparator is (considering only the fundamental frequency for V_o and excluding the 90° phase bias)

$$V_d = (\sin 2\pi F_o t + \theta_o) \times (V \sin 2\pi F_i t + \theta_i)$$

$$= \frac{V}{2} \sin (2\pi F_o t + \theta_o - 2\pi F_i t - \theta_i)$$

$$- \frac{V}{2} \sin (2\pi F_o t + \theta_o + 2\pi F_i t + \theta_i)$$

When $F_o = F_i$,

$$V_d = \frac{V}{2} \sin (\theta_o - \theta_i)$$

where $\theta_o - \theta_i = \theta_e$ (phase error). θ_e is the phase error required to change the VCO output frequency from F_n to F_o (a change $= \Delta F$) and is often called the static phase error.

Capture range is defined as the range of input frequencies (centered on the VCO center frequency, F_o) over which the PLL will lock onto the incoming signal. *Pull-in range* is the peak capture range (i.e., capture range $= 2 \times$ pull-in range). Capture and pull-in range are shown in frequency diagram form in Figure 5-8.

Loop gain.   The *loop gain* for a PLL is simply the product of the individual gains around the closed loop. In Figure 5-6a the open-loop gain is the product of the phase comparator gain, the low-pass filter gain, the op-amp gain, and the VCO gain. Mathematically, the open-loop gain is

$$K_v = (K_d)(K_f)(K_a)(K_o) \qquad \text{with amp} \tag{5-4a}$$

where

$$K_v = K_f K_o K_d \qquad \text{without amp}$$

$K_d =$ phase comparator gain (V/rad)
$K_f =$ filter gain (V/V)
$K_a =$ op-amp gain (V/V)
$K_o =$ VCO gain (Hz/V)

and

$$K_v = \text{open-loop gain}$$

$$= \left(\frac{V}{\text{rad}}\right) \left(\frac{V}{V}\right) \left(\frac{V}{V}\right) \left(\frac{Hz}{V}\right) = \frac{Hz}{\text{rad}}$$

or

$$K_v = \frac{F \text{ cycles/s}}{\text{rad}} = \frac{F \text{ cycles}}{\text{rad-s}} \times \frac{2\pi \text{rad}}{\text{cycle}} = 2\pi F s^{-1}$$

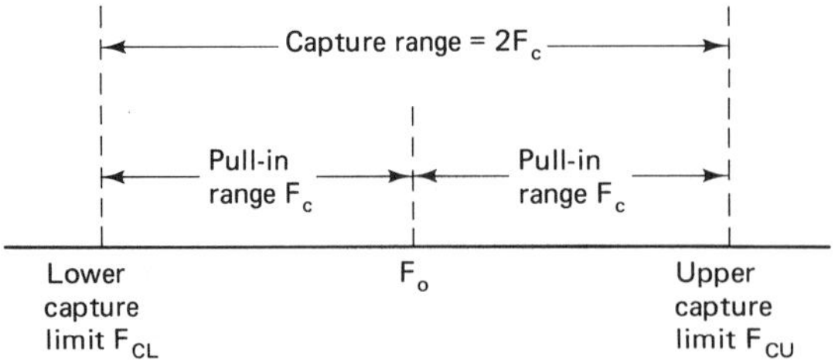

FIGURE 5-8   Capture range.

Expressed in decibels gives us

$$K_v(\text{dB}) = 20 \log k_v \qquad (5\text{-}4b)$$

From Equation 5-4 and Figure 5-6a, the following relationships are derived:

$$V_d = (\theta_e)(K_d) \qquad (5\text{-}5)$$
$$V_{out} = (V_d)(K_f)(K_a) \qquad (5\text{-}6)$$
$$\Delta F = (V_{out})(K_o) \qquad (5\text{-}7)$$

The *closed-loop gain* of a locked PLL is simply unity or 1 (0 dB). This is because there is always 100% feedback in a closed-loop PLL.

Hold-in range. The *hold-in range* for a PLL is the range of input frequencies over which the PLL will remain locked. This presumes that the PLL was initially locked. The hold-in range is often called the *tracking range*; it is the range of frequencies in which the VCO will accurately track or follow the input frequency. The hold-in range is limited by the peak-to-peak swing in V_d and is dependent on the phase comparator, op-amp, and VCO gains. From Figure 5-5c it can be seen that the phase comparator output voltage (V_d) is corrective for $\pm 90°$ ($\pm \pi/2$ rad). Beyond these limits, the polarity of V_d reverses and actually chases the VCO frequency away from the input frequency. Therefore, the maximum phase error (θ_e) that is allowed is $\pm \pi/2$ radians. Consequently, the maximum change in the VCO natural frequency is

$$\Delta F_{max} = \pm(\pi/2 \text{ rad})(K_d)(K_f)(K_a)(K_o)$$

or

$$\Delta F_{max} = \pm(\pi/2 \text{ rad})(K_v) \qquad (5\text{-}8)$$

where ΔF_{max} is the hold-in range.

EXAMPLE 5-1

For the PLL shown in Figure 5-7a, a VCO natural frequency $F_n = 200$ kHz, an input frequency $F_i = 210$ kHz, and the following circuit gains: $K_d = 0.2$ V/rad, $K_f = 1$, $K_a = 5$, and $K_o = 20$ kHz/V; determine:

    (a) The open-loop gain, K_v.
    (b) The change in frequency required to achieve lock, ΔF.
    (c) V_{out}.
    (d) V_d.
    (e) The static phase error, θ_e.
    (f) The hold-in range, ΔF_{max}.

Solution   (a) From Equation 5-4a

$$K_v = \frac{0.2 \text{ V}}{\text{rad}} \frac{1 \text{ V}}{\text{V}} \frac{5 \text{ V}}{\text{V}} \frac{20 \text{ kHz}}{\text{V}} = 20 \frac{\text{kHz}}{\text{rad}}$$

$$20 \frac{\text{kHz}}{\text{rad}} = \frac{20 \text{ kilocycles}}{\text{rad-s}} \times \frac{2\pi \text{ rad}}{\text{cycle}} = 125.6 \text{ ks}^{-1}$$

Expressed as a decibel, we have

$$K_v \text{ (dB)} = 20 \log 125.6 \text{ ks}^{-1} = 102 \text{ dB}$$

(b) $\Delta F = F_i - F_o = 210 - 200 \text{ kHz} = 10 \text{ kHz}$

(c) Rearranging Equation 5-1 gives us

$$V_{out} = \frac{\Delta F}{K_o} = \frac{10 \text{ kHz}}{20 \text{ kHz/V}} = 0.5 \text{V}$$

(d) $V_d = \dfrac{V_{out}}{(K_f)(K_a)} = \dfrac{0.5 \text{ V}}{(1)(5)} = 0.1 \text{ V}$

(e) Rearranging Equation 5-3 gives us

$$\theta_e = \frac{V_d}{K_d} = \frac{0.1 \text{ V}}{0.2 \text{ V/rad}} = 0.5 \text{ rad or } 28.65°$$

(f) $\Delta F_{max} = \dfrac{(\pi/2 \text{ rad})(20 \text{ kHz})}{\text{rad}} = \pm 31.4 \text{ kHz}$

Lock range is the range of frequencies over which the loop will stay locked onto the input signal once lock has been established. Hold-in range is half the lock range. The relationship between lock and hold-in range is shown in Figure 5-9.

The lock range is expressed in rad/s ($2\omega_L$) and is related to the open-loop voltage gain K_v ($K_v = K_f K_o K_d$ for a simple loop with LPF, phase comparator, and VCO, or $K_f K_o K_d K_a$ for a loop with an amplifier, as in Figure 5-7) as follows:

$$2\omega_L = 2K_v = 2K_f K_o K_d \text{ for a simple loop or}$$
$$2K_f K_o K_d K_a \text{ for the loop of Figure 5-7}$$

The lock range in rad/s is twice the dc voltage gain (open loop) and is independent of the LPF response. The capture range $2\omega_c$ depends on the lock range and on the LPF response, so that it changes with the type of LP filter used, and with the filter cutoff frequency. For a simple *RC* LPF, it is given by

$$2\omega_C = 2 \left(\frac{\omega_L}{RC} \right)^{1/2}$$

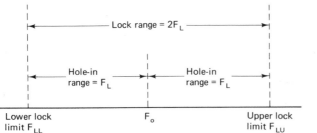

FIGURE 5-9 Lock range.

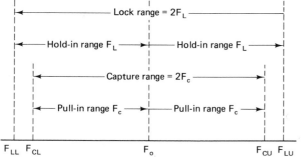

FIGURE 5-10   Capture and lock range.

The capture range is never greater than and is almost always less than the lock range. The relationships among capture, lock, hold-in, and pull-in range are shown in Figure 5-10. Note that lock range ≥ capture range and hold-in range ≥ pull-in range.

Closed-loop frequency response.   The closed-loop frequency response for an *uncompensated* (unfiltered) PLL is shown in Figure 5-11. As shown previously, the open-loop gain of a PLL for a frequency of 1 rad/s = K_v. The frequency response shown in Figure 5-11 is for the circuit and PLL parameters given in Example 5-1. It can be seen that the open-loop gain (K_v) at 1 rad/s = 102 dB, and the open-loop gain equals 1 or 0 dB at the loop cutoff frequency (ω_v). Also, the closed-loop gain is unity up to ω_v, where it drops to −3 dB and continues to roll off at 6 dB/octave (20 dB/decade). Also, $\omega_v = K_v = 125.6$ krad/s, which is the single-sided bandwidth of the uncompensated closed loop.

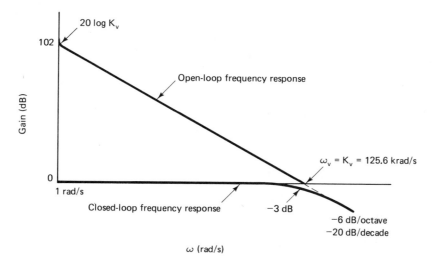

FIGURE 5-11   Frequency response for an uncompensated PLL.

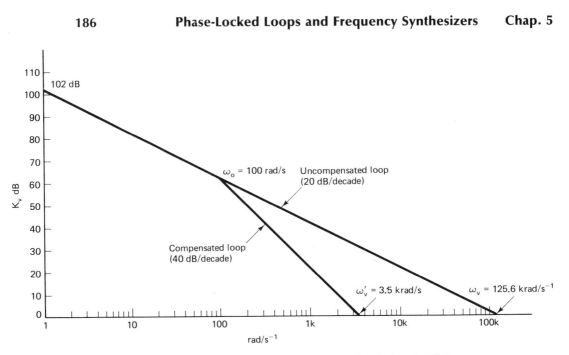

FIGURE 5-12   Loop frequency response for single-pole RC filter.

From Figure 5-11 it can be seen that the frequency response for an uncompensated PLL is identical to that of a single-pole (first-order) low-pass filter with a break frequency $\omega_c = 1$ rad/s. In essence, a PLL is a low-pass *tracking* filter that follows input frequency changes that fall within a bandwidth equal to $\pm K_v$.

If additional bandlimiting is required, a low-pass filter can be added between the phase comparator and the op-amp as shown in Figure 5-7. This filter can be either a single- or multiple-pole filter. Figure 5-12 shows the loop frequency response for a simple single-pole *RC* filter with a cutoff frequency $\omega_c = 1$ krad/s. The frequency response follows that of Figure 5-11 up to the loop filter break frequency, then the response rolls off at 12 dB/octave (40 db/decade). As a result, the compensated unity-gain frequency (ω_c') is reduced (the PLL bandwidth is reduced to approximately ± 3.5 krad/s).

EXAMPLE 5-2

Plot the frequency response for a PLL with an open-loop gain $K_v = 15$ kHz/rad ($\omega_v = 94.2$ krad/s). On the same log paper, plot the response with the addition of a single-pole loop filter with a cutoff frequency $\omega_c = 1.59$ Hz/rad (10 rad/s) and a two/pole loop filter with the same cutoff frequency.

Solution   The specified frequency response curves are shown in Figure 5-13. It can be seen that with the single-pole filter the compensated loop response $= \omega_v' = 1$ krad/s and with the two-pole filter $\omega_v'' = 200$ rad/s.

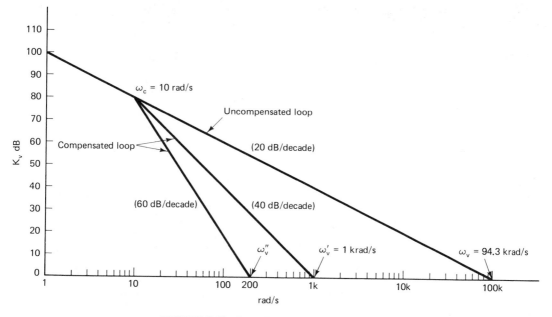

FIGURE 5-13 Loop frequency response for Example 5-2.

The bandwidth of the loop filter (or for that matter, whether a loop filter is needed) depends on the specific application. In Chapters 6, 7, and 8, several applications for PLLs are given with varying requirements. Figure 5-14 shows the specification sheets for the XR2211 integrated-circuit phase-locked loop.

FREQUENCY SYNTHESIZERS

Synthesize means to form an entity by combining parts or elements. A *frequency synthesizer* is used to generate many output frequencies through the addition, subtraction, multiplication, and division of a smaller number of fixed frequency sources. Simply stated, a frequency synthesizer is a crystal-controlled variable-frequency generator. The objective of a synthesizer is twofold. It should produce as many frequencies as possible from a minimum number of sources and each frequency should be as accurate and stable as every other frequency. The ideal frequency synthesizer can generate hundreds or even thousands of different frequencies from a single-crystal oscillator. A frequency synthesizer may be capable of simultaneously generating more than one output frequency, where each frequency is synchronous to a single reference or master oscillator frequency. Frequency synthesizers are used extensively in test and measurement equipment (audio and RF signal generators), tone-generating equipment (touch tone), remote control units (electronic tuners), and multichannel communications systems (telephony).

FSK Demodulator / Tone Decoder

GENERAL DESCRIPTION

The XR-2211 is a monolithic phase-locked loop (PLL) system especially designed for data communications. It is particularly well suited for FSK modem applications. It operates over a wide supply voltage range of 4.5 to 20 V and a wide frequency range of 0.01 Hz to 300 kHz. It can accommodate analog signals between 2 mV and 3 V, and can interface with conventional DTL, TTL, and ECL logic families. The circuit consists of a basic PLL for tracking an input signal within the pass band, a quadrature phase detector which provides carrier detection, and an FSK voltage comparator which provides FSK demodulation. External components are used to independently set center frequency, bandwidth, and output delay. An internal voltage reference proportional to the power supply provides ratio metric operation for low system performance variations with power supply changes.

The XR-2211 is available in 14 pin DTL ceramic or plastic packages specified for commercial or military temperature ranges.

FEATURES

Wide Frequency Range	0.01 Hz to 300 kHz
Wide Supply Voltage Range	4.5 V to 20 V
DTL/TTL/ECL Logic Compatibility	
FSK Demodulation, with Carrier Detection	
Wide Dynamic Range	2 mV to 3 V rms
Adjustable Tracking Range (±1% to ±80%)	
Excellent Temp. Stability	20 ppm/°C, typ.

APPLICATIONS

FSK Demodulation
Data Synchronization
Tone Decoding
FM Detection
Carrier Detection

ABSOLUTE MAXIMUM RATINGS

Power Supply	20 V
Input Signal Level	3 V rms
Power Dissipation	
Ceramic Package	750 mW
Derate above T_A = +25°C	6 mV/°C
Plastic Package	625 mW
Derate above T_A = +25°C	5.0 mW/°C

FUNCTIONAL BLOCK DIAGRAM

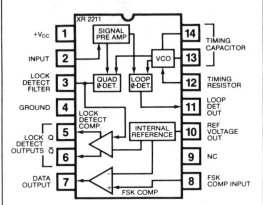

ORDERING INFORMATION

Part Number	Package	Operating Temperature
XR-2211M	Ceramic	–55°C to +125°C
XR-2211CN	Ceramic	0°C to + 75°C
XR-2211CP	Plastic	0°C to + 75°C
XR-2211N	Ceramic	–40°C to + 85°C
XR-2211P	Plastic	–40°C to + 85°C

SYSTEM DESCRIPTION

The main PLL within the XR-2211 is constructed from an input preamplifier, analog multiplier used as a phase detector, and a precision voltage controlled oscillator (VCO). The preamplifier is used as a limiter such that input signals above typically 2MV RMS are amplified to a constant high level signal. The multipling-type phase detector acts as a digital exclusive or gate. Its output (unfiltered) produces sum and difference frequencies of the input and the VCO output, f input + f input (2 f input) and f input - f input (0 Hz) when the phase detector output to remove the "sum" frequency component while passing the difference (DC) component to drive the VCO. The VCO is actually a current controlled oscillator with its nominal input current (f_0) set by a resistor (R_0) to ground and its driving current with a resistor (R_1) from the phase detector.

The other sections of the XR-2211 act to: determine if the VCO is driven above or below the center frequency (FSK comparator); produced both active high and active low outputs to indicate when the main PLL is in lock (quadrature phase detector and lock detector comparator).

EXAR Integrated Systems, Inc., 750 Palomar Avenue, Sunnyvale, CA 94086 * (408) 732-7970 * TWX 910-339-9233

FIGURE 5-14   XR-2211 Phase-locked loop. (Courtesy of EXAR Corporation.)

XR-2211

ELECTRICAL CHARACTERISTICS

Test Conditions: Test Circuit of Figure 1, $V^+ = V^- = 6V$, $T_A = +25°C$, $C = 5000$ pF, $R_1 = R_2 = R_3 = R_4 = 20$ KΩ, $R_L = 4.7$ KΩ, Binary Inputs grounded, S_1 and S_2 closed unless otherwise specified.

PARAMETERS	XR-2211/2211M			XR-2211C			UNITS	CONDITIONS
	MIN.	TYP.	MAX.	MIN.	TYP.	MAX.		
GENERAL								
Supply Voltage	4.5		20	4.5		20	V	
Supply Current		4	7		5	9	mA	$R_0 \geqslant 10$ KΩ See Fig. 4
OSCILLATOR SECTION								
Frequency Accuracy		±1	±3		±1		%	Deviation from $f_0 = 1/R_0C_0$
Frequency Stability								$R_1 = \frac{1}{2}$
Temperature		±20	±50		±20		ppm/°C	See Fig. 8.
Power Supply		0.05	0.5		0.05		%/V	$V^+ = 12 \pm 1$ V. See Fig. 7.
		0.2			0.2		%/V	$V^+ = 5 \pm 0.5$ V. See Fig. 7.
Upper Frequency Limit	100	300			300		kHz	$R_0 = 8.2$ KΩ, $C_0 = 400$ pF
Lowest Practical								
Operating Frequency			0.01		0.01		Hz	$R_0 = 2$ MΩ, $C_0 = 50$ μF
Timing Resistor, R_0								See Fig. 5.
Operating Range	5		2000	5		2000	KΩ	
Recommended Range	15		100	15		100	KΩ	See Fig. 7 and 8.
LOOP PHASE DETECTOR SECTION								
Peak Output Current	±150	±200	±300	±100	±200	±300	μA	Measured at Pin 11.
Output Offset Current		±1			±2		μA	
Output Impedance		1			1		MΩ	
Maximum Swing	±4	±5		±4	±5		V	Referenced to Pin 10.
QUADRATURE PHASE DETECTOR								Measured at Pin 3.
Peak Output Current	100	150			150		μA	
Output Impedance		1			1		MΩ	
Maximum Swing		11			11		V pp	
INPUT PREAMP SECTION								Measured at Pin 2.
Input Impedance		20			20		KΩ	
Input Signal								
Voltage Required to								
Cause Limiting		2	10		2		mV rms	
VOLTAGE COMPARATOR SECTIONS								
Input Impedance		2			2		MΩ	Measured at Pins 3 and 8.
Input Bias Current		100			100		nA	
Voltage Gain	55	70		55	70		dB	$R_L = 5.1$ KΩ
Output Voltage Low		300			300		mV	$I_C = 3$ mA
Output Leakage Current		0.01			0.01		μA	$V_O = 12$ V
INTERNAL REFERENCE								
Voltage Level	4.9	5.3	5.7	4.75	5.3	5.85	V	Measured at Pin 10.
Output Impedance		100			100		Ω	

FIGURE 5-14 (*continued*)

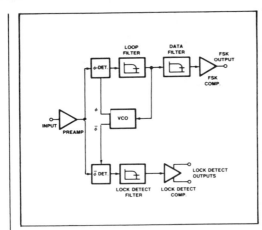

Figure 1: Functional Block Diagram of a Tone and FSK Decoding System Using XR-2211

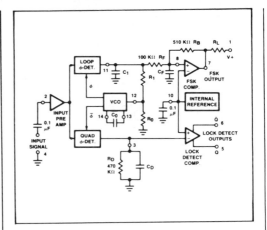

Figure 2: Generalized Circuit Connection for FSK and Tone Detection

Reference Voltage, V_R (Pin 10): This pin is internally biased at the reference voltage level, V_R: V_R = V+/2 - 650 mV. The dc voltage level at this pin forms an internal reference for the voltage levels at Pins 5, 8, 11 and 12. Pin 10 *must* be bypassed to ground with a 0.1 μF capacitor for proper operation of the circuit.

Loop Phase Detector Output (Pin 11): This terminal provides a high impedance output for the loop phase detector. The PLL loop filter is formed by R_1 and C_1 connected to Pin 11 (see Figure 2). With no input signal, or with no phase error within the PLL, the dc level at Pin 11 is very nearly equal to V_R. The peak voltage swing available at the phase detector output is equal to $\pm V_R$.

VCO Control Input (Pin 12): VCO free-running frequency is determined by external timing resistor, R_0, connected from this terminal to ground. The VCO free-running frequency, f_0, is:

$$f_0 = \frac{1}{R_0 C_o} \text{ Hz}$$

where C_0 is the timing capacitor across Pins 13 and 14. For optimum temperature stability, R_0 must be in the range of 10 KΩ to 100 KΩ see Figure 8).

This terminal is a low impedance point, and is internally biased at a dc level equal to V_R. The maximum timing current drawn from Pin 12 must be limited to $\leqslant$3 mA for proper operation of the circuit.

VCO Timing Capacitor (Pins 13 and 14): VCO frequency is inversely proportional to the external timing capacitor, C_0, connected across these terminals (see Figure 5). C_0 must be nonpolar, and in the range of 200 pF to 10 μF.

VCO Frequency Adjustment: VCO can be fine-tuned by connecting a potentiometer, R_X, in series with R_0 at Pin 12 (see Figure 9).

VCO Free-Running Frequency, f_0: XR-2211 does not have a separate VCO output terminal. Instead, the VCO outputs are internally connected to the phase detector sections of the circuit. However, for set-up or adjustment purposes, VCO free-running frequency can be measured at Pin 3 (with C_D disconnected), with no input and with Pin 2 shorted to Pin 10.

DESIGN EQUATIONS
(See Figure 2 for definition of components.)

1. VCO Center Frequency, f_0:

    $f_0 = 1/R_0 C_0$ Hz

2. Internal Reference Voltage, V_R (measured at Pin 10):

    V_R = V+/2 − 650 mV

3. Loop Low-Pass Filter Time Constant, τ:

    $\tau = R_1 C_1$

FIGURE 5-14 (*continued*)

4. Loop Damping, ζ:

$$\zeta = 1/4 \sqrt{\frac{C_0}{C_1}}$$

5. Loop Tracking Bandwidth, $\pm\Delta f/f_0$:

   $\Delta f/f_0 = R_0/R_1$

6. FSK Data Filter Time Constant, τF:

   $\tau F = R_F C_F$

7. Loop Phase Detector Conversion Gain, $K\phi$: ($K\phi$ is the differential dc voltage across Pins 10 and 11, per unit of phase error at phase detector input):

   $K\phi = -2V_R/\pi$ volts/radian

8. VCO Conversion Gain, K_0: (K_0 is the amount of change in VCO frequency, per unit of dc voltage change at Pin 11):

   $K_0 = -1/V_R C_0 R_1$ Hz/volt

9. Total Loop Gain, K_T:

   $K_T = 2\pi K\phi K_0 = 4/C_0 R_1$ rad/sec/volt

10. Peak Phase Detector Current I_A:

    $I_A = V_R$ (volts)/25 mA

APPLICATIONS INFORMATION

FSK DECODING:

Figure 9 shows the basic circuit connection for FSK decoding. With reference to Figures 2 and 9, the functions of external components are defined as follows: R_0 and C_0 set the PLL center frequency, R_1 sets the system bandwidth, and C_1 sets the loop filter time constant and the loop damping factor. C_F and R_F form a one-pole post-detection filter for the FSK data output. The resistor R_B (= 510 KΩ) from Pin 7 to Pin 8 introduces positive feedback across the FSK comparator to facilitate rapid transition between output logic states.

Recommended component values for some of the most commonly used FSK bands are given in Table 1.

Design Instructions:

The circuit of Figure 9 can be tailored for any FSK decoding application by the choice of five key circuit components: R_0, R_1, C_0, C_1 and C_F. For a given set of FSK mark and space frequencies, f_1 and f_2, these parameters can be calculated as follows:

a) Calculate PLL center frequency, f_0:

   $$f_0 = \frac{f_1 + f_2}{2}$$

b) Choose value of timing resistor R_0, to be in the range of 10 KΩ to 100 KΩ. This choice is arbitrary. The recommended value is $R_0 \equiv 20$ KΩ. The final value of R_0 is normally fine-tuned with the series potentiometer, R_X.

c) Calculate value of C_0 from design equation (1) or from Figure 6:

   $C_0 = 1/R_0 f_0$

d) Calculate R_1 to give a Δf equal to the mark space deviation:

   $R_1 = R_0 [f_0/(f_1-f_2)]$

e) Calculate C_1 to set loop damping. (See design equation no. 4.):

   Normally, $\zeta \approx 1/2$ is recommended.

   Then: $C_1 = C_0/4$ for $\zeta = 1/2$

f) Calculate Data Filter Capacitance, C_F:

   For $R_F = 100$ KΩ, $R_B = 510$ KΩ, the recommended value of C_F is:

   $C_F \approx 3/(\text{Baud Rate})$ μF

Note: All calculated component values except R_0 can be rounded to the nearest standard value, and R_0 can be varied to fine-tune center frequency, through a series potentiometer, R_X. (See Figure 9.)

FIGURE 5-14 *(continued)*

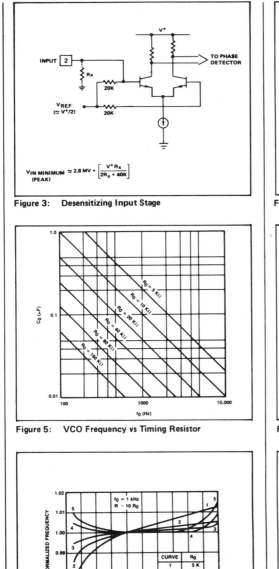

Figure 3:   Desensitizing Input Stage

Figure 5:   VCO Frequency vs Timing Resistor

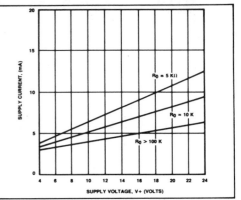

Figure 4:   Typical Supply Current vs V^+ (Logic Outputs Open Circuited).

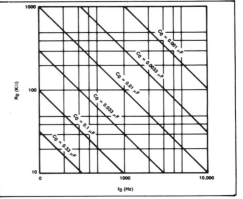

Figure 6:   VCO Frequency vs Timing Capacitor

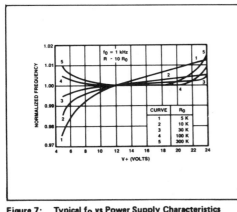

Figure 7:   Typical f_0 vs Power Supply Characteristics

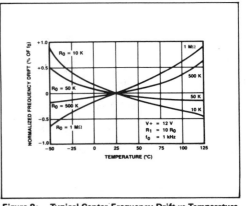

Figure 8:   Typical Center Frequency Drift vs Temperature

FIGURE 5-14 (*continued*)

192

XR-2211

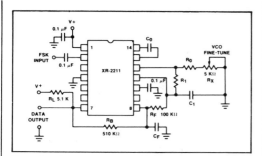

Figure 9:   Circuit Connection for FSK Decoding

Design Example:

75 Baud FSK demodulator with mark space frequencies of 1110/1170 Hz:

Step 1:   Calculate f_0: f_0 = (1110 + 1170) (1/2) = 1140 Hz

Step 2:   Choose R_0 = 20 KΩ (18 KΩ fixed resistor in series with 5 KΩ potentiometer)

Step 3:   Calculate C_0 from Figure 6: C_0 = 0.044 μF

Step 4:   Calculate R_1: R_1 = R_0 (2240/60) = 380 KΩ

Step 5:   Calculate C_1: C_1 = C_0/4 = 0.011 μF

Note: All values except R_0 can be rounded to *nearest* standard value.

Table 1.   Recommended Component Values for Commonly Used FSK Bands. (See Circuit of Figure 9.)

FSK BAND	COMPONENT VALUES	
300 Baud f_1 = 1070 Hz f_2 = 1270 Hz	C_0 = 0.039 μF C_1 = 0.01 μF R_1 = 100 KΩ	C_F = 0.005 μF R_0 = 18 KΩ
300 Baud f_1 = 2025 Hz f_2 = 2225 Hz	C_0 = 0.022 μF C_1 = 0.0047 μF R_1 = 200 KΩ	C_F = 0.005 μF R_0 = 18 KΩ
1200 Baud f_1 = 1200 Hz f_2 = 2200 Hz	C_0 = 0.027 μF C_1 = 0.01 μF R_1 = 30 KΩ	C_F = 0.0022 μF R_0 = 18 KΩ

FSK DECODING WITH CARRIER DETECT:

The lock detect section of XR-2211 can be used as a carrier detect option, for FSK decoding. The recommended circuit connection for this application is shown in Figure 10. The open collector lock detect output, Pin 6, is shorted to data output (Pin 7). Thus, data output will be disabled at "low" state, until there is a carrier within the detection band of the PPL, and the Pin 6 output goes "high," to enable the data output.

The minimum value of the lock detect filter capacitance C_D is inversely proportional to the capture range, $\pm\Delta f_C$. This is the range of incoming frequencies over which the loop can acquire lock and is always less than the tracking range. It is further limited by C_1. For most applications, $\Delta f_C > \Delta f/2$. For R_D = 470 KΩ, the approximate minimum value of C_D can be determined by:

$$C_D \ (\mu F) \geqslant 16/\text{capture range in Hz.}$$

With values of C_D that are too small, chatter can be observed on the lock detect output as an incoming signal frequency approaches the capture bandwidth. Excessively large values of C_D will slow the response time of the lock detect output.

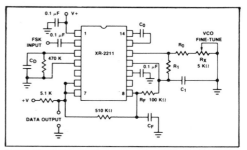

Figure 10:   External Connectors for FSK Demodulation with Carrier Detect Capability

Note: Data Output is "Low" When No Carrier is Present.

TONE DETECTION:

Figure 11 shows the generalized circuit connection for tone detection. The logic outputs, Q and $\overline{Q}$ at Pins 5 and 6 are normally at "high" and "low" logic states, respectively. When a tone is present within the detection band of the PLL, the logic state at these outputs become reversed for the duration of the input tone. Each logic output can sink 5 mA of load current.

Both logic outputs at Pins 5 and 6 are open collector type stages, and require external pull-up resistors R_{L1} and R_{L2}, as shown in Figure 11.

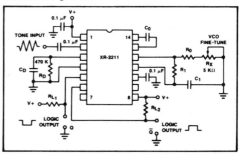

Figure 11:   Circuit Connection for Tone Detection.

FIGURE 5-14 *(continued)*

With reference to Figures 2 and 11, the functions of the external circuit components can be explained as follows: R_0 and C_0 set VCO center frequency; R_1 sets the detection bandwidth; C_1 sets the low pass-loop filter time constant and the loop damping factor. R_{L1} and R_{L2} are the respective pull-up resistors for the Q and $\bar{Q}$ logic outputs.

Design Instructions:

The circuit of Figure 11 can be optimized for any tone detection application by the choice of the 5 key circuit components: R_0, R_1, C_0, C_1 and C_D. For a given input, the tone frequency, f_S, these parameters are calculated as follows:

a) Choose R_0 to be in the range of 15 KΩ to 100 KΩ. This choice is arbitrary.

b) Calculate C_0 to set center frequency, f_0 equal to f_s (see Figure 6): $C_0 = 1/R_0 f_S$

c) Calculate R_1 to set bandwidth ±Δf (see design equation no. 5):

$$R_1 = R_0(f_0/\Delta f)$$

Note: The total detection bandwidth covers the frequency range of $f_0 \pm \Delta f$.

d) Calculate value of C_1 for a given loop damping factor;

$$C_1 = C_0/16\zeta2$$

Normally $\zeta \approx 1/2$ is optimum for most tone detector applications, giving $C_1 = 0.25\ C_0$.

Increasing C_1 improves the out-of-band signal rejection, but increases the PLL capture time.

e) Calculate value of filter capacitor C_D. To avoid chatter at the logic output, with $R_D = 470$ KΩ, C_D must be:

$$C_D(\mu F) \geqslant (16/\text{capture range in Hz})$$

Increasing C_D slows down the logic output response time.

Design Examples:

Tone detector with a detection band of 1 kHz ± 20 Hz:

a) Choose R_0 = 20 KΩ (18 KΩ in series with 5 KΩ potentiometer).

b) Choose C_0 for f_0 = 1 kHz (from Figure 6): $C_0 = 0.05$ μF.

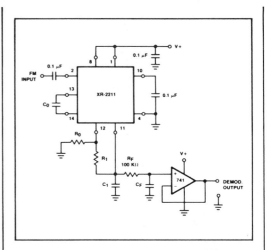

Figure 12: Linear FM Detector Using XR-2211 and an External Op Amp. (See section on Design Equation for Component Values.)

c) Calculate R_1: $R_1 = (R_0)\ (1000/20) = 1$ MΩ.
d) Calculate C_1: for $\zeta = 1/2$, $C_1 = 0.25$, $C_0 = 0.013$ μF.
e) Calculate C_D: $C_D = 16/38 = 0.42$ μF.
f) Fine-tune center frequency with 5 KΩ potentiometer, R_X.

LINEAR FM DETECTION:

XR-2211 can be used as a linear FM detector for a wide range of analog communications and telemetry applications. The recommended circuit connection for this application is shown in Figure 12. The demodulated output is taken from the loop phase detector output (Pin 11), through a post-detection filter made up of R_F and C_F, and an external buffer amplifier. This buffer amplifier is necessary because of the high impedance output at Pin 11. Normally, a non-inverting unity gain op amp can be used as a buffer amplifier, as shown in Figure 12.

The FM detector gain, i.e., the output voltage change per unit of FM deviation can be given as:

$$V_{out} = R_1\ V_R/100\ R_0 \text{ Volts/\%deviation}$$

where V_R is the internal reference voltage (V_R = V+/2 − 650 mV). For the choice of external components R_1, R_0, C_D, C_1 and C_F, see section on design equations.

FIGURE 5-14 (continued)

XR-2211

PRINCIPLES OF OPERATION

Signal Input (Pin 2): Signal is ac coupled to this terminal. The internal impedance at Pin 2 is 20 KΩ. Recommended input signal level is in the range of 10 mV rms to 3 V rms.

Quadrature Phase Detector Output (Pin 3): This is the high impedance output of quadrature phase detector and is internally connected to the input of lock detect voltage comparator. In tone detection applications, Pin 3 is connected to ground through a parallel combination of R_D and C_D (see Figure 2) to eliminate the chatter at lock detect outputs. If the tone detect section is not used, Pin 3 can be left open circuited.

Lock Detect Output, Q (Pin 5): The output at Pin 5 is at "high" state when the PLL is out of lock and goes to "low" or conducting state when the PLL is locked. It is an open collector type output and requires a pull-up resistor, R_L, to V+ for proper operation. At "low" state, it can sink up to 5 mA of load current.

Lock Detect Complement, $\overline{Q}$ (Pin 6): The output at Pin 6 is the logic complement of the lock detect output at Pin 5. This output is also an open collector type stage which can sink 5 mA of load current at low or "on" state.

FSK Data Output (Pin 7): This output is an open collector logic stage which requires a pull-up resistor, R_L, to V+ for proper operation. It can sink 5 mA of load current. When decoding FSK signals, FSK data output is at "high" or "off" state for low input frequency, and at "low" or "on" state for high input frequency. If no input signal is present, the logic state at Pin 7 is indeterminate.

FSK Comparitor Input (Pin 8): This is the high impedance input to the FSK voltage comparator. Normally, an FSK post-detection or data filter is connected between this terminal and the PLL phase detector output (Pin 11). This data filter is formed by R_F and C_F of Figure 2. The threshold voltage of the comparator is set by the internal reference voltage, V_R, available at Pin 10.

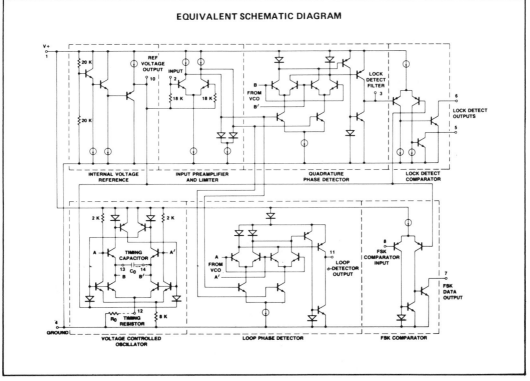

EQUIVALENT SCHEMATIC DIAGRAM

Rev. 8/83

FIGURE 5-14 (*continued*)

195

Essentially, there are two methods of frequency synthesis: direct and indirect. With *direct frequency synthesis*, multiple output frequencies are generated by mixing the outputs from two or more crystal-controlled oscillators or by dividing/multiplying the output frequency from a single-crystal oscillator. With *indirect frequency synthesis*, a feedback-controlled divider/multiplier (such as a PLL) is used to generate multiple output frequencies. Indirect frequency synthesis is slower and more susceptible to noise; however, it is less expensive and requires fewer and less complicated filters than direct synthesis.

Direct Frequency Synthesizers

Multiple-crystal frequency synthesizer.   Figure 5-15 shows a block diagram for a *multiple-crystal frequency synthesizer* that uses nonlinear mixing (heterodyning) and filtering to produce 128 different frequencies from 20 crystals and two oscillator modules. For the crystal values shown, a range of frequencies from 510 to 1890 kHz in 10-kHz steps are synthesized. A synthesizer such as this can be used to generate the carrier frequencies for the 106 AM broadcast-band stations (540 to 1600 kHz). For the switch positions shown, the 160- and 700-kHz oscillators are selected, and the outputs from the balanced mixer are their sum and difference frequencies (700 kHz $\pm$ 160

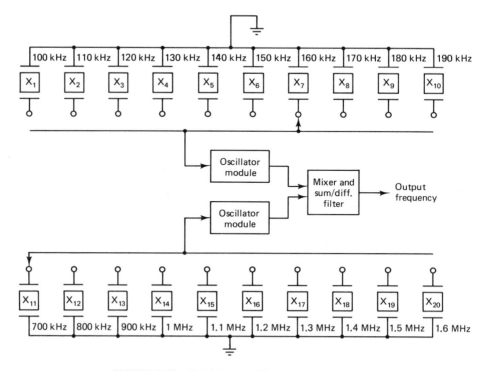

FIGURE 5-15   Multiple-crystal frequency synthesizer.

kHz = 540 and 860 kHz). The output filter is tuned to 540 kHz, which is the carrier frequency for channel 1. To generate the carrier frequency for channel 106, the 100-kHz crystal is selected with either the 1700-kHz (difference) or 1500-kHz (sum) crystal. The minimum frequency separation between output frequencies for a synthesizer is called *resolution*. The resolution for the synthesizer shown in Figure 5-15 is 10 kHz.

Single-crystal frequency synthesizer. Figure 5-16 shows a block diagram for a *single-crystal frequency synthesizer* that again uses frequency addition, multiplication, and division to generate frequencies (in 1-Hz steps) from 1 to 999,999 Hz. A 100-kHz crystal is the source for the master oscillator from which all frequencies are derived.

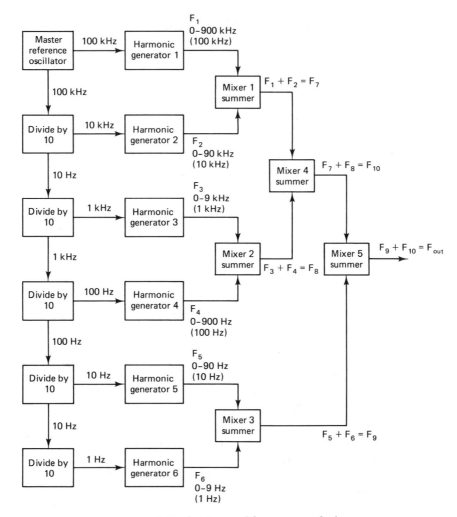

FIGURE 5-16 Single-crystal frequency synthesizer.

The master oscillator frequency is a base frequency which is repeatedly divided by 10 to generate 5 additional base frequencies (10 kHz, 1 kHz, 100 Hz, 10 Hz, and 1 Hz). Each of the base frequencies is fed to a separate harmonic generator (frequency multiplier) which consists of a nonlinear amplifier and a filter. The filter is tunable to each of the first nine harmonics of its base frequency. Therefore, the possible output frequencies for harmonic generator 1 are 100 to 900 kHz in 100-kHz steps, for harmonic generator 2 are 10 to 90 kHz in 10-kHz steps, and so on. The resolution for the synthesizer is determined by how many times the master crystal oscillator frequency is divided. For the synthesizer shown in Figure 5-16, the resolution is 1 Hz. The mixers used are balanced modulators with output filters that are tuned to the sum of the two input frequencies. For example, the harmonics selected in Figure 5-16 produce a 246,313-Hz output frequency. Table 5-1 lists the selector switch positions for each harmonic generator and the input and output frequencies from each mixer. It can be seen that the five mixers simply sum the output frequencies from the six harmonic generators with three levels of mixing (addition).

TABLE 5-1 SWITCH POSITIONS AND HARMONICS

Harmonic generator	Output frequency	Mixer	Output frequency
1	200 kHz	1	240 kHz
2	40 kHz		
3	6 kHz	2	6300 Hz
4	300 Hz		
5	10 Hz	3	13 Hz
6	3 Hz		
		4	246.3 kHz
		5	246.313 kHz

Switched crystal frequency synthesizer. In the early model citizens' band (CB) radios, each channel that a transceiver was equipped for required a separate crystal. Consequently, a transceiver equipped for 23 class D channels, required a minimum of 23 crystals. Figure 5-17 shows a block diagram for a switched crystal frequency synthesizer that is used to generate 23 RF and 5 IF frequencies (four first IFs and one second IF). As you can see, a *switched crystal frequency synthesizer* is a form of *multiple-crystal synthesizer*. Although this technique of frequency synthesis is gradually being replaced with more modern PLL frequency synthesizers, it does illustrate the concept of frequency synthesis quite well. The 28 frequencies are generated with 14 crystals; 6 crystals for the synthesizer oscillator, and 4 each in the transmit and receive oscillators. The three selector switches (S_1, S_2, and S_3) are mechanically ganged to a common tuning switch (S_1). S_1 selects the same crystal for four successive switch positions. Therefore, each crystal in the synthesizer oscillator is used to generate four successive channel frequencies except XTAL 6, which is used to generate only three (i.e., XTAL 1 is for

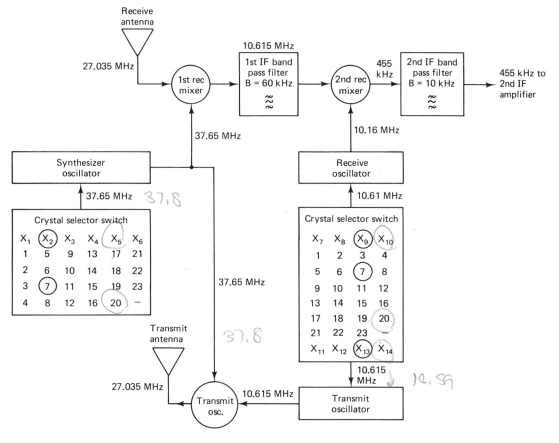

FIGURE 5-17 Switched crystal frequency synthesizer.

channels 1 to 4, XTAL 2 is for channels 5 to 8, etc.). S_2 and S_3 sequentially select the four second receive and transmit oscillator crystals, respectively, as the channel selector is switched through the 23 channel positions. Therefore, each crystal in the transmit and second receive oscillators is used for six channels except XTALs 9 and 13, which are used for only five. However, the channels are not successive (i.e., XTALs 7 and 11 are selected for channels 1, 5, 9, 13, 17, and 21; and XTALs 8 and 12 are selected for channels 2, 6, 10, 14, 18, and 22; etc.). In the receiver, the first four RF carrier frequencies combine in the first receive mixer with XTAL 1 frequency and produce four different first IF frequencies (10.635, 10.625, 10.615, and 10.595 MHz, respectively). However, for all 23 channels, the first IF mixes with the receive oscillator frequency in the second receive mixer to produce a difference frequency of 455 kHz, which is the second IF for all 23 channels. Generating the 23 transmit RF carriers is accomplished in a similar manner. For each channel, one of the six synthesizer XTAL frequencies combines in the transmit mixer with one of the four transmit XTAL frequencies to produce a different RF carrier frequency. The transmit frequency is the difference

TABLE 5-2 CRYSTAL AND SWITCH ASSIGNMENTS FOR THE TRANSCEIVER SHOWN IN FIGURE 5-17

Synthesizer oscillator			Receive oscillator			Transmit oscillator		
Channels	Crystal	Frequency	Channels	Crystal	Frequency	Channels	Crystal	Frequency
1–4	X1	37.6 MHz	1, 5, 9, 13, 17, 21	X7	10.18 MHz	1, 5, 9, 13, 17, 21	X11	10.63 MHz
5–8	X2	37.65 MHz						
9–12	X3	37.7 MHz	2, 6, 10, 14, 18, 22	X8	10.17 MHz	2, 6, 10, 14, 18, 22	X12	10.625 MHz
13–16	X4	37.75 MHz						
17–20	X5	37.8 MHz	3, 7, 11, 15, 19, 23	X9	10.16 MHz	3, 7, 11, 15, 19, 23	X13	10.615 MHz
21–23	X6	37.85 MHz						
			4, 8, 12, 16, 20	X10	10.14 MHz	4, 8, 12, 16, 20	X14	10.59 MHz

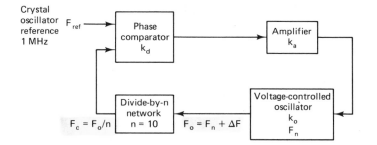

FIGURE 5-18   Single-loop PLL frequency synthesizer.

between the synthesizer and transmit oscillator frequencies. Table 5-2 lists the crystals, selector switch positions, and first and second IFs used for each of the 23 channels. The switch positions and frequencies shown in Figure 5-17 are for transmitting and receiving channel 7.

Indirect Frequency Synthesizers

Phased-locked-loop frequency synthesizers.   In recent years, PLL frequency synthesizers have rapidly become the most popular method for frequency synthesis. Figure 5-18 shows a block diagram for a simple *single-loop* PLL frequency synthesizer. The stable frequency reference is a crystal-controlled oscillator. The range of frequencies generated and the resolution are dependent on the divider network and the open-loop gain. The frequency *divider* is a divide-by-*n* circuit, where *n* is a whole integer number. The simplest form of divider circuit is a *programmable digital up-down counter* with an output frequency $F_c = F_o/n$. With this arrangement, once lock has occurred, $F_c = F_{ref}$ (where $F_c = F_n + \Delta F$) and the VCO and synthesizer output frequency $F_o = nF_{ref}$. Thus the synthesizer is essentially a times-*n* frequency multiplier. The frequency divider reduces the open-loop gain by a factor of *n*. Consequently, the other circuits around the loop must have relatively high gains. The open-loop gain for the frequency synthesizer shown in Figure 5-18 is

$$K_v = \frac{(K_d)\,(K_a)\,(K_o)}{n} \qquad (5\text{-}9a)$$

From Equation 5-9a it can be seen that as *n* changes, the open-loop gain changes in inverse proportion to *n*. A way to remedy this problem is to program the amplifier gain as well as the divider ratio. Thus the open-loop gain is

$$K_v = \frac{K_d n K_a K_o}{n} = K_d K_a K_o \qquad (5\text{-}9b)$$

For the reference frequency and divider circuit shown in Figure 5-18, the range of output frequencies is

$$F_o = nF_{\text{ref}}$$
$$= F_{\text{ref}} \text{ to } 10F_{\text{ref}}$$
$$= 1 \text{ to } 10 \text{ MHz}$$

notice $\rightarrow$

Figure 5-19 shows a block diagram for a CB *transceiver* that uses three crystal oscillators and a PLL frequency synthesizer to generate the 23 RF carriers and two IFs. The divide-by-1024 network is used to reduce the resolution of the PLL to 10 kHz and, at the same time, provide the 10.24-MHz beat frequency required to generate the 455-kHz second IF. Unlike the CB transceiver shown in Figure 5-17, the first IF is the same for all 23 channels (10.695 MHz). The programmable divider is controlled by a binary code generated in the channel selector switch. The VCO output frequency (F_o) varies from 37.66 MHz (channel 1) to 37.94 MHz (channel 23) and mixes with the receive crystal oscillator frequency (36.38 MHz) to generate a difference frequency $F_1 = 1.28$ to 1.56 MHz, which is the input frequency to the programmable divider. The programmable divider is programmed for values of n from 128 for channel 1 to

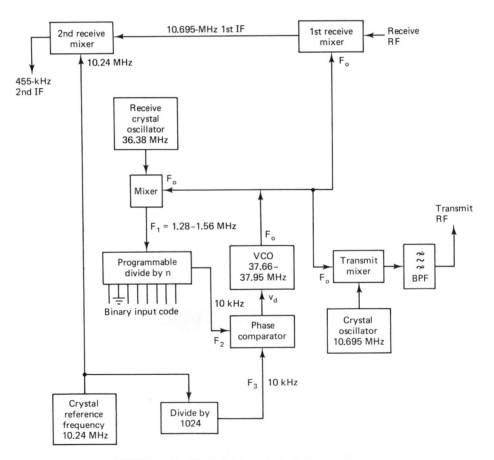

FIGURE 5-19   23-Channel synthesized CB transceiver.

156 for channel 23. Consequently, $F_2 = 10$ kHz for all 23 channels. F_2 and the crystal reference frequency (F_3) are compared in the phase comparator. The phase comparator output voltage is used to tune the VCO and lock it onto the crystal reference frequency. The VCO output frequency is also mixed with the received RF carrier to generate a 10.695-MHz difference frequency, which is the first IF. To generate the 23 transmit RF carrier frequencies, the VCO output frequency is mixed with the output frequency from crystal oscillator 3 (10.695 MHz). The output of the transmit mixer is tuned to their difference frequency. Table 5-3 lists the 23 channel carrier frequencies and their corresponding values for n and F_1.

Prescaled frequency synthesizer.   Figure 5-20 shows the block diagram for a frequency synthesizer that uses *prescaling* to achieve *fractional division*. Prescaling is also necessary for generating frequencies greater than 100 MHz because programmable counters are not available that operate at such high frequencies. The synthesizer shown in Figure 5-20 uses a *two-modulus* prescaler. The prescaler has two modes of operation. One mode provides an output for every input pulse (P) and the other mode provides an output for every $P + 1$ input pulses. Whenever the m-register contains a nonzero number, the prescaler counts in the $P + 1$ mode. Consequently, once the m and n registers have been initially loaded, the prescaler will count down $(P + 1)m$ times

TABLE 5-3

Channel	Frequency (MHz)	VCO frequency (MHz)	n	F_1 (MHz)
1	26.965	37.66	128	1.28
2	26.975	37.67	129	1.29
3	26.985	37.68	130	1.30
4	27.005	37.70	132	1.32
5	27.015	37.71	133	1.33
6	27.025	37.72	134	1.34
7	27.035	37.73	135	1.35
8	27.055	37.75	137	1.37
9	27.065	37.76	138	1.38
10	27.075	37.77	139	1.39
11	27.085	37.78	140	1.40
12	27.105	37.80	142	1.42
13	27.115	37.81	143	1.43
14	27.125	37.82	144	1.44
15	27.135	37.83	145	1.45
16	27.155	37.85	147	1.47
17	27.165	37.86	148	1.48
18	27.175	37.87	149	1.49
19	27.185	37.88	150	1.50
20	27.205	37.90	152	1.52
21	27.215	37.91	153	1.53
22	27.225	37.92	154	1.54
23	27.245	37.94	156	1.56

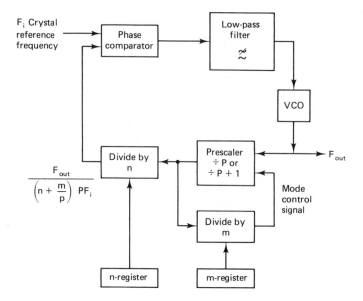

FIGURE 5-20 Frequency synthesizer using prescaling.

until the m counter goes to zero, the prescaler operates in the P mode and the n counter counts down $(n - m)$ times. At this time, both the m and n counters are reset to their initial values which have been stored in the m and n registers, respectively, and the process repeats. Mathematically, the synthesizer output frequency (F_o) is

$$F_o = \left(n + \frac{m}{P} \right) PF_i$$

Multifrequency synthesizer. Figure 5-21a shows a block diagram for a frequency synthesizer that is used to generate 12 carrier frequencies and a pilot for a multiple-channel telecommunications system. A 64-kHz pilot is received, divided by 16, then fed to the input of a PLL. The PLL generates a coherent 4-kHz base frequency from which all local carrier frequencies are derived. The 4-kHz base frequency is multiplied and heterodyned in a harmonic generator to produce the necessary output frequencies. A filter network is used to select 12 channel carrier frequencies which are all multiples of 4 kHz. In addition, the 4-kHz base frequency is used to generate a 104.08-kHz *pilot* that is synchronous with the 12 channel carrier frequencies. Each channel carrier is separated from adjacent carrier frequencies by 4 kHz. Therefore, twelve 4-kHz voice channels are stacked on top of each other in the frequency domain. The composite signal modulates a single high-frequency carrier which is transmitted as one radio channel. This technique is called *frequency-division multiplexing*. Figure 5-21b shows the composite spectrum for the 12 channel carrier frequencies and one pilot frequency generated by the frequency synthesizer shown in Figure 5-21a.

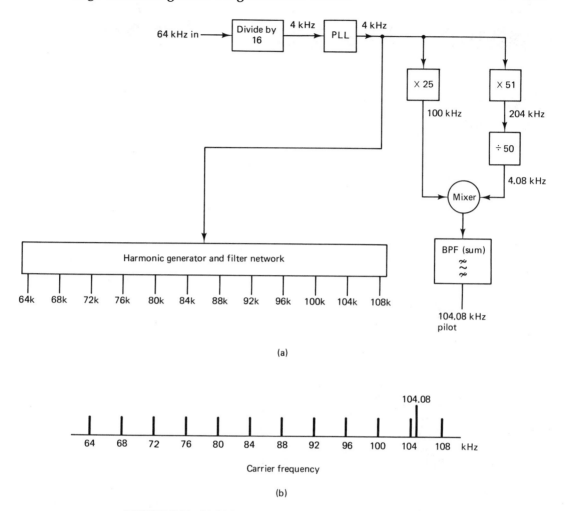

FIGURE 5-21 Multichannel communications system—frequency generation:
(a) block diagram; (b) carrier and pilot spectrum.

LARGE-SCALE INTEGRATION PROGRAMMABLE TIMERS

In recent years, the use of large-scale integration integrated circuits for frequency synthesis has increased at a tremendous rate primarily because of their simplicity and high degree of accuracy. Figure 5-22 shows the specification sheet for the XR-2240 programmable timer/counter. The XR-2240 is a monolithic controller capable of producing precision timing intervals ranging from microseconds through several days. The XR-2240 can generate 256 frequencies or pulse patterns from a single RC setting and can be synchronized with external clock signals. The XR-2240 is ideally suited for precision timing, sequential timing, pattern generation, and frequency synthesis.

Programmable Timer/Counter

GENERAL DESCRIPTION

The XR-2240 Programmable Timer/Counter is a mono-lithic controller capable of producing ultra-long time delays without sacrificing accuracy. In most applications, it provides a direct replacement for mechanical or elec-tromechanical timing devices and generates program-mable time delays from micro-seconds up to five days. Two timing circuits can be cascaded to generate time delays up to three years.

As shown in Figure 1, the circuit is comprised of an internal time-base oscillator, a programmable 8-bit counter and a control flip-flop. The time delay is set by an external R-C network and can be programmed to any value from 1 RC to 255 RC.

In astable operation, the circuit can generate 256 sepa-rate frequencies or pulse-patterns from a single RC set-ting and can be syncronized with external clock sig-nals. Both the control inputs and the outputs are com-patible with TTL and DTL logic levels.

FEATURES

Timing from micro-seconds to days
Programmable delays: 1RC to 255 RC
Wide supply range; 4V to 15V
TTL and DTL compatible outputs
High accuracy: 0.5%
External Sync and Modulation Capability
Excellent Supply Rejection: 0.2%/V

APPLICATIONS

Precision Timing   Frequency Synthesis
Long Delay Generation   Pulse Counting/Summing
Sequential Timing   A/D Conversion
Binary Pattern Generation   Digital Sample and Hold

ABSOLUTE MAXIMUM RATINGS

Supply Voltage	18V
Power Dissipation	
Ceramic Package	750 mW
Derate above +25°C	6 mw/°C
Plastic Package	625 mW
Derate above +25°C	5 mW/°C
Operating Temperature	
XR-2240M	−55°C to +125°C
XR-2240C	0°C to +70°C
Storage Temperature	−65°C to +150°C

FUNCTIONAL BLOCK DIAGRAM

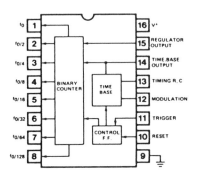

ORDERING INFORMATION

Part Number	Package	Operating Temperature
XR-2240M	Ceramic	−55°C to +125°C
XR-2240N	Ceramic	0°C to +70°C
XR-2240CN	Ceramic	0°C to +70°C
XR-2240P	Plastic	0°C to +70°C
XR-2240CP	Plastic	0°C to +70°C

SYSTEM DESCRIPTION

The XR-2240 is a combination timer/counter capable of generating accurate timing intervals ranging from mi-croseconds through several days. The time base works as an astable multivibrator with a period equal to RC. The eight bit counter can divide the time base output by any integer value from 1 to 255. The wide supply volt-age range of 4.5 to 15 V, TTL and DTL logic compatibili-ty, and 0.5% accuracy allow wide applicability. The counter may operate independently of the time base. Counter outputs are open collector and may be wire-OR connected.

The circuit is triggered or reset with positive going pulses. By connecting the reset pin (Pin 10) to one of the counter outputs, the time base will halt at timeout. If none of the outputs are connected to the reset, the circuit will continue to operate in the astable mode. Ac-tivating the trigger terminal (Pin 11) while the timebase is stopped will set all counter outputs to the low state and start the timebase.

EXAR Corporation, 750 Palomar Avenue, Sunnyvale, CA 94086 • (408) 732-7970 • TWX 910-339-9233

FIGURE 5-22   XR-2240 programmable timer/counter. (Courtesy of EXAR Corporation.)

XR-2240

ELECTRICAL CHARACTERISTICS

Test Conditions: See Figure 2, V$^+$ = 5V, T$_A$ = 25°C, R = 10 kΩ, C = 0.1 μF, unless otherwise noted.

PARAMETERS	XR-2240			XR-2240C			UNIT	CONDITIONS
	MIN	TYP	MAX	MIN	TYP	MAX		
GENERAL CHARACTERISTICS								
Supply Voltage	4		15	4		15	V	For V$^+$ < 4.5V, Short Pin 15 to Pin 16
Supply Current								
Total Circuit		3.5	6		4	7	mA	V$^+$ = 5V, V$_{TR}$ = 0, V$_{RS}$ = 5V
		12	16		13	18	mA	V$^+$ = 15V, V$_{TR}$ = 0, V$_{RS}$ = 5V
Counter Only		1			1.5		mA	See Figure 3
Regulator Output, V$_R$	4.1	4.4		3.9	4.4		V	Measured at Pin 15, V$^+$ = 5V
	6.0	6.3	6.6	5.8	6.3	6.8	V	V$^+$ = 15V, See Figure 4
TIME BASE SECTION								See Figure 2
Timing Accuracy*		0.5	2.0		0.5	5	%	V$_{RS}$ = 0, V$_{TR}$ = 5V
Temperature Drift		150	300		200		ppm/°C	V$^+$ = 5V 0°C ≤ T ≤ 75°C
		80			80		ppm/°C	V$^+$ = 15V
Supply Drift		0.05	0.2		0.08	0.3	%/V	V$^+$ ≥ 8 Volts, See Figure 11
Max. Frequency	100	130			130		kHz	R = 1 kΩ, C = 0.007 μF
Modulation Voltage								Measured at Pin 12
Level	3.00	3.50	4.0	2.80	3.50	4.20	V	V$^+$ 5V
		10.5			10.5		V	V$^+$ = 15V
Recommended Range of Timing Components								See Figure 8
Timing Resistor, R	0.001		10	0.001		10	MΩ	
Timing Capacitor, C	0.007		1000	0.01		1000	μF	
TRIGGER/RESET CONTROLS								
Trigger								Measures at Pin 11, V$_{RS}$ = 0
Trigger Threshold		1.4	2.0		1.4	2.0	V	
Trigger Current		8			10		μA	V$_{RS}$ = 0, V$_{TR}$ = 2V
Impedance		25			25		kΩ	
Response Time**		1			1		μsec.	
Reset								
Reset Threshold		1.4	2.0		1.4	2.0	V	
Reset Current		8			10		μA	V$_{TR}$ = 0, V$_{RS}$ = 2V
Impedance		25			25		kΩ	
Response Time**		0.8			0.8		μsec.	
COUNTER SECTION								See Figure 4, V$^+$ = 5V
Max. Toggle Rate	0.8	1.5			1.5		MHz	V$_{RS}$ = 0, V$_{TR}$ = 5V Measured at Pin 14
Input:								
Impedance		20			20		kΩ	
Threshold	1.0	1.4		1.0	1.4		V	
Output:								Measured at Pins 1 thru 8
Rise Time		180			180		nsec.	R$_L$ = 3k, C$_L$ = 10 pF
Fall Time		180			180		nsec.	
Sink Current	3	5		2	4		mA	V$_{OL}$ ≤ 0.4V
Leakage Current		0.01	8		0.01	15	μA	V$_{OH}$ = 15V

*Timing error solely introduced by XR-2240, measured as % of ideal time-base period of T = 1.00 RC.
**Propagation delay from application of trigger (or reset) input to corresponding state change in counter output at pin 1.

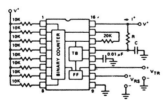

Figure 2. Generalized Test Circuit

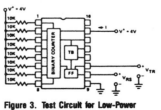

Figure 3. Test Circuit for Low-Power Operation (Time-Base Powered Down)

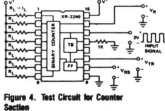

Figure 4. Test Circuit for Counter Section

FIGURE 5-22 (*continued*)

PRINCIPLES OF OPERATION

The timing cycle for the XR-2240 is initiated by applying a positive-going trigger pulse to pin 11. The trigger input actuates the time-base oscillator, enables the counter section, and sets all the counter outputs to "low" state. The time-base oscillator generates timing pulses with its period, T, equal to 1 RC. These clock pulses are counted by the binary counter section. The timing cycle is completed when a positive-going reset pulse is applied to pin 10.

Figure 5 gives the timing sequence of output waveforms at various circuit terminals, subsequent to a trigger input. When the circuit is at reset state, both the time-base and the counter sections are disabled and all the counter outputs are at "high" state.

In most timing applications, one or more of the counter outputs are connected back to the reset terminal, as shown in Figure 6, with S_1 closed. In this manner, the circuit will start timing when a trigger is applied and will automatically reset itself to complete the timing cycle when a programmed count is completed. If none of the counter outputs are connected back to the reset terminal (switch S_1 open), the circuit would operate in its astable or free-running mode, subsequent to a trigger input.

Figure 5. Timing Diagram of Output Waveforms

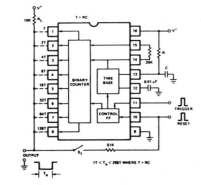

Figure 6. Generalized Circuit Connection for Timing Applications (Switch S_1 Open for Astable Operations, Closed for Monostable Operations)

PROGRAMMING CAPABILITY

The binary counter outputs (pins 1 through 8) are open-collector type stages and can be shorted together to a common pull-up resistor to form a "wired-or" connection. The combined output will be "low" as long as any one of the outputs is low. In this manner, the time delays associated with each counter output can be *summed* by simply shorting them together to a common output bus as shown in Figure 6. For example, if only pin 6 is connected to the output and the rest left open, the total duration of the timing cycle, T_0, would be 32T. Similarly, if pins 1, 5, and 6 were shorted to the output bus, the total time delay would be $T_0 = (1 + 16 + 32) T = 49T$. In this manner, by proper choice of counter terminals connected to the output bus, one can program the timing cycle to be: $1T \leq T_0 \leq 255T$, where $T = RC$.

TRIGGER AND RESET CONDITIONS

When power is applied to the XR-2240 with no trigger or reset inputs, the circuit reverts to "reset" state. Once triggered, the circuit is immune to additional trigger inputs, until the timing cycle is completed or a reset input is applied. If both the reset and the trigger controls are activated simultaneously trigger overrides reset.

DESCRIPTION OF CIRCUIT CONTROLS

COUNTER OUTPUTS (PINS 1 THROUGH 8)

The binary counter outputs are buffered "open-collector" type stages, as shown in Figure 15. Each output is capable of sinking $\approx$ 5 mA of load current. At reset condition, all the counter outputs are at high or non-conducting state. Subsequent to a trigger input, the outputs change state in accordance with the timing diagram of Figure 5.

The counter outputs can be used individually, or can be connected together in a "wired-or" configuration, as described in the Programming section.

RESET AND TRIGGER INPUTS (PINS 10 AND 11)

The circuit is reset or triggered with positive-going control pulses applied to pins 10 and 11. The threshold level for these controls is approximately two diode drops ($\approx$ 1.4V) above ground.

Minimum pulse widths for reset and trigger inputs are shown in Figure 10. Once triggered, the circuit is immune to additional trigger inputs until the end of the timing cycle.

MODULATION AND SYNC INPUT (PIN 12)

The period T of the time-base oscillator can be modulated by applying a dc voltage to this terminal (see Figure 13). The time-base oscillator can be synchronized to an external clock by applying a sync pulse to pin 12, as shown in Figure 16. Recommended sync pulse widths and amplitudes are also given in the figure.

FIGURE 5-22 (*continued*)

XR-2240

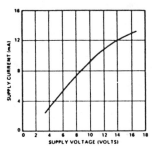

Figure 7. Supply Current vs. Supply Voltage in Reset Condition (Supply Current Under Trigger Condition is ≈ 0.7 mA less)

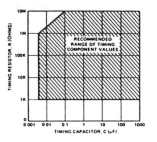

Figure 8. Recommended Range of Timing Component Values.

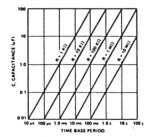

Figure 9. Time-Base Period, T, as a Function of External RC

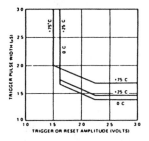

Figure 10. Minimum Trigger and Reset Pulse Widths at Pins 10 and 11

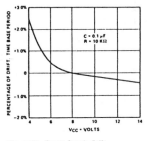

Figure 11. Power Supply Drift

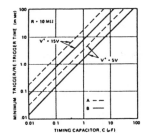

Figure 12.
A) Minimum Trigger Delay Time Subsequent to Application of Power
B) Minimum Re-trigger Time, Subsequent to a Reset Input

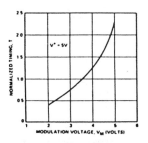

Figure 13. Normalized Change in Time-Base Period As a Function of Modulation Voltage at Pin 12

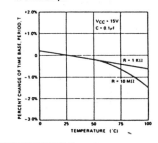

Figure 14. Temperature Drift of Time-Base Period, T

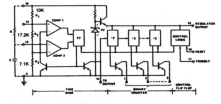

Figure 15. Simplified Circuit Diagram of XR-2240

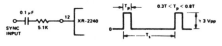

Figure 16. Operation with External Sync Signal.
(a) Circuit for Sync Input
(b) Recommended Sync Waveform

FIGURE 5-22 (*continued*)

HARMONIC SYNCHRONIZATION

Time-base can be synchronized with *integer multiples or harmonics* of input sync frequency, by setting the time-base period, T, to be an integer multiple of the sync pulse period, T_S. This can be done by choosing the timing components R and C at pin 13 such that:

$$T = RC = (T_S/m) \text{ where}$$

$$m \text{ is an integer, } 1 \leq m \leq 10.$$

Figure 17 gives the typical pull-in range for harmonic synchronization, for various values of harmonic modulus, m. For m < 10, typical pull-in range is greater than ±4% of time-base frequency.

TIMING TERMINAL (PIN 13)

The time-base period T is determined by the external R-C network connected to this pin. When the time-base is triggered, the waveform at pin 13 is an exponential ramp with a period T = 1.0 RC.

TIME-BASE OUTPUT (PIN 14)

Time-Base output is an open-collector type stage, as shown in Figure 15 and requires a 20 KΩ pull-up resistor to Pin 15 for proper operation of the circuit. At reset state, the time-base output is at "high" state. Subsequent to triggering, it produces a negative-going pulse train with a period T = RC, as shown in the diagram of Figure 5.

Time-base output is internally connected to the binary counter section and also serves as the input for the external clock signal when the circuit is operated with an external time-base.

The counter input triggers on the negative-going edge of the timing or clock pulses applied to pin 14. The trigger threshold for the counter section is ≈ +1.5 volts. The counter section can be disabled by clamping the voltage level at pin 14 to ground.
Note:
Under certain operating conditions such as high supply voltages ($V^+ > 7V$) and small values of timing capacitor (C < 0.1 μF) the pulse-width of the time-base output at pin 14 may be too narrow to trigger the counter section. This can be corrected by connecting a 300 pF capacitor from pin 14 to ground.

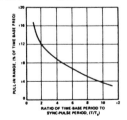

Figure 17. Typical Pull-In Range for Harmonic Synchronization

REGULATOR OUTPUT (PIN 15)

This terminal can serve as a V^+ supply to additional XR-2240 circuits when several timer circuits are cascaded (See Figure 20), to minimize power dissipation. For circuit operation with external clock, pin 15 can be used as the V^+ terminal to power-down the internal time-base and reduce power dissipation. The output current shall not exceed 10 mA.

When the internal time-base is used with $V^+ \leq 4.5V$, pin 15 should be shorted to pin 16.

APPLICATIONS INFORMATION

PRECISION TIMING (Monostable Operation)

In precision timing applications, the XR-2240 is used in its monostable or "self-resetting" mode. The generalized circuit connection for this application is shown in Figure 18.

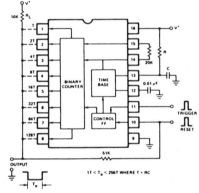

Figure 18. Circuit for Monostable Operation (T_0 = NRC where $1 \leq N \leq 255$)

The output is normally "high" and goes to "low" subsequent to a trigger input. It stays low for the time duration T_0 and then returns to the high state. The duration of the timing cycle T_0 is given as:

$$T_0 = NT = NRC$$

where T = RC is the time-base period as set by the choice of timing components at pin 13 (See Figure 9). N is an integer in the range of:

$$1 \leq N \leq 255$$

as determined by the combination of counter outputs (pins 1 through 8) connected to the output bus, as described below.

PROGRAMMING OF COUNTER OUTPUTS: The binary counter outputs (pins 1 through 8) are open-collector type stages and can be shorted together to a common pull-up resistor to form a "wired-or" connection where

FIGURE 5-22 (*continued*)

the combined output will be "low" as long as any one of the outputs is low. In this manner, the time delays associated with each counter output can be summed by simply shorting them together to a common output bus as shown in Figure 18. For example if only pin 6 is connected to the output and the rest left open, the total duration of the timing cycle, T_0, would be 32T. Similarly, if pins 1, 5, and 6 were shorted to the output bus, the total time delay would be $T_0 = (1 + 16 + 32) T = 49T$. In this manner, by proper choice of counter terminals connected to the output bus, one can program the timing cycle to be: $1T \leq T_0 \leq 255T$.

ULTRA-LONG DELAY GENERATION

Two XR-2240 units can be cascaded as shown in Figure 19 to generate extremely long time delays. In this application, the reset and the trigger terminals of both units are tied together and the time base of Unit 2 disabled. In this manner, the output would normally be high when the system is at reset. Upon application of a trigger input, the output would go to a low stage and stay that way for a total of $(265)^2$ or 65,536 cycles of the time-base oscillator.

PROGRAMMING: Total timing cycle of two cascaded units can be programmed from $T_0 = 256RC$ to $T_0 = 65,536RC$ in 256 discrete steps by selectively shorting any one or the combination of the counter outputs from Unit 2 to the output bus.

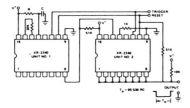

Figure 19. Cascaded Operation for Long Delay Generation

LOW-POWER OPERATION

In cascaded operation, the time-base section of Unit 2 can be powered down to reduce power consumption, by using the circuit connection of Figure 20. In this case, the V^+ terminal (pin 16) of Unit 2 is left open-circuited, and the second unit is powered from the regulator output of Unit 1, by connecting pin 15 of both units.

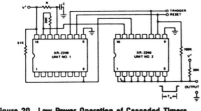

Figure 20. Low-Power Operation of Cascaded Timers

ASTABLE OPERATION

The XR-2240 can be operated in its astable or free-running mode by disconnecting the reset terminal (pin 10) from the counter outputs. Two typical circuit connections for this mode of operation are shown in Figure 21. In the circuit connection of Figure 21(a), the circuit operates in its free-running mode, with external trigger and reset signals. It will start counting and timing subsequent to a trigger input until an external reset pulse is applied. Upon application of a positive-going reset signal to pin 10, the circuit reverts back to its rest state. The circuit of Figure 21(a) is essentially the same as that of Figure 6, with the feedback switch S_1 open.

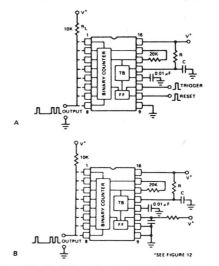

Figure 21. Circuit Connections for Astable Operation
(a) Operation with External Trigger and Reset Controls
(b) Free-running or Continuous Operation

The circuit of Figure 21(b) is designed for continuous operation. The circuit self-triggers automatically when the power supply is turned on, and continues to operate in its free-running mode indefinitely.

In astable or free-running operation, each of the counter outputs can be used individually as synchronized oscillators; or they can be interconnected to generate complex pulse patterns.

BINARY PATTERN GENERATION

In astable operation, as shown in Figure 21, the output of the XR-2240 appears as a complex pulse pattern. The waveform of the output pulse train can be determined directly from the timing diagram of Figure 5 which shows the phase relations between the counter outputs. Figure 22 shows some of these complex pulse patterns. The pulse pattern repeats itself at a rate equal to the period of the *highest* counter bit connected

FIGURE 5-22 (*continued*)

to the common output bus. The minimum pulse width contained in the pulse train is determined by the *lowest* counter bit connected to the output.

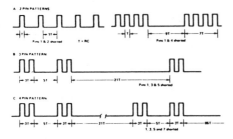

Figure 22. Binary Pulse Patterns Obtained by Shorting Various Counter Outputs

OPERATION WITH EXTERNAL CLOCK

The XR-2240 can be operated with an external clock or time-base, by disabling the internal time-base oscillator and applying the external clock input to pin 14. The recommended circuit connection for this application is shown in Figure 23. The internal time-base can be de-activated by connecting a 1 KΩ resistor from pin 13 to ground. The counters are triggered on the negative-going edges of the external clock pulse. For proper operation, a minimum clock pulse amplitude of 3 volts is required. Minimum external clock pulse width must be ≥ 1 μS.

For operation with supply voltages of 6V or less, the internal time-base section can be powered down by open-circuiting pin 16 and connecting pin 15 to V$^+$. In this configuration, the internal time-base does not draw any current, and the over-all current drain is reduced by ≈ 3 mA.

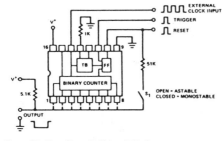

Figure 23. Operation with External Clock

FREQUENCY SYNTHESIZER

The programmable counter section of XR-2240 can be used to generate 255 discrete frequencies from a given time base setting using the circuit connection of Figure 24. The output of the circuit is a positive pulse train with a pulse width equal to T, and a period equal to (N + 1) T where N is the programmed count in the counter.

The modulus N is the *total count* corresponding to the counter outputs connected to the output bus. Thus, for example, if pins 1, 3 and 4 are connected together to the output bus, the total count is: N = 1 + 4 + 8 = 13; and the period of the output waveform is equal to (N + 1) T or 14T. In this manner, 256 different frequencies can be synthesized from a given time-base setting.

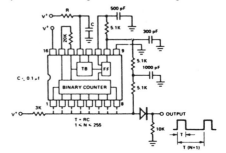

Figure 24. Frequency Synthesis from Internal Time-Base

SYNTHESIS WITH HARMONIC LOCKING: The harmonic synchronization property of the XR-2240 time-base can be used to generate a wide number of discrete frequencies from a given input reference frequency. The circuit connection for this application is shown in Figure 25. (See Figures 16 and 17 for external sync waveform and harmonic capture range.) If the time base is synchronized to (m)th harmonic of input frequency where 1 ≤ m ≤ 10, as described in the section on "Harmonic Synchronization", the frequency f_O of the output waveform in Figure 25 is related to the input reference frequency f_R as:

$$f_O = f_R \frac{m}{(N + 1)}$$

where m is the harmonic number, and N is the programmed counter modulus. For a range of 1 ≤ N ≤ 255, the circuit of Figure 25 can produce 1500 separate frequencies from a single fixed reference.

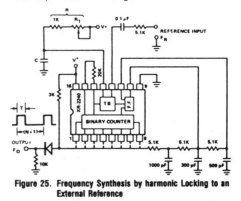

Figure 25. Frequency Synthesis by harmonic Locking to an External Reference

FIGURE 5-22 *(continued)*

212

One particular application of the circuit of Figure 25 is generating frequencies which are not harmonically related to a reference input. For example, by choosing the external R-C to set m = 10 and setting N = 5, one can obtain a 100 Hz output frequency synchronized to 60 Hz power line frequency.

STAIRCASE GENERATOR

The XR-2240 Timer/Counter can be interconnected with an external operational amplifier and a precision resistor ladder to form a staircase generator, as shown in Figure 26. Under reset condition, the output is low. When a trigger is applied, the op. amp. output goes to a high state and generates a negative going staircase of 256 equal steps. The time duration of each step is equal to the time-base period T. The staircase can be stopped at any desired level by applying a "disable" signal to pin 14, through a steering diode, as shown in Figure 26. The count is stopped when pin 14 is clamped at a voltage level less than 1.4V.

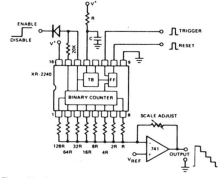

Figure 26. Staircase Generator

DIGITAL SAMPLE/HOLD

Figure 27 shows a digital sample and hold circuit using the XR-2240. The principle of operation of the circuit is similar to the staircase generator described in the previous section. When a "strobe" input is applied, the RC low-pass network between the reset and the trigger inputs of XR-2240 causes the timer to be first reset and then triggered by the same strobe input. This strobe input also sets the output of the bistable latch to a high state and activates the counter.

The circuit generates a staircase voltage at the output of the op. amp. When the level of the staircase reaches that of the analog input to be sampled, comparator changes state, activates the bistable latch and stops the count. At this point, the voltage level at the op. amp. output corresponds to the sampled analog input. Once the input is sampled, it will be held until the next strobe signal. Minimum re-cycle time of the system is ≈6 msec.

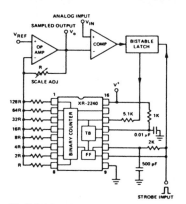

Figure 27. Digital Sample and Hold Circuit

ANALOG-TO-DIGITAL CONVERTER

Figure 28 shows a simple 8-bit A/D converter system using the XR-2240. The operation of the circuit is very similar to that described in connection with the digital sample/hold system of Figure 15. In the case of A/D conversion, the digital output is obtained in parallel format from the binary counter outputs, with the output at pin 8 corresponding to the most significant bit (MSB). The re-cycle time of the A/D converter is ≈6 msec.

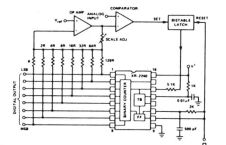

Figure 28. Analog-to-Digital Converter

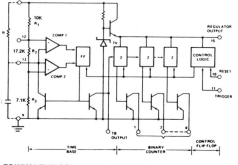

EQUIVALENT SCHEMATIC DIAGRAM

FIGURE 5-22 *(continued)*

QUESTIONS

5-1. Describe a phase-locked loop.

5-2. List the advantages of an integrated-circuit PLL over a discrete PLL.

5-3. Draw the block diagram for a PLL, list the functions of each block, and describe its basic operation.

5-4. Describe the operation of a voltage-controlled oscillator.

5-5. Describe the operation of a phase comparator.

5-6. Describe how loop acquisition is accomplished with a PLL from an initial unlocked condition until lock is achieved.

5-7. Define the following terms: *beat frequency*; *zero beat*; *acquisition time*; *open-loop gain*.

5-8. Contrast the following terms: *capture range*; *pull-in range*; *closed-loop gain*; *hold-in range*; *tracking range*; *lock range*.

5-9. Define the following terms: *uncompensated PLL*; *loop cutoff frequency*; *tracking filter*.

5-10. Define *synthesize*. What is a frequency synthesizer?

5-11. Describe direct and indirect frequency synthesis.

5-12. What is meant by the *resolution* of a frequency synthesizer?

PROBLEMS

5-1. For the VCO input-versus-output characteristics curve shown below, determine:

185 KHz (a) The frequency of operation for a −2-V input signal.

     (b) The frequency deviation for a ±2-V$_p$ input signal. $f_{high} - f_{low}$

     (c) The transfer function, K_o, for the linear portion of the curve (−3 to +3 V).

     $(200 - 60)/6 = 22.5$ KHz/v

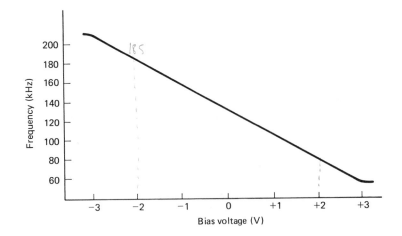

5-2. For the output voltage-versus-phase difference (θ_e) characteristic curve shown below, determine:

(a) The output voltage for a $-45°$ phase difference.
(b) The output voltage for a $+60°$ phase difference.
(c) The maximum peak output voltage.
(d) The transfer function, K_d.

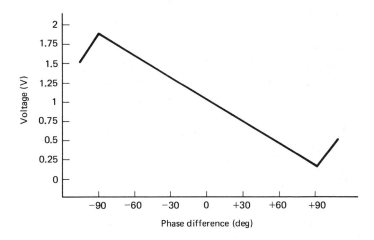

5-3. For the PLL shown in Figure 5-7a a VCO natural frequency $F_n = 150$ kHz, an input frequency $F_i = 160$ kHz, and the following circuit gains $K_d = 0.2$ V/rad, $K_f = 1$, $K_a = 4$, and $K_o = 15$ kHz/V, determine:

P. 183 - 184

(a) The open-loop gain K_v.
(b) ΔF.
(c) V_{out}.
(d) V_d.
(e) θ_e.
(f) The hold-in range, ΔF_{max}.

5-4. Plot the frequency response for a PLL with an open-loop gain $K_v = 20$ kHz/rad. On the same log paper, plot the response with a single-pole loop filter with a cutoff frequency $\omega_c = 100$ rad/s, and a two-pole loop filter with the same cutoff frequency.

5-5. Determine the change in frequency (ΔF) for a VCO with a transfer function $K_o = 2.5$ kHz/V and a dc input voltage change $\Delta V = 0.8$ V. $K_o = \frac{\Delta F}{\Delta V}$

5-6. Determine the voltage at the output of a phase comparator with a transfer function $K_d = 0.5$ V/rad and a phase error $\theta_e = 0.75$ rad.

5-7. Determine the hold-in range (ΔF_{max}) for a PLL with a loop gain $K_v = 110$ kHz/rad. $\leq \pm \frac{k}{2} \cdot K_v$

5-8. Determine the phase error necessary to produce a VCO frequency shift $\Delta F_o = 10$ kHz for a loop gain $K_v = 40$ kHz/rad.

5-9. Determine two output frequencies for the multiple-crystal frequency synthesizer shown in Figure 5-15 if crystals X8 and X18 are selected. $F_{highest} \pm F_{lower}$

5-10. Determine the output frequency from the single-crystal frequency synthesizer shown in Figure 5-16 for the following harmonics.

Harmonic Generator	Harmonic
1	6
2	4
3	7
4	1
5	2
6	6

Channel 14 **5-11.** Determine which channel the switched-crystal frequency synthesizer shown in Figure 5-17 is tuned for the following crystal selections: X4 and X8. *channel they have in common*

5-12. Determine the first and second intermediate frequencies for Problem 5-11.

10 kHz **5-13.** Determine F_c for the PLL frequency synthesizer shown in Figure 5-18 for a natural frequency $F_n = 200$ kHz, $\Delta F = 0$ Hz, and $n = 20$.

$$F_o = F_n + \Delta F$$
$$F_c = \frac{F_o}{n}$$

$$F_o = 200 \text{ kHz} + 0 \text{ Hz}$$
$$F_o = 200 \text{ kHz}$$

$$F_c = \frac{200 \text{ kHz}}{20}$$

$$\therefore F_c = 10 \text{ kHz}$$

Chapter 6

SINGLE-SIDEBAND COMMUNICATIONS SYSTEMS

INTRODUCTION

Conventional AM DSBFC communications systems, such as those discussed in Chapters 3 and 4, have several inherent and outstanding disadvantages. First, in conventional AM systems, at least two-thirds of the transmitted power is in the carrier. However, there is no information in the carrier; the information is contained in the sidebands. Also, the information contained in the upper sideband is identical to the information contained in the lower sideband. Therefore, transmitting both sidebands is redundant. Consequently, conventional AM is both power and bandwidth inefficient, which are two of the most important considerations when designing electronic communications systems.

SINGLE-SIDEBAND SYSTEMS

There are many different types of *sideband* communications systems. Some of these systems conserve bandwidth, some conserve power, and some conserve both bandwidth and power. Figure 6-1 compares the frequency spectra and relative power distributions for conventional AM and several of the more common single-sideband (SSB) systems.

AM Single-Sideband Full Carrier

AM single-sideband full carrier (SSBFC) is a form of amplitude modulation in which the carrier is transmitted at *full* power, but one of the sidebands is removed. Therefore,

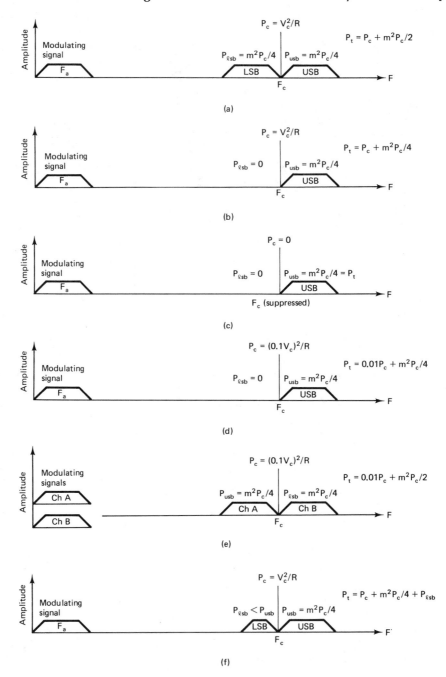

FIGURE 6-1 Single-sideband systems: (a) conventional DSBFC AM; (b) full-carrier single sideband; (c) suppressed-carrier single sideband; (d) reduced-carrier single sideband; (e) independent sideband; (f) vestigal sideband.

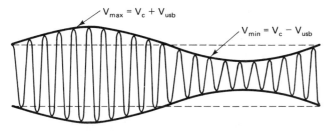

$V_{max} = V_c + V_{usb}$

$V_{min} = V_c - V_{usb}$

FIGURE 6-2 SSBFC waveform, 100% modulation.

SSBFC transmission requires only half as much bandwidth as conventional AM. The frequency spectrum and relative power distribution for SSBFC are shown in Figure 6-1b. Note that with 100% modulation, the carrier power (P_c) constitutes at least four-fifths (80%) of the total transmitted power (P_t), and only one-fifth (20%) of the total power is in the sideband. With conventional AM, two-thirds (67%) of the total power is in the carrier and one-third (33%) is in the sidebands. Therefore, although SSBFC requires less total power (P_t), it actually utilizes a smaller percentage of the total power for the information carrying portion of the signal (i.e., the sidebands). Figure 6-2 shows the envelope for a 100% modulated SSBFC wave for a single frequency-modulating signal. Comparing to Figure 3-6c shows that the SSBSC waveform is identical to that of a 50% modulated double-sideband wave. Recall from Chapter 3 that the maximum positive and negative peaks in a conventional AM envelope occur when the carrier and both of the side frequencies reach their respective peaks at the same time, and the peak change in the envelope is equal to the sum of the amplitudes of the upper and lower side frequencies. With single-sideband transmission, there is only a single side frequency (either the upper or lower) to add to the carrier. Therefore, the positive and negative excursions in the envelope are only half what they are with double-sideband transmission. Consequently, with single-sideband full-carrier transmission, the demodulated signals have only half the amplitude as a double-sideband demodulated wave. Thus a trade-off is made with SSBFC. SSBFC requires less bandwidth than DSBFC but also produces a lower amplitude-demodulated signal. If the bandwidth is halved, the total noise is reduced by 3 dB. However, if one sideband is removed, the power in the information portion of the wave is also halved. Consequently, the S/N is the same. With SSBFC, the repetition rate of the envelope is the frequency of the modulating signal. Therefore, as with double-sideband transmission, the information is contained in the shape of the modulated envelope.

AM Single-Sided Suppressed Carrier

AM single-sideband suppressed carrier (SSBSC) is a form of amplitude modulation in which the carrier is totally suppressed and one of the sidebands is removed. Therefore, SSBSC requires half as much bandwidth as conventional AM and considerably less transmitted power. The frequency spectrum and relative power distribution for SSBSC are shown in Figure 6-1c. It can be seen that the sideband power (P_{sb}) makes up 100% of the total transmitted power. Figure 6-3 shows a SSBSC waveform for a single-fre-

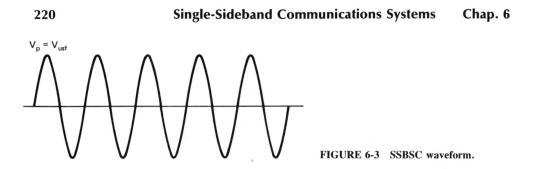

FIGURE 6-3   SSBSC waveform.

quency modulating signal. As you can see, the waveform is not an envelope; it is simply a single frequency equal to the carrier frequency plus or minus the modulating signal frequency, depending on whether the USF or LSF is suppressed.

AM Single-Sideband Reduced Carrier

AM single-sideband reduced carrier (SSBRC) is a form of amplitude modulation where one sideband is totally removed and the carrier voltage is reduced to approximately 10% of its unmodulated amplitude. Consequently, the transmitted sideband can have as much as 86% of the total transmitted power. Generally, the carrier is totally suppressed during modulation, then reinserted later at a reduced amplitude. Consequently, SSBRC is sometimes called single-sideband *reinserted* carrier. The reinserted carrier is called a *pilot* carrier and is reinserted for demodulation purposes, which is explained later in this chapter. The frequency spectrum and relative power distribution for SSBRC are shown in Figure 6-1d. The figure shows that the sideband power constitutes approximately 100% of the total transmitted power. Figure 6-4a shows the transmitted waveform for a single-frequency modulating signal when the carrier level equals the sideband level, and Figure 6-4b shows the waveform when the carrier level is less than the sideband level. Note that the repetition rate of the envelope is equal to the modulating signal frequency (F_a). To demodulate this signal with conventional peak or envelope detectors, the carrier must be separated, amplified, then reinserted in the receiver. Therefore, suppressed-carrier transmission is sometimes called *exalted* carrier because the carrier is exalted or elevated in the receiver prior to demodulation. With an exalted carrier system, the separate carrier amplification must be sufficient to raise the amplitude of the carrier to a value greater than that of the sideband signal. SSBRC requires half as much bandwidth as conventional AM and, because the carrier is transmitted at a reduced level, also conserves power.

AM Independent Sideband

AM independent sideband (ISB) is a form of amplitude modulation where a single carrier frequency is separately modulated by two independent modulating signals. In essence, ISB is a form of double-sideband transmission in which the transmitter has two single-sideband suppressed-carrier modulators. One modulator removes the upper

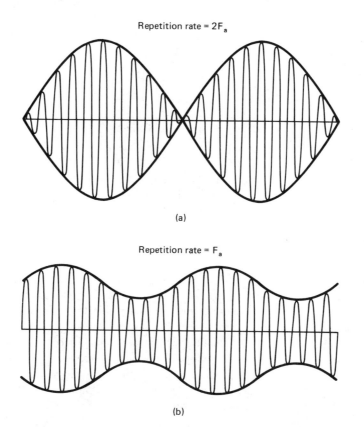

Repetition rate = $2F_a$

(a)

Repetition rate = F_a

(b)

FIGURE 6-4 **SSBRC waveform: (a) carrier level equal to the sideband level; (b) carrier level less than the sideband level.**

sideband and the other removes the lower sideband. The single-sideband output signals from the two modulators are combined to form a double-sideband signal in which the two sidebands are totally independent from each other except that they are symmetrical about a common carrier frequency. ISB transmission consists of two independent sidebands with one positioned above and one positioned below the suppressed carrier. For demodulation purposes, the carrier is generally reinserted at a reduced level. Figure 6-1e shows the frequency spectrum and power distribution for ISB, and Figure 6-5 shows the transmitted waveform for two independent single-frequency sources. The two sources are of equal frequency; therefore, the waveform is identical to a double-sideband suppressed-carrier waveform with a repetition rate equal to twice the modulating signal frequency ($2F_a$). ISB conserves both transmitted power and bandwidth as two information sources are transmitted within the frequency spectrum required for a single source with conventional AM DSBFC. ISB is one technique that is used in the United States for stereo AM transmission. One channel (the left) is transmitted in the lower sideband, and the other channel (the right) is transmitted in the upper sideband.

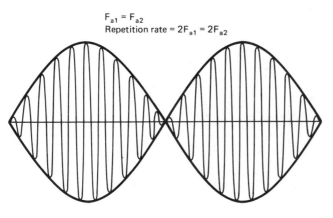

$$F_{a1} = F_{a2}$$
Repetition rate $= 2F_{a1} = 2F_{a2}$

FIGURE 6-5   ISB waveform.

AM Vestigial Sideband

AM vestigial sideband (VSB) is a form of amplitude modulation where the carrier is transmitted at full power along with one complete sideband and part of the other sideband. VSB is a form of AM double sideband where the lower frequency-modulating signals are transmitted double sideband and thus appreciate the benefit of 100% AM modulation. The higher-frequency modulating signals are transmitted single sideband and thus can achieve only 50% AM modulation. Consequently, the lower frequencies are emphasized in the demodulator and produce larger amplitude signals than the higher frequencies. The frequency spectrum and relative power distribution for VSB are shown in Figure 6-1f. Probably the most widely known VSB system is the picture portion of a commercial television broadcasting signal.

Comparison of Single-Sideband to Conventional AM

Bandwidth conservation is an obvious advantage of single-sideband transmission over conventional double-sideband AM. The preceding discussions and the frequency spectra of Figure 6-1 show that single sideband requires only half as much bandwidth as conventional AM, and also that single-sideband transmission conserves transmitted power. However, the transmitted power necessary to produce a given receive signal-to-noise ratio (*S/N*) is a convenient means of comparing the power requirements and relative performance of sideband and conventional AM systems. The receive *S/N* determines the degree of intelligibility of a received signal.

    Peak envelope power (PEP) is the rms power developed at the crest of the modulation envelope. With conventional AM, the envelope contains 1 unit of carrier power and 0.25 units of power in each sideband for a total transmitted PEP = 1.5 units. A single-sideband transmitter rated at 0.5 unit of PEP will produce the same *S/N* at the output of the receiver as 1.5 units of carrier plus sideband power from a conventional AM system. In other words, the same performance is achieved with SSB using only one-third the transmitted power. Table 6-1 compares conventional AM with single side-

TABLE 6-1   CONVENTIONAL AM VERSUS SINGLE-SIDEBAND

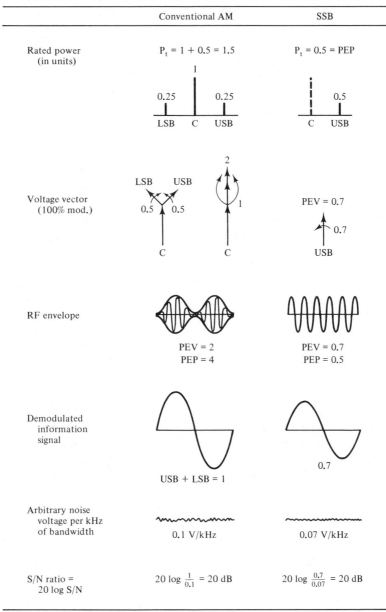

	Conventional AM	SSB
Rated power (in units)	$P_t = 1 + 0.5 = 1.5$	$P_t = 0.5 = \text{PEP}$
Voltage vector (100% mod.)		PEV = 0.7
RF envelope	PEV = 2 PEP = 4	PEV = 0.7 PEP = 0.5
Demodulated information signal	USB + LSB = 1	0.7
Arbitrary noise voltage per kHz of bandwidth	0.1 V/kHz	0.07 V/kHz
S/N ratio = 20 log S/N	$20 \log \frac{1}{0.1} = 20$ dB	$20 \log \frac{0.7}{0.07} = 20$ dB

band for a single frequency-modulating signal. The voltage vectors for the power requirements stated are shown. It can be seen that it requires 0.5 unit of voltage per sideband and 1 unit for the carrier with conventional AM for a total of 2 PEV units (peak envelope volts) and only 0.7 PEV for single sideband. The RF envelopes for AM and sideband transmission are also shown, which correspond to the voltage and power relationships outlined previously. The demodulated signal at the output of a conventional AM receiver is proportional to the quadratic sum of the upper and lower sideband signals, which equals 1 PEV unit. For sideband reception, the demodulated signal is $0.707(1) = 0.7$ PEV. If the noise voltage for conventional AM is arbitrarily chosen as 0.1 V/kHz, the noise voltage for a single-sideband waveform with half the bandwidth is 0.7 V/kHz. Consequently, the *S/N* performance for single sideband is equal to that of conventional AM.

CIRCUITS FOR GENERATING SSB

In the preceding section it was shown that with most single-sideband systems, the carrier is either totally suppressed or reduced to only a fraction of its original value. To remove the carrier from a modulated wave or to reduce its amplitude is difficult, if not impossible, without also removing a portion of each sideband (conventional notch filters simply do not have sufficient *Q*-factors). Therefore, modulator circuits that inherently remove the carrier during the modulation process have been developed. A circuit that produces a double-sideband signal with a suppressed carrier is the *balanced* modulator. The balanced modulator has rapidly become one of the most widely used circuits in electronic communications. In addition to suppressed-carrier AM systems, the balanced modulator is widely used in FM and PM systems. The balanced modulator is also used extensively in data modems and digital modulation systems such as phase shift keying (PSK) and quadrature amplitude modulation (QAM).

Balanced Ring Modulator

Figure 6-6a shows the schematic diagram for a *balanced ring modulator* constructed with diodes. Semiconductor diodes are ideally suited for use in balanced modulator circuits because they are stable, require no external power source, have a long life, and require very little maintenance. The balanced ring modulator is sometimes called a *balanced lattice modulator* or simply a *balanced modulator*. The balanced modulator has two inputs: a single frequency carrier and the modulating signal, which may be a single frequency or a complex waveform. For the balanced modulator to operate properly, the amplitude of the carrier must be sufficiently greater than the amplitude of the modulating signal (approximately six to seven times greater). This is to ensure that the carrier controls the "on" or "off" condition of the four diode switches (D_1 to D_4).

Circuit operation.   Essentially, diodes D_1 to D_4 are electronic switches that control whether the modulating signal is passed from transformer T_1 to transformer T_2 as is or with a 180° phase reversal. With the carrier (V_c) polarity as shown in Figure 6-6b, diode switches D_1 and D_2 are forward biased and "on," while diode switches D_3 and D_4 are reverse biased and "off." Consequently, the modulating signal (V_a) is transferred across the closed switches to T_2 as is. When the polarity of the carrier

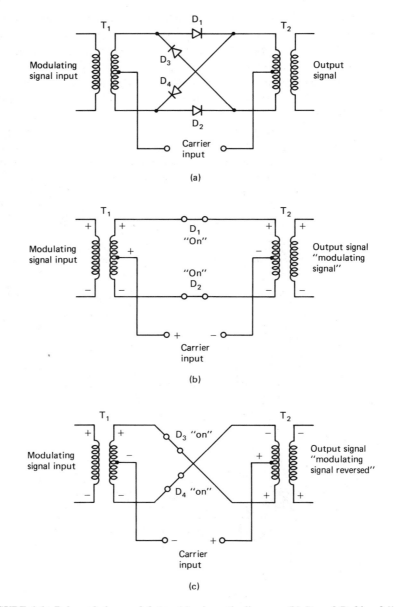

FIGURE 6-6 Balanced ring modulator: (a) schematic diagram; (b) D_1 and D_2 biased "on"; (c) D_3 and D_4 biased "on."

reverses, as shown in Figure 6-6c, diode switches D_1 and D_2 are reverse biased and "off," while diode switches D_3 and D_4 are forward biased and "on." Consequently, the modulating signal undergoes a 180° phase reversal before reaching T_2. Carrier current flows from its source to the center taps of T_1 and T_2, where it splits and goes in opposite directions through the upper and lower halves of the transformers. Thus their magnetic fields cancel in the secondary and the carrier component is removed. If the diodes are not perfectly matched or if the transformers are not tapped in their exact centers, the circuit is out of balance and the carrier is not totally suppressed. It is virtually impossible to achieve perfect balance; therefore, there is always a small carrier component in the output signal. This is commonly called *carrier leak*. The amount of carrier suppression is typically between 40 and 60 dB.

Figure 6-7 shows the balanced modulator waveforms for a single frequency-modulating signal, the carrier, the output waveform before filtering, and the output waveform after filtering. It can be seen that D_1 and D_2 conduct during the positive half-cycles of V_c, and D_3 and D_4 conduct during the negative half-cycles. The output of the modulator

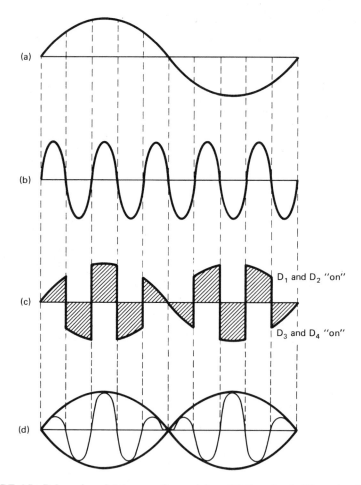

FIGURE 6-7   Balanced modulator waveforms: (a) modulating signal; (b) carrier signal; (c) output waveform before filtering; (d) output waveform after filtering.

consists of a series of RF pulses whose repetition rate is determined by the switching or RF carrier frequency and whose amplitude is controlled by the level of the modulating signal. Consequently, the output waveform takes the shape of the modulating signal except with alternating positive and negative polarities that correspond to the polarity of V_c.

Mathematical analysis.   A balanced modulator is a *product* modulator; the output signal is the product of the two input signals. The carrier frequency is multiplied by the modulating signal. Mathematically, the output of a balanced modulator for a single frequency-modulating signal is

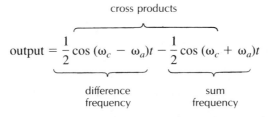

$$\text{output} = \underbrace{(\sin \omega_a t)}_{\substack{\text{modulating} \\ \text{signal}}} \times \underbrace{(\sin \omega_c t)}_{\substack{\text{carrier} \\ \text{signal}}}$$

Using the trigonometric identity for the product of two sine functions, the output is

$$\text{output} = \overbrace{\underbrace{\frac{1}{2} \cos (\omega_c - \omega_a)t}_{\substack{\text{difference} \\ \text{frequency}}} - \underbrace{\frac{1}{2} \cos (\omega_c + \omega_a)t}_{\substack{\text{sum} \\ \text{frequency}}}}^{\text{cross products}}$$

From the preceding mathematical operation, it can be seen that the output signal is simply two cosine waves; one is the sum frequency ($\omega_c + \omega_a$) and the other is the difference frequency ($\omega_c - \omega_a$). Both the carrier and the modulating signal are suppressed.

FET Push-Pull Balanced Modulator

Figure 6-8 shows a schematic diagram for a balanced modulator that uses FETs (field-effect transistors) rather than switching diodes for the nonlinear devices. A FET is a nonlinear device that exhibits square-law properties. That is, its output contains a term that is proportional to the input signal squared, which includes, of course, the second-order cross-product frequencies (i.e., the primary sum and difference frequencies). Like the diode balanced modulator, the FET modulator is a product modulator and produces only the cross-product frequencies at its output. The FET balanced modulator is similar to a standard push-pull amplifier except that the modulator circuit has two inputs (the carrier and the modulating signal).

Circuit operation.   The carrier is fed into the circuit in such a way that it is applied simultaneously and in phase to the gates of both FET amplifiers (Q_1 and Q_2). The carrier produces currents in the top and bottom halves of the output transformer that are equal in magnitude but 180° out of phase. Therefore, they cancel and no carrier component appears in the output. The modulating signal is applied to the circuit in

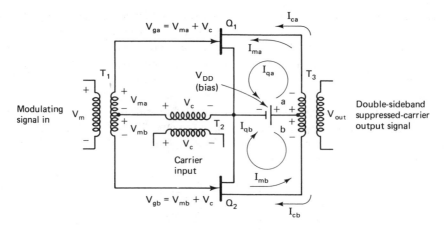

FIGURE 6-8   FET balanced modulator. For the polarities shown:

$$I_{ta} = I_{qa} + I_{ca} + I_{ma}$$
$$I_{tb} = -I_{qb} - I_{cb} + I_{mb}$$
$$I_t = I_{ma} + I_{mb} = 2I_m$$

V_{out} **is proportional to the modulating current (I_{ma} and I_{mb}).**

such a way that it is applied simultaneously to the two gates 180° out of phase. The modulating signal causes an increase in the drain current in one FET and, at the same time, causes a decrease in the drain current in the other FET.

Figure 6-9 shows the phasor diagrams for the currents produced in the output transformer of a FET balanced modulator. Figure 6-9a shows that the quiescent dc drain currents from Q_1 and Q_2 (I_{qa} and I_{qb}) pass through their respective halves of the primary winding of T_3 180° out of phase with each other. Figure 6-9a also shows that an increase in drain current due to the carrier signal (I_{ca} and I_{cb}) adds to the quiescent current in both halves of the transformer windings, producing currents (I_{ta} and I_{tb}) that are equal and simply the sum of the quiescent and carrier currents. I_{ta} and I_{tb} are equal but travel in opposite directions; consequently, they cancel each other. Figure 6-9b shows the phasor sum of the quiescent and carrier currents when the carrier currents travel in the opposite direction to the quiescent currents. The total currents in both halves of the winding are still equal in magnitude, but now they are equal to the difference between the quiescent and carrier currents. Figure 6-9c shows the phasor diagram when a current component is added due to a modulating signal. The modulating signal currents (I_{ma} and I_{mb}) produce in their respective halves of the output transformer currents that are in phase with each other. However, it can be seen that in one half of the windings the total current is equal to the sum of the dc current, the carrier current, and the modulating signal current; and in the other half of the windings the total current is equal to the difference between the dc and carrier currents and the modulating signal current. Thus the dc and carrier currents cancel in the secondary windings, while the sum and difference components remain. The continuously changing carrier and modulating signal currents produce the cross-product frequencies.

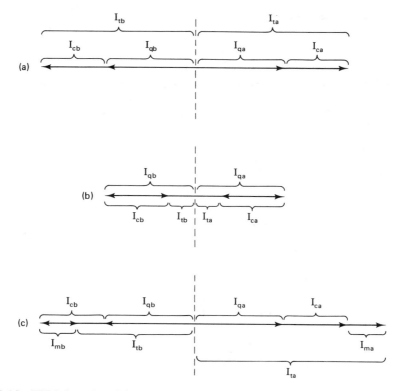

FIGURE 6-9 FET balanced modulator phasor diagrams: (a) in-phase sum of dc and carrier currents; (b) out-of-phase sum of dc and carrier currents; (c) sum of dc, carrier, and modulating signal currents.

The carrier and modulating signal polarities shown in Figure 6-8 produce an output current that is proportional to the carrier and modulating signal voltages. The carrier signal (V_c) produces a current in both FETs (I_{ca} and I_{cb}) that is in the same direction as the quiescent currents (I_{qa} and I_{qb}). The modulating signal (V_{ma} and V_{mb}) produces a current in Q_1 (I_{ma}) that is in the same direction as I_{ca} and I_{qa} and a current in Q_2 (I_{mb}) that is in the opposite direction as I_{cb} and I_{qb}. Therefore, the total current through the a-side of T_3 is $I_{ta} = I_{ca} + I_{qa} + I_{ma}$ and the total current through the b-side of T_3 is $I_{tb} = -I_{cb} - I_{qb} + I_{mb}$. Thus the net current through the primary winding of T_3 is $I_{ta} + I_{tb} = I_{ma} + I_{mb}$. For a modulating signal with the opposite polarity, the drain current in Q_2 will increase and in Q_1 it will decrease. Ignoring the quiescent dc current (I_{qa} and I_{qb}), the drain current in one FET is the sum of the carrier and modulating signal currents ($I_c + I_m$), and the drain current in the other FET is the difference between the carrier and modulating signal currents ($I_c - I_m$).

T_1 is an audio transformer while T_2 and T_3 are radio-frequency transformers. Therefore, any audio component that appears at the drain circuits of Q_1 and Q_2 is not passed on to the output. To achieve total carrier suppression, Q_1 and Q_2 must be perfectly matched and T_1 and T_3 must be tapped in their exact centers. With both the diode and FET balanced modulators, between 40 and 60 dB of carrier suppression is typical.

Balanced Bridge Modulator

Figure 6-10a shows the schematic diagram for a *balanced bridge modulator*. The operation of the bridge modulator, like that of the balanced ring modulator, is completely dependent on the switching action of diodes D_1 through D_4 under the influence of the carrier and modulating signal voltages. Again, the carrier voltage controls the "on" or "off" condition of the diodes and therefore must be appreciably larger than the modulating signal voltage.

      Circuit operation.   For the carrier polarities shown in Figure 6-10b, all four diodes are reverse biased and "off." Consequently, the audio signal voltage is transferred

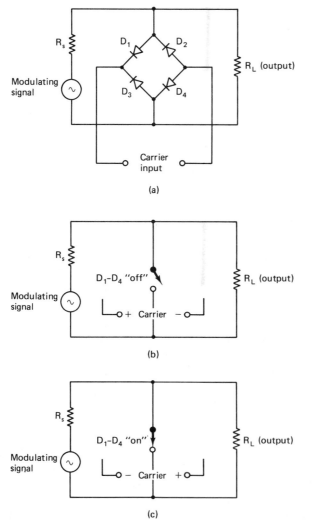

(a)

(b)

(c)

FIGURE 6-10 Balanced bridge modulator: (a) schematic diagram; (b) diodes biased "off"; (c) diodes biased "on."

directly to the load resistor (R_L). Figure 6-10c shows the equivalent circuit for a carrier with the opposite polarity. All four diodes are forward biased and "on," and the load resistor is essentially shorted out. As the carrier voltage changes from positive to negative, and vice versa, the output waveform contains a series of pulses that are comprised mainly of the upper and lower sideband frequencies.

Linear Integrated Circuit (LIC) Balanced Modulator

LIC balanced modulators are available, such as the LM1496, that can provide carrier suppression of 50 dB at 10 MHz and up to 65 dB at 500 kHz. The LM1496 balanced modulator integrated circuit is a *double-balanced modulator-demodulator* which produces an output signal that is proportional to the product of its input signals (the carrier and modulating signal). Integrated circuits are ideally suited for applications that require a balanced operation.

    Circuit operation.   Figure 6-11 shows a simplified schematic diagram for a *differential* amplifier, which is the fundamental circuit of an LIC balanced modulator because of its excellent *common-mode rejection ratio* (typically 85 dB). When a carrier signal is applied to the base of Q_1, the emitter current in Q_1 and Q_2 will vary by the

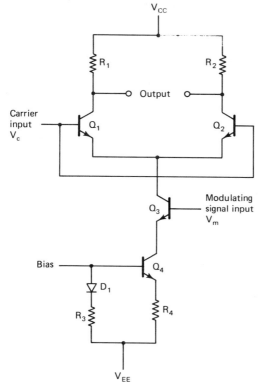

FIGURE 6-11 Differential amplifier schematic.

same amount. Because the emitter current for both Q_1 and Q_2 comes from a common current source (Q_4), any increase in Q_1's emitter current results in a corresponding decrease in Q_2's emitter current, and vice versa. Similarly, when a carrier signal is applied to the base of Q_2, the emitter currents in Q_1 and Q_2 will vary by the same amount except in opposite directions. Consequently, if the same carrier signal is fed simultaneously to the base of Q_1 and Q_2, the respective increases and decreases in the emitter currents of Q_1 and Q_2 are equal and cancel each other. Therefore, the collector currents and output voltage remain unchanged. If a modulating signal is applied to the base of Q_3, it causes a corresponding increase or decrease (depending on its polarity) in the collector currents of Q_1 and Q_2. However, the carrier and modulating signal frequencies mix in Q_1 and Q_2 and produce cross-product frequencies in the output. Therefore, the carrier and modulating signal frequencies are canceled in the balanced transistors, while the sum and difference frequencies appear at the output.

Figure 6-12a shows the schematic diagram for a typical AM DSBSC modulator using the LM1496 integrated circuit. The carrier signal is applied to pin 10, which, in conjunction with pin 8, provides an input to a quad cross-coupled differential output amplifier. This configuration is used to ensure that full-wave multiplication of the carrier and modulating signal occurs. The modulating signal is applied to pin 1, which, in conjunction with pin 4, provides a differential input to the current driving transistors for the output difference amplifiers. The 1-kΩ resistor connected between pins 2 and 3 sets the gain of the difference amplifier. The 50-kΩ potentiometer, in conjunction with V_{EE} (-8 V dc), is used to balance the bias currents for the difference amplifiers and null the carrier. Pins 6 and 12 are single-ended outputs which contain carrier and sideband components. When one of the outputs is inverted and added to the other output, the carrier is suppressed and a double-sideband suppressed carrier wave is produced. Such a process is accomplished in the op-amp subtractor circuit. The subtractor inverts the signal at the inverting ($-$) input and adds it to the signal at the noninverting ($+$) input. Thus a double-sideband suppressed carrier wave appears at the output of the op-amp. The 6.8-kΩ resistor connected to pin 5 is a bias resistor for the internal constant current supply. The specification sheet for Motorola's 1496 integrated-circuit balanced modulator chip is shown in Figure 6-12b.

SINGLE-SIDEBAND TRANSMITTERS

The transmitters used for SSBSC and SSBRC transmission are identical except that the reinserted carrier transmitters have an additional circuit which adds a low-amplitude carrier to the single-sideband waveform after suppressed carrier modulation has been performed. The reinserted carrier is called a pilot carrier. The circuit where the carrier is reinserted is called a linear summer if it is a resistive network and a *hybrid coil* if the SSB waveform and the pilot are inductively combined in a transformer bridge. There are three techniques that are commonly used for single-sideband generation: the *filter method*, the *phase shift method*, and the so-called *"third method."*

SSB Transmitter: Filter Method

Figure 6-13 shows a block diagram for a SSB transmitter that uses balanced modulators to suppress the carrier and bandpass filters to suppress the unwanted sideband. The modulating signal is an audio spectrum that extends from 0 to 5 kHz. In the first balanced modulator (balanced modulator 1), the modulating signal mixes with a low-frequency (LF) 100-kHz carrier to produce a single-sideband suppressed-carrier spectrum that is centered around the suppressed 100-kHz IF carrier. Bandpass filter 1 (BPF 1) is tuned to a 5-kHz bandwidth centered around 102.5 kHz, which is the center frequency of the upper sideband spectrum. The pilot or reduced amplitude carrier is added to the single-sideband waveform in the carrier reinsertion stage, which is simply a linear summer. The summer is a simple adder circuit that combines the 100-kHz pilot carrier to the 100- to 105-kHz upper sideband spectrum. Thus the output of the summer is a SSBRC waveform. (If suppressed-carrier transmission is used, the carrier pilot and summer circuit are omitted.) The low-frequency IF is converted to the final operating frequency through a series of frequency translations. First, the SSBRC waveform is mixed in balanced modulator 2 with a 2-MHz medium-frequency carrier (MF). The output is a double-sideband suppressed-carrier signal where the upper and lower sidebands each contain the original SSBRC signal spectrum. The upper and lower sidebands are

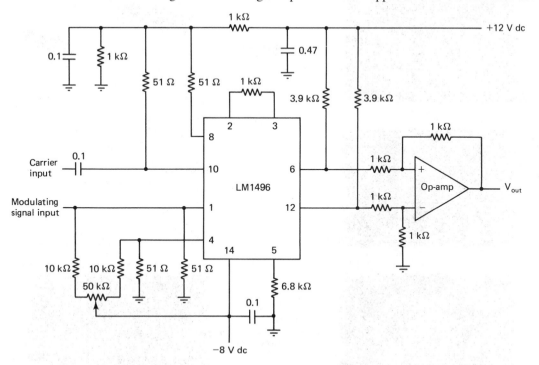

FIGURE 6-12   AM DSBSC modulator using the LM1496 linear integrated circuit: (a) schematic diagram; (b) 1496 specification sheets. (Copyright Motorola, Inc., 1982. Used by permission.)

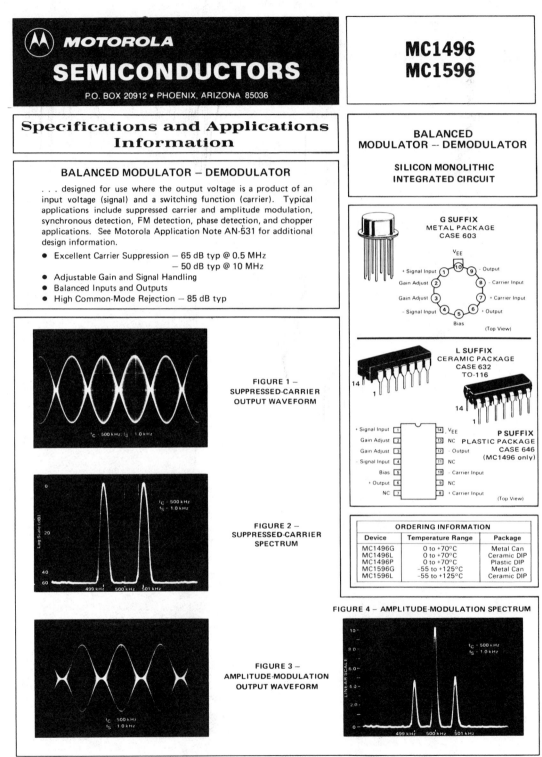

MOTOROLA SEMICONDUCTORS

P.O. BOX 20912 • PHOENIX, ARIZONA 85036

MC1496
MC1596

Specifications and Applications Information

BALANCED MODULATOR — DEMODULATOR

. . . designed for use where the output voltage is a product of an input voltage (signal) and a switching function (carrier). Typical applications include suppressed carrier and amplitude modulation, synchronous detection, FM detection, phase detection, and chopper applications. See Motorola Application Note AN-531 for additional design information.

- Excellent Carrier Suppression — 65 dB typ @ 0.5 MHz
  — 50 dB typ @ 10 MHz
- Adjustable Gain and Signal Handling
- Balanced Inputs and Outputs
- High Common-Mode Rejection — 85 dB typ

BALANCED MODULATOR -- DEMODULATOR

SILICON MONOLITHIC INTEGRATED CIRCUIT

G SUFFIX
METAL PACKAGE
CASE 603

V_{EE}

+ Signal Input (1) (10) - Output
Gain Adjust (2) (9) - Carrier Input
Gain Adjust (3) (8) + Carrier Input
- Signal Input (4) (7) + Output
Bias (5) (6)

(Top View)

L SUFFIX
CERAMIC PACKAGE
CASE 632
TO-116

P SUFFIX
PLASTIC PACKAGE
CASE 646
(MC1496 only)

Pin		Pin	
+ Signal Input	1	14	V_{EE}
Gain Adjust	2	13	NC
Gain Adjust	3	12	- Output
- Signal Input	4	11	NC
Bias	5	10	- Carrier Input
+ Output	6	9	NC
NC	7	8	+ Carrier Input

(Top View)

ORDERING INFORMATION

Device	Temperature Range	Package
MC1496G	0 to +70°C	Metal Can
MC1496L	0 to +70°C	Ceramic DIP
MC1496P	0 to +70°C	Plastic DIP
MC1596G	-55 to +125°C	Metal Can
MC1596L	-55 to +125°C	Ceramic DIP

FIGURE 1 — SUPPRESSED-CARRIER OUTPUT WAVEFORM

f_C = 500 kHz; f_S = 1.0 kHz

FIGURE 2 — SUPPRESSED-CARRIER SPECTRUM

f_C = 500 kHz
f_S = 1.0 kHz

499 kHz   500 kHz   501 kHz

FIGURE 3 — AMPLITUDE-MODULATION OUTPUT WAVEFORM

f_C = 500 kHz
f_S = 1.0 kHz

FIGURE 4 — AMPLITUDE-MODULATION SPECTRUM

f_C = 500 kHz
f_S = 1.0 kHz

499 kHz   500 kHz   501 kHz

DS 9132 R2

FIGURE 6-12(b)

MAXIMUM RATINGS* (T_A = +25°C unless otherwise noted)

Rating	Symbol	Value	Unit
Applied Voltage ($V_6 - V_7$, $V_8 - V_1$, $V_9 - V_7$, $V_9 - V_8$, $V_7 - V_4$, $V_7 - V_1$, $V_8 - V_4$, $V_6 - V_8$, $V_2 - V_5$, $V_3 - V_5$)	ΔV	30	Vdc
Differential Input Signal	$V_7 - V_8$ $V_4 - V_1$	+5.0 $\pm(5 + I_5 R_e)$	Vdc
Maximum Bias Current	I_5	10	mA
Thermal Resistance, Junction to Air Ceramic Dual In-Line Package Plastic Dual In-Line Package Metal Package	$R_{\theta JA}$	 180 100 200	°C/W
Operating Temperature Range MC1496 MC1596	T_A	 0 to +70 −55 to +125	°C
Storage Temperature Range	T_{stg}	−65 to +150	°C

ELECTRICAL CHARACTERISTICS* (V_{CC} = +12 Vdc, V_{EE} = −8.0 Vdc, I_5 = 1.0 mAdc, R_L = 3.9 kΩ, R_e = 1.0 kΩ, T_A = +25°C unless otherwise noted) (All input and output characteristics are single-ended unless otherwise noted.)

Characteristic	Fig	Note	Symbol	MC1596 Min	MC1596 Typ	MC1596 Max	MC1496 Min	MC1496 Typ	MC1496 Max	Unit
Carrier Feedthrough V_C = 60 mV(rms) sine wave and f_C = 1.0 kHz offset adjusted to zero f_C = 10 MHz	5	1	V_{CFT}	 − −	 40 140	 − −	 − −	 40 140	 − −	µV(rms)
V_C = 300 mVp-p square wave: offset adjusted to zero f_C = 1.0 kHz offset not adjusted f_C = 1.0 kHz				 − −	 0.04 20	 0.2 100	 − −	 0.04 20	 0.4 200	mV(rms)
Carrier Suppression f_S = 10 kHz, 300 mV(rms) f_C = 500 kHz, 60 mV(rms) sine wave f_C = 10 MHz, 60 mV(rms) sine wave	5	2	V_{CS}	 50 −	 65 50	 − −	 40 −	 65 50	 − −	dB k
Transadmittance Bandwidth (Magnitude) (R_L = 50 ohms) Carrier Input Port, V_C = 60 mV(rms) sine wave f_S = 1.0 kHz, 300 mV(rms) sine wave Signal Input Port, V_S = 300 mV(rms) sine wave $|V_C|$ = 0.5 Vdc	8	8	BW_{3dB}	 − −	 300 80	 − −	 − −	 300 80	 − −	MHz
Signal Gain V_S = 100 mV(rms), f = 1.0 kHz; $|V_C|$ = 0.5 Vdc	10	3	A_{VS}	2.5	3.5	−	2.5	3.5	−	V/V
Single-Ended Input Impedance, Signal Port, f = 5.0 MHz Parallel Input Resistance Parallel Input Capacitance	6	−	 r_{ip} c_{ip}	 − −	 200 2.0	 − −	 − −	 200 2.0	 − −	 kΩ pF
Single-Ended Output Impedance, f = 10 MHz Parallel Output Resistance Parallel Output Capacitance	6	−	 r_{op} c_{op}	 − −	 40 5.0	 − −	 − −	 40 5.0	 − −	 kΩ pF
Input Bias Current $I_{bS} = \frac{I_1 + I_4}{2}$; $I_{bC} = \frac{I_7 + I_8}{2}$	7	−	 I_{bS} I_{bC}	 − −	 12 12	 25 25	 − −	 12 12	 30 30	µA
Input Offset Current $I_{ioS} = I_1 - I_4$; $I_{ioC} = I_7 - I_8$	7	−	 $|I_{ioS}|$ $|I_{ioC}|$	 − −	 0.7 0.7	 5.0 5.0	 − −	 0.7 0.7	 7.0 7.0	µA
Average Temperature Coefficient of Input Offset Current (T_A = −55°C to +125°C)	7	−	$|TC_{Iio}|$	−	2.0	−	−	2.0	−	nA/°C
Output Offset Current ($I_6 - I_9$)	7	−	$|I_{oo}|$	−	14	50	−	14	80	µA
Average Temperature Coefficient of Output Offset Current (T_A = −55°C to +125°C)	7	−	$|TC_{Ioo}|$	−	90	−	−	90	−	nA/°C
Common-Mode Input Swing, Signal Port, f_S = 1.0 kHz	9	4	CMV	−	5.0	−	−	5.0	−	Vp-p
Common-Mode Gain, Signal Port, f_S = 1.0 kHz, $|V_C|$ = 0.5 Vdc	9	−	ACM	−	−85	−	−	−85	−	dB
Common-Mode Quiescent Output Voltage (Pin 6 or Pin 9)	10	−	V_o	−	8.0	−	−	8.0	−	Vdc
Differential Output Voltage Swing Capability	10	−	V_{out}	−	8.0	−	−	8.0	−	Vp-p
Power Supply Current $I_6 + I_9$ I_{10}	7	6	 I_{CC} I_{EE}	 − −	 2.0 3.0	 3.0 4.0	 − −	 2.0 3.0	 4.0 5.0	mAdc
DC Power Dissipation	7	5	P_D	−	33	−	−	33	−	mW

*Pin number references pertain to this device when packaged in a metal can. To ascertain the corresponding pin numbers for plastic or ceramic packaged devices refer to the first page of this specification sheet.

Ⓜ **MOTOROLA** *Semiconductor Products Inc.*

FIGURE 6-12(b) *(continued)*

GENERAL OPERATING INFORMATION *

Note 1 — Carrier Feedthrough

Carrier feedthrough is defined as the output voltage at carrier frequency with only the carrier applied (signal voltage = 0).

Carrier null is achieved by balancing the currents in the differential amplifier by means of a bias trim potentiometer (R_1 of Figure 5).

Note 2 — Carrier Suppression

Carrier suppression is defined as the ratio of each sideband output to carrier output for the carrier and signal voltage levels specified.

Carrier suppression is very dependent on carrier input level, as shown in Figure 22. A low value of the carrier does not fully switch the upper switching devices, and results in lower signal gain, hence lower carrier suppression. A higher than optimum carrier level results in unnecessary device and circuit carrier feedthrough, which again degenerates the suppression figure. The MC1596 has been characterized with a 60 mV(rms) sinewave carrier input signal. This level provides optimum carrier suppression at carrier frequencies in the vicinity of 500 kHz, and is generally recommended for balanced modulator applications.

Carrier feedthrough is independent of signal level, V_S. Thus carrier suppression can be maximized by operating with large signal levels. However, a linear operating mode must be maintained in the signal-input transistor pair — or harmonics of the modulating signal will be generated and appear in the device output as spurious sidebands of the suppressed carrier. This requirement places an upper limit on input-signal amplitude (see Note 3 and Figure 20). Note also that an optimum carrier level is recommended in Figure 22 for good carrier suppression and minimum spurious sideband generation.

At higher frequencies circuit layout is very important in order to minimize carrier feedthrough. Shielding may be necessary in order to prevent capacitive coupling between the carrier input leads and the output leads.

Note 3 — Signal Gain and Maximum Input Level

Signal gain (single-ended) at low frequencies is defined as the voltage gain,

$$A_{VS} = \frac{V_o}{V_S} = \frac{R_L}{R_e + 2r_e} \text{ where } r_e = \frac{26 \text{ mV}}{I_5 \text{ (mA)}}$$

A constant dc potential is applied to the carrier input terminals to fully switch two of the upper transistors "on" and two transistors "off" ($V_C = 0.5$ Vdc). This in effect forms a cascode differential amplifier.

Linear operation requires that the signal input be below a critical value determined by R_E and the bias current I_5.

$$V_S \leq I_5 R_E \text{ (Volts peak)}$$

Note that in the test circuit of Figure 10, V_S corresponds to a maximum value of 1 volt peak.

Note 4 — Common-Mode Swing

The common-mode swing is the voltage which may be applied to both bases of the signal differential amplifier, without saturating the current sources or without saturating the differential amplifier itself by swinging it into the upper switching devices. This swing is variable depending on the particular circuit and biasing conditions chosen (see Note 6).

Note 5 — Power Dissipation

Power dissipation, P_D, within the integrated circuit package should be calculated as the summation of the voltage-current products at each port, i.e. assuming $V_9 = V_6$, $I_5 = I_6 = I_9$ and ignoring

base current, $P_D = 2 I_5 (V_6 - V_{10}) + I_5 (V_5 - V_{10})$ where subscripts refer to pin numbers.

Note 6 — Design Equations

The following is a partial list of design equations needed to operate the circuit with other supply voltages and input conditions. See Note 3 for R_e equation.

A. Operating Current

The internal bias currents are set by the conditions at pin 5. Assume:

$$I_5 = I_6 = I_9$$

$$I_B \ll I_C \text{ for all transistors}$$

then:

$$R_5 = \frac{V^- - \phi}{I_5} - 500 \ \Omega \quad \text{where: } R_5 \text{ is the resistor between pin } 5 \text{ and ground}$$
$$\phi = 0.75 \text{ V at } T_A = +25^\circ C$$

The MC1596 has been characterized for the condition $I_5 = 1.0$ mA and is the generally recommended value.

B. Common-Mode Quiescent Output Voltage

$$V_6 = V_9 = V^+ - I_5 R_L$$

Note 7 — Biasing

The MC1596 requires three dc bias voltage levels which must be set externally. Guidelines for setting up these three levels include maintaining at least 2 volts collector-base bias on all transistors while not exceeding the voltages given in the absolute maximum rating table;

$$30 \text{ Vdc} \geq [(V_6, V_9) - (V_7, V_8)] \geq 2 \text{ Vdc}$$

$$30 \text{ Vdc} \geq [(V_7, V_8) - (V_1, V_4)] \geq 2.7 \text{ Vdc}$$

$$30 \text{ Vdc} \geq [(V_1, V_4) - (V_5)] \geq 2.7 \text{ Vdc}$$

The foregoing conditions are based on the following approximations:

$$V_6 = V_9, \quad V_7 = V_8, \quad V_1 = V_4$$

Bias currents flowing into pins 1, 4, 7, and 8 are transistor base currents and can normally be neglected if external bias dividers are designed to carry 1.0 mA or more.

Note 8 — Transadmittance Bandwidth

Carrier transadmittance bandwidth is the 3-dB bandwidth of the device forward transadmittance as defined by:

$$y_{21C} = \frac{i_o \text{ (each sideband)}}{v_s \text{ (signal)}} \Bigg|_{V_O = 0}$$

Signal transadmittance bandwidth is the 3-dB bandwidth of the device forward transadmittance as defined by:

$$y_{21S} = \frac{i_o \text{ (signal)}}{v_s \text{ (signal)}} \Bigg|_{V_c = 0.5 \text{ Vdc, } V_O = 0}$$

---

*Pin number references pertain to this device when packaged in a metal can. To ascertain the corresponding pin numbers for plastic or ceramic packaged devices refer to the first page of this specification sheet.

(M) MOTOROLA *Semiconductor Products Inc.*

FIGURE 6-12(b) *(continued)*

Note 9 — Coupling and Bypass Capacitors C_1 and C_2

Capacitors C_1 and C_2 (Figure 5) should be selected for a reactance of less than 5.0 ohms at the carrier frequency.

Note 10 — Output Signal, V_o

The output signal is taken from pins 6 and 9, either balanced or single-ended. Figure 12 shows the output levels of each of the two output sidebands resulting from variations in both the carrier and modulating signal inputs with a single-ended output connection.

Note 11 — Negative Supply, V_{EE}

V_{EE} should be dc only. The insertion of an RF choke in series with V_{EE} can enhance the stability of the internal current sources.

Note 12 — Signal Port Stability

Under certain values of driving source impedance, oscillation may occur. In this event, an RC suppression network should be connected directly to each input using short leads. This will reduce the Q of the source-tuned circuits that cause the oscillation.

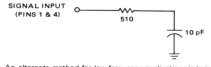

An alternate method for low-frequency applications is to insert a 1 k-ohm resistor in series with the inputs, pins 1 and 4. In this case input current drift may cause serious degradation of carrier suppression.

TEST CIRCUITS

FIGURE 5 — CARRIER REJECTION AND SUPPRESSION

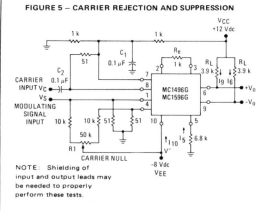

NOTE: Shielding of input and output leads may be needed to properly perform these tests.

FIGURE 6 — INPUT-OUTPUT IMPEDANCE

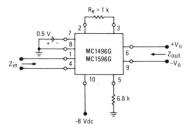

FIGURE 7 — BIAS AND OFFSET CURRENTS   FIGURE 8 — TRANSCONDUCTANCE BANDWIDTH

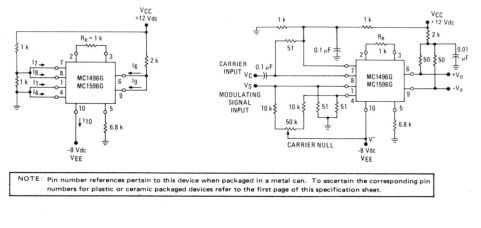

NOTE: Pin number references pertain to this device when packaged in a metal can. To ascertain the corresponding pin numbers for plastic or ceramic packaged devices refer to the first page of this specification sheet.

Ⓜ **MOTOROLA** *Semiconductor Products Inc.*

FIGURE 6-12(b) *(continued)*

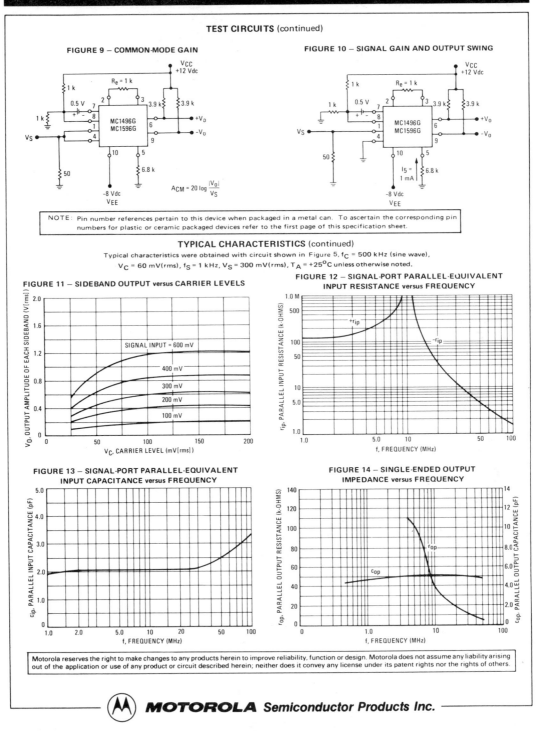

TEST CIRCUITS (continued)

FIGURE 9 – COMMON-MODE GAIN

FIGURE 10 – SIGNAL GAIN AND OUTPUT SWING

$A_{CM} = 20 \log \frac{|V_0|}{V_S}$

NOTE: Pin number references pertain to this device when packaged in a metal can. To ascertain the corresponding pin numbers for plastic or ceramic packaged devices refer to the first page of this specification sheet.

TYPICAL CHARACTERISTICS (continued)

Typical characteristics were obtained with circuit shown in Figure 5, f_C = 500 kHz (sine wave), V_C = 60 mV(rms), f_S = 1 kHz, V_S = 300 mV(rms), T_A = +25°C unless otherwise noted.

FIGURE 11 – SIDEBAND OUTPUT versus CARRIER LEVELS

FIGURE 12 – SIGNAL-PORT PARALLEL-EQUIVALENT INPUT RESISTANCE versus FREQUENCY

FIGURE 13 – SIGNAL-PORT PARALLEL-EQUIVALENT INPUT CAPACITANCE versus FREQUENCY

FIGURE 14 – SINGLE-ENDED OUTPUT IMPEDANCE versus FREQUENCY

MOTOROLA *Semiconductor Products Inc.*

FIGURE 6-12(b) *(continued)*

238

TYPICAL CHARACTERISTICS (continued)

Typical characteristics were obtained with circuit shown in Figure 5, f_C = 500 kHz (sine wave),
V_C = 60 mV(rms), f_S = 1 kHz, V_S = 300 mV(rms), T_A = +25°C unless otherwise noted.

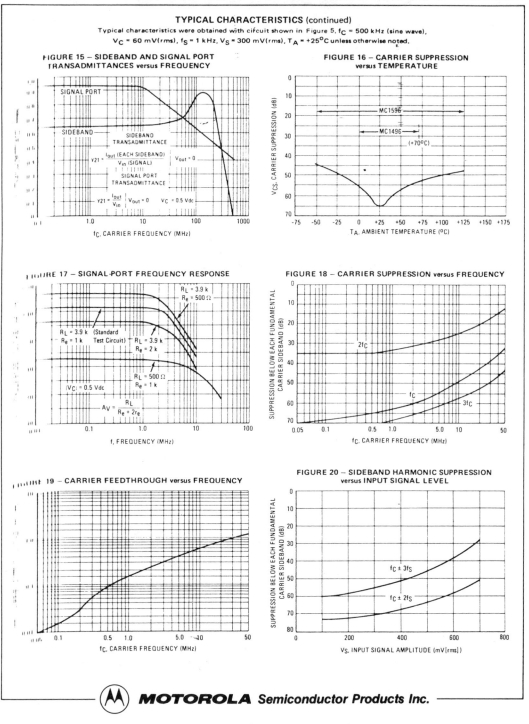

FIGURE 15 – SIDEBAND AND SIGNAL PORT
TRANSADMITTANCES versus FREQUENCY

FIGURE 16 – CARRIER SUPPRESSION
versus TEMPERATURE

FIGURE 17 – SIGNAL-PORT FREQUENCY RESPONSE

FIGURE 18 – CARRIER SUPPRESSION versus FREQUENCY

FIGURE 19 – CARRIER FEEDTHROUGH versus FREQUENCY

FIGURE 20 – SIDEBAND HARMONIC SUPPRESSION
versus INPUT SIGNAL LEVEL

Ⓜ **MOTOROLA** *Semiconductor Products Inc.*

FIGURE 6-12(b) *(continued)*

239

TYPICAL CHARACTERISTICS (continued)

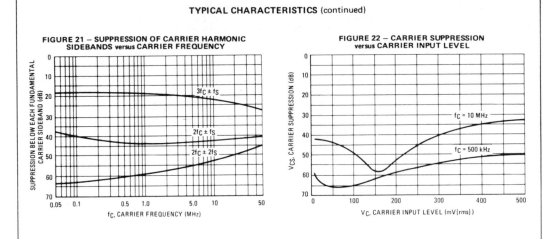

FIGURE 21 – SUPPRESSION OF CARRIER HARMONIC
SIDEBANDS versus CARRIER FREQUENCY

FIGURE 22 – CARRIER SUPPRESSION
versus CARRIER INPUT LEVEL

OPERATIONS INFORMATION

The MC1596/MC1496, a monolithic balanced modulator circuit, is shown in Figure 23.

This circuit consists of an upper quad differential amplifier driven by a standard differential amplifier with dual current sources. The output collectors are cross-coupled so that full-wave balanced multiplication of the two input voltages occurs. That is, the output signal is a constant times the product of the two input signals.

Mathematical analysis of linear ac signal multiplication indicates that the output spectrum will consist of only the sum and difference of the two input frequencies. Thus, the device may be used as a balanced modulator, doubly balanced mixer, product detector, frequency doubler, and other applications requiring these particular output signal characteristics.

The lower differential amplifier has its emitters connected to the package pins so that an external emitter resistance may be used. Also, external load resistors are employed at the device output.

Signal Levels

The upper quad differential amplifier may be operated either in a linear or a saturated mode. The lower differential amplifier is operated in a linear mode for most applications.

For low-level operation at both input ports, the output signal will contain sum and difference frequency components and have an amplitude which is a function of the product of the input signal amplitudes.

For high-level operation at the carrier input port and linear operation at the modulating signal port, the output signal will contain sum and difference frequency components of the modulating signal frequency and the fundamental and odd harmonics of the carrier frequency. The output amplitude will be a constant times the modulating signal amplitude. Any amplitude variations in the carrier signal will not appear in the output.

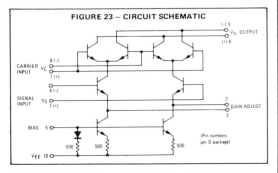

FIGURE 23 – CIRCUIT SCHEMATIC

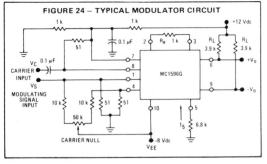

FIGURE 24 – TYPICAL MODULATOR CIRCUIT

NOTE: Pin number references pertain to this device when packaged in a metal can. To ascertain the corresponding pin numbers for plastic or ceramic packaged devices refer to the first page of this specification sheet.

 MOTOROLA *Semiconductor Products Inc.*

FIGURE 6-12(b) *(continued)*

OPERATIONS INFORMATION (continued)

The linear signal handling capabilities of a differential amplifier are well defined. With no emitter degeneration, the maximum input voltage for linear operation is approximately 25 mV peak. Since the upper differential amplifier has its emitters internally connected, this voltage applies to the carrier input port for all conditions.

Since the lower differential amplifier has provisions for an external emitter resistance, its linear signal handling range may be adjusted by the user. The maximum input voltage for linear operation may be approximated from the following expression:

$$V = \left(I_5\right)\left(R_E\right) \text{volts peak}.$$

This expression may be used to compute the minimum value of R_E for a given input voltage amplitude.

The gain from the modulating signal input port to the output is the MC1596/MC1496 gain parameter which is most often of interest to the designer. This gain has significance only when the lower differential amplifier is operated in a linear mode, but this includes most applications of the device.

As previously mentioned, the upper quad differential amplifier may be operated either in a linear or a saturated mode. Approximate gain expressions have been developed for the MC1596/MC1496 for a low-level modulating signal input and the following carrier input conditions:

1) Low-level dc
2) High-level dc
3) Low-level ac
4) High-level ac

These gains are summarized in Table 1, along with the frequency components contained in the output signal.

NOTES:
1. Low-level Modulating Signal, V_M, assumed in all cases. V_C is Carrier Input Voltage.
2. When the output signal contains multiple frequencies, the gain expression given is for the output amplitude of each of the two desired outputs, $f_C + f_M$ and $f_C - f_M$.
3. All gain expressions are for a single-ended output. For a differential output connection, multiply each expression by two.
4. R_L = Load resistance.
5. R_E = Emitter resistance between pins 2 and 3.
6. r_e = Transistor dynamic emitter resistance, at +25°C;

$$r_e \approx \frac{26 \text{ mV}}{I_5 \text{ (mA)}}$$

7. K = Boltzmann's Constant, T = temperature in degrees Kelvin, q = the charge on an electron.

$$\frac{KT}{q} \approx 26 \text{ mV at room temperature}$$

FIGURE 25 – TABLE 1
VOLTAGE GAIN AND OUTPUT FREQUENCIES

Carrier Input Signal (V_C)	Approximate Voltage Gain	Output Signal Frequency(s)
Low-level dc	$\dfrac{R_L \, V_C}{2(R_E + 2r_e)\left(\frac{KT}{q}\right)}$	f_M
High-level dc	$\dfrac{R_L}{R_E + 2r_e}$	f_M
Low-level ac	$\dfrac{R_L \, V_C(\text{rms})}{2\sqrt{2}\left(\frac{KT}{q}\right)(R_E + 2r_e)}$	$f_C \pm f_M$
High-level ac	$\dfrac{0.637 \, R_L}{R_E + 2r_e}$	$f_C \pm f_M$, $3f_C \pm f_M$, $5f_C \pm f_M$, . . .

APPLICATIONS INFORMATION

Double sideband suppressed carrier modulation is the basic application of the MC1596/MC1496. The suggested circuit for this application is shown on the front page of this data sheet.

In some applications, it may be necessary to operate the MC1596/MC1496 with a single dc supply voltage instead of dual supplies. Figure 26 shows a balanced modulator designed for operation with a single +12 Vdc supply. Performance of this circuit is similar to that of the dual supply modulator.

AM Modulator

The circuit shown in Figure 27 may be used as an amplitude modulator with a minor modification.

All that is required to shift from suppressed carrier to AM operation is to adjust the carrier null potentiometer for the proper amount of carrier insertion in the output signal.

However, the suppressed carrier null circuitry as shown in Figure 27 does not have sufficient adjustment range. Therefore, the modulator may be modified for AM operation by changing two resistor values in the null circuit as shown in Figure 28.

Product Detector

The MC1596/MC1496 makes an excellent SSB product detector (see Figure 29).

This product detector has a sensitivity of 3.0 microvolts and a dynamic range of 90 dB when operating at an intermediate frequency of 9 MHz.

The detector is broadband for the entire high frequency range. For operation at very low intermediate frequencies down to 50 kHz the 0.1 μF capacitors on pins 7 and 8 should be increased to 1.0 μF. Also, the output filter at pin 9 can be tailored to a specific intermediate frequency and audio amplifier input impedance.

As in all applications of the MC1596/MC1496, the emitter resistance between pins 2 and 3 may be increased or decreased to adjust circuit gain, sensitivity, and dynamic range.

This circuit may also be used as an AM detector by introducing carrier signal at the carrier input and an AM signal at the SSB input.

The carrier signal may be derived from the intermediate frequency signal or generated locally. The carrier signal may be introduced with or without modulation, provided its level is sufficiently high to saturate the upper quad differential amplifier. If the carrier signal is modulated, a 300 mV(rms) input level is recommended.

Ⓜ MOTOROLA *Semiconductor Products Inc.*

FIGURE 6-12(b) *(continued)*

APPLICATIONS INFORMATION (continued)

Doubly Balanced Mixer

The MC1596/MC1496 may be used as a doubly balanced mixer with either broadband or tuned narrow band input and output networks.

The local oscillator signal is introduced at the carrier input port with a recommended amplitude of 100 mV(rms).

Figure 30 shows a mixer with a broadband input and a tuned output.

Frequency Doubler

The MC1596/MC1496 will operate as a frequency doubler by introducing the same frequency at both input ports.

Figures 31 and 32 show a broadband frequency doubler and a tuned output very high frequency (VHF) doubler, respectively.

Phase Detection and FM Detection

The MC1596/MC1496 will function as a phase detector. High-level input signals are introduced at both inputs. When both inputs are at the same frequency the MC1596/MC1496 will deliver an output which is a function of the phase difference between the two input signals.

An FM detector may be constructed by using the phase detector principle. A tuned circuit is added at one of the inputs to cause the two input signals to vary in phase as a function of frequency. The MC1596/MC1496 will then provide an output which is a function of the input signal frequency.

NOTE: Pin number references pertain to this device when packaged in a metal can. To ascertain the corresponding pin numbers for plastic or ceramic packaged devices refer to the first page of this specification sheet.

TYPICAL APPLICATIONS

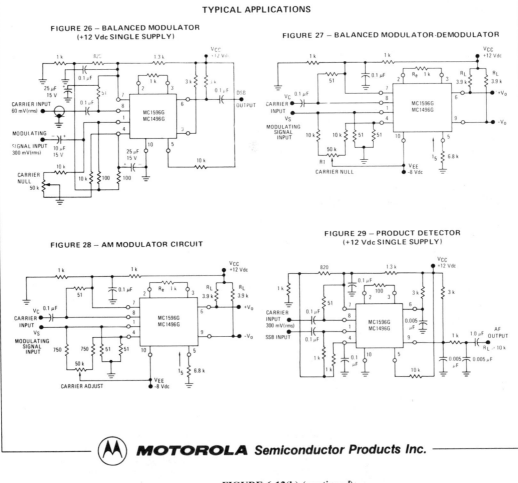

FIGURE 26 – BALANCED MODULATOR
(+12 Vdc SINGLE SUPPLY)

FIGURE 27 – BALANCED MODULATOR-DEMODULATOR

FIGURE 28 – AM MODULATOR CIRCUIT

FIGURE 29 – PRODUCT DETECTOR
(+12 Vdc SINGLE SUPPLY)

MOTOROLA *Semiconductor Products Inc.*

FIGURE 6-12(b) *(continued)*

TYPICAL APPLICATIONS (continued)

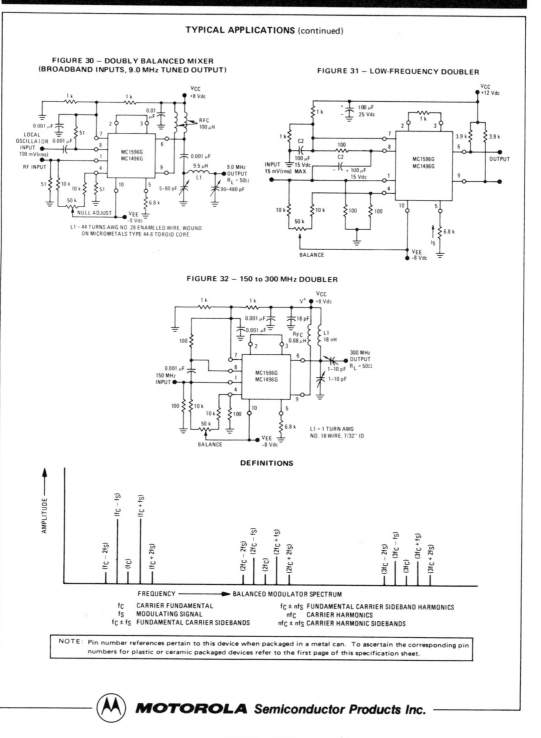

FIGURE 30 – DOUBLY BALANCED MIXER
(BROADBAND INPUTS, 9.0 MHz TUNED OUTPUT)

L1 = 44 TURNS AWG NO. 28 ENAMELED WIRE, WOUND
ON MICROMETALS TYPE 44-6 TOROID CORE.

FIGURE 31 – LOW-FREQUENCY DOUBLER

FIGURE 32 – 150 to 300 MHz DOUBLER

L1 = 1 TURN AWG
NO. 18 WIRE, 7/32'' ID

DEFINITIONS

BALANCED MODULATOR SPECTRUM

f_C	CARRIER FUNDAMENTAL
f_S	MODULATING SIGNAL
$f_C \pm f_S$	FUNDAMENTAL CARRIER SIDEBANDS

$f_C \pm nf_S$	FUNDAMENTAL CARRIER SIDEBAND HARMONICS
nf_C	CARRIER HARMONICS
$nf_C \pm nf_S$	CARRIER HARMONIC SIDEBANDS

NOTE: Pin number references pertain to this device when packaged in a metal can. To ascertain the corresponding pin
numbers for plastic or ceramic packaged devices refer to the first page of this specification sheet.

MOTOROLA *Semiconductor Products Inc.*

FIGURE 6-12(b) *(continued)*

OUTLINE DIMENSIONS

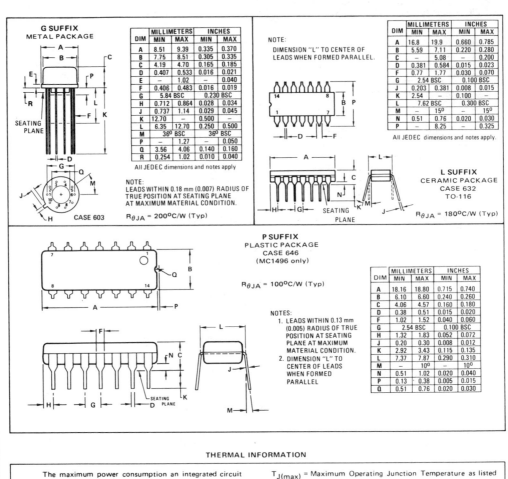

G SUFFIX
METAL PACKAGE

DIM	MILLIMETERS		INCHES	
	MIN	MAX	MIN	MAX
A	8.51	9.39	0.335	0.370
B	7.75	8.51	0.305	0.335
C	4.19	4.70	0.165	0.185
D	0.407	0.533	0.016	0.021
E	–	1.02	–	0.040
F	0.406	0.483	0.016	0.019
G	5.84 BSC		0.230 BSC	
H	0.712	0.864	0.028	0.034
J	0.737	1.14	0.029	0.045
K	12.70	–	0.500	–
L	6.35	12.70	0.250	0.500
M	36° BSC		36° BSC	
P	–	1.27	–	0.050
Q	3.56	4.06	0.140	0.160
R	0.254	1.02	0.010	0.040

All JEDEC dimensions and notes apply

SEATING PLANE

CASE 603

NOTE:
LEADS WITHIN 0.18 mm (0.007) RADIUS OF TRUE POSITION AT SEATING PLANE AT MAXIMUM MATERIAL CONDITION.

$R_{\theta JA} = 200°C/W$ (Typ)

NOTE:
DIMENSION "L" TO CENTER OF LEADS WHEN FORMED PARALLEL.

DIM	MILLIMETERS		INCHES	
	MIN	MAX	MIN	MAX
A	16.8	19.9	0.660	0.785
B	5.59	7.11	0.220	0.280
C	–	5.08	–	0.200
D	0.381	0.584	0.015	0.023
F	0.77	1.77	0.030	0.070
G	2.54 BSC		0.100 BSC	
J	0.203	0.381	0.008	0.015
K	2.54	–	0.100	–
L	7.62 BSC		0.300 BSC	
M	–	15°	–	15°
N	0.51	0.76	0.020	0.030
P	–	8.25	–	0.325

All JEDEC dimensions and notes apply.

L SUFFIX
CERAMIC PACKAGE
CASE 632
TO-116

$R_{\theta JA} = 180°C/W$ (Typ)

SEATING PLANE

P SUFFIX
PLASTIC PACKAGE
CASE 646
(MC1496 only)

$R_{\theta JA} = 100°C/W$ (Typ)

NOTES:
1. LEADS WITHIN 0.13 mm (0.005) RADIUS OF TRUE POSITION AT SEATING PLANE AT MAXIMUM MATERIAL CONDITION.
2. DIMENSION "L" TO CENTER OF LEADS WHEN FORMED PARALLEL

DIM	MILLIMETERS		INCHES	
	MIN	MAX	MIN	MAX
A	18.16	18.80	0.715	0.740
B	6.10	6.60	0.240	0.260
C	4.06	4.57	0.160	0.180
D	0.38	0.51	0.015	0.020
F	1.02	1.52	0.040	0.060
G	2.54 BSC		0.100 BSC	
H	1.32	1.83	0.052	0.072
J	0.20	0.30	0.008	0.012
K	2.92	3.43	0.115	0.135
L	7.37	7.87	0.290	0.310
M	–	10°	–	10°
N	0.51	1.02	0.020	0.040
P	0.13	0.38	0.005	0.015
Q	0.51	0.76	0.020	0.030

SEATING PLANE

THERMAL INFORMATION

The maximum power consumption an integrated circuit can tolerate at a given operating ambient temperature, can be found from the equation:

$$P_{D(T_A)} = \frac{T_{J(max)} - T_A}{R_{\theta JA}(Typ)} \geq V_I\,I_S \cdot V_O\,I_O$$

Where: $P_{D(T_A)}$ = Power Dissipation allowable at a given operating ambient temperature.

$T_{J(max)}$ = Maximum Operating Junction Temperature as listed in the Maximum Ratings Section

T_A = Maximum Desired Operating Ambient Temperature

$R_{\theta JA}(Typ)$ = Typical Thermal Resistance Junction to Ambient

I_S = Total Supply Current

MOTOROLA Semiconductor Products Inc.

FIGURE 6-12(b) (*continued*)

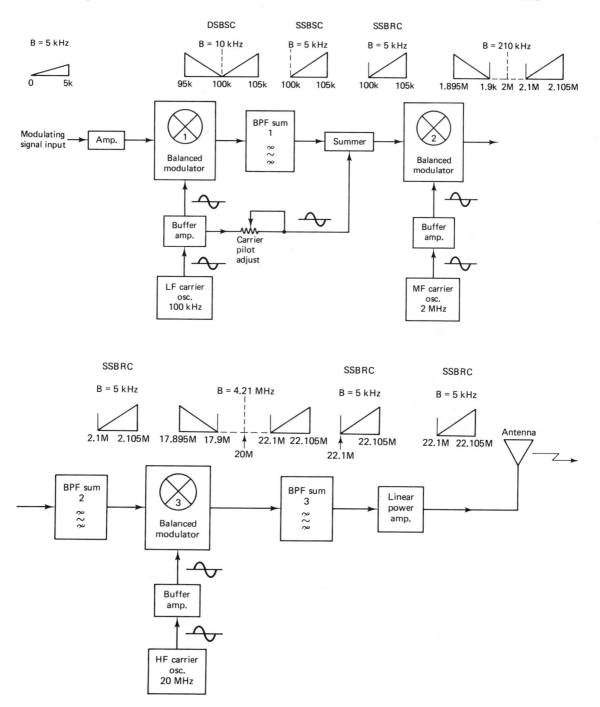

FIGURE 6-13 Single-sideband transmitter: filter method.

separated by a 200-kHz frequency band that is void of information. BPF 2 is centered on 2.1025 MHz with a 5 kHz bandwidth. Therefore, the output of BPF 2 is once again a single-sideband reduced-carrier waveform. Its spectrum comprises a reduced 2.1-MHz second IF carrier and a 5-kHz upper sideband. The output of BPF 2 is mixed with a 20-MHz high-frequency (HF) carrier in balanced modulator 3. The output is a double-sideband suppressed-carrier signal where the upper and lower sidebands again each contain the original SSBRC signal spectrum. The sidebands are separated by a 4.2-MHz frequency band that is void of information. BPF 3 is centered on 22.1025 MHz with a 5 kHz bandwidth. Therefore, the output of BPF 3 is once again a single-sideband reduced-carrier waveform with a reduced 22.1-MHz RF carrier and a 5-kHz upper sideband. The output waveform is amplified in the linear power amplifier and transmitted.

In the transmitter just described, the original modulating signal spectrum was up-converted in three modulation steps to a final carrier frequency of 22.1 MHz and a single upper sideband that extended to 22.105 MHz. After each up-conversion (frequency translation), the desired sideband is separated from the double-sideband spectrum with a BPF. The same final output spectrum can be produced with a single heterodyning process: one balanced modulator, one bandpass filter, and a single HF carrier supply. Figure 6-14 shows the block diagram and output spectrum for a single-step up-conversion transmitter. The output of the balanced modulator is a double-sideband spectrum centered around a suppressed carrier frequency of 22.1 MHz. To separate the 5-kHz upper sideband from the composite spectrum, a multiple-pole BPF with an extremely high Q is required. A BPF that meets this criterion is in itself difficult to construct but suppose that this were a multiple-channel transmitter and the carrier frequency were tunable; then the BPF must also be tunable. Constructing a tunable BPF in the MHz range with a passband of only 5 kHz is beyond economic and engineering feasibility. The only BPF in the transmitter shown in Figure 6-13 that had to separate sidebands that were immediately adjacent was BPF 1. To construct a multiple-pole, steep-skirted BPF at 100 kHz is a relatively simple task, as only a moderate Q is required. The sidebands separated by BPF 2 are 200 kHz apart; thus a low-Q filter with gradual roll-off characteristics can be used with no danger of passing any portion of the undesired sideband. BPF 3 separates sidebands that are 4.2 MHz apart. If multiple channels are used and the HF carrier is tunable, a single broadband filter can be used for BPF 3 with no danger of the undesired sideband leaking through the filter. For single-channel operation, the single conversion transmitter is the simplest design, but for multiple-channel operation, the three-conversion system is more practical. Figure 6-14b and c show the output spectrum and filtering requirements for both methods.

Single-sideband filters.   It is evident that filters are an essential part of a single-sideband system. Transmitters as well as receivers have requirements for highly selective networks for limiting both the signal and noise spectrums. Conventional LC filters do not have a high enough Q for most sideband transmitters. Therefore, most filters used for sideband generation are constructed from either *crystal* or *ceramic* material, or they use *mechanical* filters.

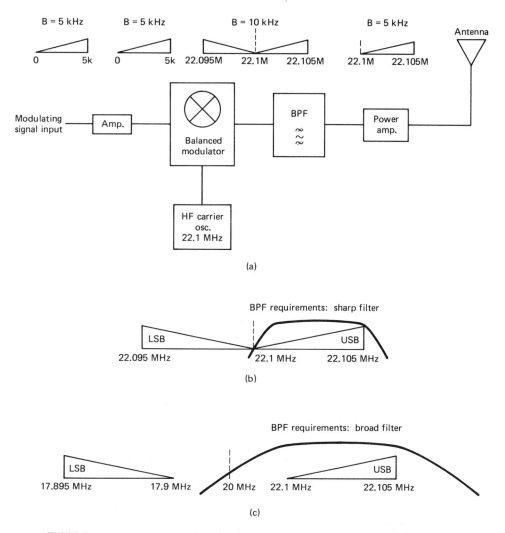

FIGURE 6-14 Single conversion SSBSC transmitter—filter method: (a) block diagram; (b) output spectrum and filtering requirements for a single-conversion transmitter; (c) output spectrum and filtering requirements for a three-conversion transmitter.

Crystal filters. The *crystal lattice filter* is commonly used in single-sideband systems. The schematic diagram for a typical crystal lattice bandpass filter is shown in Figure 6-15a. The lattice comprises two sets of matched crystal pairs (X_1 and X_2, X_3 and X_4) connected between tuned input and output transformers. Crystals X_1 and X_2 are series connected, while X_3 and X_4 are connected in parallel. X_1 and X_2 are cut to operate at the filter lower cutoff frequency, while X_3 and X_4 are cut to operate at the upper cutoff frequency. The input and output transformers are tuned to the center of the desired passband; this tends to spread the difference between the series and parallel

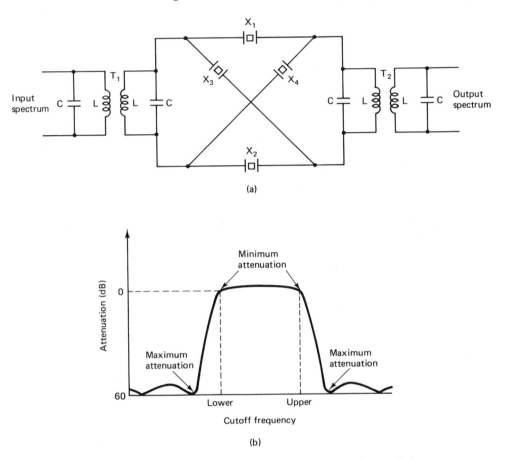

(a)

(b)

FIGURE 6-15   Crystal lattice filter: (a) schematic diagram; (b) characteristic curve.

resonant crystal frequencies. When the reactances of the bridge arms are equal and have the same sign (inductive or capacitive), the input signals propagating through the two possible paths cancel each other. When the reactances are equal with opposite signs (one inductive and one capacitive), the maximum signal is transmitted through the network. Figure 6-15b shows a typical characteristic curve for a crystal lattice filter. Crystal filters are available with a Q as high as 100,000. The filter shown in Figure 6-15a is a single-element filter. However, in order for a crystal filter to adequately pass a specific band and reject all other frequencies, at least two elements are required. Typical insertion losses for crystal filters are between 1.5 and 3 dB.

Ceramic filter. Ceramic filters are made from lead zinconate–titanate, which exhibits the piezoelectric effect. Therefore, they operate quite similar to crystal filters except that ceramic filters do not have as high a Q-factor. Typically, Q values for ceramic filters go up to about 2000. Ceramic filters are less expensive, smaller, and more rugged than their crystal counterparts. However, ceramic filters have more loss than crystal filters. The insertion loss for ceramic filters is typically between 2 and 4 dB.

Mechanical filters. A *mechanical filter* is a *mechanically resonant* device. It receives electrical energy, converts it to mechanical vibrations, then converts the vibrations back to electrical energy at its output. Essentially, there are four elements that comprise a mechanical filter: an input transducer that converts the input electrical energy to mechanical vibrations, a series of mechanical resonant metal disks that vibrate at the desired resonant frequency, a coupling rod that couples the metal disks together, and an output transducer that converts the mechanical vibrations back to electrical energy. Figure 6-16 shows the electrical equivalent circuit for a mechanical filter. The series resonant circuits (L and C) represent the metal disks, coupling capacitor C_1 represents the coupling rod, and R represents the matching mechanical loads. The resonant frequency of the filter is determined by the series LC disks and C_1 determines the bandwidth. Mechanical filters are more rugged than either ceramic or crystal filters and have comparable frequency response characteristics. However, mechanical filters are larger and heavier and therefore are impractical for mobile communications equipment.

SSB Transmitter: Phase-Shift Method

With the phase-shift method of single-sideband generation, the undesired sideband is canceled in the output of the modulator; therefore, sharp filtering is unnecessary. Figure 6-17 shows a block diagram for a SSB transmitter that uses the phase-shift method to eliminate the upper sideband. Essentially, there are two separate double-sideband modulators (balanced modulators 1 and 2). The modulating signal and carrier are applied to the two modulators 90° out of phase. The outputs from the two balanced modulators are double-sideband suppressed carrier signals with the proper phase such that when they are combined in the linear summer the upper sidebands cancel.

Phasor representation. The phasors shown in Figure 6-17 illustrate how the upper sideband is eliminated by rotating both the carrier and the modulating signal 90° prior to modulation. The output phasor from balanced modulator 1 shows the relative position and direction of rotation of the upper (ω_u) and lower (ω_l) side frequencies to the suppressed carrier (ω_c). The output of balanced modulator 2 is essentially the same phasor except that the phase of the carrier and the modulating signal are each rotated 90° from the reference. The output of the linear summer shows the sum of the output

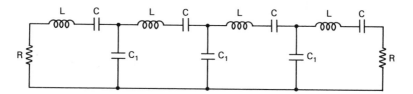

FIGURE 6-16 Mechanical filter equivalent circuit.

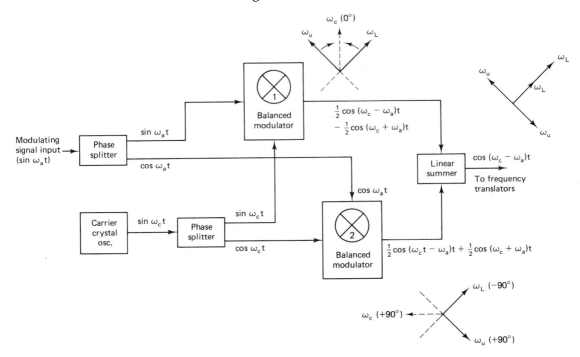

FIGURE 6-17   SSB transmitter: phase-shift method.

phasors from the two balanced modulators. The two phasors for the lower sideband are in phase and reinforce, whereas the phasors for the upper sideband are 180° out of phase and cancel. Consequently, the upper sideband is removed at the output of the linear summer.

Mathematical analysis.   In Figure 6-17 the input modulating signal ($\sin \omega_a t$) is fed directly to balanced modulator 1 and shifted 90° (to $\cos \omega_a t$) and fed to balanced modulator 2. The low-frequency carrier ($\sin \omega_c t$) is also fed directly to balanced modulator 1 and shifted 90° and fed to balanced modulator 2. The balanced modulators are product modulators and their outputs are expressed mathematically as

$$\begin{aligned}
\text{output from} \\
\text{balanced modulator 1} &= (\sin \omega_a t) \times (\sin \omega_c t) \\
&= \tfrac{1}{2} \cos (\omega_c - \omega_a)t - \tfrac{1}{2} \cos (\omega_c + \omega_a)t
\end{aligned}$$

$$\begin{aligned}
\text{output from} \\
\text{balanced modulator 2} &= (\cos \omega_a t) \times (\cos \omega_c t) \\
&= \tfrac{1}{2} \cos (\omega_c - \omega_a)t + \tfrac{1}{2} \cos (\omega_c + \omega_a)t
\end{aligned}$$

and the output from the linear summer is

$$\frac{\frac{1}{2} \cos (\omega_c - \omega_a)t - \frac{1}{2} \cos (\omega_c + \omega_a)t}{\frac{+ \frac{1}{2} \cos (\omega_c - \omega_a)t + \frac{1}{2} \cos (\omega_c + \omega_a)t}{\underbrace{\cos (\omega_c - \omega_a)t}} \qquad \text{canceled}}$$

$$\underbrace{\cos (\omega_c - \omega_a)t}_{\substack{\text{lower sideband} \\ \text{(difference frequencies)}}}$$

SSB Transmitter: The Third Method

The so-called *third method* of single-sideband generation, developed by D. K. Weaver in the 1950s, is similar to the phase shift method described previously in that it uses phase shifting and summing to cancel the undesired sideband. However, it has an advantage in that the information signal is initially modulated onto an audio subcarrier, thus eliminating the need for a *wideband* phase shifter (which is difficult to build in practice). The block diagram for a third-method SSB modulator is shown in Figure 6-18. Notice that all of the inputs to the phase shifters are single-frequencies (ω_o, $\omega_o + 90°$, ω_c, and $\omega_c + 90°$). The input audio mixes with the audio subcarrier in balanced modulators

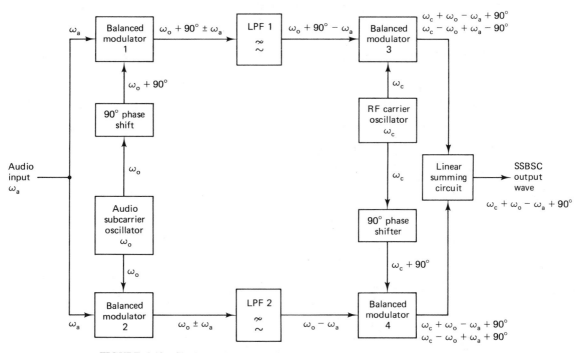

FIGURE 6-18 Single-sideband suppressed-carrier modulator: the "third method."

1 and 2, which are supplied with quadrature (90° out of phase) subcarrier signals. The output from balanced modulator 2 contains the upper and lower sidebands ($\omega_o \pm \omega_a$), while the output from balanced modulator 1 contains the upper and lower sidebands, each shifted in phase 90° ($\omega_o + 90° \pm \omega_a$). The upper sidebands are removed by their respective low-pass filters, which have an upper cutoff frequency equal to that of the suppressed audio subcarrier. The output from LPF 1 ($\omega_o + 90° - \omega_a$) is mixed with the RF carrier (ω_c) in balanced modulator 3, and the output from LPF 1 ($\omega_o - \omega_a$) is mixed with a 90° phase-shifted RF carrier ($\omega_c + 90°$) in balanced modulator 4. The RF carriers are, of course, suppressed in balanced modulators 3 and 4. Therefore, the output from balanced modulator 3 ($\omega_c + \omega_o - \omega_a + 90°$) + ($\omega_c - \omega_o + \omega_a - 90°$) is combined in the linear summer with the output from balanced modulator 4 ($\omega_c + \omega_o - \omega_a + 90°$) + ($\omega_c - \omega_o + \omega_a + 90°$). The output from the summer is

$$
\frac{
\begin{array}{l}
(\omega_c + \omega_o - \omega_a + 90°) + (\omega_c - \omega_o + \omega_a - 90°) \\
+ (\omega_c + \omega_o - \omega_a + 90°) + (\omega_c - \omega_o + \omega_a + 90°)
\end{array}
}{
(\omega_c + \omega_o - \omega_a + 90°) \qquad \text{canceled}
}
$$

The final RF output frequency is $F_c + F_o - F_a$, which is essentially the lower sideband of RF carrier $F_c + F_o$. The +90° offset phase is an absolute phase shift and is, consequently, insignificant. If the RF upper sideband is desired, simply interchange the carrier inputs to balanced modulators 3 and 4, in which case the final RF carrier is $F_c - F_o$.

Independent Sideband Transmitter

Figure 6-19 shows a block diagram for an independent sideband transmitter with three stages of modulation. The transmitter uses the filter method to produce two independent single-sideband channels (channel A and channel B). The two channels are combined; then a pilot carrier is reinserted. The composite ISB reduced carrier waveform is up-converted to RF with two additional stages of frequency translation. There are two 0- to 5-kHz information signals (channels A and B) that originate from two independent sources. The channel A information spectrum modulates a 100-kHz LF carrier in balanced modulator A. The output from balanced modulator A passes through BPF A, which is tuned to the lower sideband (95 to 100 kHz). The channel B information spectrum modulates the same 100-kHz LF carrier in balanced modulator B. The output from balanced modulator B passes through BPF B, which is tuned to the upper sideband (100 to 105 kHz). The two SSB spectrums are combined in a hybrid network to form a composite ISB suppressed carrier spectrum (95 to 105 kHz) The LF carrier (100 kHz) is reinserted in the linear summer to form an ISB reduced carrier waveform. The ISB spectrum is mixed with a 2.7-MHz MF carrier in balanced modulator 3. The output from balanced modulator 3 passes through BPF 3 to produce an ISB reduced carrier spectrum that extends from 2.795 to 2.805 MHz with a reduced 2.8-MHz pilot carrier. Balanced modulator 4, BPF 4, and the HF carrier translate the MF spectrum to an RF ISB spectrum that extends from 27.795 to 27.8 MHz (channel A) and 27.8 to 27.805 MHz (channel B) with a 27.8-MHz reduced carrier.

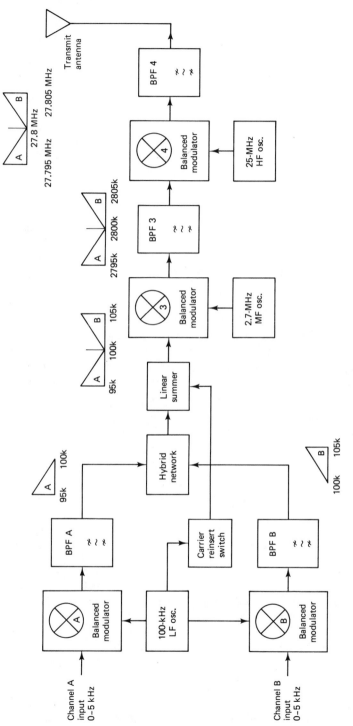

FIGURE 6-19   ISB transmitter: block diagram.

253

SINGLE-SIDEBAND RECEIVERS

Single-Sideband BFO Receiver

Figure 6-20 shows the block diagram for a simple noncoherent single-sideband receiver. The received RF is selected, amplified, then mixed down to IF for further amplification and band reduction. The output from the IF stage is heterodyned (beat) with the output from a beat frequency oscillator (BFO). The difference between the IF carrier frequency and the BFO frequency is the original information signal spectrum. Demodulation is accomplished simply by mixing and filtering the received signal with locally generated carriers. The receiver is noncoherent because the RF local oscillator and BFO frequencies are not synchronized to the transmitter local oscillators. Consequently, any difference between the transmit and receive carrier frequencies produces a frequency offset error in the demodulated information spectrum. For example, if the receive local oscillator is 100 Hz above its designated frequency and the BFO is 50 Hz above its designated frequency, the restored information is offset 150 Hz from its original input frequency spectrum. Fifty hertz or more offset is distinguishable by a normal listener as a tonal variation.

Mathematical analysis. The RF mixer and second detector shown in Figure 6-20 are product detectors (i.e., balanced modulators). Like the balanced modulators in the transmitter, their outputs are the product of their inputs. Essentially, the only difference between a product modulator and a product detector is that in a product modulator the input is tuned to a low-frequency modulating signal and the output is tuned to a high-frequency modulated signal. With a product detector, the input is tuned to a high-frequency modulated signal and the output is tuned to a low-frequency difference signal. With both the modulator and the detector, the single frequency carrier is the switching signal. In a receiver, the received input signal, which is a suppressed RF carrier and

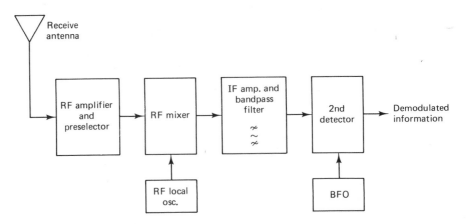

FIGURE 6-20   Noncoherent BFO SSB receiver.

one sideband, is multiplied (mixed) with the HF local oscillator frequency to generate an SSB IF difference signal at its output. The output from the second product detector is the sum and difference frequencies between the SSB IF and the beat frequency. The difference frequency band is the original input information spectrum. Mathematically, the output from the second product detector is

$$\text{output} = [\cos{(\omega_c - \omega_a)t}] \times (\sin{\omega_b t})$$

where

$$\omega_c = \text{suppressed IF carrier frequency}$$
$$\omega_a = \text{modulating signal frequencies}$$
$$\omega_c - \omega_a = \text{IF sideband frequencies}$$
$$\omega_b = \text{BFO frequency}$$

and

$$\omega_c = \omega_b$$

The trigonometric identity for the product of a cosine and sine wave of different frequencies is

$$(\cos{A}) \times (\sin{B}) = \tfrac{1}{2}\sin{(A + B)} + \tfrac{1}{2}\sin{(A - B)}$$

Therefore, the output is

$$\text{output} = \cos{(\omega_c - \omega_a)t} \times \sin{\omega_b t}$$
$$= \tfrac{1}{2}\sin{(\omega_c - \omega_a + \omega_b)t} + \tfrac{1}{2}\sin{(\omega_c - \omega_a - \omega_b)t}$$
$$\text{output} = \underbrace{\tfrac{1}{2}\sin{(2\omega_c + \omega_a)t}}_{\text{filtered off}} + \underbrace{\tfrac{1}{2}\sin{\omega_a t}}_{\substack{\text{original input} \\ \text{spectrum}}}$$

EXAMPLE 6-1

For the BFO receiver shown in Figure 6-20, an RF received signal of 30 to 30.005 MHz, an RF local oscillator frequency of 20 MHz, an IF frequency band of 10 to 10.005 MHz, and a BFO frequency of 10 MHz, determine:
(a) The demodulated first IF frequency band and demodulated information spectrum.
(b) The demodulated information spectrum when the RF local oscillator drifts down 0.001%.

Solution (a) The output from the RF mixer is

$$\text{RF mixer output} = F_{rf} - F_{lo}$$
$$F_{ip} = (30 \text{ to } 30.005 \text{ MHz}) - 20 \text{ MHz} = 10 \text{ to } 10.005$$

The demodulated information spectrum is

$$\text{demodulated output} = F_{if} - F_b$$
$$= (10 \text{ to } 10.005 \text{ MHz}) - 10 \text{ MHz} = 0 \text{ to } 5 \text{ kHz}$$

(b) A 0.001% drift would cause the corresponding decrease in the RF local oscillator frequency:

$$\text{Rf local oscillator drift} = (0.00001)(30 \text{ MHz}) = 300 \text{ Hz}$$

Therefore, the output from the RF mixer is

$$\begin{aligned}\text{RF mixer output} &= F_{rf} - F_{lo} \\ &= (30 \text{ to } 30.005 \text{ MHz}) - 19.9997 \text{ MHz} = 10.0003 \text{ to } 10.0053 \text{ MHz}\end{aligned}$$

and

$$\begin{aligned}\text{demodulated output} &= F_{if} - F_b f_o \\ &= (10.0003 \text{ to } 10.0053 \text{ MHz}) - 10 \text{ MHz} = 300 \text{ to } 5300 \text{ Hz}\end{aligned}$$

The 0.001% drift in the RF local oscillator frequency caused a corresponding 300-Hz shift or offset in the demodulated information spectrum.

Coherent SSB BFO Receiver

Figure 6-21 shows a block diagram for a *coherent SSB BFO receiver*. This receiver is identical to the receiver shown in Figure 6-19 except that the carrier and BFO frequencies are synchronized to the transmit carrier oscillators. The carrier recovery circuit is a narrowband PLL that tracks and removes the suppressed carrier from the composite SSBRC received spectrum and uses the recovered pilot to regenerate coherent receive carrier frequencies in the synthesizer. The synthesizer circuit generates the RF local oscillator and BFO frequencies. The carrier recovery circuit tracks the received carrier pilot. Therefore, minor changes in the transmit carrier frequencies are compensated for in the receiver and frequency offset error is eliminated. If the coherent receiver shown

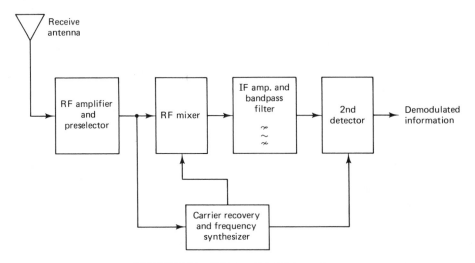

FIGURE 6-21 Coherent SSB BFO receiver.

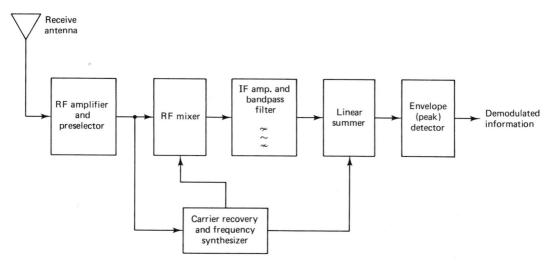

FIGURE 6-22   Single-sideband envelope detection receiver.

in Figure 6-21 had been used in Example 6-1, the receiver oscillators would have been locked onto the received pilot carrier and unable to drift independently.

Single-Sideband Envelope Detection Receiver

Figure 6-22 shows the block diagram for a single-sideband receiver that uses synchronous carriers and envelope detection to demodulate the received signals. The reduced carrier pilot is detected, separated from the composite receive spectrum, and regenerated in the carrier recovery circuit. The regenerated pilot is divided and used as the stable frequency source for a frequency synthesizer, which supplies the receiver with frequency coherent carriers (the receiver carrier oscillators are synchronized to the transmit carrier oscillators). The received RF is mixed down to IF in the first product detector. A regenerated IF carrier is added to the IF spectrum in the linear summer, which produces a SSB full carrier envelope. The SSBFC waveform is demodulated with a conventional peak detector to produce the original input signal spectrum.

Multichannel Pilot Carrier SSB Receiver

Figure 6-23 shows a block diagram for a multichannel pilot carrier SSB receiver that uses a PLL carrier recovery circuit and a frequency synthesizer. The RF input range extends from 4 to 30 MHz, and the VCO natural frequency is coarsely adjusted with an external channel selector switch over a frequency range of 6 to 32 MHz. The VCO frequency tracks above the incoming RF by 2 MHz, which is the first IF. A 1.8-MHz beat frequency sets the second IF to 200 kHz.

Circuit operation.   The VCO frequency is coarsely set with the channel selector switch. The output frequency from the VCO mixes with the incoming RF in the first

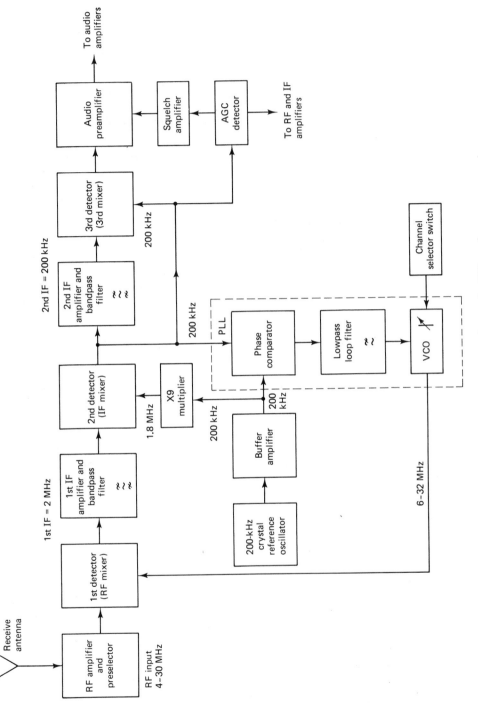

FIGURE 6-23 Multichannel pilot carrier SSB receiver.

detector to produce a first IF difference frequency of 2 MHz. The first IF mixes with the 1.8-MHz beat frequency to produce a 200-kHz second IF. The PLL locks onto the 200-kHz pilot and produces a dc correction voltage that fine tunes the VCO. The second IF is beat down to audio in the third detector, which is passed on to the audio preamplifier for processing. The AGC detector produces an AGC voltage that is proportional to the amplitude of the 200-kHz pilot. The AGC voltage is fed back to the RF and/or IF amplifiers to adjust their gains proportionate to the received pilot level and to the squelch circuit to turn the audio preamplifier off in the absence of a received pilot. The PLL compares the 200-kHz pilot to a stable crystal-controlled reference. Consequently, although the receiver carrier supply is not directly synchronized to the transmit oscillators, the first and second IFs are, thus eliminating any frequency offset in the demodulated audio spectrum.

Single-Sideband Measurements

As mentioned previously, single-sideband transmitters are rated in peak envelope power (PEP) and peak envelope volts (PEV) rather than simply rms power and voltage. For a single frequency-modulating signal, the modulated output signal with SSB is not an envelope as with conventional AM, but rather, a continuous single frequency tone. A single frequency is not representative of a typical modulating signal spectrum. Therefore, for test purposes, a *two-frequency* test signal is used for the modulating signal where the two tones have equal amplitudes. Figure 6-24a shows the envelope produced in a

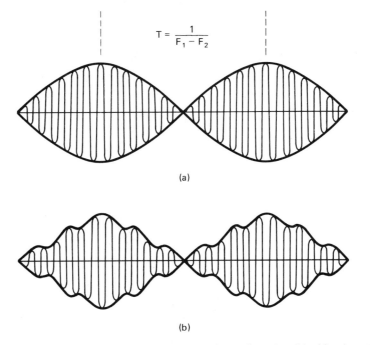

(a)

(b)

FIGURE 6-24 Two-tone SSB test signal: (a) without reinserted carrier; (b) with reinserted carrier.

SSB modulator for a two-tone test signal. The envelope is the vector sum of the two equal amplitude side frequencies and is similar to a conventional AM envelope except that the repetition rate is equal to the difference between the two tone frequencies. Figure 6-24b shows the envelope for a two-tone test signal when a low-amplitude pilot carrier is added. The envelope has basically the same shape except with the addition of a low-amplitude sine-wave ripple at the carrier frequency.

A two-tone SSB envelope is an important consideration because it is from this envelope that the output power for a SSB transmitter is determined. The PEP for a SSB transmitter is analogous to the total output power from a conventional AM transmitter. The rated PEP is the rms output power measured at the peak of the envelope when the input is a two-tone test signal and the two tones are equal in amplitude. With such an output signal, the actual power dissipated in the load is equal to half the PEP. Therefore, the voltage developed across the load is

$$e_{\text{total}} = e_1^2 + e_2^2$$

where e_1 and e_2 are the rms voltages of the two test tones. Therefore

$$\text{PEP} = \frac{e_1^2 + e_2^2}{R} \quad \text{W}$$

and since $e_1 = e_2$.

$$\text{PEP} = \frac{(2e)^2}{R}$$

$$= \frac{4e^2}{R} \quad \text{W} \tag{6-1}$$

However, the average power dissipated in the load is equal to the sum of the powers of the two tones:

$$P_{\text{ave}} = \frac{e_1^2}{R} + \frac{e_2^2}{R} = \frac{2e^2}{R} \quad \text{W} \tag{6-2}$$

and

$$P_{\text{ave}} = \frac{\text{PEP}}{2} \quad \text{W} \tag{6-3}$$

Two equal-amplitude test tones are used for the test signal for the following reasons:

1. One tone produces a continuous single-frequency output that does not produce intermodulation.
2. A single-frequency output signal is not analogous to normal conversation.
3. More than two tones makes analysis impractical.
4. Two tones of equal amplitude place a more demanding requirement on the transmitter than is likely to occur during normal operation.

EXAMPLE 6-2

For a two-tone test signal of 1.5 and 3 kHz and a carrier frequency of 100 kHz:
(a) Determine the output frequency spectrum.
(b) For $e_1 = e_2 = 10 \text{ V}_p$ and a load resistance of 50 Ω, determine the PEP and average output power.

Solution   (a) The output spectrum contains the two upper side frequencies

$$\text{USF}_1 = 100 \text{ kHz} + 1.5 \text{ kHz} = 101.5 \text{ kHz}$$

$$\text{USF}_2 = 100 \text{ kHz} + 3 \text{ kHz} = 103 \text{ kHz}$$

(b) Substituting into Equation 6-1 yields

$$\text{PEP} = \frac{4(3.535)^2}{50} = 1 \text{ W}$$

Substituting into Equation 6-2 gives us

$$P_{\text{ave}} = \frac{2(3.535)^2}{50} = \tfrac{1}{2} \text{ W}$$

QUESTIONS

6-1. Describe AM SSBFC. Compare SSBFC to conventional AM.

6-2. Describe AM SSBSC. Compare SSBSC to conventional AM.

6-3. Describe AM SSBRC. Compare SSBRC to conventional AM.

6-4. What is a pilot carrier?

6-5. What is an exalted carrier?

6-6. Describe AM ISB. Compare ISB to conventional AM.

6-7. Describe AM VSB. Compare VSB to conventional AM.

6-8. Define *peak envelope power*.

6-9. Draw the schematic diagram and describe the operation of a balanced ring modulator.

6-10. What is a product modulator?

6-11. Draw the schematic diagram and describe the operation of a FET push-pull balanced modulator.

6-12. Draw the schematic diagram and describe the operation of a balanced bridge modulator.

6-13. What are the advantages of an LIC integrated-circuit balanced modulator over a discrete balanced modulator?

6-14. Draw the block diagram for an SSB transmitter using the filter method. Describe its operation.

6-15. Contrast crystal, ceramic, and mechanical filters.

6-16. Draw the schematic diagram for an SSB transmitter using the phase shift method. Describe its operation.

6-17. Draw the schematic diagram for an SSB transmitter using the "third method." Describe its operation.

6-18. Draw the block diagram for an independent sideband transmitter. Describe its operation.

6-19. What is a product detector?

6-20. What is the difference between a product detector and a product modulator?

6-21. What is the difference between a noncoherent and a coherent receiver?

6-22. Draw the block diagram for a noncoherent single-sideband BFO receiver. Describe its operation.

6-23. Draw the block diagram for a coherent single-sideband BFO receiver. Describe its operation.

6-24. Draw the block diagram for an envelope detection receiver. Describe its operation.

6-25. Draw the block diagram for a coherent multichannel pilot carrier SSB receiver. Describe its operation.

6-26. Why is a two-tone test signal used for making PEP measurements?

PROBLEMS

6-1. For the balanced ring modulator shown in Figure 6-6a, carrier input frequency $F_c = 400$ kHz, and modulating signal frequency spectrum $F_a = 0$ to 4 kHz; determine:
  (a) The output frequency range. $396 - 404$ kHz
  (b) The output frequency for a single frequency input $F_a = 2.8$ kHz. 397.2 and 402.8 kHz

6-2. For the LIC balanced modulator circuit shown in Figure 6-12, carrier input frequency $F_c = 200$ kHz, and modulating signal frequency spectrum $F_a = 0$ to 3 kHz; determine:
  (a) The output frequency range.
  (b) The output frequency for a single frequency input $F_a = 1.2$ kHz.

6-3. For the SSB transmitter shown in Figure 6-13, LF carrier frequency $F_{1f} = 100$ kHz, MF carrier frequency $F_{mf} = 4$ MHz, HF carrier frequency $F_{hf} = 30$ MHz, and audio input frequency spectrum $F_a = 0$ to 5 kHz:
  (a) Sketch the frequency spectrums for the following points: balanced modulator 1 out, BPF 1 out, summer out, balanced modulator 2 out, BPF 2 out, balanced modulator 3 out, and BPF 3 out.
  (b) For a single frequency input $F_a = 1.5$ kHz, determine the translated frequency for the following points: BPF 1 out, BPF 2 out, and BPF 3 out.

6-4. Repeat Problem 6-3 except change the LF carrier frequency to 500 kHz. Which transmitter has the more stringent filtering requirements?

6-5. For the SSB transmitter shown in Figure 6-14a, audio input frequency spectrum $F_a = 0$ to 3 kHz and HF carrier frequency $F_{hf} = 28$ MHz:
  (a) Sketch the output frequency spectrum. $28 - 28.003$ MHz
  (b) For a single-frequency input $F_a = 2.2$ kHz, determine the output frequency. 28.0022 MHz

6-6. Repeat Problem 6-5 except change the audio input frequency spectrum to $F_a = 300$ to 5000 Hz.

6-7. For the SSB transmitter shown in Figure 6-17, carrier input frequency $F_c = 500$ kHz, and input frequency spectrum $F_a = 0$ to 4 kHz:
  (a) Sketch the frequency spectrum at the output of the linear summer. 496 500
  (b) For a single audio input frequency $F_a = 3$ kHz, determine the output frequency.
  497 kHz

6-8. Repeat Problem 6-7 except change the carrier input frequency to 400 kHz and the input frequency spectrum to F_a = 300 to 5000 Hz.

6-9. For the ISB transmitter shown in Figure 6-19, channel A input frequency spectrum F_a = 0 to 4 kHz, channel B input spectrum F_b = 0 to 4 kHz, LF carrier frequency F_{1f} = 200 kHz, MF carrier frequency F_{mf} = 4 MHz, and HF carrier frequency F_{hf} = 32 MHz:

P. 511

(a) Sketch the frequency spectrums for the following points: balanced modulator A out, BPF A out, balanced modulator B out, BPF B out, hybrid network out, linear summer out, balanced modulator 3 out, BPF 3 out, balanced modulator 4 out, and BPF 4 out.

(b) For A-channel input frequency F_a = 2.5 kHz and B-channel input frequency F_b = 3 kHz, determine the frequency components at the following points: BPF A out, BPF B out, BPF 3 out, and BPF 4 out.

6-10. Repeat Problem 6-9 except change the channel A input frequency spectrum to 0 to 10 kHz and the channel B input frequency spectrum to 0 to 6 kHz.

6-11. For the SSB receiver shown in Figure 6-20, RF input frequency F_{rf} = 35.602 MHz, RF local oscillator frequency F_{lo} = 25 MHz, and a 2-kHz modulating signal frequency, determine the IF frequency and the BFO frequency. $F_{IF} = F_{rf} - F_{lo}$, $BFO = F_{IF} - F_{mod.\,signal}$

IF : 10.602 MHz
BFO = 10.6 MHz

6-12. For the multichannel pilot carrier SSB receiver shown in Figure 6-23, crystal oscillator frequency F_{co} = 300 kHz, first IF frequency F_{1if} = 3.3 MHz, RF input frequency F_{rf} = 23.303 MHz, and modulating signal frequency F_a = 3 kHz, determine the following: VCO output frequency, multiplication factor, second IF frequency.

6-13. For a two-tone test signal of 2 and 3 kHz and a carrier frequency of 200 kHz:

(a) Determine the output frequency spectrum.

(b) For $e_1 = e_2$ = 12 V_p and a load resistor R_L = 50 Ω, determine the PEP and average power.

Chapter 7

ANGLE MODULATION TRANSMISSION

INTRODUCTION

As stated previously, there are three properties of an analog signal that can be varied: its amplitude, its frequency, or its phase. Chapters 3, 4, and 6 dealt with amplitude modulation. This chapter (as well as Chapter 8) deals with both frequency modulation (FM) and phase modulation (PM). FM and PM are both forms of *angle* modulation. Unfortunately, both forms of angle modulation are often referred to simply as FM when, actually, there is a distinct (although subtle) difference between the two. Angle modulation was first introduced in 1931 as an alternative to amplitude modulation. It was suggested that an angle-modulated wave was less susceptible to noise than AM and, consequently, could improve the performance of radio communications. Major E. H. Armstrong developed the first successful FM radio system in 1936, and in 1939 the first regularly scheduled broadcasting of FM signals began in Alpine, New Jersey.

ANGLE MODULATION

Angle modulation results whenever the phase angle (θ) of a sinusoidal wave is varied with respect to time. An angle-modulated wave is expressed mathematically as

$$M(t) = V_c \cos [\omega_c t + \theta(t)] \qquad (7\text{-}1)$$

where

$M(t)$ = angle-modulated carrier

V_c = peak carrier amplitude (V)

$$\omega_c = \text{carrier frequency, } 2\pi F_c$$

$$\theta(t) = \text{angle modulation (rad)}$$

With angle modulation it is necessary that $\theta(t)$ be a prescribed function of the modulating signal. Therefore, if $V(t)$ is the modulating signal, the angle modulation is expressed mathematically as

$$\theta(t) = F[V(t)] \tag{7-2}$$

where $V(t)$ is the modulating signal $= V_a \sin \omega_a t$.

In essence, the difference between FM and PM lies in which property (the frequency or the phase) is varied directly by the modulating signal. Whenever the frequency of a carrier is varied, the phase is also varied, and vice versa. Therefore, FM and PM must both occur whenever either form of angle modulation is performed. If the frequency of the carrier is varied directly in accordance with the modulating signal, FM results. If the phase of the carrier is varied directly in accordance with the modulating signal, PM results. Therefore, direct FM is indirect PM and direct PM is indirect FM.

Figure 7-1 shows the waveform for a sinusoidal carrier where the frequency is changing in respect to time. Whenever the period (T) of a sine wave is changed, both its frequency and phase change, and if the changes are continuous, the wave is no longer a single frequency. It can be shown that the resultant waveform comprises the original carrier frequency (sometimes called the *carrier rest frequency*) and an infinite number of side frequencies. The change in frequency is called the frequency deviation (ΔF) and the change in phase is called the phase deviation $(\Delta \theta)$. Frequency deviation is the relative displacement of the carrier frequency, and phase deviation is the relative *angular displacement* of the carrier in respect to a reference phase.

Figure 7-2 shows a sinusoidal carrier in which the frequency (F) is changed (*deviated*) over a period of time (t). After t seconds, the frequency has changed ΔF hertz and is now $F - \Delta F$. The phase has also changed $\Delta \theta$ radians.

Mathematical Analysis

The difference between FM and PM is more easily understood by defining the following four terms with reference to Equation 7-1: instantaneous phase, instantaneous phase deviation, instantaneous frequency, and instantaneous frequency deviation.

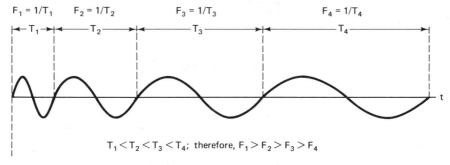

FIGURE 7-1 Frequency changing with time.

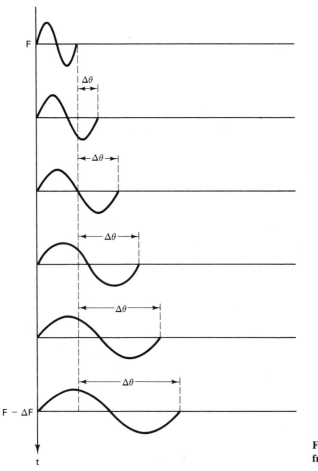

FIGURE 7-2   Phase changing with frequency.

Instantaneous phase. The *instantaneous phase* is the precise phase of the carrier at a given time and is expressed mathematically as

$$\text{instantaneous phase} = \omega_c t + \theta(t) \qquad \text{rad} \qquad (7\text{-}3)$$

where

$$\omega_c t = \left(2\pi \, \frac{\text{rad}}{\text{cycle}} \right) \left(F \, \frac{\text{cycles}}{\text{second}} \right) (t \text{ seconds}) = 2\pi F t \qquad \text{rad}$$

$$\theta(t) = \text{rad}$$

Instantaneous phase deviation. The *instantaneous phase deviation* is the instantaneous change in the phase of the carrier at a given time and is expressed mathematically as

$$\text{instantaneous phase deviation} = \theta(t) \qquad \text{rad} \qquad (7\text{-}4)$$

Instantaneous frequency. The *instantaneous frequency* of an angle-modulated carrier is the precise frequency of the carrier at a given time and is defined as the first time derivative of the instantaneous phase. In terms of Equation 7-3, the instantaneous frequency is expressed mathematically as

$$\begin{array}{c}\text{instantaneous}\\\text{frequency}\end{array} = \frac{d}{dt}[\omega_c t + \theta(t)] = \omega_c + \theta'(t) \qquad \text{rad/s} \qquad (7\text{-}5)$$

where

$$\omega_c = \left(2\pi \frac{\text{rad}}{\text{cycle}}\right)\left(F \frac{\text{cycles}}{\text{second}}\right) = 2\pi F \qquad \text{rad/s}$$

$$\theta'(t) = \text{rad/s}$$

Instantaneous frequency deviation. The *instantaneous frequency deviation* is the instantaneous change in the frequency of the carrier and is defined as the first time derivative of the instantaneous phase deviation. Therefore, the instantaneous phase deviation is the first integral of the instantaneous frequency deviation. In terms of Equation 7-4, the instantaneous frequency deviation is expressed mathematically as[*]

$$\begin{array}{c}\text{instantaneous}\\\text{frequency deviation}\end{array} = \theta'(t) \qquad \text{rad/s} \qquad (7\text{-}6)$$

For a modulating signal, $V(t)$, the phase and frequency modulation are

$$\text{phase modulation} = \theta(t) = KV(t) \qquad (7\text{-}7)$$

$$\text{frequency modulation} = \theta'(t) = K_1 V(t) \qquad (7\text{-}8)$$

where K and K_1 are constants which are, in fact, equal to the *deviation sensitivities* of the modulators. The deviation sensitivity is the input-versus-output transfer function for the modulator. The deviation sensitivity for a PM modulator is

$$K = \text{rad/V}$$

and for an FM modulator

$$K_1 = 2\pi F \text{ rad/V}$$

or,

$$= \frac{2\pi F \text{ rad/V}}{2\pi \text{ rad}} = \text{Hz/V}$$

Phase modulation is the first integral of the frequency modulation. Therefore, from Equations 7-7 and 7-8,

$$\text{PM} = \theta(t) = \int \theta'(t) = \int K_1 V(t) = K_1 \int V(t) \qquad (7\text{-}9)$$

---

[*] Note that a prime is used to denote the first derivative.

For a modulating signal $V_a \cos \omega_a t$ and substituting into Equation 7-1 yields

$$\text{phase modulation} = V_c \cos \omega_c t + KV_a \cos \omega_a t)$$

$$\text{frequency modulation} = V_c \cos \left(\omega_c t + \frac{K_1 V_a}{\omega_a} \sin \omega_a t \right)$$

The preceding mathematical relationships are summarized in Table 7-1. Also, the FM and PM waves that result when the modulating signal is a single frequency (sinusoidal) are shown.

FM and PM Waveforms

Figure 7-3 illustrates both frequency and phase modulation of a sinusoidal carrier by a single frequency-modulating signal. It can be seen that the FM and PM waveforms are identical except for their time relationship (phase). Thus it is impossible to distinguish an FM waveform from a PM waveform without knowing the characteristics of the modulating signal. Figure 7-3a shows the unmodulated carrier and Figure 7-3b the modulating signal (both sine waves). Figure 7-3e shows the first derivative of the modulating signal (a cosine wave). Figure 7-3c shows the frequency-modulated wave whose instantaneous frequency is proportional to the modulating signal. Note that the frequency deviation is maximum at the positive and negative peaks of the modulating signal and minimum at the *zero crossings* (the frequency deviation is proportional to the modulating signal—more specifically, to the amplitude of the modulating signal). Also, note in Figure 7-3d that the frequency deviation is maximum at the zero crossings of the modulating signal and minimum at the positive and negative peaks (the frequency deviation is proportional to the slope of the modulating signal). Therefore, for the phase-modulated waveform, the frequency deviation is proportional to the waveform shown in Figure 7-3e, which is a cosine wave (the first derivative of the modulating signal). Note that the amplitudes of both the FM and PM waveforms remain constant. Therefore, the following can be concluded:

1. With frequency modulation, the instantaneous frequency is proportional to the modulating signal and the instantaneous phase is proportional to the first time integral of the modulating signal.

TABLE 7-1   EQUATIONS FOR PHASE- AND FREQUENCY-MODULATED CARRIERS

Type of modulation	Modulating signal	Angle-modulated wave, $M(t)$
(a) Phase	$V(t)$	$V_c \cos [\omega_c t + KV(t)]$
(b) Frequency	$V(t)$	$V_c \cos [\omega_c t + K_1 \int V(t)\, dt]$
(c) Phase	$V_a \cos \omega_a t$	$V_c \cos (\omega_c t + KV_a \cos \omega_a t)$
(d) Frequency	$-V_a \sin \omega_a t$	$V_c \cos \left(\omega_c t + \dfrac{K_1 V_a}{\omega_a} \cos \omega_a t \right)$
(e) Frequency	$V_a \cos \omega_a t$	$V_c \cos \left(\omega_c t + \dfrac{K_1 V_a}{\omega_a} \sin \omega_a t \right)$

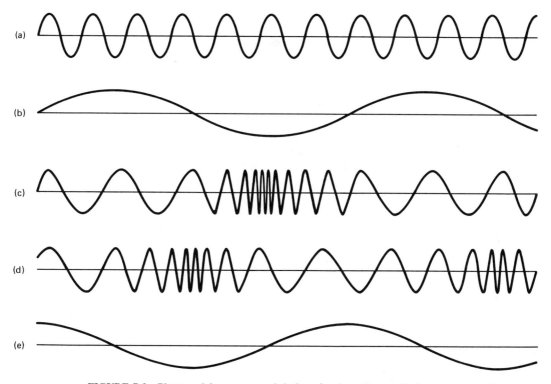

FIGURE 7-3 Phase and frequency modulation of a sine-wave carrier by a sine-wave signal: (a) unmodulated carrier; (b) modulating signal; (c) frequency-modulated wave; (d) phase-modulated wave; (e) first derivative of the modulating signal.

2. With phase modulation, the instantaneous phase is proportional to the modulating signal and the instantaneous frequency is proportional to the first time derivative (waveform slope) of the modulating signal.

Frequency deviation. *Frequency deviation* (ΔF) is the change in frequency that occurs in the carrier when it is acted on by a modulating signal. From Figure 7-3 it can be seen that ΔF is proportional to the amplitude of the modulating signal V_a, and the rate at which the frequency change occurs is equal to the modulating signal frequency F_a. Frequency deviation is typically given as a peak frequency shift in hertz. The peak-to-peak frequency deviation is sometimes called *carrier swing*. Frequency deviation is a function of the deviation sensitivity of the modulator and the modulating signal amplitude. Mathematically, ΔF is

$$\Delta F = K_1 V_a \qquad 2\pi F \; rad \qquad (7\text{-}10a)$$

or

$$\Delta F = \frac{K_1 V_a}{2\pi \; rad} \qquad F \; \text{Hz} \qquad (7\text{-}10b)$$

carrier swing = 2 ΔF

where

$$V_a = \text{peak modulating signal amplitude (V)}$$

$$K_1 = \text{deviation sensitivity } (2\pi F \text{ rad/V})$$

If the deviation sensitivity is given in Hz/V,

$$\Delta F = K_1 V_a = \text{Hz} \qquad (7\text{-}10\text{c})$$

Modulation index.   Comparing the expressions for the angle-modulated wave types (c), (d), and (e) of Table 7-1 shows that the formula for either a frequency- or a phase-modulated wave with a sinusoidal modulating signal can be written in the general form

$$M(t) = V_c \cos \left[\omega_c t + \theta(t) \right]$$

or

$$M(t) = V_c \cos \left(\omega_c t + m \cos \omega_a t \right) \qquad (7\text{-}11)$$

where

$$\theta(t) = m \cos \omega_a t = \text{instantaneous phase deviation}$$

$$m = \text{modulation index}$$

and

$$m = K V_a \text{ for phase modulation} \qquad (7\text{-}12)$$

$$m = \frac{K_1 V_a}{\omega_a} \text{ for frequency modulation} \qquad (7\text{-}13)$$

where

$$V_a = \text{peak amplitude of the modulating signal}$$

Therefore for **PM**

$$m = (\text{rad/V})(V_a) = \text{rad}$$

and for **FM**

$$m = \frac{(2\pi F \text{ rad/V})(V_a)}{2\pi F_a \text{rad}} = \frac{\Delta F}{F_a} \text{ (unitless)}$$

$\theta(t)$ is the instantaneous phase deviation, and from Equation 7-11 it can be seen that $\theta(t)$ is a function of $\cos \omega_a t$. Therefore, the phase deviation varies at a rate equal to ω_a (the modulating signal frequency) and reaches a maximum peak value of $K V_a$ for PM and $K_1 V_a / \omega_a$ for FM. Therefore, for PM, m is the peak phase deviation in radians and is called the *modulation index*. For FM, the modulation index is a unitless ratio and is used to describe the depth of modulation. From the preceding relationships it can be seen that for phase modulation m is independent of the frequency of the modulating signal. However, for frequency modulation, m is inversely proportional to the frequency of the modulating signal. Also, for both phase and frequency modulation, the modulation index is directly proportional to the deviation sensitivity (K or K_1) and the modulating signal amplitude (V_a).

With FM it is more common to express the modulation index as the peak frequency deviation divided by the modulating signal frequency. Mathematically, modulation index is

$$m = \frac{\Delta F}{F_a} \tag{7-14}$$

where

K_1 = deviation sensitivity (Hz/V)

$\Delta F = K_1 V_a$ = peak frequency deviation (Hz)

F_a = modulating signal frequency (Hz)

EXAMPLE 7-1

(a) Determine the peak frequency deviation (ΔF) and modulation index (m) for an FM modulator with a deviation sensitivity K_1 = 5 kHz/V and a modulating signal $V(t) = 2 \cos (2\pi 2000t)$.

(b) Determine the peak phase deviation (m) for a PM modulator with a deviation sensitivity K = 2.5 rad/V and a modulating signal $V(t) = 2 \cos (2\pi 2000t)$.

Solution   (a) From Equation 7-10c,

$$\Delta F = K_1 V_a = \left(\frac{5000 \text{ Hz}}{\text{V}}\right)(2 \text{ V}) = 10 \text{ kHz}$$

From Equation 7-14,

$$m = \frac{\Delta F}{F_a} = \frac{10 \text{ kHz}}{2 \text{ kHz}} = 5$$

(b) From Equation 7-12,

$$m = KV_a = \left(\frac{2.5 \text{ rad}}{\text{V}}\right)(2 \text{ V}) = 5 \text{ rad}$$

In Example 7-1 the modulation index for the frequency-modulated carrier is equal to the peak phase deviation of the phase-modulated carrier (5). If the amplitude of the modulating signal is changed, both the FM modulation index and the peak phase deviation will change proportionally. If the frequency of the modulating signal changes, the FM modulation index changes inversely proportional. However, the phase deviation is unaffected by changes in the modulating signal frequency. Therefore, under identical conditions, FM and PM are indistinguishable for a single modulating frequency; however, when the frequency changes, the PM modulation index remains constant, whereas the FM modulation index increases as the modulating frequency is reduced, and vice versa.

Percent modulation.   With FM, *percent modulation* is simply the ratio of the frequency deviation actually produced to the maximum frequency deviation allowed by law stated in percent form. Mathematically, percent modulation is

$$\% \text{ modulation} = \frac{\Delta F \text{ (actual)}}{\Delta F \text{ (maximum)}} \times 100 \qquad (7\text{-}15)$$

For example, in the United States the FCC (in Canada the DDC), limits the frequency deviation for FM broadcast band transmitters to ± 75 kHz. If a given modulating signal produces ± 50 kHz of frequency deviation, the percent modulation is

$$\% \text{ modulation} = \frac{50 \text{ kHz}}{75 \text{ kHz}} \times 100 = 67\%$$

Phase and Frequency Modulators and Demodulators

A *phase modulator* is a circuit in which the carrier is varied in such a way that its instantaneous phase is proportional to the modulating signal. The unmodulated carrier is a single-frequency sinusoid and is commonly called the *rest* frequency. A *frequency modulator* (often called a *frequency deviator*) is a circuit in which the carrier is varied in such a way that its instantaneous phase is proportional to the integral of the modulating wave. Therefore, with a frequency modulator if the modulating wave $V(t)$ is differentiated prior to being applied to the modulator, the instantaneous phase deviation is proportional to the integral of $V'(t)$ or, in other words, proportional to $V(t)$ $[\int V'(t) = V(t)]$. Similarly, an FM modulator that is preceded by a differentiator produces an output wave in which the phase deviation is proportional to the modulating wave and is, therefore, equivalent to a phase modulator. Several other equivalences are possible. For example, a frequency demodulator followed by an integrator is equivalent to a phase demodulator. Four common equivalences are listed below and illustrated in Figure 7-4.

  1. PM modulator = differentiator + FM modulator
  2. PM demodulator = FM demodulator + integrator
  3. FM modulator = integrator + PM modulator
  4. FM demodulator = PM demodulator + differentiator

Frequency Analysis of FM and PM Signals

With angle modulation, the frequency components of the modulated wave are much more complexly related to the frequency components of the modulating signal than with amplitude modulation. In a frequency or a phase modulator, a single-frequency modulating signal produces an infinite number of side frequencies, where each side frequency is displaced from the carrier by an integral multiple of the modulating signal

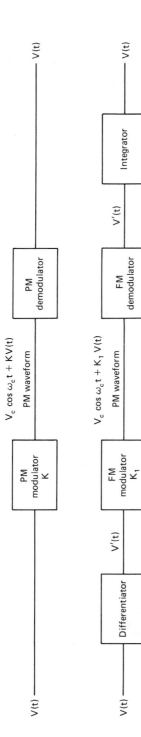

FIGURE 7-4 Frequency and phase modulation and demodulation.

273

frequency. However, most of the side frequencies are negligibly small in amplitude and can be ignored when using a variety of criteria.

Modulation by a single-frequency sinusoid.   Frequency analysis of an angle-modulated wave with a single-frequency sinusoid produces a peak phase deviation of m radians, where m is the modulation index. Again, from Equation 7-11 and for a modulating frequency equal to ω_a, $M(t)$ is written as

$$M(t) = V_c \cos (\omega_c t + m \cos \omega_a t)$$

From Equation 7-11, the individual frequency components that make up the modulated wave are not obvious. However, *Bessel function identities* are available that may be applied directly. One such identity is

$$\cos (\alpha + m \cos \beta) = \sum_{n=-\infty}^{\infty} J_n(m) \cos \left(\alpha + n\beta + \frac{n\pi}{2} \right) \qquad (7\text{-}16)$$

$J_n(m)$ is the Bessel function of the first kind of nth order and of argument m. If Identity 7-16 is applied to Equation 7-11, $M(t)$ may be rewritten as

$$M(t) = V_c \sum_{n=-\infty}^{\infty} J_n(m) \cos \left(\omega_c t + n\omega_a t + \frac{n\pi}{2} \right) \qquad (7\text{-}17)$$

Expanding Equation 7-17 for the first four terms yields

$$M(t) = V_c \{ J_0(m) \cos \omega_c t + J_1(m) \cos \left[(\omega_c + \omega_a)t + \frac{\pi}{2} \right]$$

$$+ J_1(m) \cos \left[(\omega_c - \omega_a)t + \frac{\pi}{2} \right] - J_2(m) \cos \left[(\omega_c + 2\omega_a)t \right] \qquad (7\text{-}18)$$

$$- J_2(m) \cos \left[(\omega_c - 2\omega_a)t \right] + \cdots$$

Equations 7-17 and 7-18 show that a single-frequency modulating signal produces an infinite number of sets of side frequencies, each displaced from the carrier by integral multiples of the modulating frequency. A sideband set includes an upper and a lower side frequency ($\pm F_a$, $\pm 2F_a$, $\pm 3F_a$, etc.). The successive sets of sidebands are called the first-order sidebands, second-order sidebands, and so on; and their magnitudes are determined by the coefficients $J_1(m)$, $J_2(m)$, and so on, respectively. Table 7-2 shows the Bessel functions of the first kind for several values of m. The values shown for J_n are relative to the amplitude of the unmodulated carrier. For example, $J_2 = 0.35$ indicates that the amplitude of the second set of side frequencies is equal to 35% of the unmodulated carrier amplitude ($0.35V_c$). It can be seen that the amplitude of the higher-order sidebands rapidly becomes insignificant as the modulation index decreases below unity. For larger values of m, the value of $J_n(m)$ starts to decrease rapidly as soon as $n = m$. As the modulation index increases from zero, the magnitude of the carrier $J_0(m)$ decreases.

TABLE 7-2 BESSEL FUNCTIONS OF THE FIRST KIND, $J_n(m)$

m_f	J_0	J_1	J_2	J_3	J_4	J_5	J_6	J_7	J_8	J_9	J_{10}	J_{11}	J_{12}	J_{13}	J_{14}
0.00	1.00	—	—	—	—	—	—	—	—	—	—	—	—	—	—
0.25	0.98	0.12	—	—	—	—	—	—	—	—	—	—	—	—	—
0.5	0.94	0.24	0.03	—	—	—	—	—	—	—	—	—	—	—	—
1.0	0.77	0.44	0.11	0.02	—	—	—	—	—	—	—	—	—	—	—
1.5	0.51	0.56	0.23	0.06	0.01	—	—	—	—	—	—	—	—	—	—
2.0	0.22	0.58	0.35	0.13	0.03	—	—	—	—	—	—	—	—	—	—
2.4	0	0.52	0.43	0.20	0.06	0.02	—	—	—	—	—	—	—	—	—
2.5	−0.05	0.50	0.45	0.22	0.07	0.02	0.01	—	—	—	—	—	—	—	—
3.0	−0.26	0.34	0.49	0.31	0.13	0.04	0.01	—	—	—	—	—	—	—	—
4.0	−0.40	−0.07	0.36	0.43	0.28	0.13	0.05	0.02	—	—	—	—	—	—	—
5.0	−0.18	−0.33	0.05	0.36	0.39	0.26	0.13	0.05	0.02	—	—	—	—	—	—
6.0	0.15	−0.28	−0.24	0.11	0.36	0.36	0.25	0.13	0.06	0.02	—	—	—	—	—
7.0	0.30	0.00	−0.30	−0.17	0.16	0.35	0.34	0.23	0.13	0.06	0.02	—	—	—	—
8.0	0.17	0.23	−0.11	−0.29	−0.10	0.19	0.34	0.32	0.22	0.13	0.06	0.03	—	—	—
9.0	−0.09	0.25	0.14	−0.18	−0.27	−0.06	0.20	0.33	0.31	0.21	0.12	0.06	0.03	0.01	—
10.0	−0.25	0.05	0.25	0.06	−0.22	−0.23	−0.01	0.22	0.32	0.29	0.21	0.12	0.06	0.03	0.01

For m equal to approximately 2.4, $J_0(m) = 0$ and the carrier vanishes (this is called the *first carrier null*). This property is often used to determine the modulation index or set the deviation for an FM modulator. The carrier reappears as m is increased beyond 2.4. When m reaches 5.4, the carrier once again vanishes (this is called the *second carrier null*). The carrier goes to zero at periodic intervals as the index of modulation is further increased. Figure 7-5 shows the curves for the relative amplitudes of the carrier and several sets of side frequencies for values of m up to 10. It can be seen that the amplitude of both the carrier and the side frequencies vary at a periodic rate that is a damped sine wave. The negative values for $J(m)$ simply indicate the relative phase of that side-frequency set.

In Table 7-2 only the significant side frequencies are listed. A side frequency is not considered significant unless it has an amplitude equal to or greater than 1% of the unmodulated carrier amplitude ($J_n \geq 0.01$). From Table 7-2 it can be seen that as m increases, the number of significant sidebands also increases. Therefore, the bandwidth of an angle-modulated wave is directly proportional to the modulation index.

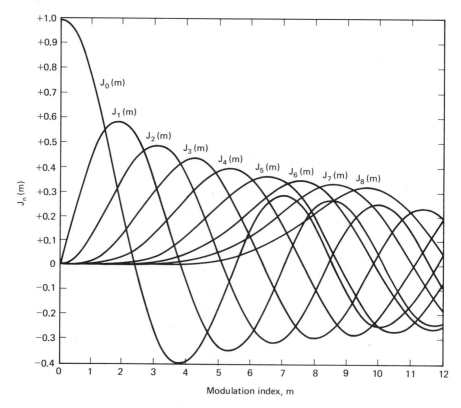

FIGURE 7-5 $J_n(m)$ versus m.

EXAMPLE 7-2

For an FM modulator with a modulation index $m = 1$, a modulating signal $V(t) = V_a \sin (2\pi 1000t)$ and an unmodulated carrier $v_c = 10 \sin (2\pi 500kt)$:
   (a) Determine the number of sets of significant side frequencies.
   (b) Determine their amplitudes.
   (c) Draw the frequency spectrum.

Solution (a) From Table 7-2 a modulation index of 1 yields a reduced carrier component and three sets of side frequencies.
   (b) Their amplitudes are
$$(1) = .77$$
$$J_0 = 0.77(10) = 7.7 \text{ V}_p \qquad carrier$$
$$J_1 = 0.44(10) = 4.4 \text{ V}_p$$
$$J_2 = 0.11(10) = 1.1 \text{ V}_p \qquad \bigg\} \text{ 3 sets of sidebands}$$
$$J_3 = 0.02(10) = 0.2 \text{ V}_p$$

(c) The frequency spectrum is shown in Figure 7-6.

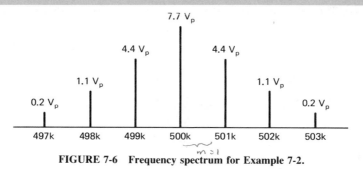

$m = 1$

FIGURE 7-6   Frequency spectrum for Example 7-2.

If the FM modulator in Example 7-2 were replaced with a PM modulator and the same carrier and modulating signal frequencies were used, a peak phase deviation $m = 1$ rad produces exactly the same frequency spectrum.

Bandwidth Requirements for Angle-Modulated Waves

In 1922, J. R. Carson mathematically proved that frequency modulation cannot be accommodated in a narrower bandwidth than AM. From the preceding discussion and Example 7-2, it can be seen that the bandwidth of an angle-modulated wave is a function of the modulating signal frequency and the modulation index. With angle modulation, multiple sets of sidebands are produced and, consequently, the bandwidth can be significantly wider than that of an AM wave with the same modulating signal. The modulator output waveform in Example 7-2 requires 6 kHz of bandwidth to pass the carrier and the significant side frequencies. A conventional AM modulator would require only 2 kHz of bandwidth, and a single-sideband system only 1 kHz.

Angle-modulated waveforms are generally classified as either *low*, *medium*, or *high* index. For the low-index case, the peak phase deviation (modulation index) is less than 1 rad, and the high-index case is when the peak phase deviation is greater than 10 rad. Modulation indices between 1 and 10 are classified as medium index. From Table 7-2 it can be seen that with low-index angle modulation, most of the signal information is carried by the first set of sidebands and the minimum bandwidth required is approximately equal to twice the highest modulating signal frequency. For a high-index signal, a method of determining the bandwidth called the quasi-stationary approach is used. With this approach, it is assumed that the modulating signal is changing very slowly. For example, for an FM modulator with a deviation sensitivity $K_1 = 2 \pi 2000$ rad per volt-second and a 1-V_p modulating signal, the peak frequency deviation $\Delta F = 2000$ Hz. If the rate of change of the frequency of the modulating signal is very slow, the bandwidth is determined by the peak-to-peak frequency deviation. Therefore, for large indices, the minimum bandwidth required is equal to the peak-to-peak frequency deviation or twice the peak frequency deviation.

Thus for low-index modulation, the minimum bandwidth is approximated by

$$B = 2F_a \qquad (7\text{-}19)$$

and for high-index modulation, the minimum bandwidth is approximated by

$$B = 2(\Delta F) \qquad (7\text{-}20)$$

The actual bandwidth required to pass all of the significant sidebands for an angle-modulated wave is equal to two times the product of the highest modulating signal and the number of significant sidebands determined from the table of Bessel functions. Mathematically, the rule for determining the bandwidth for an angle-modulated wave using the Bessel table is

$$B = 2(n \times F_a) \qquad (7\text{-}21)$$

where

> n = number of significant sidebands
> F_a = highest modulating signal frequency

In an unpublished memorandum dated August 28, 1939, Carson established a general rule to estimate the bandwidth for all angle-modulated systems regardless of the modulation index. This rule is called Carson's rule. Simply stated, Carson's rule approximates the minimum bandwidth of an angle-modulated wave as twice the sum of the peak frequency deviation and the highest modulating signal frequency. Mathematically, Carson's rule is

$$B = 2(\Delta F + F_a) \qquad (7\text{-}22)$$

where

ΔF = peak frequency modulation
F_a = highest modulating signal frequency

Carson's rule is an approximation and gives transmission bandwidths that are narrower than the bandwidths actually determined using the Bessel table and Equation 7-21. The actual bandwidth required is a function of the modulating signal waveform and the quality of transmission desired.

EXAMPLE 7-3

For an FM modulator with a peak frequency deviation ΔF = 10 kHz, a modulating signal frequency F_a = 10 kHz, and a 500-kHz carrier:
    (a) Determine the actual minimum bandwidth from the Bessel function table.
    (b) Determine the approximate minimum bandwidth using Carson's rule.
    (c) Plot the output frequency spectrum.

Solution   (a) Substituting into Equation 7-14 yields

$$m = \frac{\Delta F}{F_a} = \frac{10,000}{10,000} = 1$$

From Table 7-2 a modulation index of 1 yields three sets of significant sidebands. Substituting into Equation 7-21, the minimum bandwidth is

$$B = 2(3 \times 10,000) = 60 \,\text{kHz}$$

(b) Substituting into Equation 7-22, the minimum bandwidth is

$$B = 2(10 \,\text{kHz} + 10 \,\text{kHz}) = 40 \,\text{kHz}$$

(c) The output frequency spectrum is shown in Figure 7-7.

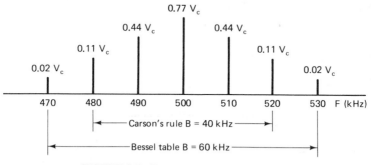

FIGURE 7-7   **Frequency spectrum for Example 7-3.**

From Example 7-3 it can be seen that the minimum bandwidth determined from Carson's rule is less than the actual minimum bandwidth required to pass all of the significant sideband sets as defined in the Bessel table. Therefore, a system that was designed using Carson's rule would have a poorer performance than a system designed using the Bessel table. For modulation indices above 5, Carson's rule is a close approximation to the actual minimum bandwidth required.

Deviation ratio.   For a given FM system, the minimum bandwidth is greatest when the maximum frequency deviation is produced by the maximum modulating signal frequency (i.e., the highest modulating frequency occurs with the maximum amplitude allowed). By definition, *deviation ratio* is the *worst-case* modulation index and is equal to the maximum peak frequency deviation divided by the maximum modulating signal frequency. The worst-case modulation index produces the widest FM output frequency spectrum. Mathematically, the deviation ratio is

$$\text{DR} = \frac{\Delta F_{max}}{F_{a(max)}} \tag{7-23}$$

where

$$\text{DR} = \text{deviation ratio}$$
$$F_{a(max)} = \text{maximum modulating signal frequency (Hz)}$$
$$\Delta F_{max} = \text{maximum peak frequency deviation (Hz)}$$

For example, for the commercial FM broadcast band, the maximum frequency deviation set by the FCC is 75 kHz and the maximum modulating signal frequency is 15 kHz. Therefore, the deviation ratio for FM broadcasting is 75 kHz/15 kHz = 5. This does not mean that whenever a modulation index of 5 occurs, the widest bandwidth also occurs. It means that whenever a modulation index of 5 occurs for the maximum modulating signal frequency, the widest bandwidth occurs.

EXAMPLE 7-4

(a) Determine the deviation ratio and bandwidth for the worst-case (widest-bandwidth) modulation index for an FM broadcast-band transmitter.
(b) Determine the deviation ratio and maximum bandwidth for an equal modulation index with only half the peak frequency deviation and modulating signal frequency.

Solution   (a) $\text{DR} = \dfrac{75 \text{ kHz}}{15 \text{ kHz}} = 5 \; \simeq \; m$

From Table 7-2 a modulation index of 5 produces eight sets of significant sidebands. Substituting into Equation 7-21 yields $B = 2(n \times F_a)$ $P. \; 278$
$$B = 2(8 \times 15,000) = 240 \text{ kHz}$$

(b) For a 37.5-kHz frequency deviation and a modulating signal frequency $F_a = 7.5$ kHz, the modulation coefficient is

all values ½ of part a. $\dfrac{37.5 \text{ kHz}}{7.5 \text{ kHz}} = 5$

and the bandwidth is

$$B = 2(8 \times 7500) = 120 \text{ kHz}$$

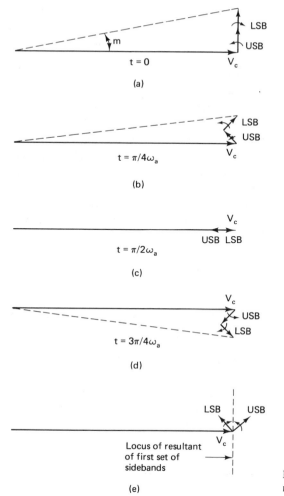

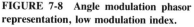

FIGURE 7-8 Angle modulation phasor representation, low modulation index.

It would seem that a higher modulation index (i.e., more sideband sets) with a lower modulating frequency would generate a wider bandwidth because there are considerably more sideband sets generated, but remember, the sidebands are closer together. For example, a 1-kHz modulating signal that produces 10 kHz of frequency deviation has a modulation index $m = 10$ and produces 14 sets of sidebands. However, the sidebands are displaced from each other by only 1 kHz, and therefore the total bandwidth $B = 2(14 \times 1000) = 28,000$ Hz.

Phasor Representation of Angle Modulation

As with AM, an angle-modulated wave can be shown with phasors. The phasor diagram for a low-index-angle modulated wave with a single-frequency modulating signal is shown in Figure 7-8. For this special case ($m < 1$) only the first set of sideband pairs is considered and the phasor diagram is very similar to that of an AM wave except for a phase reversal of one of the side frequencies. The resultant vector has an amplitude that is close to unity at all times and a peak phase deviation of m radians. It is important to note that if the higher-order terms were included, the vector would have no amplitude variations. The dashed line shown in Figure 7-8e is the locus of the resultant formed by the carrier and the first set of side frequencies.

Figure 7-9 shows the phasor diagram for a high-index-angle modulated wave with five sets of side frequencies (for simplicity, the vectors for only two sets are shown). The resultant vector is the sum of the carrier component and the significant side frequencies with their magnitudes adjusted according to the Bessel table. Each side frequency is shifted an additional 90° from the preceding one. The locus of the resultant five-component approximation is curved and closely follows the signal locus. By definition, the locus is a segment of the circle with radius equal to the amplitude of the unmodulated carrier. It should be noted that the resultant signal amplitude and, consequently, the signal power remains constant.

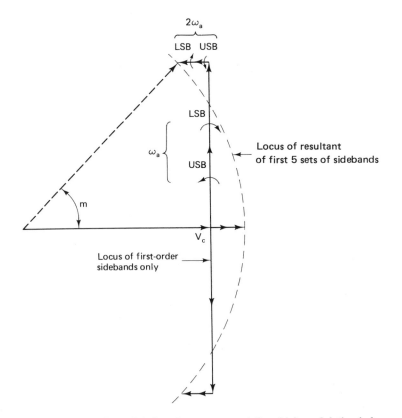

FIGURE 7-9   Angle modulation phasor representation, high modulation index.

Average Power of an Angle-Modulated Wave

The total power in an angle-modulated wave is equal to the power of the unmodulated carrier. Therefore, with angle modulation (unlike AM) the power that was originally in the unmodulated carrier is redistributed among the carrier and its sidebands. The rms power of an angle-modulated wave is independent of the modulating signal and equal to the rms power of the unmodulated carrier. Mathematically, the rms power in the unmodulated carrier is simply

$$P_c = \frac{V_c^2}{2R} \text{ Wrms} \tag{7-24a}$$

where

P_c = carrier power (watts rms)
V_c = unmodulated carrier voltage (volts peak)
R = load resistance

and peak power is

$$P_c = \frac{V_c^2}{R} \tag{7-24b}$$

and the total rms power in an angle-modulated carrier is

$$P_t = \frac{M^2(t)}{R} \tag{7-25a}$$

or

$$P_t = \frac{V_c^2}{R} \cos^2[\omega_c t + \theta(t)] \tag{7-25b}$$

$$= \frac{V_c^2}{R} \left\{ \frac{1}{2} + \frac{1}{2} \cos[2\omega_c t + 2\theta(t)] \right\} \tag{7-25c}$$

In Equation 7-25c, the second term consists of an infinite number of sinusoidal side-frequency components about a frequency equal to twice the unmodulated carrier frequency ($2F_c$). Consequently, the average value of the second term is zero and the rms power of the modulated wave is

$$P = \frac{V_c^2}{2R} \text{ Wrms} \tag{7-26}$$

Note that Equations 7-24 and 7-26 are identical, so that the average power of the modulated carrier is equal to the power of the unmodulated carrier. The modulated carrier power is the sum of the powers of the carrier and the side frequencies. Therefore, the total modulated carrier power is

$$P_t = P_c + P_1 + P_2 + P_3 + P_n \tag{7-27a}$$

$$= \frac{V_c^2}{R} + \frac{2(V_1)^2}{R} + \frac{2(V_2)^2}{R} + \frac{2(V_3)^2}{R} + \frac{2(V_n)^2}{R} \tag{7-27b}$$

where

P_c = carrier power
P_1 = power in first set of sidebands
P_2 = power in second set of sidebands
P_3 = power in third set of sidebands
P_n = power in nth set of sidebands

EXAMPLE 7-5

(a) Determine the peak unmodulated carrier power for the FM modulator and conditions given in Example 7-2 (assume a load resistance $R_L = 50\ \Omega$).
(b) Determine total peak angle-modulated wave power.

Solution   (a) Substituting into Equation 7-24 yields

$$P_c = \frac{10^2}{50} = 2 \text{ W peak}$$

(b) Substituting into Equation 7-27b gives us

$$P_t = \frac{7.7^2}{50} + \frac{2(4.4)^2}{50} + \frac{2(1.1)^2}{50} + \frac{2(0.2)^2}{50}$$

$$= 1.1858 + 0.7744 + 0.0484 + 0.0016$$

$$= 2.0102 \text{ W peak}$$

The results of (a) and (b) are not exactly equal because the values given in the Bessel table have been rounded off. However, the results are close enough to illustrate that the power in the modulated wave is equal to the power of the unmodulated carrier.

Noise and Angle Modulation

When random white noise with a constant spectral density is added to an FM signal, it produces an unwanted deviation of the carrier frequency. The magnitude of the unwanted modulation depends on the relative amplitude of the noise in respect to the carrier. When this unwanted carrier deviation is demodulated, it becomes noise if it has frequency components that fall into the modulating signal spectrum. The spectral shape of the demodulated noise depends on whether an FM or a PM demodulator is used. The noise voltage at the output of a PM demodulator is constant with frequency, whereas the noise voltage at the output of an FM demodulator increases linearly with frequency. This is commonly called the FM *noise triangle* and is shown in Figure 7-10. From Figure 7-10 it can be seen that the demodulated noise voltage is inherently higher for the higher-modulating signal frequencies.

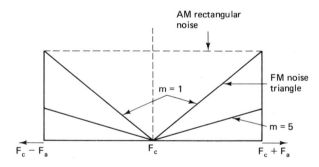

FIGURE 7-10   FM noise triangle.

PM due to an interfering sinusoid.   Figure 7-11a shows phase modulation caused by a single-frequency noise signal. The noise component V_n is separated in frequency from the signal component V_c by frequency ω_n. This is shown in Figure 7-11b. Assuming that $V_c > V_n$, the peak phase deviation due to an interfering single-frequency sinusoid is

$$\Delta\theta(\text{peak}) = \frac{V_n}{V_c} \quad \text{rad} \tag{7-28}$$

Figure 7-11c shows the result of *limiting* the amplitude of the composite FM signal on noise. (Limiting is commonly used in angle-modulation receivers and is explained in Chapter 8.) It can be seen that the single-frequency noise signal has been transposed into a noise sideband pair each with amplitude $V_n/2$. These sidebands are coherent; therefore, the peak phase deviation is still V_n/V_c radians. However, the unwanted amplitude variations have been removed, which reduces the total power but does not reduce the interference in the demodulated signal due to the unwanted phase deviation. Because the phase modulation is sinusoidal, the rms phase deviation due to the interfering noise signal is the peak phase deviation divided by $\sqrt{2}$. Mathematically, the rms phase deviation is

$$\Delta\theta(\text{rms}) = \frac{V_n}{V_c\sqrt{2}} \tag{7-29}$$

FM due to an interfering sinusoid.   As stated previously and from Equation 7-5, the instantaneous frequency deviation $\Delta F(t)$ is the first time derivative of the instantaneous phase deviation $\theta(t)$. When the carrier component is much larger than the interfering noise voltage, the instantaneous phase deviation is approximately

$$\theta(t) = \frac{V_n}{V_c}\sin(\omega_n t + \theta_n) \quad \text{rad} \tag{7-30}$$

and taking the first derivative, we obtain

$$\Delta F(t) = \frac{V_n}{V_c}\omega_n(\cos \omega_n t + \theta_n) \quad \text{rad/s} \tag{7-31}$$

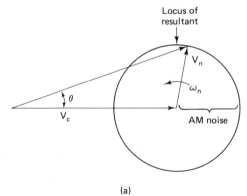

(a)

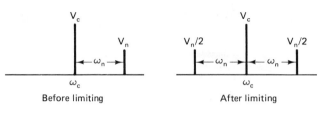

(b)

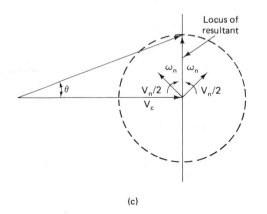

(c)

**FIGURE 7-11   Interfering sinusoid of
noise: (a) before limiting; (b) frequency
spectrum; (c) after limiting.**

Therefore, peak frequency deviation is

$$\Delta F(\text{peak}) = \frac{V_n}{V_c}\,\omega_n \qquad \text{rad/s} \qquad (7\text{-}32\text{a})$$

$$= \frac{V_n}{V_c}\,F_n \qquad \text{Hz} \qquad (7\text{-}32\text{b})$$

and the rms frequency deviation is

$$\Delta F(\text{rms}) = \frac{V_n}{V_c \sqrt{2}} \omega_n \qquad \text{rad} \tag{7-33a}$$

$$= \frac{V_n}{V_c \sqrt{2}} F_n \qquad \text{Hz} \tag{7-33b}$$

Rearranging Equation 7-14, it can be seen that the peak frequency deviation (ΔF) is a function of the modulating frequency and the modulation index. Therefore, for a noise modulating frequency F_n, the peak frequency deviation is

$$\Delta F(\text{peak}) = mF_n \tag{7-34}$$

where m = peak phase deviation

From Equation 7-34 it can be seen that the further the noise frequency is displaced from the carrier frequency, the larger the frequency deviation. Therefore, noise frequencies that produce components at the high end of the modulating signal spectrum produce more frequency deviation for the same phase deviation than frequencies that fall at the low end. FM demodulators generate an output voltage that is proportional to the frequency deviation and of frequency equal to the difference between the carrier and the interfering signal. Therefore, high-frequency noise components produce more demodulated noise than do low-frequency components.

The signal-to-noise ratio at the output of an FM demodulator due to unwanted frequency deviation from an interfering sinusoid is the ratio of the peak frequency deviation due to the information signal to the peak frequency deviation due to the interfering signal.

$$\frac{S}{N} = \frac{\Delta F(\text{signal})}{\Delta F(\text{noise})} \tag{7-35}$$

EXAMPLE 7-6

For an angle-modulated carrier $V_c = 6 \cos(2\pi 110 \text{ MHz}t)$ with 75 kHz of frequency deviation due to the information signal and a single-frequency interfering signal $V_n = 0.3 \cos(2\pi 109.985 \text{ MHz}t)$.
   (a) Determine the frequency of the demodulated interference signal.
   (b) Determine the peak and rms phase and frequency deviations due to the interfering signal.
   (c) Determine the signal-to-noise ratio at the output of the demodulator.

Solution   (a) The frequency of the noise interference is

$$F_c - F_n = 110 \text{ MHz} = 109.985 \text{ MHz} = 15 \text{ kHz}$$

(b) Substituting into Equation 7-28 yields

$$\theta(\text{peak}) = \frac{0.3}{6} = 0.05 \text{ rad}$$

and from Equation 7-29 the rms phase deviation is

$$\theta(\text{rms}) = \frac{0.3}{6\sqrt{2}} = 0.03535 \text{ rad}$$

Substituting into Equation 7-32b yields

$$\Delta F(\text{peak}) = \left(\frac{0.3}{6}\right) 15 \text{ kHz} = 750 \text{ Hz}$$

Substituting into Equation 7-33b gives us

$$\Delta F(\text{rms}) = \left(\frac{0.3}{6\sqrt{2}}\right) 15 \text{ kHz} = 530 \text{ Hz}$$

(c) The *S/N* ratio of the interfering tone and the carrier is the ratio of the peak carrier amplitude to the peak noise amplitude, or

$$\frac{6}{0.3} = 20$$

The *S/N* ratio after demodulation is found by substituting into Equation 7-35:

$$\frac{S}{N} = \frac{75 \text{ kHz}}{750 \text{ Hz}} = 100$$

There is an *S/N* improvement of 100/20 = 5 or 10 log 5 = 7 dB.

Angle modulation due to random noise.   Random noise comprises an infinite number of frequencies, of equal amplitude, and of arbitrary phase. It is convenient to analyze system noise on a per-hertz basis and to consider a band of noise N hertz wide to be made up of N uniformly spaced sinusoids. If the total peak noise voltage and, consequently, the total noise power is assumed to be relatively small compared to the carrier, the phase modulation is approximated as

$$\theta(t) = \sum_{N=-1}^{N} \frac{V_n}{V_c} \sin(\omega_n t + \theta_n) \qquad (7\text{-}36)$$

From Equation 7-36 it can be seen that the total phase modulation due to random noise is equal to the summation of the phase modulation components that are produced by each individual noise component. If these individual noise components are large compared to the carrier, intermodulation distortion occurs. Because phase relationships that exist between the noise sinusoids are not known, the peak phase modulation they produce cannot be defined. Therefore, the total rms voltage is

$$\text{total rms voltage} = \sum_{N=1}^{N} \frac{V_n^2}{2} \qquad (7\text{-}37\text{a})$$

$$= \frac{V_n}{2} N \qquad (7\text{-}37\text{b})$$

where $V_n/2$ is the rms amplitude of each sinusoid. The total rms phase deviation is

$$\theta(t)_{\text{total}} = \frac{(V_n/\sqrt{2})N}{V_c} \quad \text{rad} \tag{7-38a}$$

$$= \frac{V_n N}{V_c \sqrt{2}} \quad \text{rad} \tag{7-38b}$$

EXAMPLE 7-7

Determine the total rms phase deviation produced by a 10-kHz band of random noise with a peak voltage $V_n = 0.1$ μV and a carrier $V_c = 1 \sin(2\pi 10$ MHz $t)$.

Solution Substituting into Equation 7-38b gives us

$$\theta(t)_{\text{total}} = \frac{(0.1 \text{ μV})(10,000)}{\sqrt{2}} = 0.0007 \text{ rad}$$

Preemphasis and Deemphasis

The noise triangle shown in Figure 7-10 showed that, with FM, noise at the higher modulating frequencies is inherently greater in amplitude than noise at the lower frequencies. This includes both single-frequency interference and random noise. Therefore, assuming that the amplitudes of all of the information signals are equal, a nonuniform *S/N* is evident and the higher modulating frequencies suffer a greater *S/N* degradation. This is shown in Figure 7-12a. It can be seen that the *S/N* is lower at the high-frequency ends of the triangle. *Preemphasis* is artificially boosting the amplitude of the high-frequency modulating signals prior to modulation, and *deemphasis* is simply the opposite action and is performed in the receiver after demodulation to restore the original modulating signal voltage spectrum. Figure 7-12b shows the effects on the signal-to-noise ratio using pre- and deemphasis. The figure shows that pre- and deemphasis produces a uniform *S/N* ratio throughout the modulating signal spectrum.

A preemphasis network is a high-pass filter (i.e., a differentiator) and a deemphasis network is a low-pass filter (i.e., an integrator). Figure 7-13a shows the schematic diagrams for an active preemphasis network and a passive deemphasis network. Their corresponding frequency response curves are shown in Figure 7-13b. A preemphasis network provides a constant increase in the amplitude of the modulating signal with increase in modulating frequency. With FM, approximately 12 dB of improvement in noise performance is achieved using pre- and deemphasis. The break frequency (where pre- and deemphasis begins) is determined by the RC or L/R time constant of the network. Mathematically, the break frequency is

$$F = \frac{1}{2\pi RC} \quad \text{or} \quad \frac{1}{2\pi L/R} \tag{7-39}$$

The networks shown in Figure 7-13 are for the FM broadcast band, which uses a 75-μs time constant; therefore, the break frequency is approximately 2.12 kHz. The

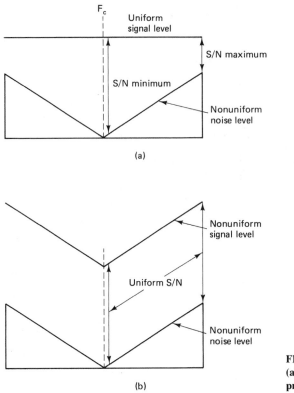

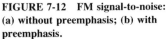

FIGURE 7-12   FM signal-to-noise:
(a) without preemphasis; (b) with
preemphasis.

FM transmission of the audio portion of commercial television broadcasting uses a 50-μs time constant. Because the noise at the output of a PM demodulator is constant with frequency, phase modulators do not require preemphasis networks.

FREQUENCY MODULATION TRANSMISSION

Direct FM

Direct FM is angle modulation in which the frequency of the carrier is varied (deviated) directly by the modulating signal. With direct FM, the instantaneous frequency deviation is directly proportional to the amplitude of the modulating signal. Figure 7-14 shows a schematic diagram for a simple (although highly impractical) direct FM generator. The tank circuit (L and C_m) is the frequency-determining section for a standard LC oscillator. The capacitor microphone is a *transducer* that converts *acoustical energy* to *mechanical energy* which is used to vary the distance between the plates of C_m and, consequently, change its capacitance. As C_m is varied, the resonant frequency is varied. Consequently, the oscillator output frequency varies directly with the input from the external sound

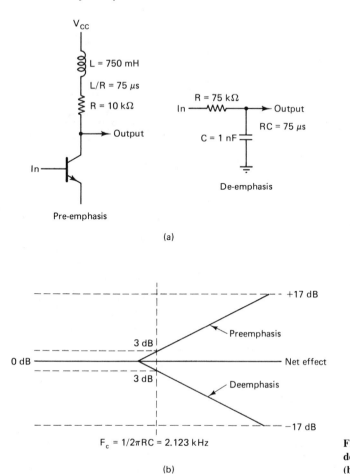

(a)

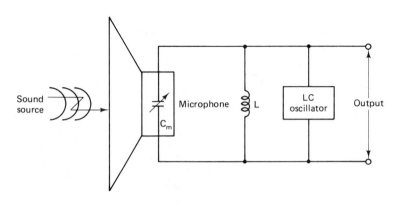

$F_c = 1/2\pi RC = 2.123$ kHz

(b)

FIGURE 7-13 Preemphasis and deemphasis: (a) schematic diagrams; (b) attenuation curves.

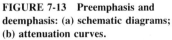

FIGURE 7-14 Simple direct FM modulator.

source. This is direct FM generation because the oscillator frequency is directly changed (deviated) by the modulating signal and the magnitude of the frequency change is proportional to the amplitude of the modulating signal.

Varactor diode modulators.   Figure 7-15 shows the schematic diagram for a more practical direct FM generator that uses a varactor diode to deviate the frequency of a crystal oscillator. R_1 and R_2 develop a dc bias that reverse biases varactor diode VD_1 and determines the center or rest frequency for the oscillator. The external audio modulating signal voltage adds to and subtracts from the dc bias which changes the varactor's capacitance and thus the oscillator's frequency. Positive alternations of V_a increase the reverse bias across VD_1, which decreases its capacitance and increases the frequency of oscillation. Negative alternations of V_a decrease the reverse bias across VD_1, increase its capacitance, and decrease the frequency of oscillation. Varactor diode FM modulators are extremely popular because they are simple to use, reliable, and have the stability of a crystal oscillator. However, because a crystal is used, the peak frequency deviation is limited to relatively small values. Consequently, they are used primarily for low-index applications, such as two-way mobile FM radio.

Figure 7-16 shows a simplified schematic diagram for a voltage-controlled oscillator (VCO) FM generator. Again, a varactor diode is used to transform changes in the modulating signal to changes in frequency. The center or rest frequency for the modulator is

$$F_c = \frac{1}{2\pi \sqrt{LC}}$$ (7-40)

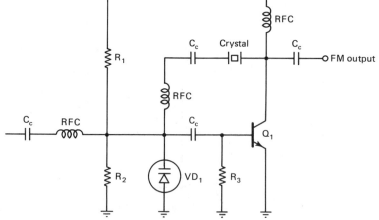

FIGURE 7-15   Varactor diode direct FM modulator.

where

L = primary inductance
C = varactor diode capacitance

With a modulating signal the frequency is

$$F_m = \frac{1}{2\pi \sqrt{L(C + \Delta C)}} \tag{7-41}$$

where ΔC is the change in varactor diode capacitance due to the modulating signal. The change in frequency with modulation (ΔF) is

$$\Delta F = F_c - F_m \tag{7-42}$$

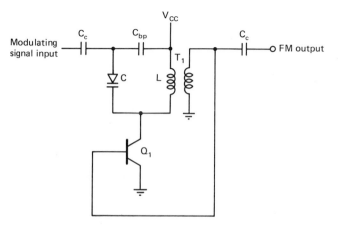

FIGURE 7-16   Varactor diode VCO FM modulator.

Reactance modulator.   Figure 7-17a shows a schematic diagram for a *reactance modulator* using a JFET as the active device. This circuit configuration is called a reactance modulator because the JFET looks like a variable-reactance load to the LC tank circuit. The modulating signal varies the reactance of Q_1, which causes a corresponding change in the resonant frequency of the oscillator tank circuit.

Figure 7-17b shows the ac equivalent circuit. R_1, R_2, R_3, R_4, R_E, and R_C provide the dc bias for Q_1. However, R_E is bypassed by C_E and is, therefore, omitted from the ac equivalent circuit. Circuit operation is as follows. Assuming an ideal JFET ($i_g = 0$)

$$v_g = i_R \tag{7-43}$$

where

$$i = \frac{v}{R - jX_c} \tag{7-44}$$

Therefore,

$$v_g = \frac{v}{R - jX_c} R \tag{7-45}$$

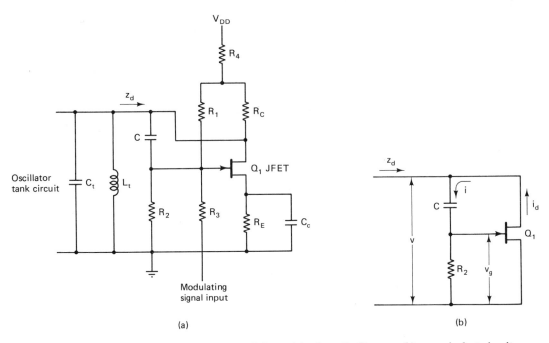

FIGURE 7-17 JFET reactance modulator: (a) schematic diagram; (b) ac equivalent circuit.

and the JFET drain current i_d is

$$i_d = g_m v_g = g_m \frac{v}{R - jX_c} R \qquad (7\text{-}46)$$

where g_m is the JFET transconductance, and the impedance between the drain and ground is

$$z_d = \frac{e}{id} = g_m \frac{v/v}{R - jX_c} R \qquad (7\text{-}47\text{a})$$

Rearranging Equation 7-47a yields

$$z_d = \frac{R - jX_c}{g_m R} = \frac{1}{g_m} - \frac{jX_c}{g_m R} \qquad (7\text{-}47\text{b})$$

Assuming that $R \ll X_c$,

$$z_d = -j\frac{X_c}{g_m R} = \frac{-j}{2\pi F g_m RC} \qquad (7\text{-}47\text{c})$$

$g_m RC$ is equivalent to a variable capacitance and is inversely proportional to resistance (R), the angular frequency of the modulating signal $(2\pi F)$, and the transconductance (g_m) of Q_1, which varies with the gate-to-source voltage. When a modulating signal is applied to R_3, the gate-to-source voltage is varied accordingly, causing a proportional change in g_m. As a result, the equivalent circuit impedance (z_d) is a function of the modulating signal. Therefore, the resonant frequency of the oscillator tank circuit is a function of the amplitude of the modulating signal and the rate at which it deviates is equal to $2\pi F$. Interchanging R_2 and C causes the variable reactance to be inductive rather than capacitive but does not affect the output FM waveform. The maximum frequency deviation for a reactance modulator is approximately 5 kHz.

Linear integrated-circuit direct FM generator. *Linear integrated-circuit* (LIC) *FM modulators* generate a high-quality output waveform that is stable, accurate, and directly proportional to the input modulating signal. Figure 7-18 shows a simplified schematic diagram for a LIC direct FM generator. The VCO center frequency is determined by the external resistor and capacitor (R and C). The input modulating signal deviates the center frequency, which produces an FM output waveform. The analog multiplier and sine shaper convert the VCO square-wave output to a sine wave, and the unity-gain amplifier provides a buffered output. The modulator output frequency is

$$FM = (F_o + \Delta F)N \qquad (7\text{-}48)$$

where $\Delta F = V_a K$.

Figure 7-19 shows the specification sheets for the XR-2206 *monolithic function generator*. The 2206 can generate either a sine or a triangular FM output waveform. The 2206 can be used for either sweep frequency operation, frequency shift keying (FSK), or direct FM generation.

Indirect FM

Indirect FM is angle modulation in which the frequency of the carrier is changed indirectly by the modulating signal. Indirect FM is accomplished by directly deviating the phase of the carrier and is therefore a form of phase modulation. The instantaneous phase of an indirect FM carrier is directly proportional to the modulating signal amplitude. Therefore, the instantaneous frequency is proportional to the integral of the modulating signal.

Figure 7-20 shows a schematic diagram for an indirect FM modulator (henceforth referred to as a phase modulator). The phase modulator comprises a varactor diode VD_1 in series with an inductive network (tunable coil L_1 and resistor R_1). The combined series–parallel network appears as a series resonant circuit to the output frequency from the crystal oscillator. A modulating signal applied to VD_1 changes its capacitance and,

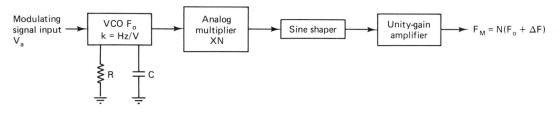

FIGURE 7-18   LIC direct FM generator: simplified schematic diagram.

Monolithic Function Generator

GENERAL DESCRIPTION

The XR-2206 is a monolithic function generator integrated circuit capable of producing high quality sine, square, triangle, ramp, and pulse waveforms of high-stability and accuracy. The output waveforms can be both amplitude and frequency modulated by an external voltage. Frequency of operation can be selected externally over a range of 0.01 Hz to more than 1 MHz.

The circuit is ideally suited for communications, instrumentation, and function generator applications requiring sinusoidal tone, AM, FM, or FSK generation. It has a typical drift specification of 20 ppm/°C. The oscillator frequency can be linearly swept over a 2000:1 frequency range, with an external control voltage, having a very small affect on distortion.

FEATURES

Low-Sine Wave Distortion	.5%, Typical
Excellent Temperature Stability	20 ppm/°C, Typical
Wide Sweep Range	2000:1, Typical
Low-Supply Sensitivity	0.01%V, Typical
Linear Amplitude Modulation	
TTL Compatible FSK Controls	
Wide Supply Range	10V to 26V
Adjustable Duty Cycle	1% to 99%

APPLICATIONS

Waveform Generation
Sweep Generation
AM/FM Generation
V/F Conversion
FSK Generation
Phase-Locked Loops (VCO)

ABSOLUTE MAXIMUM RATINGS

Power Supply	26V
Power Dissipation	750 mW
Derate Above 25°C	5 mW/°C
Total Timing Current	6 mA
Storage Temperature	−65°C to +150°C

FUNCTIONAL BLOCK DIAGRAM

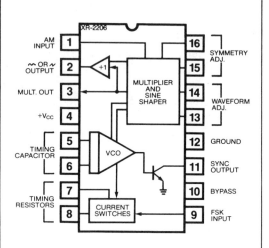

ORDERING INFORMATION

Part Number	Package	Operating Temperature
XR-2206M	Ceramic	-55°C to +125°C
XR-2206N	Ceramic	0°C to +70°C
XR-2206P	Plastic	0°C to +70°C
XR-2206CN	Ceramic	0°C to +70°C
XR-2206CP	Plastic	0°C to +70°C

SYSTEM DESCRIPTION

The XR-2206 is comprised of four functional blocks; a voltage-controlled oscillator (VCO), an analog multiplier and sine-shaper; a unity gain buffer amplifier; and a set of current switches.

The VCO actually produces an output frequency porportional to an input current, which is produced by a resistor from the timing terminals to ground. The current switches route one of the timing pins current to the VCO controlled by an FSK input pin, to produce an output frequency. With two timing pins, two discrete output frequencies can be independently produced for FSK Generation Applications.

EXAR Integrated Systems, Inc., 750 Palomar Avenue, Sunnyvale, CA 94086 * (408) 732-7970 * TWX 910-339-9233

FIGURE 7-19 Specification sheet for XR-2206 monolithic function generator. (Courtesy of EXAR Corporation.)

XR-2206

ELECTRICAL CHARACTERISTICS

Test Conditions: Test Circuit of Figure 1, $V^+ = 12V$, $T_A = 25°$, $C = 0.01\,\mu F$, $R_1 = 100\,k\Omega$, $R_2 = 10\,k\Omega$, $R_3 = 25\,k\Omega$ unless otherwise specified. S_1 open for triangle, closed for sine wave.

PARAMETER	XR-2206M			XR-2206C			UNIT	CONDITIONS
	MIN.	TYP.	MAX.	MIN.	TYP.	MAX.		
GENERAL CHARACTERISTCS								
Single Supply Voltage	10		26	10		26	V	
Split-Supply Voltage	±5		±13	±5		±13	V	
Supply Current		12	17		14	20	mA	$R_1 \geqslant 10\,k\Omega$
OSCILLATOR SECTION								
Max. Operating Frequency	0.5	1		0.5	1		MHz	$C = 1000\,pF$, $R_1 = 1\,k\Omega$
Lowest Practical Frequency		0.01			0.01		Hz	$C = 50\,\mu F$, $R_1 = 2\,M\Omega$
Frequency Accuracy		±1	±4		±2		% of f_o	$f_o = 1/R_1 C$
Temperature Stability		±10	±50		±20		ppm/°C	$0°C \leqslant T_A \leqslant 75°C$,
								$R_1 = R_2 = 20\,k\Omega$
Supply Sensitivity		0.01	0.1		0.01		%/V	$V_{LOW} = 10V$, $V_{HIGH} = 20V$,
								$R_1 = R_2 = 20\,k\Omega$
Sweep Range	1000:1	2000:1			2000:1		$f_H = f_L$	f_H @ $R_1 = 1\,k\Omega$
								f_L @ $R_1 = 2\,M\Omega$
Sweep Linearity								
10:1 Sweep		2			2		%	$f_L = 1\,kHz$, $f_H = 10\,kHz$
1000:1 Sweep		8			8		%	$f_L = 100\,Hz$, $f_H = 100\,kHz$
FM Distortion		0.1			0.1		%	±10% Deviation
Recommended Timing								
Components								
Timing Capacitor: C	0.001		100	0.001		100	μF	See Figure 4.
Timing Resistors: R_1 & R_2	1		2000	1		2000	$k\Omega$	
Triangle Sine Wave Output								See Note 1, Figure 2.
Triangle Amplitude		160			160		mV/kΩ	Figure 1, S_1 Open
Sine Wave Amplitude	40	60	80		60		mV/kΩ	Figure 1, S_1 Closed
Max. Output Swing		6			6		V p-p	
Output Impedance		600			600		Ω	
Triangle Linearity		1			1		%	
Amplitude Stability		0.5			0.5		dB	For 1000:1 Sweep
Sine Wave Amplitude Stability		4800			4800		ppm/°C	See Note 2.
Sine Wave Distortion								
Without Adjustment		2.5			2.5		%	$R_1 = 30\,k\Omega$
With Adjustment		0.4	1.0		0.5	1.5	%	See Figures 6 and 7.
Amplitude Modulation								
Input Impedance	50	100		50	100		$k\Omega$	
Modulation Range		100			100		%	
Carrier Suppression		55			55		dB	
Linearity		2			2		%	For 95% modulation
Square-Wave Output								
Amplitude		12			12		V p-p	Measured at Pin 11.
Rise Time		250			250		nsec	$C_L = 10\,pF$
Fall Time		50			50		nsec	$C_L = 10\,pF$
Saturation Voltage		0.2	0.4		0.2	0.6	V	$I_L = 2\,mA$
Leakage Current		0.1	20		0.1	100	μA	$V_{11} = 26V$
FSK Keying Level (Pin 9)	0.8	1.4	2.4	0.8	1.4	2.4	V	See section on circuit controls
Reference Bypass Voltage	2.9	3.1	3.3	2.5	3	3.5	V	Measured at Pin 10.

Note 1: Output amplitude is directly proportional to the resistance, R_3, on Pin 3. See Figure 2.
Note 2: For maximum amplitude stability, R_3 should be a positive temperature coefficient resistor.

FIGURE 7-19 (*continued*)

XR-2206

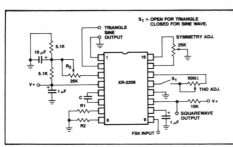

Figure 1:   Basic Test Circuit.

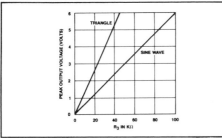

Figure 2:   Output Amplitude as a Function of the Resistor, R$_3$, at Pin 3.

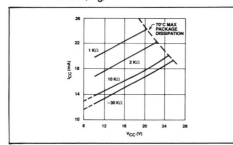

Figure 3:   Supply Current versus Supply Voltage, Timing, R.

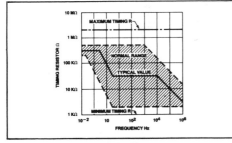

Figure 4:   R versus Oscillation Frequency.

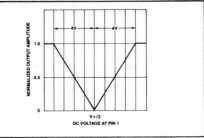

Figure 5:   Normalized Output Amplitude versus DC Bias at AM Input (Pin 1).

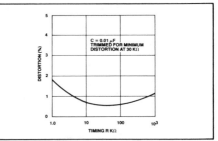

Figure 6:   Trimmed Distortion versus Timing Resistor.

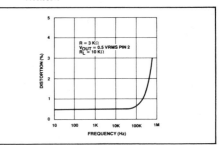

Figure 7:   Sine Wave Distortion versus Operating Frequency with Timing Capacitors Varied.

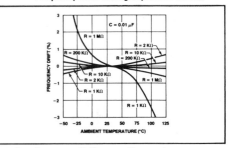

Figure 8:   Frequency Drift versus Temperature.

FIGURE 7-19 (*continued*)

298

XR-2206

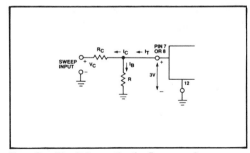

Figure 9:   Circuit Connection for Frequency Sweep.

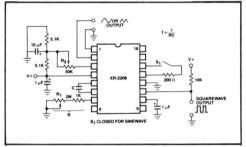

Figure 10: Circuit for Sine Wave Generation without
External Adjustment. (See Figure 2 for
Choice of R_3.)

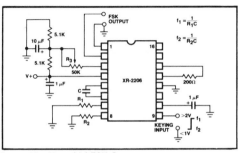

Figure 12: Sinusoidal FSK Generator.

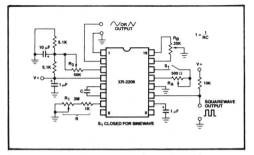

Figure 11: Circuit for Sine Wave Generation with
Minimum Harmonic Distortion. (R_3
Determines Output Swing — See Figure 2.)

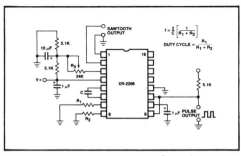

Figure 13: Circuit for Pulse and Ramp Generation.

FIGURE 7-19 (*continued*)

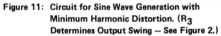

Frequency-Shift Keying:

The XR-2206 can be operated with two separate timing resistors, R_1 and R_2, connected to the timing Pin 7 and 8, respectively, as shown in Figure 12. Depending on the polarity of the logic signal at Pin 9, either one or the other of these timing resistors is activated. If Pin 9 is open-circuited or connected to a bias voltage $\geqslant 2V$, only R_1 is activated. Similarly, if the voltage level at Pin 9 is $\leqslant 1V$, only R_2 is activated. Thus, the output frequency can be keyed between two levels, f_1 and f_2, as:

$$f_1 = 1/R_1C \text{ and } f_2 = 1/R_2C$$

For split-supply operation, the keying voltage at Pin 9 is referenced to V^-.

Output DC Level Control:

The dc level at the output (Pin 2) is approximately the same as the dc bias at Pin 3. In Figures 10, 11 and 12, Pin 3 is biased midway between V^+ and ground, to give an output dc level of $\approx V^+/2$.

APPLICATIONS INFORMATION

Sine Wave Generation

Without External Adjustment:

Figure 10 shows the circuit connection for generating a sinusoidal output from the XR-2206. The potentiometer, R_1 at Pin 7, provides the desired frequency tuning. The maximum output swing is greater than $V^+/2$, and the typical distortion (THD) is <2.5%. If lower sine wave distortion is desired, additional adjustments can be provided as described in the following section.

The circuit of Figure 10 can be converted to split-supply operation, simply by replacing all ground connections with V^-. For split-supply operation, R_3 can be directly connected to ground.

With External Adjustment:

The harmonic content of sinusoidal output can be reduced to $\approx 0.5\%$ by additional adjustments as shown in Figure 11. The potentiometer, R_A, adjusts the sine-shaping resistor, and R_B provides the fine adjustment for the waveform symmetry. The adjustment procedure is as follows:

1. Set R_B at midpoint, and adjust R_A for minimum distortion.

2. With R_A set as above, adjust R_B to further reduce distortion.

Triangle Wave Generation

The circuits of Figures 10 and 11 can be converted to triangle wave generation, by simply open-circuiting Pin 13 and 14 (i.e., S_1 open). Amplitude of the triangle is approximately twice the sine wave output.

FSK Generation

Figure 12 shows the circuit connection for sinusoidal FSK signal operation. Mark and space frequencies can be independently adjusted, by the choice of timing resistors, R_1 and R_2; the output is phase-continuous during transitions. The keying signal is applied to Pin 9. The circuit can be converted to split-supply operation by simply replacing ground with V^-.

Pulse and Ramp Generation

Figure 13 shows the circuit for pulse and ramp waveform generation. In this mode of operation, the FSK keying terminal (Pin 9) is shorted to the square-wave output (Pin 11), and the circuit automatically frequency-shift keys itself between two separate frequencies during the positive-going and negative-going output waveforms. The pulse width and duty cycle can be adjusted from 1% to 99%, by the choice of R_1 and R_2. The values of R_1 and R_2 should be in the range of 1 kΩ to 2 MΩ.

FIGURE 7-19 *(continued)*

XR-2206

PRINCIPLES OF OPERATION

Description of Controls

Frequency of Operation:

The frequency of oscillation, f_0, is determined by the external timing capacitor, C, across Pin 5 and 6, and by the timing resistor, R, connected to either Pin 7 or 8. The frequency is given as:

$$f_0 = \frac{1}{RC} \text{ Hz}$$

and can be adjusted by varying either R or C. The recommended values of R, for a given frequency range, are shown in Figure 4. Temperature stability is optimum for $4 \text{ k}\Omega < R < 200 \text{ k}\Omega$. Recommended values of C are from 1000 pF to 100 μF.

Frequency Sweep and Modulation:

Frequency of oscillation is proportional to the total timing current, I_T, drawn from Pin 7 or 8:

$$f = \frac{320 I_T \text{ (mA)}}{C \text{ } (\mu F)} \text{ Hz}$$

Timing terminals (Pin 7 or 8) are low-impedance points, and are internally biased at +3V, with respect to Pin 12. Frequency varies linearly with I_T, over a wide range of current values, from 1 μA to 3 mA. The frequency can be controlled by applying a control voltage, V_C, to the activated timing pin as shown in Figure 9. The frequency of oscillation is related to V_C as:

$$f = \frac{1}{RC} \left[1 + \frac{R}{R_C} \left(1 - \frac{V_C}{3} \right) \right] \text{ Hz}$$

where V_C is in volts. The voltage-to-frequency conversion gain, K, is given as:

$$K = \partial f / \partial V_C = -\frac{0.32}{R_C C} \text{ Hz/V}$$

CAUTION: For safe operation of the circuit, I_T should be limited to $\leqslant 3$ mA.

Output Amplitude:

Maximum output amplitude is inversely proportional to the external resistor, R_3, connected to Pin 3 (see Figure 2). For sine wave output, amplitude is approximately 60 mV peak per kΩ of R_3; for triangle, the peak amplitude is approximately 160 mV peak per kΩ of R_3. Thus, for example, $R_3 = 50 \text{ k}\Omega$ would produce approximately ± 3V sinusoidal output amplitude.

Amplitude Modulation:

Output amplitude can be modulated by applying a dc bias and a modulating signal to Pin 1. The internal impedance at Pin 1 is approximately 100 kΩ. Output amplitude varies linearly with the applied voltage at Pin 1, for values of dc bias at this pin, within ± 4 volts of $V^+/2$ as shown in Figure 5. As this bias level approaches $V^+/2$, the phase of the output signal is reversed, and the amplitude goes through zero. This property is suitable for phase-shift keying and suppressed-carrier AM generation. Total dynamic range of amplitude modulation is approximately 55 dB.

CAUTION: AM control must be used in conjunction with a well-regulated supply, since the output amplitude now becomes a function of V^+.

EQUIVALENT SCHEMATIC DIAGRAM

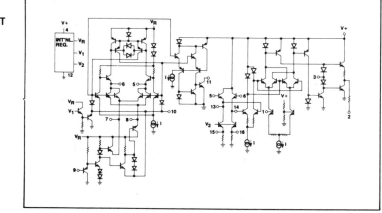

FIGURE 7-19 *(continued)*

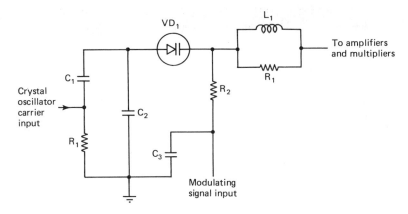

FIGURE 7-20   Indirect FM modulator schematic diagram.

consequently, the phase angle of the impedance seen by the RF carrier varies, which results in a corresponding phase shift in the carrier. The phase shift is directly proportional to the amplitude of the modulating signal. An advantage of indirect FM is that a buffered crystal oscillator is used for the carrier source. Consequently, indirect FM transmitters are more frequency stable than their direct counterparts. A disadvantage is that the capacitance-versus-voltage characteristic of a varactor diode is nonlinear. In fact, it closely resembles a square-root function. Consequently, to minimize distortion of the PM waveform, the amplitude of the modulating signal must be kept small, which, of course, limits the phase deviation (modulation index) to small values.

FM Communication Systems

Before FM transmitters are examined, it is helpful to have a basic understanding of the services allocated by the FCC for FM transmission, their specifications, and limitations.

Commercial FM broadcast band.   The commercial FM broadcast band is a wideband FM communications system that comprises one hundred 200-kHz channels that occupy the frequency band from 88 to 108 MHz. As stated previously, the maximum frequency deviation for a commercial FM broadcast-band transmitter is 75 kHz, with a maximum modulating signal frequency of 15 kHz. These requirements yield a deviation ratio of 5, which produces eight sets of significant sidebands. The sidebands are separated by 15 kHz, which produces a bandwidth of 240 kHz. It appears that *adjacent channel interference* would be a problem because the 240-kHz bandwidth determined using Bessel approximations exceeds the allocated 200-kHz bandwidth. However, from the Bessel table, it can be seen that the voltage components for the two highest sideband sets are 0.05 and 0.02, respectively, which are considered insignificant. Also, using Carson's approximation, the bandwidth is 2(75 kHz + 15 kHz), or only 180 kHz. The FCC gets around this problem by leaving every other channel assignment vacant in a given metropolitan area, and it is unlikely that any area would have the need for more than 50 FM broadcast stations. Figure 7-21 shows the frequency spectrum allocated by the FCC for commercial FM broadcasting and the worst-case frequency spectrum for a single station's transmission.

Two-way FM mobile communications. The FCC has allocated several frequency bands for the purpose of FM mobile communications, which includes such services as mobile telephone, local law enforcement, civil defense, marine, and fire department communications systems. The primary purpose for two-way FM radio is voice communications. Therefore, the bandwidth allocated for two-way FM is typically only 10 to 30 kHz. The maximum frequency deviation is approximately 5 kHz, and the highest modulating signal frequency is typically 3 kHz. Two-way mobile FM radio is *narrowband FM* (NBFM). That is, the modulation indices are low enough so that only one or two pairs of significant sidebands are generated and the bandwidth resembles that of a conventional AM system. NBFM is sometimes called *low-index FM* because low deviation ratios are used.

Several frequency bands used for two-way FM mobile radio communications are: 25 to 50 MHz, 152 to 162 MHz, 450 to 470 MHz, and 806 to 947 MHz. Cellular radio is a mobile FM telephone communications system and is described in Chapter 8.

Television broadcasting. The sound portion of commercial television broadcasting uses frequency modulation. A 4.5-MHz *subcarrier* is frequency modulated, then up-converted to either VHF or UHF for transmission. The FM subcarrier is transmitted 4.5 MHz above the picture carrier. A more detailed description of television broadcasting is given in Chapter 12. The maximum frequency deviation allowed by the FCC for television broadcasting is 25 kHz, and the maximum modulating signal frequency is 15 kHz. Therefore, the deviation ratio DR = 25 kΩ/15 kΩ or 1.67 and the maximum

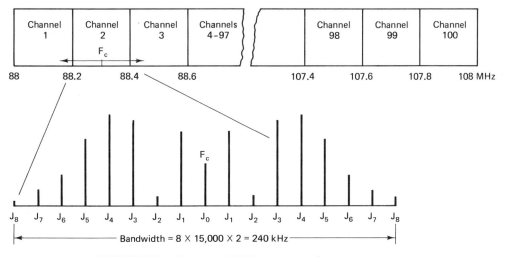

FIGURE 7-21   Commercial FM broadcasting frequency spectrum.

bandwidth using Carson's rule is $B = 2(25 \text{ kHz} + 15 \text{ kHz}) = 80 \text{ kHz}$. Figure 7-22 shows the output frequency spectrum for channel 10 of the commercial television broadcast band.

Microwave and satellite radio communications.   FM microwave and satellite systems are gradually being replaced with higher-capacity digital systems. However, due to the rapid increase in the demand for long-distance telecommunications, the existing FM systems will be with us for some time. With both microwave and satellite FM radio, an intermediate frequency (typically between 60 and 80 MHz) is frequency modulated by a wideband signal spectrum, then up-converted in frequency to the gigahertz range for transmission. The modulating signal frequency spectrum can extend from a few kilohertz up to several megahertz. Consequently, low-index NBFM must be used to minimize the transmission bandwidth. Keep in mind that narrowband FM simply means that low modulation indices are used and generally only one set of significant sidebands is generated. However, the bandwidth can still be quite wide, depending on the bandwidth of the modulating signal spectrum.

Direct FM Transmitter

Crosby direct FM transmitter.   Figure 7-23 shows the block diagram for a *Crosby direct FM transmitter* with an *automatic frequency control* (AFC) loop. The frequency modulator can be either a reactance modulator or a VCO. The carrier frequency is the unmodulated output frequency from the master oscillator (F_c). It is important to note that the carrier source is not a crystal, and if it were not for the AFC loop, this type of transmitter could not meet the frequency-stability requirements set by the FCC for commercial FM broadcasting. For the transmitter shown in Figure 7-23, the center frequency of the master oscillator is 5.1 MHz, which is multiplied by 18 in three steps ($3 \times 2 \times 3$) to produce a transmit carrier frequency $F_t = 91.8$ MHz. At this time there are three aspects of frequency conversion that should be noted. First, when the frequency of a frequency-modulated carrier is multiplied, its frequency and phase deviation are multiplied as well. Second, the rate at which the carrier is deviated (i.e., the

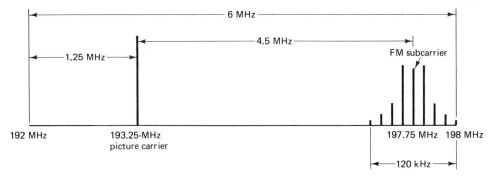

FIGURE 7-22   RF spectrum channel 10, commercial TV broadcasting.

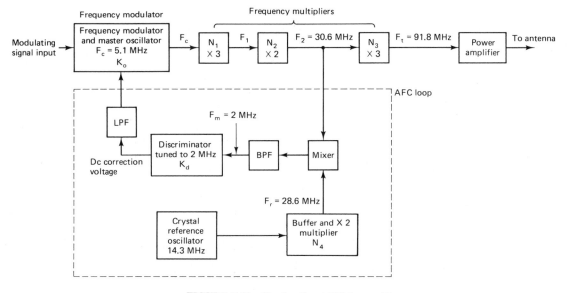

FIGURE 7-23 Crosby direct FM transmitter.

modulating signal frequency) is unaffected by the multiplication process. Therefore, the modulation index is also multiplied, Third, when a frequency-modulated carrier is heterodyned with another frequency in a mixer, the carrier can be either up- or down-converted, depending on the passband of the output filter. However, the frequency deviation, phase deviation, and rate of change are unaffected by the heterodyning process. Therefore, for the transmitter shown in Figure 7-23, the frequency and phase deviation at the output of the modulator are also multiplied by 18. To achieve 75 kHz of frequency deviation at the antenna, the deviation at the output of the modulator is

$$\Delta F = \frac{75 \text{ kHz}}{18} = 4166.7 \text{ Hz}$$

and the modulation index is

$$m = \frac{4166.7 \text{ Hz}}{F_a}$$

For a maximum modulating signal frequency $F_a = 15$ kHz,

$$m = \frac{4166.7 \text{ Hz}}{15 \text{ kHz}} = 0.2778$$

Thus the modulation index at the antenna is

$$m = 0.2778(18) = 5$$

which is the deviation ratio for commercial FM broadcasting.

AFC loop. The purpose of the *AFC loop* is to achieve near-crystal stability of the transmit carrier frequency without using a crystal as the master oscillator. With AFC, the carrier is beat against a crystal source in a mixer, down-converted in frequency, then fed back to the input of a *frequency discriminator*. A discriminator is a frequency-selective device whose output is a dc voltage that is proportional to the difference between the input frequency and its resonant frequency (discriminator operation is explained in Chapter 8). For the transmitter shown in Figure 7-23, the output from the doubler $F_2 = 30.6$ MHz, which is mixed with a crystal-controlled reference frequency $F_r = 28.6$ MHz to produce a difference frequency $F_m = 2$ MHz. The discriminator is a relatively high-Q (narrow band) tuned circuit that reacts only to frequencies near 2 MHz. Therefore, the discriminator responds to long-term, low-frequency changes in the carrier center frequency due to master oscillator drift and does not respond to the frequency deviation caused by the modulating signal. If the discriminator responded to the frequency deviation produced by the modulating signal, the feedback loop would cancel the deviation and thus remove the modulation from the FM wave (this is called ''wipe off''). The dc correction voltage is added to the modulating signal to automatically adjust the master oscillator's center frequency to compensate for the low-frequency drift.

EXAMPLE 7-8

Use the transmitter model shown in Figure 7-23 to answer the following questions. For a total frequency multiplication of ×20 and a transmit carrier frequency $F_t = 88.8$ MHz, determine:

 (a) The master oscillator center frequency.
 (b) The maximum frequency deviation at the output of the modulator.
 (c) The deviation ratio at the output of the modulator for a maximum modulating frequency $F_a = 15$ kHz.
 (d) The deviation ratio at the antenna.

Solution   (a) $F_c = \dfrac{F_t}{N_1 N_2 N_3} = \dfrac{88.8 \text{ MHz}}{20} = 4.43 \text{ MHz}$

(b) $\Delta F = \dfrac{\Delta F_t}{N_1 N_2 N_3} = \dfrac{75 \text{ kHz}}{20} = 3750 \text{ Hz}$

(c) $\text{DR} = \dfrac{\Delta F(\text{maximum})}{F_a(\text{maximum})} = \dfrac{3750 \text{ Hz}}{15 \text{ kHz}} = 0.25$

(d) $\text{DR} = 0.25 \times 20 = 5$

Automatic frequency control. Because the Crosby transmitter uses either a VCO or a reactance oscillator to generate the carrier frequency, it is more susceptible to *frequency drift* due to temperature change, power supply fluctuations, and so on, than if it were a crystal oscillator. As stated in Chapter 2, the stability of an oscillator is often given in *parts per million* (ppm) per degree celsius. For example, for the transmitter shown in Figure 7-23, an oscillator stability of ±40 ppm could produce ±204 Hz (5.1

MHz × 40/Hz/million) of frequency drift per degree celsius at the output of the master oscillator. This corresponds to ±3672 Hz drift at the antenna (18 × 204), which far exceeds the ±2 kHz maximum set by the FCC for commercial FM broadcasting. Although an AFC circuit does not totally eliminate frequency drift, it can substantially reduce it. Assuming a rock-stable crystal reference oscillator and a perfectly tuned discriminator, the frequency drift at the output of the second multiplier without feedback is

$$\text{open-loop drift} = dF_{ol} = N_1 N_2 dF_c \qquad (7\text{-}49)$$

and the closed-loop drift is

$$\text{closed-loop drift} = dF_{cl} = dF_{ol} - N_1 N_2 k_d k_o dF_{cl} \qquad (7\text{-}50)$$

Therefore,

$$dF_{cl} + N_1 N_2 k_d k_o dF_{cl} = dF_{ol}$$

and

$$dF_{cl}(1 + N_1 N_2 k_d k_o) = dF_{ol}$$

Thus

$$dF_{cl} = \frac{dF_{ol}}{1 + N_1 N_2 k_d k_o} \qquad (7\text{-}51)$$

where

K_d = discriminator transfer function (V/Hz)
K_o = master oscillator transfer function (Hz/V)

From Equation 7-51 it can be seen that the frequency drift at the output of the second multiplier and consequently at the input to the discriminator is reduced by a factor of $1 + N_1 N_2 k_d k_o$ when the AFC loop is closed. The carrier frequency drift is multiplied by the AFC loop gain and fed back to the master oscillator as a dc correction voltage. Of course, the total frequency error cannot be canceled because then there would not be any error voltage to feed back.

EXAMPLE 7-9

Use the transmitter block diagram and values given in Figure 7-23 to answer the following questions. Determine the reduction in frequency drift at the antenna for a transmitter without AFC compared to a transmitter with AFC. Use a VCO stability = 200 ppm, K_o = 10 kHz/V, and K_d = 2 V/kHz.

Solution With the feedback loop open, the master oscillator output frequency is

$$F_c = 5.1 \text{ MHz} + (200 \text{ ppm})(5.1 \text{ MHz}) = 5,101,020 \text{ Hz}$$

and the frequency at the output of the second multiplier is

$$F_2 = N_1 N_2 F_c = (5,101,020)(6) = 30,606,120 \text{ Hz}$$

Thus the frequency drift is

$$dF_2 = 30,606,120 - 30,600,000 = 6120 \text{ Hz}$$

Therefore, the antenna transmit frequency is

$$F_t = (30,606,120)(3) = 91.81836 \text{ MHz}$$

which is 18.36 kHz above the assigned frequency and well out of limits.

    With the feedback loop closed, the frequency drift at the output of the second multiplier is reduced by a factor of $1 + N_1 N_2 K_o K_d$ or

$$1 + (2)(3) \left(\frac{10 \text{ kHz}}{\text{V}} \right) \left(\frac{2 \text{ V}}{\text{kHz}} \right) = 121$$

Therefore,

$$dF_2 = \frac{6120}{121} = 51 \text{ Hz}$$

Thus

$$F_2 = 30,600,051$$

and the antenna transmit frequency is

$$F_t = 30,600,051 \times 3 = 91,800,153 \text{ Hz}$$

The frequency drift at the antenna has been reduced from 18,360 Hz to 153 Hz, which is now well within the ±2 kHz FCC requirements.

    The preceding discussion and Example 7-9 assumed a perfectly stable crystal reference oscillator and a perfectly tuned discriminator. In actuality, both the discriminator and the reference crystal are subject to drift, and the worst-case situation is when they both drift in the same direction as the VCO. The drift characteristics for a typical discriminator are on the order of 100 ppm. Perhaps now you can see why the output frequency from the second multiplier was mixed down to a relatively low frequency (2 MHz) prior to being fed to the discriminator. For a discriminator tuned to 2 MHz with a stability of 100 ppm, the maximum discriminator drift is

$$dF_d = 100 \text{ ppm} \times 2 \text{ MHz} = 200 \text{ Hz}$$

    If the 30.6-MHz signal were fed directly into the discriminator, the maximum drift would be

$$dF_d = 100 \text{ ppm} \times 30.6 \text{ MHz} = 3060 \text{ Hz}$$

Frequency drift due to discriminator instability is multiplied by the AFC open-loop gain. Therefore, the change in the second multiplier output frequency due to discriminator drift is

$$dF_2 = dF_d N_1 N_2 K_o K_d \qquad (7\text{-}52)$$

Similarly, the crystal reference oscillator can drift and also contribute to the total frequency change at the output of the second multiplier. The drift due to crystal instability is multiplied by 2 before entering the mixer; therefore,

$$dF_2 = N_4 dF_o N_1 N_2 K_o K_d \tag{7-53}$$

and the maximum open-loop frequency drift at the output of the second multiplier is

$$dF_2(\text{total}) = N_1 N_2 K_o K_d (dF_c + dF_d + N_4 dF_o) \tag{7-54}$$

Phase-locked-loop FM transmitter.   Figure 7-24 shows a *wideband* FM transmitter that uses a phase-locked loop to achieve crystal stability from a VCO master oscillator and, at the same time, generate a high-index, wideband FM signal at its output. The output frequency from the VCO is divided by *N* and fed back to the PLL phase comparator, where it is compared to a stable crystal reference frequency. The phase comparator generates a dc voltage that is proportional to the difference in the two frequencies. The dc correction voltage is added to the modulating signal and applied to the VCO input. The correction voltage adjusts the VCO center frequency to its proper value. Again, the LPF prevents changes in the VCO output frequency due to the modulating signal from being converted to dc, fed back to the VCO, and wiping out the modulation. The LPF also prevents the loop from locking onto a sideband frequency.

PM from FM.   As stated previously and shown in Figure 7-4, an FM modulator preceded by a differentiator generates a PM waveform. If the transmitters shown in Figures 7-23 and 7-24 are preceded by a preemphasis network, which is a differentiator (high-pass filter), an interesting situation occurs. For a 75-μs time constant, the amplitude of frequencies above 2.12 kHz are emphasized through differentiation. Therefore, for

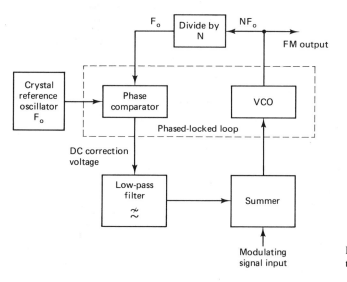

FIGURE 7-24   Phase-locked-loop FM transmitter.

modulating frequencies below 2.12 kHz, the output waveform is proportional to the modulating signal, and for frequencies above 2.12 kHz, the output waveform is proportional to the derivative of the input signal. In other words, frequency modulation occurs for frequencies below 2.12 kHz and phase modulation occurs for frequencies above 2.12 kHz. Because the gain of a differentiator increases with frequency above the break frequency (2.12 kHz), and since the frequency deviation is proportional to the modulating signal amplitude, the frequency deviation also increases with frequency above 2.12 kHz. From Equation 7-15 it can be seen that if ΔF and F_a increase proportionately, the modulation index remains constant, which is, of course, a characteristic of PM.

Indirect FM Transmitter

Armstrong indirect FM transmitter.   With indirect FM, the phase of the carrier is deviated directly by the modulating signal, which indirectly deviates the carrier frequency. Figure 7-25 shows the block diagram for a *wideband Armstrong indirect FM transmitter*. The carrier source is a crystal oscillator. Therefore, the stability requirements set by the FCC can be achieved without the use of AFC.

With an Armstrong transmitter, a relatively low-frequency subcarrier is phase shifted 90° and fed to a balanced modulator, where it is mixed with the input modulating signal (F_a). The output from the balanced modulator is a double-sideband suppressed

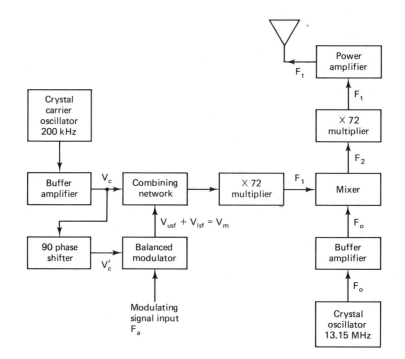

FIGURE 7-25   Armstrong indirect FM transmitter.

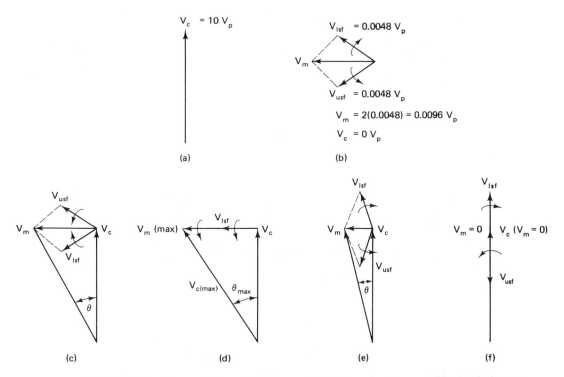

FIGURE 7-26 Phasor addition of V_c, V_{usf}, and V_{lsf}: (a) carrier phasor; (b) sideband phasors; (c) – (f) progressive phasor addition. Part (d) shows the peak phase shift.

carrier waveform which is combined with the original carrier to produce a low-index phase-modulated waveform. Figure 7-26a shows the phasor for the original carrier (V_c), and Figure 7-26b shows the phasors for the sideband components of the suppressed carrier waveform (V_{usf} and V_{lsf}). Because the suppressed carrier (V_c') is 90° out of phase with V_c, the upper and lower sidebands combine to produce a component (V_m) that is always in quadrature (at right angles) with V_c. Figure 7-26c through f show the progressive phasor addition of V_c, V_{usf}, and V_{lsf}. It can be seen that the output from the combining network is a phase-modulated wave where the deviation rate is equal to the modulating signal frequency, F_a, and the peak phase deviation (modulation index) is

$$\theta = m = \arctan \frac{V_m}{V_c} \qquad (7\text{-}55\text{a})$$

For very small angles, the tangent is approximately equal to the angle; therefore,

$$\theta = m \approx \frac{V_m}{V_c} \qquad (7\text{-}55\text{b})$$

EXAMPLE 7-10

For the phase-shifted carrier (V_c'), upper side frequency (V_{usf}), and lower side frequency (V_{lsb}) components shown in Figure 7-26; determine:

    (a) The peak carrier phase shift in both radians and degrees.

    (b) The frequency deviation for modulating signal frequency $F_a = 15$ kHz.

Solution   (a) The peak phase deviation is equal to the modulation index; therefore, from Equation 7-55a,

$$m = \theta = \arctan \frac{V_m}{V_c} \tag{7-55a}$$

$$= \arctan \frac{0.0096}{10} = 0.00553°$$

$$= (0.00553°) \frac{\pi r}{180°} = 0.000965 \text{ rad}$$

(b) Rearranging Equation 7-14 gives us

$$\Delta F = mF_a = (0.000965)(15,000) = 14.47 \text{ Hz}$$

From the phasor diagrams shown in Figure 7-26, it can be seen that the carrier amplitude is varied, which produces unwanted amplitude modulation in the output waveform, and V_c(max) occurs when V_{usf} and V_{lsf} are in phase with each other and with V_c. The maximum phase deviation that can be produced with this type of modulator is approximately 1.67 mrad. Therefore, from Equation 7-14 and for a maximum modulating signal frequency $F_a = 15$ kHz, the maximum frequency deviation is

$$\Delta F_{max} = (0.00167)(15 \text{ kHz}) = 25 \text{ Hz}$$

From the preceding discussion it is evident that the modulation index at the output of the combining network is insufficient to produce a wideband FM spectrum and therefore must be multiplied considerably before being transmitted. For the transmitter shown in Figure 7-25, a 200-kHz phase-modulated carrier with a peak phase deviation $\theta = 0.000965$ rad and a frequency deviation $\Delta F = 14.17$ Hz is produced at the output of the combining network. To achieve 75-kHz frequency deviation, the waveform must be multiplied by 5184 (75,000/14.47). However, this would produce a transmit carrier frequency $F_t = 5184 \times 200$ kHz $= 1036.8$ MHz, which is well beyond the commercial FM broadcast band limits. It is apparent that multiplication by itself is inadequate. Therefore, a combination of multiplying and mixing is used to develop the desired transmit carrier frequency with 75-kHz frequency deviation. The waveform at the output of the combining network is multiplied by 72, producing the following output signal:

$$F_1 = 72 \times 200 \text{ kHz} = 14.4 \text{ MHz}$$

$$m = 72 \times 0.000965 = 0.06948 \text{ rad}$$

$$\Delta F = 72 \times 14.47 = 1042 \text{ Hz}$$

The output from the first multiplier is mixed with a 13.15-MHz crystal controlled frequency (F_o) to produce a difference signal (F_2) with the following characteristics:

$$F_2 = 14.4 \text{ MHz} - 13.15 \text{ MHz} = 1.25 \text{ MHz}$$

$$m = 0.06948 \text{ rad}$$

$$\Delta F = 1042 \text{ Hz}$$

Note that only the carrier frequency is affected by the heterodyning process. The output from the mixer is once again multiplied by 72, to produce a transmit signal with the following characteristics:

$$F_t = 1.25 \text{ MHz} \times 72 = 90 \text{ MHz}$$

$$m = 0.06948 \times 72 = 5.0 \text{ rad}$$

$$\Delta F = 1042 \text{ Hz} \times 72 = 75 \text{ kHz}$$

In the preceding example, the carrier was multiplied by a factor of 450, while the frequency deviation and modulation index were multiplied by a factor of 5184.

With the Armstrong transmitter, the phase of the carrier is directly modulated in the combining network, producing indirect frequency modulation. The magnitude of the phase deviation is directly proportional to the amplitude of the modulating signal but independent of its frequency. Therefore, for changes in the modulating signal frequency, the modulation index remains constant and the frequency deviation changes proportionately. For example, for the transmitter show in Figure 7-25, if the modulating signal amplitude is held constant while its frequency is decreased to 5 kHz, the modulation index remains at 5 while the frequency deviation is reduced to $\Delta F = 5 \times 5 \text{ kHz} = 25$ kHz.

FM from PM. As stated previously and shown in Figure 7-4, a PM modulator preceded by an integrator generates an FM waveform. If the PM transmitter shown in Figure 7-25 is preceded by a low-pass filter (which is an integrator) FM results. The low-pass filter is simply a $1/F$ filter, which is commonly called a *predistorter* or a *frequency correction* network.

FM Versus PM

From a purely theoretical viewpoint, the difference between FM and PM is quite simple: the modulation index for FM is defined differently than for PM. With PM, the phase deviation is directly proportional to the amplitude of the modulating signal and independent of its frequency. With FM, the frequency deviation is directly proportional to the amplitude of the modulating signal and inversely proportional to its frequency.

Considering FM as a form of phase modulation, the larger the frequency deviation, the larger the phase deviation. Therefore, the latter depends, at least to a certain extent, on the amplitude of the modulating signal, just as with PM. With PM, the modulation index is proportional to the amplitude of the modulating signal voltage only, whereas with FM the modulation index is also inversely proportional to the modulating signal frequency. If FM transmissions are received on a PM receiver, the bass frequencies have considerably more phase deviation than a PM modulator would have given them. Since the output from a PM demodulator is proportional to the phase deviation (modulation index), the signal appears excessively bass-boosted. Alternatively (and this is the more practical situation), PM demodulated by an FM receiver appears lacking in bass frequencies.

QUESTIONS

7-1. Define *angle modulation*.

7-2. Define *direct FM*; *indirect FM*.

7-3. Define *direct PM*; *indirect PM*.

7-4. Define *frequency deviation*; *phase deviation*.

7-5. Define *instantaneous phase*; *instantaneous phase deviation*; *instantaneous frequency*; *instantaneous frequency deviation*.

7-6. Define *deviation sensitivity* for a frequency modulator and for a phase modulator.

7-7. Describe the relationship between the instantaneous carrier frequency and the modulating signal for FM.

7-8. Describe the relationship between the instantaneous carrier phase and the modulating signal for PM.

7-9. Describe the relationship between frequency deviation and the modulating signal.

7-10. Define *carrier swing*.

7-11. Define *modulation index* for FM and for PM.

7-12. Describe the relationship between modulation index and the modulating signal for FM; for PM.

7-13. Define *percent modulation*.

7-14. Describe the difference between a frequency and a phase modulator.

7-15. How can a frequency modulator be converted to a phase modulator; a phase modulator to a frequency modulator?

7-16. How many sets of side frequencies are produced when a carrier is frequency modulated by a single input frequency?

7-17. What are the requirements for a side frequency to be significant?

7-18. Define a *low*, a *medium*, and a *high modulation index*.

7-19. Describe the significance of the Bessel table.

7-20. State Carson's general rule for determining the bandwidth for an angle-modulated wave.

7-21. Define *deviation ratio*.

7-22. Describe the relationship between the power in the unmodulated carrier and the power in the modulated wave for FM.

7-23. Describe the FM noise triangle.

7-24. What effect does limiting have on the composite FM waveform?

7-25. Define *preemphasis*; *deemphasis*.

7-26. Describe a preemphasis network; a deemphasis network.

7-27. Describe the operation of a varactor diode FM generator.

7-28. Describe the operation of a reactance FM modulator.

7-29. Describe the operation of a linear integrated-circuit FM modulator.

7-30. List four communications services that use frequency modulation.

7-31. Draw the block diagram for a Crosby direct FM transmitter and describe its operation.

7-32. What is the purpose of an AFC loop? Why is one required for the Crosby system?

7-33. Draw the block diagram for a phase-locked-loop FM transmitter and describe its operation.

7-34. Draw the block diagram for an Armstrong indirect FM transmitter and describe its operation.

7-35. Contrast FM and PM.

PROBLEMS

7-1. If a frequency modulator produces 5 kHz of frequency deviation for a 10-Vp modulating signal, determine the deviation sensitivity. D_{vvv}, $\leq 00 \, Hz/v$

7-2. If a phase modulator produces 2 rad of frequency deviation for a 5-Vp modulating signal, determine the deviation sensitivity.

7-3. Determine (a) the peak frequency deviation, (b) the carrier swing, and (c) the modulation index for an FM modulator with deviation sensitivity $K_1 = 4$ kHz/V and modulating signal $v_a = 10 \sin (2\pi 2000t)$. P, 269, 271

7-4. Determine the peak phase deviation for a PM modulator with deviation sensitivity $K = 1.5$ rad/V and modulating signal $v_a = 2 \sin (2\pi 1000t)$.

7-5. Determine the percent modulation for a television broadcast station with a maximum frequency deviation $\Delta F = 50$ kHz when the modulating signal produces 40-kHz frequency deviation at the antenna. $\%$ modulation $= \dfrac{\Delta F \; actual}{\Delta F \; maximum} \cdot 100$

7-6. From the Bessel table determine the number of sets of side frequencies produced for the following modulation indices: 0.5, 1.0, 2.0, 5.0, and 10.0.

7-7. For an FM modulator with modulation index $m = 2$, modulating signal $V(t) = v_a \sin (2\pi 2000t)$, and unmodulated carrier $v_c = 8 \sin (2\pi 800kt)$:
(a) Determine the number of sets of significant sidebands.
(b) Determine their amplitudes.   P.277
(c) Draw the frequency spectrum.

7-8. For an FM transmitter with 60-kHz carrier swing, determine the frequency deviation. If the amplitude of the modulating signal decreases by a factor of 2, determine the new frequency deviation.

7-9. For a given input signal, an FM broadcast transmitter has a frequency deviation $\Delta F = 20$ kHz. Determine the frequency deviation if the amplitude of the modulating signal increases by a factor of 2.5. $\Delta F = K_1 V_a$ so it would increase by fact. 2.5 (50 kHz)

7-10. An FM transmitter has a rest frequency $F_c = 96$ MHz and a deviation sensitivity $K_1 = 4$ kHz/V. Determine the frequency deviation for a modulating signal $V(t) = 8 \sin (2\pi 2000t)$. Determine the modulation index.

P. 280 **7-11.** Determine the deviation ratio and worst-case bandwidth for an FM system with a maximum frequency deviation $\Delta F = 25$ kHz and maximum modulating signal frequency $F_a = 12.5$ kHz.

7-12. For an FM modulator with 40-kHz frequency deviation and a modulating signal frequency $F_a = 10$ kHz, determine the bandwidth using both the Bessel table and Carson's rule.

$P = \dfrac{V_c^2}{R}$ **7-13.** For an FM modulator with an unmodulated carrier voltage $V_c = 20$ V$_p$, a modulation index $m = 1$, and a load resistance $RL = 10\ \Omega$; determine the power in the modulated carrier and each significant side frequency, and sketch the power spectrum for the modulated wave. (see answers)

7-14. For an angle-modulated carrier $v_c = 2\cos(2\pi 200\text{MHz }t)$ with 50 kHz of frequency deviation due to the modulating signal and a single-frequency interfering signal $v_n = 0.5\cos(2\pi 200.01$ MHz$)$, determine:
(a) The frequency of the demodulated interference signal.
(b) The peak and rms phase and frequency deviation due to the interfering signal.
(c) The S/N ratio at the output of the demodulator.

P. 285 **7-15.** Determine the total rms phase deviation produced by a 5-kHz band of random noise with a peak voltage $V_n = 0.08$ V and a carrier $v_c = 1.5\sin(2\pi 40\text{MHz }t)$. $\dfrac{V_n}{V_c\sqrt{2}} = \dfrac{.08}{1.5(\sqrt{2})} = .0377\varsigma$

7-16. For a Crosby direct FM transmitter similar to the one shown in Figure 7-23 with the following parameters, determine:
(a) The frequency deviation at the output of the VCO and the power amplifier.
(b) The modulation index at the same two points.
(c) The bandwidth at the output of the power amplifier.

$$N_1 = \times 3$$
$$N_2 = \times 3$$
$$N_3 = \times 2$$

Crystal reference oscillator frequency = 13 MHz

Reference multiplier = ×3

VCO deviation sensitivity $K_1 = 450$ Hz/V

Modulating signal $v_a = 3\sin(2\pi 5kt)$

VCO rest frequency $F_c = 4.5$ MHz

Discriminator resonant frequency $F_d = 1.5$ MHz

7-17. For an Armstrong indirect FM transmitter similar to the one shown in Figure 7-25 with the following parameters, determine:
(a) The modulation index at the output of the combining network and the power amplifier.
(b) The frequency deviation at the same two points.
(c) The transmit carrier frequency.

CN$_1$ $m = .0036$
P.A. $m = 7.2$
CN $\delta f = 7.2$ Hz
PA $\delta f = 14.4$ kHz

$F_c = 90$ MHz

Crystal carrier oscillator = 210 kHz

Crystal reference oscillator = 10.2 MHz

Sideband voltage $V_m = 0.018$ V$_p$

Carrier input to combiner $V_c = 5\ V_p$

First multiplier $= \times 40$

Second multiplier $= \times 50$

Modulating signal frequency $F_a = 2$ kHz

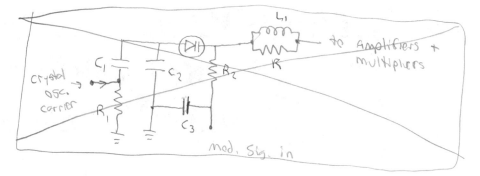

7-17)

a)   C.N. $m = \dfrac{V_m}{V_c}$

    P.A. $m = mF_a$ $\left(m \times 1\text{st mult.} \times 2\text{nd mult.}\right)$

b) C.N. $\Delta F = mF_a$
   P.A $\Delta F = 2\Delta F$

c)   ?

Chapter 8

ANGLE MODULATION RECEIVERS AND SYSTEMS

INTRODUCTION

The receivers used for receiving angle-modulated signals are essentially identical to those that are used for conventional AM or SSB reception, except for the method used to extract the audio information from the composite IF waveform. In FM receivers, the output of the audio detector is proportional to the frequency deviation at its input. With PM receivers, the output of the audio detector is proportional to the phase deviation at its input. Because frequency and phase modulation both occur with either angle modulation system, FM signals can be demodulated by a PM receiver, and vice versa. Therefore, the circuits used to demodulate FM and PM signals are both described under the heading ''FM receivers.''

In conventional AM, the modulating signal is impressed onto the carrier in the form of amplitude variations. However, noise introduced in the system also produces changes in the amplitude of the envelope. Therefore, the noise cannot be removed from the composite waveform without also removing a portion of the information. With angle modulation, the information is impressed onto the carrier in the form of frequency or phase variations. Therefore, with angle modulation receivers, amplitude variations caused by noise can be removed from the composite waveform simply by *limiting* (*clipping*) the peaks of the envelope prior to detection. With angle modulation, an improvement in the signal-to-noise ratio is achieved during the demodulation process; thus system performance in the presence of noise can be improved by limiting. In essence, this is the major advantage of angle modulation over conventional AM.

FM RECEIVERS

Figure 8-1 shows a simplified block diagram for a double-conversion superheterodyne FM receiver. As you can see, it is very similar to a conventional AM receiver. The RF, mixer, and IF stages are identical to those used in AM receivers, although FM receivers generally have more IF amplification. Also, due to the inherent noise suppression characteristics of FM receivers, RF amplifiers are often not required. However the audio detector stage in an FM receiver is quite different from those used in AM receivers. The envelope (peak) detector used in conventional AM receivers is replaced by a *limiter*, *frequency discriminator*, and *deemphasis network*. The limiter and deemphasis network contribute to the improvement in the *S/N* ratio that is achieved in the audio demodulator stage. For FM broadcast band receivers, the first IF is a relatively high frequency (generally, 10.7 MHz) for good image frequency rejection, and the second IF is a relatively low frequency (generally, 455 kHz) to reduce the possibility that the amplifiers will oscillate.

FM Demodulators

FM demodulators are frequency-dependent circuits that produce an output voltage that is directly proportional to the instantaneous frequency at its input ($V_o = \Delta F K$, where $K = $ V/Hz and is the transfer function for the demodulator, and ΔF is the difference between the input frequency and the center frequency of the demodulator). There are several circuits that are used for demodulating FM signals. The most common are the *slope detector*, *Foster–Seeley discriminator*, *ratio detector*, *PLL demodulator*, and *quadrature detector*. The slope detector, Foster–Seeley discriminator, and ratio detector are all forms of *tuned circuit frequency discriminators*. Tuned circuit frequency discriminators convert FM to AM, then demodulate the AM waveform with a conventional peak detector. Also, most frequency discriminators require a 180° phase inverter, an adder circuit, and one or more frequency-dependent circuits.

Slope detector. Figure 8-2a shows the schematic diagram for a *single-ended slope detector*, which is the simplest form of tuned circuit frequency discriminator. The single-ended slope detector has the most nonlinear voltage-versus-frequency characteristics and is, therefore, seldom used. However, its circuit operation is basic to all tuned circuit frequency discriminators.

In Figure 8-2a, tuned circuit (L_a and C_a) produces an output voltage that is proportional to the input frequency. The maximum output voltage occurs at its resonant frequency (F_o), and its output rolls off proportionately as the input frequency is deviated above and below F_o. The circuit is designed so that the IF center frequency (F_c) falls in the center of the most linear portion of the voltage-versus-frequency curve (Figure 8-2b). As the IF deviates above F_c, the output voltage increases and, as the IF deviates below F_c, the output voltage decreases. Therefore, the tuned circuit converts frequency variations to amplitude variations (i.e., FM-to-AM conversion). D_i, C_i, and R_i make up a simple peak detector which converts the amplitude variations to an output voltage

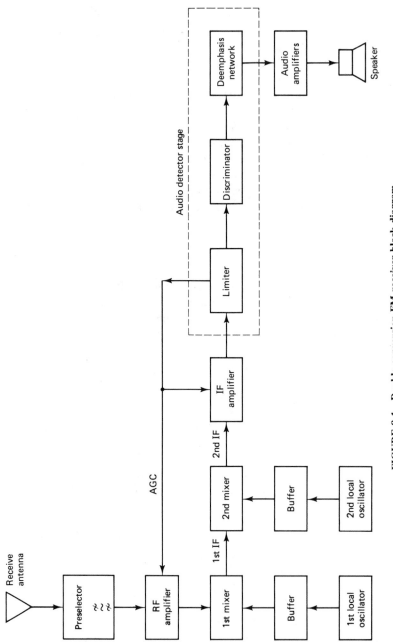

FIGURE 8-1 Double-conversion FM receiver block diagram.

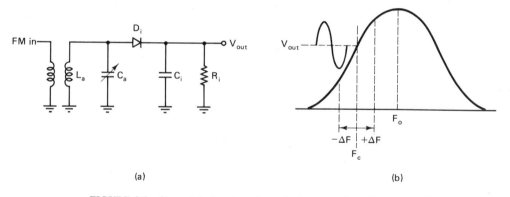

FM in — V_{out} — L_a — C_a — D_i — C_i — R_i

V_{out} — F_o — $-\Delta F$ $+\Delta F$ — F_c

(a)   (b)

FIGURE 8-2 Slope detector: (a) schematic diagram; (b) voltage-versus-frequency curve.

that varies at a rate equal to that of the input frequency variations and whose amplitude is proportional to the magnitude of the frequency changes.

Balanced slope detector.   Figure 8-3a shows the schematic diagram for a *balanced slope detector*. A single-ended slope detector is a tuned circuit frequency discriminator, and a balanced slope detector is simply two slope detectors connected in parallel and fed 180° out of phase. The phase inversion is accomplished by center tapping the tuned secondary windings of transformer T_1. In Figure 8-3a, tuned circuits (L_a, C_a, and L_b, C_b) perform the FM-to-AM conversion, and balanced peak detectors (D_1, C_1, R_1 and D_2, C_2, R_2) remove the information from the AM waveform. The top tuned circuit (L_a and C_a) is tuned to a frequency (F_a) which is above the IF center frequency (F_o) by approximately $1.33 \times \Delta F$ (for the FM broadcast band this is approximately 1.33×75 kHz = 100 kHz). The lower tuned circuit (L_b and C_b) is tuned to a frequency (F_b) that is below the IF center frequency by an equal amount. Circuit operation is quite simple. The output voltage from each tuned circuit is proportional to the input frequency and each output is rectified by its respective peak detector. Therefore, the closer the input frequency is to the tank circuit resonant frequency, the greater the tank circuit output voltage. The IF center frequency falls exactly halfway between the resonant frequencies of the two tuned circuits. Therefore, at the IF center frequency, the output voltages from the two tank circuits are equal in amplitude but opposite in polarity. Consequently, the rectified output voltages across R_1 and R_2, when added, produce a differential output voltage $V_{out} = 0$ V. When the IF is deviated above resonance, the top tuned circuit produces a higher output voltage than the lower tank circuit and V_{out} goes positive. When the IF is deviated below resonance, the output voltage from the lower tank circuit is larger and V_{out} goes negative. The output voltage-versus-frequency response curve is shown in Figure 8-3b. Although the slope detector is probably the simplest FM detector, it has several disadvantages, which include poor linearity, difficulty in tuning, and lack of provision for limiting. Because limiting is not provided, a slope detector will demodulate amplitude as well as frequency variations and consequently, must be preceded by a limiter stage. A balanced slope detector is aligned by injecting

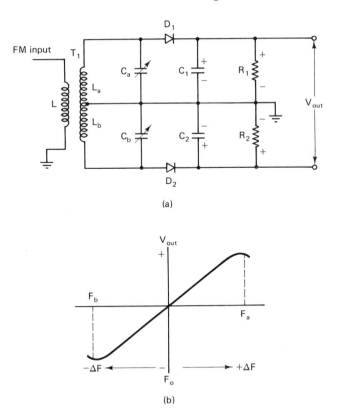

FM input

(a)

V_{out}

F_b

F_a

$-\Delta F$ $+\Delta F$

F_o

(b)

FIGURE 8-3 Balanced slope detector: (a) schematic diagram; (b) voltage-versus-frequency response curve.

an unmodulated IF center frequency and tuning C_a and C_b for zero volts out. Then F_a and F_b are alternately injected while C_a and C_b are tuned for maximum and equal output voltages with opposite polarities.

Foster–Seeley discriminator.   A *Foster–Seeley discriminator* (sometimes called a *phase-shift discriminator*) is a tuned circuit frequency discriminator whose operation is very similar to that of the balanced slope detector. The schematic diagram for a Foster–Seeley discriminator is shown in Figure 8-4a. The capacitance values for C_c, C_1, and C_2 are chosen such that they are short circuits for the IF center frequency. Therefore, the right side of L_3 is ac ground and the IF signal (V_{in}) is fed directly (in phase) across L_3 (VL_3). The incoming IF is inverted $180°$ by transformer T_1 and divided equally between L_a and L_b. At the resonant frequency of the secondary tank circuit (the IF center frequency), the secondary current (I_s) is in phase with the total secondary voltage (V_s) and $180°$ out of phase with VL_3. Also, due to loose coupling, the primary of T_1 is essentially an inductor, so that the primary current I_p is $90°$ out of phase with V_{in}, and since magnetic induction depends on primary current, the voltage induced in the secondary is $90°$ out of phase with V_{in} (VL_3). Therefore, VL_a and VL_b are $180°$ out of phase with each other and in quadrature or $90°$ out of phase with VL_3. The voltage

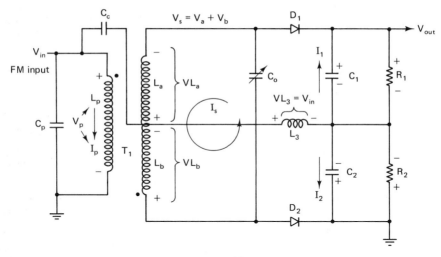

(a)

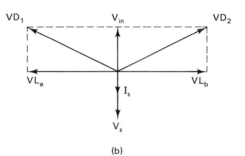

(b)

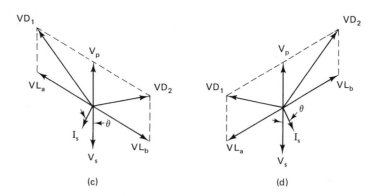

(c)   (d)

FIGURE 8-4 Foster–Seeley discriminator: (a) schematic diagram; (b) vector diagram, $F_{in} = F_o$; (c) vector diagram, $F_{in} > F_o$; (d) vector diagram, $F_{in} < F_o$.

across the top diode VD_1 is the vector sum of VL_3 and VL_a, and the voltage across the bottom diode VD_2 is the vector sum of VL_3 and VL_b. The corresponding vector diagrams are shown in Figure 8-4b. It can be seen that the voltages across D_1 and D_2 are equal. Therefore, at resonance, I_1 and I_2 are equal and C_1 and C_2 charge to equal voltages with opposite polarities. Consequently, $V_{out} = VC_1 - VC_2 = 0$ V. When the IF is deviated above resonance $(X_L > X_C)$ and the secondary tank circuit impedance becomes inductive and the secondary current lags the secondary voltage by some angle θ, which is proportional to the frequency deviation. The corresponding phasor diagram is shown in Figure 8-4c. The figure shows that the vector sum of the voltage across D_1 is greater than the voltage across D_2. Consequently, C_1 charges while C_2 discharges and V_{out} goes positive. When the IF is deviated below resonance $(X_L < X_C)$, the secondary current leads the secondary voltage by some angle θ, which is again proportional to the change in frequency. The corresponding phasors are shown in Figure 8-4d. It can be seen that the vector sum of the voltage across D_1 is now less than the voltage across D_2. Consequently, C_1 discharges while C_2 charges and V_{out} goes negative. A Foster–Seeley discriminator is tuned by injecting an unmodulated IF center frequency and tuning C_o for zero volts out.

The preceding discussion and Figure 8-4 show that the output voltage from a Foster–Seeley discriminator is directly proportional to the magnitude and the direction of the frequency deviation. Figure 8-5 shows a typical voltage-versus-frequency response curve for a Foster–Seeley discriminator. For obvious reasons, it is often called an *S-curve*. It can be seen that the output voltage-versus-frequency deviation curve is more linear than that of a slope detector, and because there is only one tank circuit, it is easier to tune. For distortionless demodulation, the frequency deviation should be restricted to the linear portion of the secondary tuned circuit frequency response curve. Like the slope detector, a Foster–Seeley discriminator responds to amplitude variations and therefore must be preceded by a limiter.

Ratio detector.   The *ratio detector* has one major advantage over the slope detector and the Foster–Seeley discriminator; a ratio detector is relatively immune to amplitude variations in the input signal. Figure 8-6a shows the schematic diagram for a ratio detector. Like the Foster–Seeley discriminator, a ratio detector has a single tuned circuit

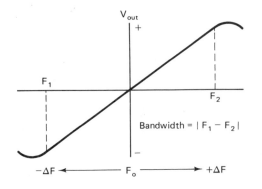

FIGURE 8-5 Discriminator voltage-versus-frequency response curve.

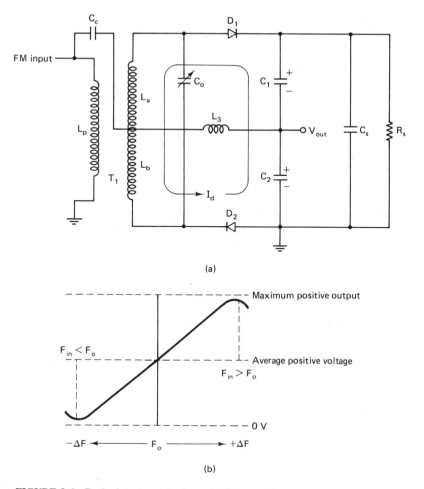

(a)

(b)

FIGURE 8-6 Ratio detector: (a) schematic diagram; (b) voltage-versus-frequency response curve.

in the transformer secondary. Therefore, the operation of a ratio detector is very similar to that of the Foster–Seeley discriminator. In fact, the voltage vectors for D_1 and D_2 are identical to those for the Foster–Seeley discriminator shown in Figure 8-4. However, with the ratio detector, one diode is reversed (D_2) and current (I_d) can flow around the outermost loop of the circuit. Therefore, after several cycles of the input signal, shunt capacitor C_s charges to approximately the peak voltage across the secondary windings of T_1. The reactance of C_s is low and R_s simply provides a dc path for diode current. Therefore, the time constant for R_s and C_s is sufficiently long so that rapid changes in the amplitude of the input signal due to thermal noise or other interfering signals are shorted to ground and therefore have no effect on the average voltage across C_s. Consequently, C_1 and C_2 charge and discharge proportional to frequency changes in the input signal and are relatively immune to amplitude variations. Also, the output from a ratio detector is taken with respect to ground, and for the diode polarities shown in Figure 8-6a, the average output voltage is positive. At resonance, the output voltage is divided equally between C_1 and C_2 and redistributed as the input frequency is deviated above and below resonance. Therefore, changes in V_{out} are due to the changing ratio of the voltages across C_1 and C_2, while the total voltage is clamped by C_s.

Figure 8-6b shows the output response curve for the ratio detector shown in Figure 8-6a. It can be seen that at resonance V_{out} is not equal to 0 V, but rather, to one-half of the voltage across the secondary windings of T_1. Because a ratio detector is relatively immune to amplitude variations, it is often selected over a discriminator. However, a discriminator produces a more linear output voltage-versus-frequency response curve.

PLL FM demodulator. Since the development of the LSI linear integrated circuit, FM demodulation can be accomplished quite simply with a phase-locked loop. Although the operation of a PLL is quite involved, the operation of a *PLL FM demodulator* is probably the simplest and easiest to understand. A PLL frequency demodulator requires no tuned circuits and automatically compensates for carrier shift due to instability of the transmit oscillator.

In Chapter 5 a detailed description of PLL operation was given. It was shown that after frequency lock had occurred, the VCO tracked frequency changes in the input signal by maintaining a phase error at the input to the phase comparator. Therefore, if the PLL input is a deviated FM signal and the VCO natural frequency is equal to the IF center frequency, the dc correction voltage fed back to the input of the VCO is proportional to the frequency deviation and is thus the demodulated information signal. If the IF is sufficiently limited prior to the PLL and the loop is properly compensated, the PLL loop gain is constant and equal to K_v. Therefore, the demodulated signal can be taken directly from the output of the internal buffer and is mathematically given as

$$V_{out} = \Delta F k_d K_a \qquad (8\text{-}1)$$

Figure 8-7 shows a schematic diagram for an FM demodulator using the XR-2212 PLL, and Figure 8-8 shows the manufacturer's specification sheets for the XR-2212. R_o and C_o are coarse adjustments for setting the VCO's free-running frequency. R_x is for fine tuning, and R_F and R_C set the internal op-amp voltage gain (K_a). The PLL closed-loop frequency response should be compensated to allow unattentuated demodulation of the entire information signal bandwidth. The PLL op-amp buffer provides voltage gain and current drive stability.

Quadrature FM demodulator. A *quadrature FM demodulator* (sometimes called a *coincidence detector*) extracts the original information signal from the composite IF waveform by multiplying the two quadrature (90° out of phase) signals. A quadrature detector uses a 90° phase shifter, a single tuned circuit, and a product detector to demodulate FM signals. The 90° phase shifter produces a signal that is in quadrature with the received IF. The tuned circuit converts frequency variations to phase variations, and the product detector multiplies the received IF by the phase-shifted signal.

Figure 8-9 shows a simplified schematic diagram for an FM quadrature detector.

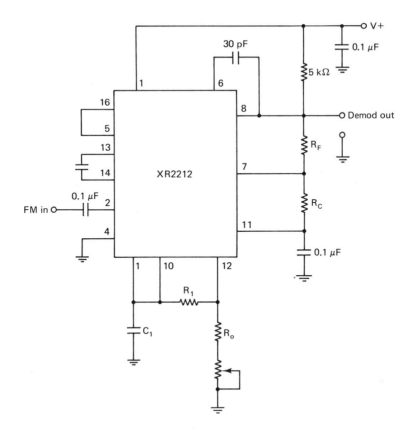

FIGURE 8-7   PLL FM demodulator.

C_i is a high reactance capacitor that, when placed in series with tank circuit (R_o, L_o, and C_o), produces a 90° phase shift at the IF center frequency. The tank circuit is tuned to the IF center frequency and produces an additional phase shift that is proportional to frequency deviation. The IF input signal (V_i) is multiplied by the quadrature signal (V_o) in the product detector and produces an output signal that is proportional to the frequency deviation. At resonance, the tank circuit impedance is resistive. However, frequency variations in the IF produce an additional positive or negative phase shift. Therefore, the product detector output voltage is proportional to the phase difference between the two input signals and is expressed mathematically as

$$V_{out} \propto V_o V_i$$
$$\propto (V_o \cos \omega_o t) \times (V_i \sin \omega_i t + \phi)$$

Substituting into the trigonometric identity for the product of a sine and a cosine wave of equal frequency gives us

$$V_{out} \propto V_i \sin (2\omega_i t + \phi) + V_o \sin \phi$$

XR-2212

Precision Phase-Locked Loop

GENERAL DESCRIPTION

The XR-2212 is an ultra-stable monolithic phase-locked loop (PLL) system especially designed for data communications and control system applications. Its on board reference and uncommitted operational amplifier, together with a typical temperature stability of better than 20 ppm/°C, make it ideally suited for frequency synthesis, FM detection, and tracking filter applications. The wide input dynamic range, large operating voltage range, large frequency range, and ECL, DTL, and TTL compatibility contribute to the usefulness and wide applicability of this device.

FEATURES

Quadrature VCO Outputs
Wide Frequency Range   0.01 Hz to 300 kHz
Wide Supply Voltage Range   4.5V to 20V
DTL/TTL/ECL Logic Compatibility
Wide Dynamic Range   2 mV to 3 Vrms
Adjustable Tracking Range (±1% to ±80%)
Excellent Temp. Stability   20 ppm/°C, Typ.

APPLICATIONS

Frequency Synthesis
Data Synchronization
FM Detection
Tracking Filters
FSK Demodulation

ABSOLUTE MAXIMUM RATINGS

Power Supply   18V
Input Signal Level   3 Vrms
Power Dissipation
Ceramic Package: 750 mW
     Derate Above $T_A = +25°C$ 6 mW/°C
Plastic Package: 625 mW
     Derate Above $T_A = +25°C$ 5 mW/°C

FUNCTIONAL BLOCK DIAGRAM

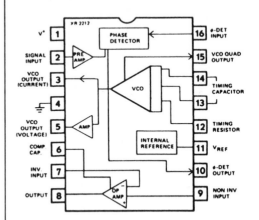

ORDERING INFORMATION

Part Number	Package	Operating Temperature
XR-2212M	Ceramic	−55°C to +125°C
XR2212CN	Ceramic	0°C to +70°C
XR-2212CP	Plastic	0°C to +70°C
XR-2212N	Ceramic	−40°C to +85°C
XR-2212P	Plastic	−40°C to +85°C

SYSTEM DESCRIPTION

The XR-2212 is a complete PLL system with buffered inputs and outputs, an internal reference, and an uncommited op amp. Two VCO outputs are pinned out; one sources current, the other sources voltage. This enables operation as a frequency synthesizer using an external programmable divider. The op amp section can be used as an audio preamplifier for FM detection or as a high speed sense amplifier (comparator) for FSK demodulation. The center frequency, bandwidth, and tracking range of the PLL are controlled independantly by external components. The PLL output is directly compatible with MOS, DTL, ECL, and TTL logic families as well as microprocessor peripheral systems.

The precision PLL system operates over a supply voltage range of 4.5 V to 20 V, a frequency range of 0.01 Hz to 300 kHz, and accepts input signals in the range of 2 mV to 3 Vrms. Temperature stability of the VCO is typically better than 20 ppm/°C.

 EXAR Corporation, 750 Palomar Avenue, Sunnyvale, CA 94086 • (408) 732-7970 • TWX 910-339-9233

FIGURE 8-8   Specification sheet for XR-2212 PLL. (Courtesy of EXAR Corporation.)

XR-2212

ELECTRICAL CHARACTERISTICS

Test Conditions: $V^+ = +12V$, $T_A = +25°C$, $R_0 = 30$ kΩ, $C_0 = 0.033$ µF, unless otherwise specified. See Figure 2 for component designation.

PARAMETERS	XR-2212/2212M			XR-2212C			UNITS	CONDITIONS
	MIN	TYP	MAX	MIN	TYP	MAX		
GENERAL								
Supply Voltage	4.5		15	4.5		15	V	
Supply Current		6	10		6	12	mA	$R_0 \geq 10$ KΩ. See Fig. 4
OSCILLATOR SECTION								
Frequency Accuracy		±1	±3		±1		%	Deviation from $f_0 = 1/R_0C_0$ $R_1 = \infty$
Frequency Stability								
Temperature		±20	±50		±20		ppm/°C	See Fig. 8.
Power Supply		0.05	0.5		0.05		%/V	$V^+ = 12 \pm 1$ V. See Fig. 7.
		.2			.2		%/V	$V^+ = 5 \pm 0.5$ V. See Fig. 7.
Upper Frequency Limit	100	300			300		kHz	$R_0 = 8.2$ KΩ, $C_0 = 400$ pF
Lowest Practical Operating Frequency		0.01			0.01		Hz	$R_0 = 2$ MΩ, $C_0 = 50$ µF See Fig. 5.
Timing Resistor, R_0								
Operating Range	5		2000	5		2000	KΩ	
Recommended Range	15		100	15		100	KΩ	See Fig. 7 and 8.
OSCILLATOR OUTPUTS								
Voltage Output								Measured at Pin 5.
Positive Swing, V_{OH}		11			11		V	
Negative Swing, V_{OL}	.8	.4			.5		V	
Current Sink Capability		1			1		mA	
Current Output								Measured at Pin 3.
Peak Current Swing	100	150			150		µA	
Output Impedance		1			1		MΩ	
Quadrature Output								Measured at Pin 15.
Output Swing		0.6			0.6		V	
DC Level		0.3			0.3		V	Referenced to Pin 11.
Output Impedance		3			3		KΩ	
LOOP PHASE DETECTOR SECTION								Measured at Pin 10.
Peak Output Current	±150	±200	±300	±100	±200	±300	µA	
Output Offset Current		±1			±2		µA	
Output Impedance		1			1		MΩ	
Maximum Swing	±4	±5		±4	±5		V	Referenced to Pin 11.
INPUT PREAMP SECTION								Measured at Pin 2.
Input Impedance		20			20		KΩ	
Input Signal to Cause Limiting		2	10		2		mVrms	
OP AMP SECTION								
Voltage Gain	55	70		55	70		dB	$R_L = 5.1$ KΩ, $R_F = \infty$
Input Bias Current		0.1	1		0.1	1	µA	
Offset Voltage		±5	±20		±5	±20	mV	
Slew Rate		2			2		V/µsec	
INTERNAL REFERENCE								Measured at Pin 11.
Voltage Level	4.9	5.3	5.7	4.75	5.3	5.85	V	
Output Impedance		100			100		Ω	

FIGURE 8-8 (*continued*)

329

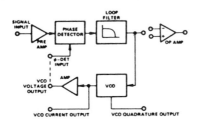

Figure 1. Functional Block Diagram of XR-2212 Precision PLL System

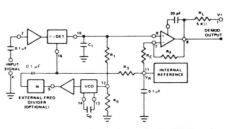

Figure 2. Generalized Circuit Connection for FM Detection, Signal Tracking or Frequency Synthesis

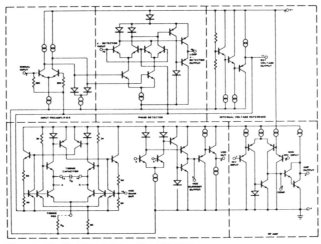

Figure 3. Simplified Circuit Schematic of XR-2212

TYPICAL CHARACTERISTICS

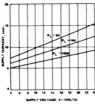

Figure 4. Typical Supply Current vs V+ (Logic Outputs Open Circuited)

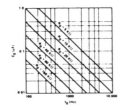

Figure 5. VCO Frequency vs Timing Resistor

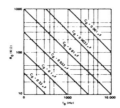

Figure 6. VCO Frequency vs Timing Capacitor

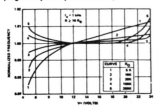

Figure 7. Typical f_0 vs Power Supply Characteristics

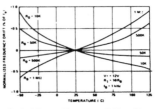

Figure 8. Typical Center Frequency Drift vs Temperature

FIGURE 8-8 (*continued*)

330

XR-2212

DESCRIPTION OF CIRCUIT CONTROLS

Signal Input (Pin 2): Signal is ac coupled to this terminal. The internal impedance at Pin 2 is 20 KΩ. Recommended input signal level is in the range of 10 mV to 5V peak-to-peak.

VCO Current Output (Pin 3): This is a high impedance (MΩ) current output terminal which can provide ± 100 μA drive capability with a voltage swing equal to V+. This output can directly interface with CMOS or NMOS logic families.

VCO Voltage Output (Pin 5): This terminal provides a low-impedance (≈ 50Ω) buffered output for the VCO. It can directly interface with low-power Schottley TTL. For interfacing with standard TTL circuits, a 750Ω pull-down resistor from pin 5 to ground is required. For operation of the PLL without an external divider, pin 5 can be dc coupled to pin 16.

Op Amp Compensation (Pin 6): The op amp section is frequency compensated by connecting an external capacitor from pin 6 to the amplifier output (pin 8). For unity-gain compensation a 20 pF capacitor is recommended.

Op Amp Inputs (Pins 7 and 9): These are the inverting and the non-inverting inputs for the op amp section. The common-mode range of the op amp inputs is from + 1V to (V+ − 1.5) volts.

Op Amp Output (Pin 8): The op amp output is an open-collector type gain stage and requires a pull-up resistor, R_L, to V+ for proper operation. For most applications, the recommended value of R_L is in 5 kΩ to 10 kΩ range.

Phase Detector Output (Pin 10): This terminal provides a high-impedance output for the loop phase-detector. The PLL loop filter is formed by R_1 and C_1 connected to Pin 10 (see Figure 2). With no input signal, or with no phase-error within the PLL, the dc level at Pin 10 is very nearly equal to V_R. The peak voltage swing available at the phase detector output is equal to ±V_R.

Reference Voltage, V_R (Pin 11): This pin is internally biased at the reference voltage level, V_R:V_R = V+/2— 650 mV. The dc voltage level at this pin forms an internal reference for the voltage levels at pins 10, 12 and 16. Pin 1 *must* be bypassed to ground with a 0.1 μF capacitor, for proper operation of the circuit.

VCO Control Input (Pin 12): VCO free-running frequency is determined by external timing resistor, R_0, connected from this terminal to ground. For optimum temperature stability, R_0 must be in the range of 10 KΩ to 100 KΩ (see Figure 8).

VCO Frequency Adjustment: VCO can be fine-tuned by connecting a potentiometer, R_X, in series with R_0 at Pin 12 (see Figure 10).

This terminal is a low-impedance point, and is internally biased at a dc level equal to V_R. The maximum timing current drawn from Pin 12 must be limited to ≤3 mA for proper operation of the circuit.

VCO Timing Capacitor (Pins 13 and 14): VCO frequency is inversely proportional to the external timing capacitor, C_0, connected across these terminals (see Figure 5). C_0 must be nonpolar, and in the range of 200 pF to 10 μF.

VCO Quadrature Output (Pin 15): The low-level (≈ 0.6 Vpp) output at this pin is at quadrature phase (i.e. 90° phase-offset) with the other VCO outputs at pins 3 and 5. The dc level at pin 15 is approximately 300 mV above V_R. The quadrature output can be used with an external multiplier as a "lock detect" circuit. In order not to degrade oscillator performance, the output at pin 15 must be buffered with an external high-impedance low-capacitance amplifier. When not in use, pin 15 should be left open-circuited.

Phase Detector Input (Pin 16): Voltage output of the VCO (pin 5) or the output of an external frequency divider is connected to this pin. The dc level of the sensing threshold for the phase detector is referenced to V_R. If the signal is capacitively coupled to pin 16, then this pin must be biased from pin 11, through an external resistor, R_B (R_B ≈ 10 KΩ). The peak voltage swing applied to pin 16 *must not* exceed (V+ − 1.5) volts.

PHASE-LOCKED LOOP PARAMETERS:

Transfer Characteristics:

Figure 9 shows the basic frequency to voltage characteristics of XR-2212. With no input signal present, filtered phase detector output voltage is approximately equal to the internal reference voltage, V_R, at pin 11. The PLL can track an input signal over its tracking bandwidth, shown in the figure. The frequencies f_{TL} and f_{TH} represent the lower and the upper edge of the tracking range, f_0 represents the VCO center frequency.

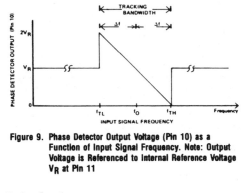

Figure 9. Phase Detector Output Voltage (Pin 10) as a Function of Input Signal Frequency. Note: Output Voltage is Referenced to Internal Reference Voltage V_R at Pin 11

Design Equations:

(See Figure 2 and Figure 9 for definition of components.)

1. VCO Center Frequency, f_0: $f_0 = 1/R_0C_0$ Hz

FIGURE 8-8 *(continued)*

2. Internal Reference Voltage, V_R (measured at Pin 11)

$$V_R = V + /2 - 650 \text{ mV}$$

3. Loop Low-Pass Filter Time Constant, τ: $\quad \tau = R_1 C_1$

4. Loop Damping, ζ: $\quad \zeta = 1/4 \sqrt{\dfrac{NC_0}{C_1}}$

where N is the external frequency divider modular (See 2). If no divider is used, $N = 1$.

5. Loop Tracking Bandwidth, $\pm \Delta f/f_0$: $\quad \Delta f/f_0 = R_0/R_1$

6. Phase Detector Conversion Gain, K_ϕ: (K_ϕ is the differential dc voltage across Pins 10 and 11, per unit of phase error at phase-detector input) $K_\phi = -2V_R/\pi$ volts/radian

7. VCO Conversion Gain, K_0: (K_0 is the amount of change in VCO frequency, per unit of dc voltage change at Pin 10. It is the reciprocal of the slope of conversion characteristics shown in Figure 9). $K_0 = -1/V_R C_0 R_1$ Hz/volt

8. Total Loop Gain, K_T:

$$K_T = 2\pi K_\phi K_0 = 4/C_0 R_1 \text{ rad/sec/volt}$$

9. Peak Phase-Detector Current, I_A; available at pin 10.

$$I_A = V_R \text{ (volts)}/25 \text{ mA}$$

APPLICATION INFORMATION

FM DEMODULATION:

XR-2212 can be used as a linear FM demodulator for both narrow-band and wide-band FM signals. The generalized circuit connection for this application is shown in Figure 10, where the VCO output (pin 5) is directly connected to the phase detector input (pin 16). The demodulated signal is obtained at phase detector output (pin 10). In the circuit connection of Figure 10, the op amp section of XR-2212 is used as a buffer amplifier to provide both additional voltage amplification as well as current drive capability. Thus, the demodulated output signal available at the op amp output (pin 8) is fully buffered from the rest of the circuit.

In the circuit of Figure 10, $R_0 C_0$ set the VCO center frequency, R_1 sets the tracking bandwidth, C_1 sets the low-pass filter time constant. Op amp feedback resistors R_F and R_C set the voltage gain of the amplifier section.

Design Instructions:

The circuit of Figure 10 can be tailored to any FM demodulation application by a choice of the external components R_0, R_1, R_C, R_F, C_0 and C_1. For a given FM center frequency and frequency deviation, the choice of these components can be calculated as follows, using the design equations and definitions given on page 1-34, 1-35 and 1-36.

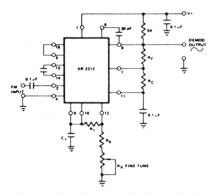

Figure 10. Circuit Connection for FM Demodulation

a) Choose VCO center frequency f_0 to be the same as FM carrier frequency.

b) Choose value of timing resistor R_0, to be in the range of 10 KΩ to 100 KΩ. This choice is arbitrary. The recommended value is $R_0 \cong 20$ KΩ. The final value of R_0 is normally fine-tuned with the series potentiometer, R_X.

c) Calculate value of C_0 from design equation (1) or from Figure 6:

$$C_0 = 1/R_0 f_0$$

d) Choose R_1 to determine the tracking bandwidth, Δf (see design equation 5). The tracking bandwidth, Δf, should be set significantly wider than the maximum input FM signal deviation, Δf_{SM}. Assuming the tracking bandwidth to be "N" times larger than Δf_{SM}, one can re-unite design equation 5 as:

$$\frac{\Delta f}{f_0} = \frac{R_0}{R_1} = N \frac{\Delta f_{SM}}{f_0}$$

Table I lists recommended values of N, for various values of the maximum deviation of the input FM signal.

% Deviation of FM Signal ($\Delta f_{SM}/f_0$)	Recommended value of Bandwidth Ratio, N ($N = \Delta f/\Delta f_{SM}$)
1% or less	10
1 to 3%	5
1 to 5%	4
5 to 10%	3
10 to 30%	2
30 to 50%	1.5

TABLE I

Recommended values of bandwidth ratio, N, for various values of FM signal frequency deviation. (Note: N is the ratio of tracking bandwidth Δf to max. signal frequency deviation, Δf_{SM}).

FIGURE 8-8 (*continued*)

XR-2212

e) Calculate C_1 to set loop damping (see design equation 4). Normally, $\zeta = 1/2$ is recommended. Then, $C_1 = C_0/4$ for $\zeta = 1/2$.

f) Calculate R_C and R_F to set peak output signal amplitude. Output signal amplitude, V_{out}, is given as:

$$V_{out} = \left(\frac{\Delta f_{SM}}{f_0}\right)(V_R)\left(\frac{R_1}{R_0}\right)\left[\frac{R_C + R_F}{R_C}\right]$$

In most applications, $R_F = 100 \text{ K}\Omega$ is recommended; then R_C, can be calculated from the above equation to give desired output swing. The output amplifier can also be used as a unity-gain voltage follower, by open circuiting R_C (i.e., $R_C = \infty$).

Note: All calculated component values except R_0 can be rounded-off to the nearest standard value, and R_0 can be varied to fine-tune center frequency, through a series potentiometer, R_X. (See Figure 10.)

Design Example:

Demodulator for FM signal with 67 kHz carrier frequency with ± 5 kHz frequency deviation. Supply voltage is +12V and required peak output swing is ± 4 volts.

Step a) f_0 is chosen as 67 kHz.

Step b) Choose $R_0 = 20 \text{ K}\Omega$ (18 KΩ fixed resistor in series with 5 KΩ potentiometer).

Step c) Calculate C_0; from design Eq. (1).

$C_0 = 746$ pF

Step d) Calculate R_1. For given FM deviation, $\Delta f_{SM}/f_0 = 0.0746$, and N = 3 from Table I.

Then:

$R_0/R_1 = (3)(0.0746) = 0.224$

or:

$R_1 = 89.3 \text{ K}\Omega$.

Step e) Calculate $C_1 = (C_0/4) = 186$ pF.

Step f) Calculate R_C and R_F to get ± 4 volts peak output swing: Let $R_F = 100 \text{ K}\Omega$. Then,

$R_C = 80.6 \text{ K}\Omega$.

Note: All values except R_0 can be rounded-off to nearest standard value.

FREQUENCY SYNTHESIS

Figure 11 shows the generalized circuit connection for frequency synthesis. In this application an external frequency divider is connected between the VCO output (pin 5) and the phase-detector input (pin 16). When the circuit is in lock, the two signals going into the phase-detector are at the same frequency, or $f_S = f_1/N$ where

N is the modulus of the external frequency divider. Conversely, the VCO output frequency, f_1 is equal to $N f_S$.

In the circuit configuration of Figure 11, the external timing components, R_0 and C_0, set the VCO free-running frequency; R_1 sets the tracking bandwidth and C1 sets the loop damping, i.e., the low-pass filter time constant (see design equations).

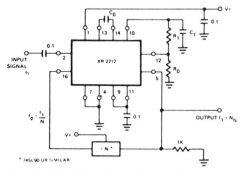

Figure 11. Circuit Connection for Frequency Synthesizer

The total tracking range of the PLL (see Figure 9), should be chosen to accommodate the lowest and the highest frequency, f_{max} and f_{min}, to be synthesized. A recommended choice for most applications is to choose a tracking half-bandwidth Δf, such that:

$$\Delta f \approx f_{max} - f_{min}.$$

If a fixed output frequency is desired, i.e. N and f_S are fixed, then a $\pm 10\%$ tracking bandwidth is recommended. Excessively large tracking bandwidth may cause the PLL to lock on the harmonics of the input signals; and the small tracking range increases the "lock-up" or acquisition time.

If a variable input frequency and a variable counter modulus N is used, then the maximum and the minimum values of output frequency will be:

$$f_{max} = N_{max}(f_S)_{max} \text{ and } f_{min} = N_{min}(f_S)_{min}.$$

Design Instructions:

For a given performance requirement, the circuit of Figure 11 can be optimized as follows:

a) Choose center frequency, f_0, to be equal to the output frequency to be synthesized. If a range of output frequencies is desired, set f_0 to be at mid-point of the desired range.

b) Choose timing resistor R_0 to be in the range of 15 KΩ to 100 KΩ. This choice is arbitrary. R_0 can be fine tuned with a series potentiometer, R_X.

c) Choose timing capacitor, C_0 from Figure 6 or Equation 1.

FIGURE 8-8 (*continued*)

d) Calculate R_1 to set tracking bandwidth (see Figure 9, and design equation 5). If a range of output frequencies are desired, set R_1 to get:

$$\Delta f = f_{max} - f_{min}.$$

If a single fixed output frequency is desired, set R_1 to get:

$$\Delta f = 0.1\ f_0.$$

e) Calculate C_1 to obtain desired loop damping. (See design equation 4). For most applications, $\zeta = 1/2$ is recommended, thus:

$$C_1 = NC_0/4$$

Note: All component values except R_0 can be rounded-off to nearest standard value.

FIGURE 8-8 (*continued*)

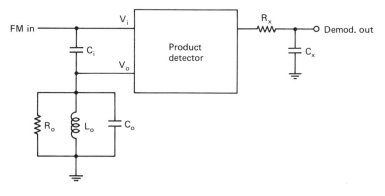

FIGURE 8-9   Quadrature FM demodulator.

The second harmonic ($2\omega_i$) is filtered off, leaving

$$V_{\text{out}} \propto V_o \sin \phi$$

where $\phi = \tan^{-1} \rho Q$ and

$\rho \approx 2\pi F/F_o$ (fractional frequency deviation)
Q = tank circuit quality factor

Amplitude Limiters and FM Thresholding

The vast majority of terrestrial FM communications systems use conventional noncoherent demodulation because most standard frequency discriminators use envelope detection to remove the intelligence from the FM waveform. Unfortunately, envelope detectors (including ratio detectors) will demodulate incidental amplitude variations in the IF. Transmission noise and interference add to the signal and thus produce unwanted amplitude variations. Also, FM is generally accompanied by small amounts of residual amplitude modulation. In the receiver, the unwanted AM and random noise interference are demodulated along with the signal and produce unwanted distortion in the recovered information signal. The noise is more prevalent at the peaks of the FM envelope and relatively insignificant during the zero crossings. A limiter is a circuit that produces a constant-amplitude output for all input signals above a prescribed minimum input level; which is often called either the *threshold*, *quieting*, or *capture* level. Limiters are required in most FM receivers because many of the demodulators that were discussed earlier in this chapter will demodulate AM as well as FM. With amplitude limiters, the *S/N* ratio at the output of the demodulator (*postdetection S/N*) can be improved by as much as 20 or more dB over the input (*predetection*) *S/N*.

Essentially, an amplitude limiter is an additional IF amplifier that is overdriven. Limiting begins when the IF signal is sufficiently large that it drives the amplifier both into saturation and cutoff. Figure 8-10 shows the input and output waveforms for a typical limiter. In Figure 8-10b, it can be seen that for IF signals that are below threshold, the noise is not reduced, and for IF signals above threshold, there is a large reduction

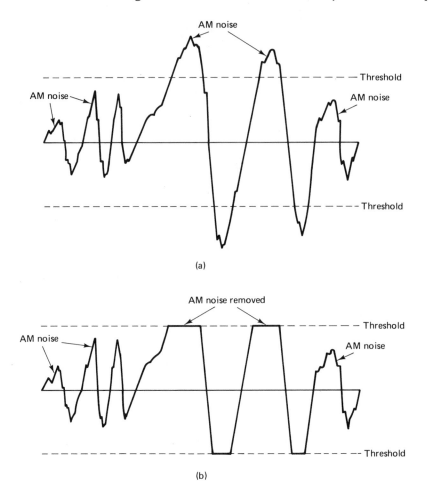

(a)

(b)

FIGURE 8-10 Amplitude limiter input and output waveforms: (a) input waveform; (b) output waveform.

in the noise level. The purpose of the limiter is to remove all amplitude variations from the IF signal.

Figure 8-11a shows the output from a limiter when the noise is greater than the signal (i.e., the noise has captured the limiter). The irregular widths of the serrations are caused by noise impulses saturating the limiter. Figure 8-11b shows the limiter output when the signal is sufficiently greater than the noise (i.e., the signal has captured the limiter). The sinusoidal signal peaks have the limiter so far into saturation that the weaker noise is totally masked. The improvement in the *S/N* ratio is called *FM thresholding*, *FM quieting*, or the *FM capture effect*. There are three criteria that must be satisfied before FM thresholding can occur. They are:

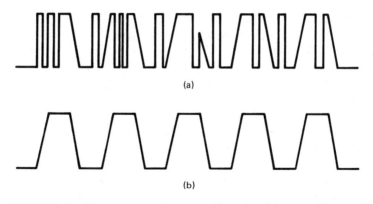

(a)

(b)

FIGURE 8-11   Limiter output: (a) captured by noise; (b) captured by signal.

1. The predetection *S/N* must be 10 dB or greater.
2. The IF signal must be sufficiently amplified to overdrive the limiter.
3. The signal must have a modulation index $m \geq 1$.

Figure 8-12 shows typical FM thresholding curves for a low modulation index signal ($m = 1$) and a high modulation index signal ($m = 4$). The output voltage from

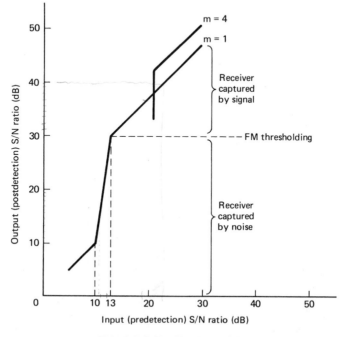

FIGURE 8-12   FM thresholding.

an FM detector is proportional to m^2. Therefore, doubling m increases the S/N ratio by 6 dB. The quieting ratio for $m = 1$ is an input $S/N = 13$ dB, and 22 dB for $m = 4$. For S/N ratios below threshold, the receiver is said to be captured by the noise, and for S/N ratios above threshold, the receiver is said to be captured by the signal. Figure 8-12 shows that IF signals at the input to the limiter with 13 or more dB of S/N undergo 17 dB of S/N improvement.

Limiter circuits.   Figure 8-13a shows a schematic diagram for a single-stage limiter with a built-in output filter. This configuration is commonly called a *bandpass limiter/amplifier* (BPL). A BPL is essentially a class A biased tuned IF amplifier, and for limiting and FM quieting to occur, requires an IF input signal sufficient to drive it into saturation and cutoff. The output tank circuit is tuned to the IF center frequency. Filtering removes the harmonic and intermodulation distortion present in the rectangular pulses due to hard limiting. This is shown in Figure 8-14. If resistor R_2 were removed entirely, the amplifier would be biased for class C operation, which is also appropriate for this type of circuit. Figure 8-13b shows limiter action for the circuit shown in Figure 8-13a. It can be seen that for small signals (below the threshold voltage) no

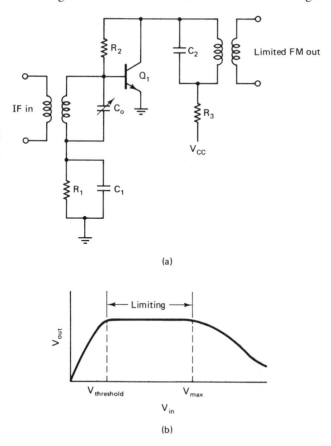

(a)

(b)

FIGURE 8-13 Single-stage tuned limiter: (a) schematic diagram; (b) limiter action.

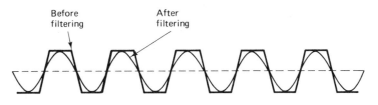

FIGURE 8-14 Filtered limiter output.

limiting occurs. When V_{in} reaches $V_{threshold}$, limiting begins, and for input amplitudes above V_{max}, there is actually a decrease in V_{out} with V_{in}. This is because with high input drive levels, the collector current pulses are sufficiently narrow that they actually develop less tank circuit power. The problem of overdriving the limiter can be rectified by incorporating AGC into the circuit.

When two limiter stages are used, it is called *double* limiting; three stages, *triple* limiting; and so on. Figure 8-15 shows a three-stage *cascoded limiter* without a built-in filter. This type of limiter must be followed by either a ceramic or crystal filter to remove the nonlinear distortion. The limiter shown has three *RC*-coupled limiter stages that are dc series connected to reduce the current drain. Cascoded amplifiers combine several of the advantages of common-emitter and common-gate amplifiers. Cascoding amplifiers also decreases the thresholding level and thus improves the quieting capabilities of the stage. The effects of double and triple limiting are shown in Figure 8-16.

EXAMPLE 8-1

For an FM receiver with a 200-kHz bandwidth, a noise figure = 8 dB, and an input noise temperature T = 100 K; determine the minimum receive carrier power to achieve a postdetection *S/N* = 37 dB. Use the receiver block diagram shown in Figure 8-1 as the receiver model and the FM thresholding curve shown in Figure 8-12.

Solution From Figure 8-12 it can be seen that to achieve a post detection *S/N* = 37 dB, the *S/N* at the input to the limiter must be at least

$$37 \text{ dB} - 17 \text{ dB} = 20 \text{ dB}$$

Therefore, for a receiver noise figure = 8 dB, the *S/N* ratio at the receiver input must be at least

$$20 \text{ dB} + 8 \text{ dB} = 28 \text{ dB}$$

The receiver input noise power, *N*, is

$$10 \log \frac{KTB}{0.001} = 10 \log \frac{(1.38 \times 10^{-23})(100)(200{,}000)}{0.001}$$

$$N = -125.6 \text{ dBm}$$

Therefore, the minimum receive carrier power is

$$-125.6 \text{ dBm} + 28 \text{ dB} = -97.6 \text{ dBm}$$

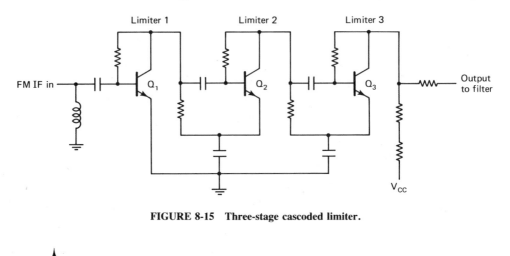

FIGURE 8-15   Three-stage cascoded limiter.

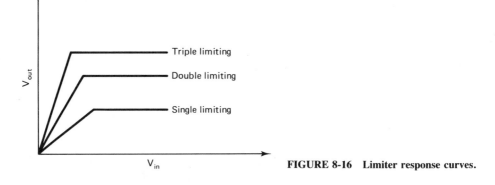

FIGURE 8-16   Limiter response curves.

FM SYSTEMS

FM Stereo

Until 1961, all commercial FM broadcast-band transmissions were *monophonic*. That is, a single 50-Hz to 15-kHz audio channel made up the entire voice and music information spectrum. This single audio channel modulated a high-frequency carrier and was transmitted through a 200-kHz-bandwidth FM channel. With mono transmission, each speaker assembly at the receiver reproduces exactly the same information. It is possible to separate the information in the frequency domain with special speakers, such as *woofers* for low frequencies and *tweeters* for high frequencies. However, it is impossible to separate monophonic sound *spatially*. The entire information signal sounds as though it is coming from the same direction (i.e., from a *point source*—there is no directivity to the sound). In 1961, the FCC authorized *stereophonic* transmission for the commercial FM broadcast band. With stereophonic transmission, the information signal is spatially divided into two 50-Hz to 15-kHz audio channels (a left and a right). Music that originated on the left side is reproduced only on the left speaker, and music that originated on the right side is reproduced only on the right speaker. Therefore, with stereophonic transmission, it is possible to reproduce music with a unique directivity and spatial dimension that before was possible only with live entertainment (i.e., from an *extended* source). Also, with stereo transmission, it is possible to separate music or sound by *tonal quality*, such as percussion, strings, horns, and so on.

A primary concern of the FCC before authorizing stereophonic transmission was its compatibility with monophonic receivers. Stereo transmission was not to affect mono reception. Also, monophonic receivers must be able to receive stereo transmission as monaural without any perceptible degradation in program quality. In addition, stereophonic receivers were to receive stereo programming with nearly perfect separation (40 dB or more) between the left and right channels.

The original FM audio spectrum is shown in Figure 8-17a. The audio channel extended from 50 Hz to 15 kHz. In 1955, the FCC approved *subcarrier* transmission under the Subsidiary Communications Authorization (SCA). SCA is used to broadcast uninterrupted music to private subscribers, such as department stores, restaurants, medical offices, and so on, equipped with special SCA receivers. Originally, the SCA subcarrier ranged from 25 to 75 kHz, but has since been standardized at 67 kHz. The subcarrier and its associated sidebands become part of the total signal that modulates the main

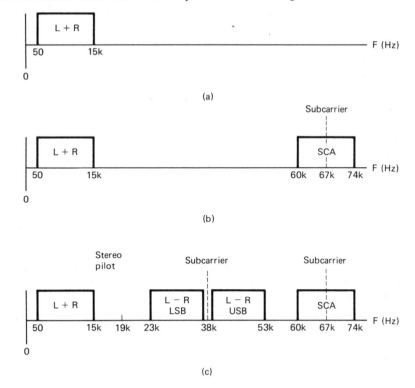

FIGURE 8-17 FM baseband spectrum: (a) prior to 1955; (b) prior to 1961; (c) since 1961.

carrier. At the receiver, the subcarrier is demodulated along with the primary channel but, of course, cannot be heard because of its high frequency. The process of stacking two or more independent channels in the frequency domain then modulating a single carrier is called frequency-division multiplexing (Chapter 1). With FM stereophonic broadcasting, three voice or music channels are frequency-division multiplexed onto a single FM carrier. Figure 8-17b shows the FM baseband spectrum prior to 1961 (baseband comprises the total modulating signal spectrum). The primary audio channel remained at 50 Hz to 15 kHz, while an additional SCA audio channel is translated to the 60- to 74-kHz passband. The SCA subcarrier may be AM single- or double-sideband transmission, or FM with a maximum modulating signal frequency of 7 kHz. However, the SCA modulation of the main carrier is low-index narrowband FM and, consequently, is a much lower-quality transmission than the primary FM channel. The total frequency deviation remained at 75 kHz with 90% (67.5 kHz) reserved for the primary channel and 10% (7.5 kHz) reserved for SCA.

Figure 8-17c shows the FM baseband spectrum since 1961. It comprises the original 50-Hz to 15-kHz audio channel plus two additional audio channels frequency-division multiplexed into a composite baseband signal. The three channels are (1) the left (L) plus the right (R) audio channels (i.e., the L + R stereo channel), (2) the left plus the inverted right audio channels (i.e., the L − R stereo channel), and (3) the SCA subcarrier and its associated sidebands. The L + R stereo channel occupies the 0- to 15-kHz passband, and the 23- to 53-kHz passband is used with stereophonic transmission for carrying the L − R stereo channel. The L − R signal amplitude modulates a 38-kHz subcarrier and produces a double-sideband suppressed carrier signal that occupies the 23- to 53-kHz passband. SCA transmissions occupy the 60- to 74-kHz spectrum. The information contained in the L + R and L − R stereo channels is identical except for their phase. With this scheme, mono receivers can demodulate the total baseband spectrum, but only the 50-Hz to 15-kHz L + R stereo channel is amplified and fed to the speakers. Stereophonic receivers must provide additional demodulation of the 23- to 53-kHz L − R stereo channel, then separate the left and right audio information and feed them to their respective speakers. Again, the SCA subcarrier is demodulated by all FM receivers, although only those with special SCA equipment demodulate the subcarrier to audio.

With stereo transmission, the maximum frequency deviation is still 75 kHz; 7.5 kHz (10%) is reserved for SCA transmission and another 7.5 kHz (10%) is reserved for the 19 kHz stereo pilot, which is explained in the next section. This leaves 60 kHz of frequency deviation for the actual stereophonic transmission of the L + R and L − R stereo channels. However, the L + R and L − R channels are not necessarily limited to 30 kHz deviation each. A rather simple but unique technique is used to *interleave* the two channels such that at times either the L + R or the L − R channel may deviate the main carrier 60 kHz by themselves. However, the total deviation will never exceed 60 kHz. This interleaving technique is explained later in the chapter.

FM stereo generation.   Figure 8-18 shows a simplified block diagram for a stereo FM transmitter. The L and R audio channels are combined in a matrix network

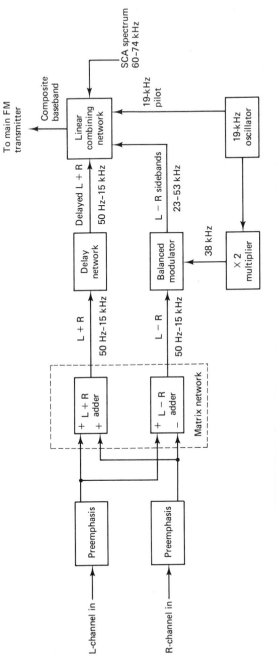

FIGURE 8-18 Stereo FM transmitter using frequency-division multiplexing.

to produce the L + R and L − R stereo channels. The L − R audio channel modulates a 38-kHz subcarrier and produces a 23- to 53-kHz L − R stereo channel. Because there is a time delay introduced in the L − R signal path as it propagates through the balanced modulator, the L + R channel must be artificially delayed somewhat to maintain phase integrity with the L − R channel for demodulation purposes. Also, for demodulation purposes, a 19-kHz subharmonic of the 38-kHz suppressed carrier is added to the baseband signal. The 19-kHz pilot is transmitted rather than the 38-kHz carrier because it is considerably more difficult to recover the 38-kHz carrier in the receiver. The composite baseband signal is fed to the FM transmitter, where it modulates the main carrier.

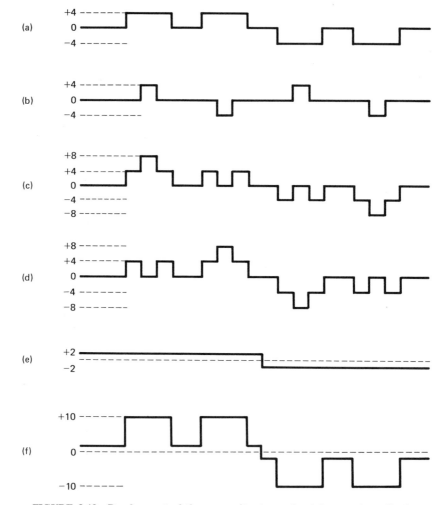

FIGURE 8-19 Development of the composite stereo signal for equal amplitude L and R signals: (a) L audio signal; (b) R audio signal; (c) L + R stereo channel; (d) L − R stereo channel; (e) SCA + 19-kHz pilot; (f) composite baseband waveform.

L + R and L − R channel interleaving. Figure 8-19 shows the development of the composite stereo signal for equal-amplitude L and R channel signals. For illustration purposes, rectangular waveforms are shown. Table 8-1 is a tabular summary of the individual and total signal voltages for Figure 8-19. Note that the L − R audio channel does not appear in the composite waveform. The L − R channel modulates the 38-kHz subcarrier to form the L − R sidebands, which are part of the composite spectrum.

For the FM modulator in this example, it is assumed that 10 V of baseband signal will produce 75 kHz of frequency deviation of the main carrier, and the SCA and 19-kHz pilot polarities shown are for maximum frequency deviation. The L and R channels are each limited to a maximum value of 4 V; 1 V is for SCA, and 1 V is for the 19-kHz stereo pilot. Therefore, 8 V is left for the L + R and L − R stereo channels. Figure 8-19 shows the L, R, L + R, L − R channels, the SCA and 19-kHz pilot, and the composite stereo waveform. It can be seen that the L + R and L − R stereo channels interleave and never produce more than 8 V of total amplitude and therefore never produce more than 60 kHz of frequency deviation. The total composite baseband never exceeds 10 V (75 kHz deviation).

Figure 8-20 shows the development of the composite stereo waveform for unequal values of the L and R signals. Again, it can be seen that the composite stereo waveform never exceeds 10 V or 75 kHz of frequency deviation. For the first set of waveforms, it appears that the sum of the L + R and L − R waveforms completely cancels. Actually, this is not true; it only appears that way because rectangular waveforms are used in this example.

FM stereo reception. FM stereo receivers are identical to standard FM receivers up to the output of the audio detector stage. The output of the discriminator is the total baseband spectrum that was shown in Figure 8-17c.

Figure 8-21 shows a simplified block diagram for an FM receiver giving both mono and stereophonic audio outputs. In the mono signal processing, the L + R stereo channel, which contains all of the original information from both the L and R audio channels, is simply filtered, amplified, then fed to both the L and R speakers. In the stereo portion of the signal processor, the baseband signal is fed to a stereo demodulator where the L and R channels are separated and fed to their respective speakers. The

TABLE 8-1   COMPOSITE FM VOLTAGES

L	R	L + R	L − R	SCA and pilot	Total
0	0	0	0	2	2
4	0	4	4	2	10
0	4	4	−4	2	2
4	4	8	0	2	10
4	−4	0	8	2	10
−4	4	0	−8	−2	−10
−4	−4	−8	0	−2	−10

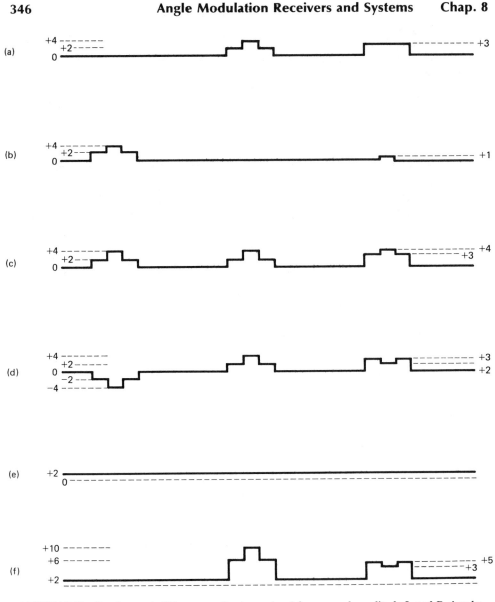

FIGURE 8-20   Development of the composite stereo signal for unequal amplitude L and R signals:
(a) L audio signal; (b) R audio signal; (c) L + R stereo channel; (d) L− R stereo channel;
(e) SCA + 19-kHz pilot; (f) composite baseband waveform.

L + R and L − R stereo channels and the 19-kHz pilot are separated from the composite
baseband signal with filters. The 19-kHz pilot is filtered off with a high-Q bandpass
filter, multiplied by 2, amplified, and then fed to the L − R demodulator. The L + R
stereo channel is filtered off by a low-pass filter with an upper cutoff frequency of 15
kHz. The L − R double-sideband signal is separated with a broadly tuned bandpass

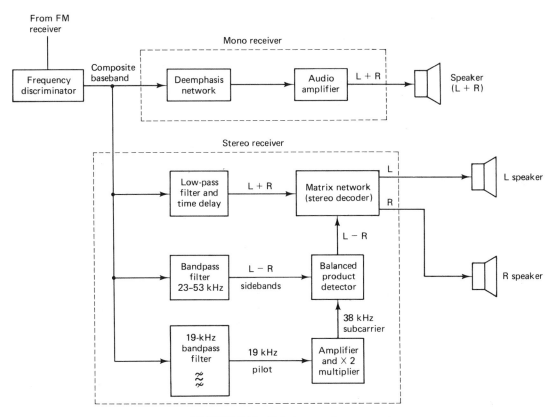

FIGURE 8-21   FM stereo and mono receiver.

filter, then mixed with the recovered 38-kHz carrier in a balanced modulator to produce the L − R audio information. The matrix network combines the L + R and L − R signals in such a way to separate the L and R audio information signals, which are fed to their respective deemphasis networks and speakers.

Figure 8-22 shows the block diagram for the stereo matrix decoder. The L − R audio channel is added directly to the L + R audio channel. The output from the adder is

$$
\begin{array}{r}
L + R \\
+ \ \underline{(L - R)} \\
2L
\end{array}
$$

The L − R audio channel is inverted, then added to the L + R audio channel. The output from the adder is

$$
\begin{array}{r}
L + R \\
- \ \underline{(L - R)} \\
2R
\end{array}
$$

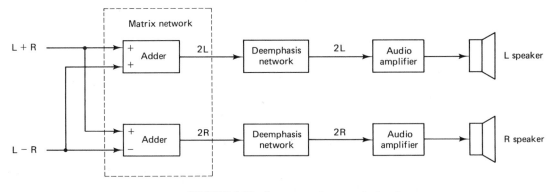

FIGURE 8-22 **Stereo matrix network decoder.**

Large-scale integration stereo demodulator.   Figure 8-23 shows the specification sheet for an XR-1310 stereo demodulator. The XR-1310 is a monolithic FM stereo demodulator which uses phase locked techniques to derive the right and left audio channels from the composite stereo signal. The XR-1310 uses no external LC tank circuits for tuning, and alignment is accomplished with a single potentiometer. The XR-1310 features simple noncritical tuning, excellent channel separation, low distortion, and a wide dynamic range.

Two-Way FM Radio Communications

Two-way FM radio communications is used extensively for *public safety mobile* communications, such as police and fire department calls and emergency medical services. There are three primary frequency bands allocated by the FCC for two-way FM radio communications: 132 to 174 MHz, 450 to 470 MHz, and 806 to 947 MHz. The maximum frequency deviation for two-way FM transmitters is typically 5 kHz, and the maximum modulating signal frequency is 3 kHz. These values give a deviation ratio of 1.67 and a maximum Bessel bandwidth of approximately 24 kHz. However, the allocated FCC channel spacing is 30 kHz. Two-way FM radio is half duplex, which supports two-way communications but not simultaneously; only one party can transmit at a time. Transmissions are initiated by closing a *push-to-talk* (PTT) switch, which turns on the transmitter and shuts off the receiver. During idle conditions, the transmitter is shut off and the receiver is turned on to allow monitoring of the radio channel for transmissions from other parties.

Two-way FM transmitter.   The simplified block diagram for a *modular integrated circuit* two-way FM transmitter is shown in Figure 8-24. This is actually a direct PM transmitter. PM is generally used because direct FM transmitters do not have the stability necessary to meet FC standards without using AFC. The transmitter shown is a four-channel unit that operates in the 150- to 174-MHz band. The channel selector switch applies power to one of four crystal oscillator modules that operate at

STEREO DEMODULATOR

GENERAL DESCRIPTION

The XR-1310 is a unique FM stereo demodulator which uses phase-locked techniques to derive the right and left audio channels from the composite signal. Using a phase-locked loop to regenerate the 38 kHz subcarrier, it requires no external L-C tanks for tuning. Alignment is accomplished with a single potentiometer.

FEATURES

Requires No Inductors
Low External Part Count
Simple, Noncritical Tuning by Single Potentiometer Adjustment
Internal Stereo/Monaural Switch with 100 mA Lamp Driving Capability
Wide Dynamic Range: 600 mV (RMS) Maximum Composite Input Signal
Wide Supply Voltage Range: 8 to 14 Volts
Excellent Channel Separation
Low Distortion
Excellent SCA Rejection

ORDERING INFORMATION

Part Number	Package	Operating Temperature
XR-1310CP	Plastic	−40°C to +85°C

FUNCTIONAL BLOCK DIAGRAM   March 1982

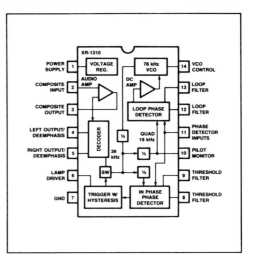

APPLICATIONS

FM Stereo Demodulation

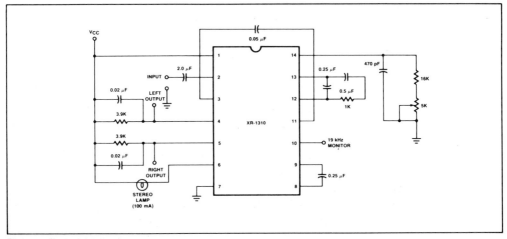

Figure 1. Typical Application

EXAR INTEGRATED SYSTEMS, INC.
750 Palomar Avenue, Sunnyvale, CA 94088   (408) 732-7970   TWX 910-339-9233   3-82 Rev. 3

FIGURE 8-23   XR-1310 stereo demodulator. (Courtesy of EXAR Corporation.)

ELECTRICAL CHARACTERISTICS

Test Conditions: Unless otherwise noted; $V_{CC}^* = +12$ Vdc, $T_A = +25°C$, 560 mV (RMS) (2.8 Vp-p) standard multiplex composite signal with L or R channel only modulated at 1.0 kHz and with 100 mV (RMS) (10 % pilot level), using circuit of Figure 1.

PARAMETERS	MIN.	TYP.	MAX.	UNIT
Maximum Standard Composite Input Signal (0.5 % THD)	2.8			Vp-p
Maximum Monaural Input Signal (1.0 % THD)	2.8			Vp-p
Input Impedance		50		kΩ
Stereo Channel Separation (50 Hz — 15 kHz)	30	40		dB
Audio Output Voltage (desired channel)		485		mV (RMS)
Monaural Channel Balance (pilot tone "off")			1.5	dB
Total Harmonic Distortion		0.3		%
Ultrasonic Frequency Rejection 19 kHz 38 kHz		34.4 45		dB
Inherent SCA Rejection (f = 67 kHz; 9.0 kHz beat note measured with 1.0 kHz modulation "off")		80		dB
Stereo Switch Level (19 kHz input for lamp "on") Hysteresis	13	 6	20	mV (RMS) dB
Capture Range (permissable tuning error of internal oscillator, reference circuit values of Figure 1)		±3.5		%
Operating Supply Voltage (loads reduced to 2.7 kΩ for 8.0-volt operation)	8.0		14	Vdc
Current Drain (lamp "off")		13		mAdc

*Symbols conform to JEDEC Engineering Bulletin No. 1 when applicable.

ABSOLUTE MAXIMUM RATINGS

($T_A = +25°C$ unless otherwise noted)

		Power Dissipation (package limitation) Derate above $T_A = +25°C$	625 mW 5.0 mW/°C
Power Supply Voltage	14 V	Operating Temperature	−40 to +85°C
Lamp Current (nominal rating, 12 V lamp)	75 mA	Range (Ambient) Storage Temperature Range	 −65 to +150°C

FIGURE 8-23 (*continued*)

frequencies between 12.5 and 14.5 MHz, depending on the specific channel assignment. The oscillator frequency is temperature compensated by the compensation module to ensure a stability of ±0.0002%. The phase modulator uses a varactor diode which is modulated by the audio signal at the output of the audio limiter. The audio signal amplitude is limited to ensure that the transmitter is not overdeviated. The modulated IF carrier is amplified, then multiplied by 12 to produce the desired RF carrier frequency. The RF is further amplified and filtered prior to transmission. The electronic push-to-talk module ensures that power is applied to the transmitter only when a transmission is in progress and that the receiver is on only when the transmitter is off. An electronic PTT is used rather than a simple mechanical switch, to reduce the static noise associated with *contact bounce* in mechanical switches. Keying the PTT applies dc power to the selected transmit oscillator module and the RF power amplifiers. Figure 8-25 shows the schematic diagram for a typical PTT module. Keying the PTT switch grounds the base of Q_1, causing it to conduct and turn off Q_2. With Q_2 off, V_{CC} is applied to the

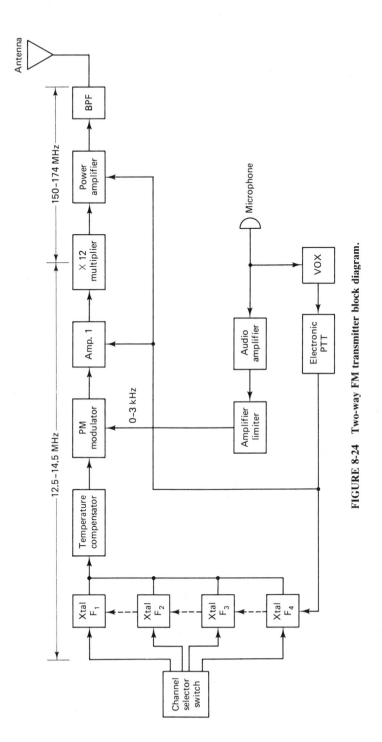

FIGURE 8-24 Two-way FM transmitter block diagram.

351

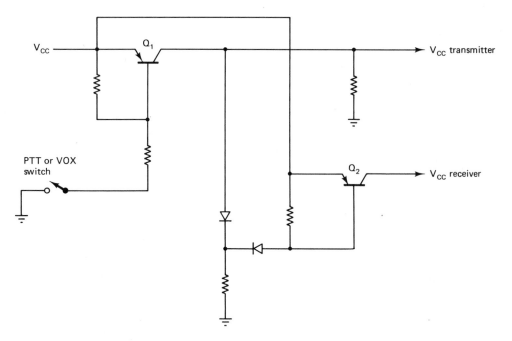

FIGURE 8-25 **Electronic PTT schematic diagram.**

transmitter and removed from the receiver. With the PTT switch released, Q_1 shuts off, removing V_{CC} from the transmitter and Q_2 is turned on applying V_{CC} to the receiver. Transmitters equipped with VOX (*voice-operated transmitter*) are automatically keyed each time the operator speaks into the microphone, regardless of whether or not the PTT button is depressed. Transmitters equipped with VOX require an external microphone. The schematic diagram for a typical VOX module is shown in Figure 8-26. Audio signal power in the 400- to 600-Hz passband is filtered and amplified by Q_1, Q_2, and Q_3. The output from Q_3 is rectified and used to turn on Q_4, which places a ground on the PTT circuit, enabling the transmitter and disabling the receiver. With no audio input signal, Q_4 is off and the PTT pin is open, disabling the transmitter and enabling the receiver.

Two-way FM receiver.   The block diagram for a typical two-way FM receiver is shown in Figure 8-27. Again, this is a four-channel integrated-circuit modular receiver with four separate crystal oscillator modules. Whenever the receiver is on, one of the four oscillator modules is activated, depending on the position of the channel selector switch. The oscillator frequency is again temperature compensated, then multiplied by 9. The output from the multiplier is applied to the mixer, where it beats with the incoming RF to produce a 20-MHz IF. This receiver uses low-side injection and the crystal oscillator frequency is determined as follows:

$$\text{crystal frequency} = \frac{\text{RF frequency} - 20\ \text{MHz}}{9}$$

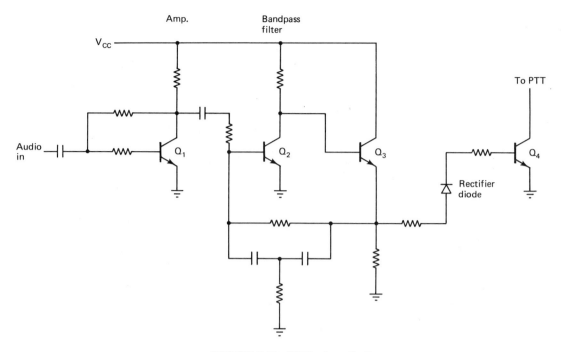

FIGURE 8-26 VOX schematic diagram.

The IF signal is filtered, amplified, limited, and then applied to the frequency discriminator for demodulation. The output from the discriminator is amplified, then applied to the speaker. A typical noise amplifier/squelch circuit is shown in Figure 8-28. The squelch circuit is keyed by out-of-band noise at the output of the audio amplifier. With no receive RF, the AGC causes the gain of the IF amplifiers to increase, which increases the receiver noise in the 3- to 5-kHz band. Whenever excessive noise is present, the audio amplifier is turned off and the receiver is quieted. The input bandpass filter passes the 3- to 5-kHz noise signal, which is amplified and rectified. The rectified output voltage determines the off/on condition of squelch switch, Q_3. When Q_3 is on, V_{CC} is applied to the receive audio amplifier. When Q_3 is off, V_{CC} is removed from the audio amplifier, quieting the receiver. R_x is a squelch sensitivity adjustment.

Mobile Telephone Service

Mobile telephone service is best described by explaining the differences between it and two-way mobile radio. As stated previously, mobile radio is half duplex and all transmissions (unless scrambled) can be heard by any listener tuned to that channel. Mobile telephone is full duplex and operates much the same as the wireline telephone service provided by local telephone companies. Mobile telephone permits two-way simultaneous transmission and, for privacy, each mobile unit is assigned a unique telephone number.

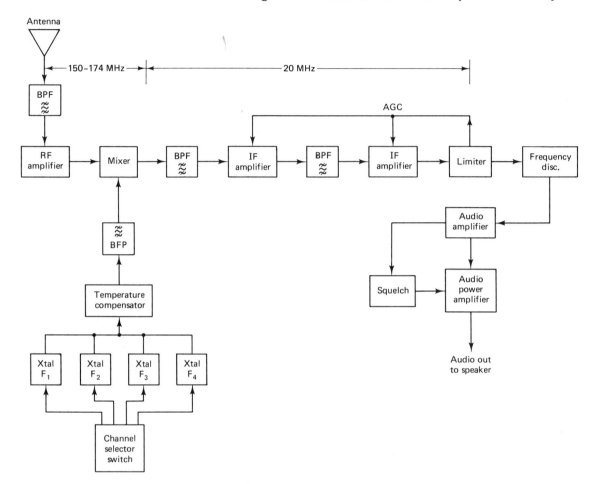

FIGURE 8-27   **Two-way FM receiver block diagram.**

Historical perspective. The first mobile telephone system in the United States was established in 1946 in St. Louis, Missouri. The FCC allocated six 60-kHz mobile telephone channels in the 150-MHz frequency range. In 1947, a public mobile telephone system was established along the highway between New York City and Boston that operated in the 35- to 40-MHz frequency range. In 1949, the FCC authorized six additional mobile channels to *radio common carriers*, which are defined as companies that do not provide public wireline telephone service but do interconnect to the public telephone network and provide equivalent nonwireline telephone service. The FCC later increased the number of channels from 6 to 11 by reducing the bandwidth to 30 kHz and spacing the new channels between the old ones. In 1950, the FCC added 12 new channels in the 450-MHz band. Until 1964, mobile telephone systems operated only in the manual mode; a special mobile telephone operator handled every call to and from each *mobile unit*. In 1964, *automatic channel selection systems* were placed in service

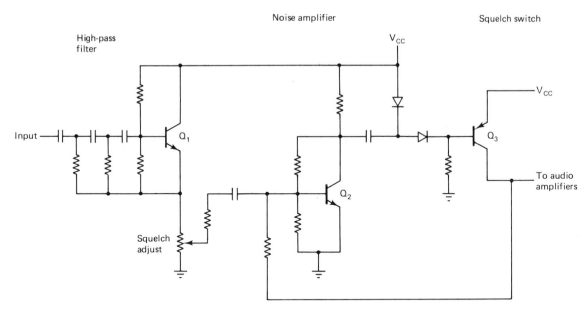

FIGURE 8-28   Squelch circuit.

for mobile telephone systems. This eliminated the need for push-to-talk operation and allowed customers to direct dial their calls without the aid of an operator. *Automatic call completion* was extended to the 450-MHz band in 1969 and *improved mobile telephone systems* (IMTS) became the United States standard mobile telephone service. Presently, there are more than 160,000 mobile telephone service (MTS) subscribers nationwide. MTS uses FM radio channels to establish communication links between mobile telephone and central base station transceivers, which are linked to the local telephone exchange via normal metallic telephone lines. Most MTS systems serve an area approximately 40 mi in diameter and each channel operates similar to a party line. Each channel may be assigned to several subscribers, but only one at a time. If the subscriber's preassigned channel is busy, the subscriber must wait until it is idle before he or she can either place or receive a call.

The growing demand for the overcrowded mobile telephone frequency spectrum prompted the FCC to issue Docket 18262, which inquired into a means for providing a higher spectrum efficiency. In 1971, AT&T submitted a proposal on the technical feasibility of providing efficient use of the mobile telephone frequency spectrum. AT&T's report, entitled *High Capacity Mobile Phone Service*, outlined the principles of cellular radio.

Cellular radio.   *Cellular radio* corrects many of the problems of traditional two-way mobile telephone service and creates a totally new environment for both mobile radio and traditional wireline telephone service. The key concepts of cellular radio were uncovered by researchers at Bell Telephone Laboratories in 1947. It was determined that by subdividing a relatively large geographical area into smaller sections called *cells*, a concept called *frequency reuse* could be employed to increase dramatically the capacity of a mobile telephone channel (reuse is explained later in the chapter). In addition, integrated-circuit technology and microprocessors have recently enabled complex radio and logic circuits to be used in electronic switching machines to store programs that provide faster and more efficient call processing.

In 1974, the FCC allocated an additional 40 MHz of bandwidth for cellular radio service (825 to 845 MHz and 870 to 890 MHz). These frequency bands were formerly allocated to UHF television channels 70 to 83. In 1975, AT&T was granted the first license to operate a developmental cellular radio service in Chicago and AT&T subsequently formed the *Advanced Mobile Phone Service* (AMPS). The following year *American Radio Telephone Service* (ARTS) was granted authorization from the FCC to install a second developmental system in the Baltimore/Washington, D.C., area.

The basic cellular radio concept is quite simple. The FCC defined geographic cellular radio coverage areas based on 1980 census figures. With the cellular concept, each area is further divided into *hexagonal cells* that fit together to form a honeycomb pattern. The hexagon shape was chosen because it provides the most effective transmission by approximating a circular pattern while eliminating gaps present between adjacent circles. The number of cells per system is not defined by the FCC and has been left to the provider to establish in accordance with anticipated traffic patterns. Each geographic mobile service area is allocated 666 cellular radio channels. Each transceiver within a covered area has a fixed subset of the 666 available radio channels based on expected traffic flow.

Figure 8-29 shows a simplified cellular telephone system which includes all of the basic components necessary for cellular radio communications. There is a radio-frequency transceiver located at the physical center of each cell. Cellular radio uses several moderately powered transceivers over a relatively wide service area, as opposed to MTS, which uses a single high-powered transceiver at high elevation. Also, each cell contains a computerized cell-site controller to handle all cell-site control functions. All cell sites are connected to an electronic switching center through a leased four-wire metallic telephone line. The electronic switching center also provides access to the public switched telephone network. Generally, each cell can accommodate up to 70 different channels simultaneously. Within a cell, each channel can support only one mobile telephone user at a time. Channels are dynamically assigned and dedicated to a single user for the duration of the call, and any user may be assigned to any channel. Called *frequency reuse*, this allows a cellular telephone system in a single area to handle considerably more than the 666 available channels. Thus cellular radio makes more efficient use of the available frequency spectrum than does traditional MTS service.

As a car moves away from the transceiver in the center of a cell, the received signal begins to decrease. When the signal strength is reduced to a predetermined level, the electronic switching center locates the cell in the *honeycomb* that is receiving the strongest signal from the mobile unit and transfers the mobile unit to the transceiver in the new cell. The transfer includes converting the call to an available frequency within the new cell's allocated channel subset. This transfer is called a *handoff* and is completely

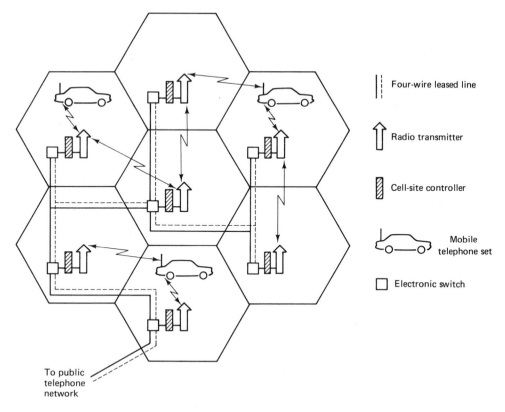

Four-wire leased line

Radio transmitter

Cell-site controller

Mobile
telephone set

Electronic switch

To public
telephone
network

FIGURE 8-29 Typical cellular telephone system.

transparent to the subscriber (i.e., the subscriber does not know that his facility has been switched). The transfer takes approximately 0.2 s, which is imperceptible to voice telephone users. However, this delay may be disruptive to data communications.

There are five primary components of a cellular radio system. They are the *electronic switching center*, a *controller*, a *radio transceiver*, *system interconnections*, and *mobile telephone units*.

The electronic switching center. The electronic switching center is a digital telephone exchange and is the heart of the system. The switch performs two essential functions: (1) it controls the switching between the public telephone network and the cell sites for all wireline-to-mobile, mobile-to-wireline, and mobile-to-mobile calls; and (2) it processes data received from the cell-site controllers concerning mobile unit status, diagnostic data, and bill compiling information.

The cell site. Each cell contains one cell-site controller that operates under the direction of the switching center. The cell-site controller manages each of the radio channels at the site, supervises calls, turns the radio transmitter and receiver on and off, injects data onto the control and user channels, and performs diagnostic tests on the cell-site equipment.

System interconnections. Four-wire leased telephone lines are generally used to connect the switching centers to each of the cell sites. There is one dedicated four-wire trunk for each of the cell's user channels. Also, there must be at least one four-wire circuit to connect the switch to the cell-site controller as a control channel.

Mobile units. A mobile telephone unit consists of a control unit, a radio transceiver, a logic unit, and a mobile antenna. The control unit houses all of the user interfaces, including a handset. The transceiver uses a frequency synthesizer to tune into any designated cellular system channel. The logic unit interprets subscriber actions and system commands and manages the transceiver and control units.

QUESTIONS

8-1. Describe the differences between an AM receiver and an FM receiver.

8-2. Draw the schematic diagram for a single-ended slope detector and describe its operation.

8-3. Draw the schematic diagram for a doubled-ended slope detector and describe its operation.

8-4. Draw the schematic diagram for a Foster–Seeley discriminator and describe its operation.

8-5. Draw the schematic diagram for a ratio detector and describe its operation.

8-6. Describe the operation of a PLL FM demodulator.

8-7. Draw the schematic diagram for a quadrature FM demodulator and describe its operation.

8-8. Contrast the advantages and disadvantages of the FM demodulators described in Questions 8-2 through 8-7.

8-9. What is the purpose of a limiter in an FM receiver?

8-10. Describe FM thresholding.

8-11. Describe the operation of an FM stereo transmitter; an FM stereo receiver.

8-12. Draw the block diagram for a two-way FM transmitter and describe its operation.

8-13. Draw the block diagram for a two-way FM receiver and describe its operation.

8-14. Describe the operation of an electronic PTT circuit.

8-15. Describe the operation of a cellular radio system.

PROBLEMS

8-1. Determine the minimum input S/N ratio required for a receiver with 15 dB of FM improvement, a noise figure F = 4 dB, and a desired postdetection signal-to-noise ratio = 33 dB.

8-2. For an FM receiver with a 100 kHz bandwidth, a noise figure F = 6 dB, and an input noise temperature T = 200°C; determine the minimum receive carrier power to achieve a postdetection S/N = 40 dB. Use the receiver block diagram shown in Figure 8-1 as the receiver model and the FM thresholding curve shown in Figure 8-12.

$F_{lo} = 92.75 + 10.7$

$IF = 2IF + 92.75$

8-3. For an FM receiver tuned to 92.75 MHz using high-side injection and a first IF of 10.7 MHz, determine the image frequency. The local oscillator frequency.

8-4. For an FM receiver with an input frequency deviation $\Delta F = 40$ kHz and a transfer ratio $K = 0.01$ V/kHz, determine V_{out}.

8-5. For the balanced slope detector shown in Figure 8-3a, a center frequency $F_c = 20.4$ MHz, and a maximum input frequency deviation $\Delta F = 50$ kHz, determine the upper and lower cutoff frequencies for the tuned circuit. $F_{upper} = 1.33(50k) + 20.4 mHz$, $F_{lower} = 20.4 mHz - 1.33(50k)$

8-6. For the Foster–Seeley discriminator shown in Figure 8-4, $VC_1 = 1.2$ V and $VC_2 = 0.8$ V, determine V_{out}.

8-7. For the ratio detector shown in Figure 8-6, $VC_1 = 1.2$ V and $VC_2 = 0.8$ V, determine V_{out}.

8-8. For an FM demodulator with an FM improvement factor equal to 23 dB and an input (predetection) signal-to-noise ratio $S_i/N_i = 26$ dB, determine the postdetection S/N.

8-9. From Figure 8-12, determine the approximate FM improvement factor for an input $S/N = 10.5$ dB and $m = 1$.

8-1) $\dfrac{S}{N}_{min\ input} = \dfrac{S}{N}_{Postdetection} - \dfrac{S}{N}_{improvement} + NF$

8-3) $IF = 2IF_{1st} + 97.75\ mHz$

8-5) $F_{upper} = 1.33(50\,kHz) + 20.4\ mHz$
$F_{lower} = 20.4\ mHz - 1.33(50\,kHz)$

8-7) $V_{out} = VC_2\ (Pg. 325)$

8-9) ? $Fig\ 8-12$ (guess)

Chapter 9

TRANSMISSION LINES

INTRODUCTION

A *transmission line* is a *metallic conductor system* that is used to transfer electrical energy from one point to another. More specifically, a transmission line is two or more conductors separated by an insulator, such as a pair of wires or a system of wire pairs. A transmission line can be as short as a few inches or it can span several thousand miles. Transmission lines can be used to propagate dc or low-frequency ac (such as 60 cycle power and audio signals); they can also be used to propagate very high frequencies (such as intermediate and radio-frequency signals). When propagating low-frequency signals, transmission-line behavior is rather simple and quite predictable. However, when propagating high-frequency signals, the characteristics of transmission lines become more involved and their behavior somewhat peculiar to a student of lumped constant circuits and systems.

TRANSVERSE ELECTROMAGNETIC WAVES

Propagation of electrical power along a transmission line occurs in the form of *transverse electromagnetic* (TEM) *waves*. A wave is an *oscillatory motion*. The vibration of a particle excites similar vibrations in nearby particles. A TEM wave propagates primarily in the nonconductor (dielectric) that separates the two conductors of a transmission line. Therefore, a wave travels or propagates itself through a medium. A transverse wave is a wave in which the direction of displacement is perpendicular to the direction of propagation. A surface wave of water is a transverse wave. A wave in which the displacement is in the direction of propagation is called a *longitudinal wave*. Sound waves are longitudinal. An electromagnetic (EM) wave is a wave produced by the acceleration of an electric charge. In a conductor, current and voltage are always accompanied by an electric (E) and a magnetic (H) field in the adjoining region of space. Figure 9-1a shows the spatial relationships between the E and H fields of an electromagnetic wave. Figure 9-1b shows the cross-sectional views of the E and H fields that

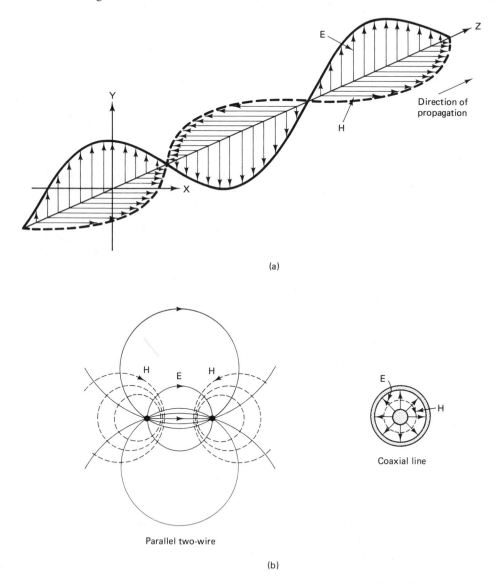

(a)

(b)

FIGURE 9-1 (a) Spatial and (b) cross-sectional views showing the relative displacement of the E and H fields on a transmission line.

surround a parallel two-wire and a coaxial line. It can be seen that the E and H fields are perpendicular to each other (at $90°$ angles) at all points. This is referred to as *space quadrature*. Electromagnetic waves that travel along a transmission line from the source toward the load are called *incident waves*, and those that travel from the load back toward the source are called *reflected waves*.

Characteristics of Electromagnetic Waves

Wave velocity.   Waves travel at various speeds, depending on the type of wave and the characteristics of the propagation medium. Sound waves travel at approximately 1100 f/s in the normal atmosphere. Electromagnetic waves travel much faster. In free space (a vacuum), TEM waves travel at the speed of light, $c = 186,283$ statute mi/s or 299,793 m/s, rounded off to 186,000 mi/s and 3×10^8 m/s. However, in air (such as the earth's atmosphere), TEM waves travel slightly more slowly, and along a transmission line, electromagnetic waves travel considerably more slowly.

Frequency and wavelength.   The oscillations of an electromagnetic wave are periodic and repetitious. Therefore, they are characterized by a frequency. The rate at which the periodic wave repeats is its frequency. The distance of one cycle occurring in space is called the *wavelength* and is determined from the following fundamental equation:

$$\text{distance} = \text{velocity} \times \text{time} \tag{9-1}$$

If the time for one cycle is substituted into Equation 9-1, we get the length of one cycle, which is called the wavelength and whose symbol is the Greek lowercase letter lambda (λ).

$$\lambda = \text{velocity} \times \text{period}$$

$$= v \times T$$

and since $T = 1/F$,

$$\lambda = \frac{v}{F} \tag{9-2}$$

for free-space propagation, $v = c$; therefore, the length of one cycle is

$$\lambda = \frac{c}{F} = \frac{3 \times 10^8 \text{ m/s}}{F \text{ cycles/s}} = \frac{\text{meters}}{\text{cycle}} \tag{9-3}$$

Figure 9-2 shows a graph of the displacement and velocity of a longitudinal wave as it propagates along a transmission line from a source to a load. The horizontal (x) axis is distance and the vertical (y) axis is displacement. One wavelength is the distance covered by one cycle of the wave. It can be seen that the wave moves to the right or propagates down the line with time. If a voltmeter is placed at any stationary point on the line, the voltage measured will fluctuate from 0 to maximum positive, back to zero, to maximum negative, back to zero again, and then the cycle repeats.

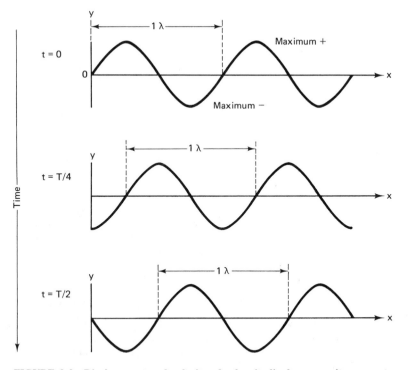

FIGURE 9-2   Displacement and velocity of a longitudinal wave as it propagates down a transmission line.

TYPES OF TRANSMISSION LINES

Transmission lines can be generally classified as balanced or unbalanced. With a *balanced transmission line*, both conductors carry current and the current in each wire is 180° out of phase with the current in the other wire. With an *unbalanced line*, one wire is at ground potential, while the other wire carries all of the current. Both conductors in a balanced line carry current and the signal currents are equal magnitude with respect to electrical ground but travel in opposite directions. Currents that flow in opposite directions in a balanced wire pair are called *metallic circuit currents*. Currents that flow in the same direction are called *longitudinal currents*. A balanced pair has the advantage that most noise interference is induced equally in both wires, producing longitudinal currents that cancel in the load. Figure 9-3 shows the results of metallic and longitudinal currents on a balanced transmission line. It can be seen that longitudinal currents (generally produced by static interference) cancel in the load. Balanced transmission lines can be connected to unbalanced loads, and vice versa, with special transformers called *baluns* (*bal*anced to *un*balanced).

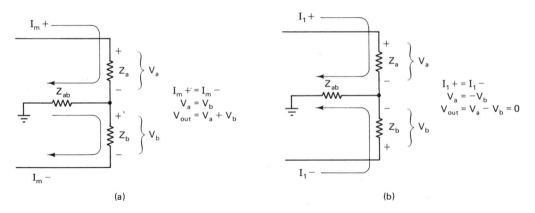

FIGURE 9-3 Results of metallic and longitudinal currents on a balanced transmission line: (a) metallic currents due to signal voltages; (b) longitudinal currents due to noise voltages.

Parallel-Conductor Transmission Lines

Open-wire transmission line. An *open-wire transmission line* is a *two-wire parallel conductor* and is shown in Figure 9-4a. It consists simply of two parallel wires, closely spaced and separated by air. Nonconductive spacers are placed at periodic intervals for support and to keep the distance between the conductors constant. The distance between the two conductors is generally between 2 and 6 in. The dielectric is simply the air between and around the two conductors in which the TEM wave propagates. The only real advantage of this type of transmission line is its simple construction. Because there is no shielding, radiation losses are high and it is susceptible to noise pickup. These are the primary disadvantages of an open-wire transmission line. Therefore, open-wire transmission lines are normally operated in the balanced mode.

Twin-lead. *Twin-lead* is another form of two-wire parallel conductor transmission line and is shown in Figure 9-4b. Twin-lead is often called *ribbon cable*. Twin-lead is essentially the same as an open-wire transmission line except that the spacers between the two conductors are replaced with a continuous solid dielectric. This assures uniform spacing along the entire cable, which is a desirable characteristic for reasons that are explained later in the chapter. Typically, the distance between the two conductors is $\frac{5}{16}$ in. Common dielectric materials are Teflon and polyethylene.

Twisted-pair cable. A *twisted-pair cable* is formed by twisting together two insulated conductors. Pairs are often stranded in *units*, and the units are then cabled into *cores*. The cores are covered with various types of *sheaths*, depending on their intended use. Neighboring pairs are twisted with different *pitch* (twist length) to reduce interference between pairs due to mutual induction. The *primary constants* of twisted-pair cable are its electrical parameters (resistance, inductance, capacitance, and conductance), which are subject to variations with the physical environment such as temperature,

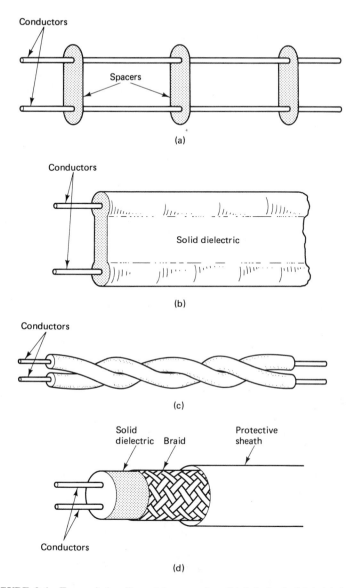

FIGURE 9-4 Transmission lines: (a) open-wire; (b) twin-lead; (c) twisted pair; (d) shielded pair.

moisture, and mechanical stress and are dependent on manufacturing deviations. A twisted-pair cable is shown in Figure 9-4c.

Shielded cable pair. To reduce radiation losses and interference, parallel two-wire transmission lines are often enclosed in a conductive metal *braid*. The braid is connected to ground and acts like a shield. The braid also prevents signals from radiating

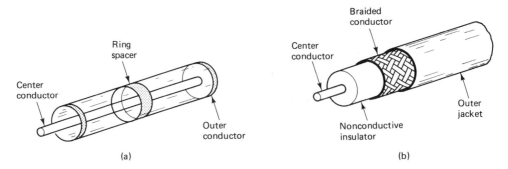

FIGURE 9-5 Concentric or coaxial transmission lines; (a) rigid air filled; (b) solid flexible line.

beyond its boundaries and keeps electromagnetic interference from reaching the signal conductors. A shielded parallel wire pair is shown in Figure 9-4d. It consists of two parallel wire conductors separated by a solid dielectric material. The entire structure is enclosed in a braided conductive tube, then covered with a protective plastic coating.

Concentric or Coaxial Transmission Lines

Parallel-conductor transmission lines are suitable for low-frequency applications. However, at high frequencies, their radiation and dielectric losses, as well as their susceptibility to external interference, are excessive. Therefore, *coaxial conductors* are used extensively for high-frequency applications, to reduce losses and to isolate transmission paths. The basic coaxial cable consists of a center conductor surrounded by a *concentric* (uniform distance from the center) *outer conductor*. At relatively high operating frequencies, the coaxial outer conductor provides excellent shielding against external interference. However, at lower frequencies, the shielding is ineffective. Also, a coaxial cable's outer conductor is generally grounded, which limits its use to unbalanced applications.

Essentially, there are two types of coaxial cables: *rigid air filled* or *solid flexible* lines. Figure 9-5a shows a rigid air coaxial line. It can be see that the center conductor is surrounded coaxially by a tubular outer conductor and the insulating material is air. The outer conductor is physically isolated and separated from the center conductor by a spacer, which is generally made of Pyrex, polystyrene, or some other nonconductive material. Figure 9-5b shows a solid flexible coaxial cable. The outer conductor is braided, flexible, and coaxial to the center conductor. The insulating material is a solid nonconductive polyethylene material that provides both support and electrical isolation between the inner and outer conductors. The inner conductor is a flexible copper wire that can either be solid or hollow.

Rigid air-filled coaxial cables are relatively expensive to manufacture, and to minimize losses, the air insulator must be relatively free of moisture. Solid coaxial cables have lower losses, are easier to construct, and are easier to install and maintain. Both types of coaxial cables are relatively immune to external radiation, radiate little themselves, and can operate at higher frequencies than can their parallel-wire counterparts.

The basic disadvantages of coaxial transmission lines is that they are expensive and must be used in the unbalanced mode.

TRANSMISSION-LINE EQUIVALENT CIRCUIT

Uniformly Distributed Lines

The characteristics of a transmission line are determined by its electrical properties, such as wire conductivity and insulator dielectric constant, and its physical properties; such as wire diameter and conductor spacing. These properties, in turn, determine the primary electrical constants: series dc resistance (R), series inductance (L), shunt capacitance (C), and shunt conductance (G). Resistance and inductance occur along the line, whereas capacitance and conductance occur between the two conductors. The primary constants are uniformly distributed throughout the length of the line and are therefore commonly called distributed parameters. To simplify analysis, distributed parameters are commonly *lumped* together per a given unit length to form an artificial electrical model of the line. For example, series resistance is generally given in ohms per mile or kilometer.

Figure 9-6 shows the electrical equivalent circuit for a metallic two-wire transmission line showing the relative placement of the various lumped parameters. The conductance between the two wires is shown in reciprocal form and given as a shunt leakage resistance (R_s).

Transmission Characteristics

The transmission characteristics of a transmission line are called secondary constants and are determined from the four primary constants. The secondary constants are characteristic impedance and propagation constant.

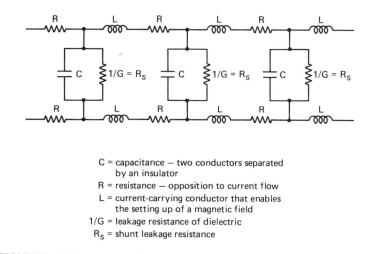

C = capacitance — two conductors separated
      by an insulator
R = resistance — opposition to current flow
L = current-carrying conductor that enables
      the setting up of a magnetic field
1/G = leakage resistance of dielectric
R_s = shunt leakage resistance

FIGURE 9-6 Two-wire parallel transmission line — electrical equivalent circuit.

Characteristic impedance. For maximum power transfer from the source to the load (i.e., no reflected energy), a transmission line must be terminated in a purely resistive load equal to the *characteristic impedance* of the line. The characteristic impedance (Z_o) of a transmission line is a complex ac quantity which is expressed in ohms, is totally independent of both length and frequency, and cannot be measured directly. Characteristic impedance (which is sometimes called *surge impedance*) is defined as the impedance seen looking into an infinitely long line or the impedance seen looking into a finite length of line which is terminated in a purely resistive load equal to the characteristic impedance of the line. A transmission line stores energy in its distributed inductance and capacitance. If the line is infinitely long, it can store energy indefinitely; energy from the source is entering the line and none is returned. Therefore, the line acts like a resistor that dissipates all of the energy. An infinite line can be simulated if a finite line is terminated in a purely resistive load equal to Z_o; all of the energy that enters the line from the source is dissipated in the load (this assumes a totally lossless line).

Figure 9–7 shows a single section of a transmission line terminated in a load Z_L that is equal to Z_o. The impedance seen looking into a line of n such sections is determined from the following expression:

$$Z_o^2 = Z_1 Z_2 + \frac{ZL^2}{n} \tag{9-4}$$

where n is the number of sections. For an infinite number of sections, ZL^2/n approaches 0 if

$$\lim \frac{ZL^2}{n} \bigg|_{n \to \infty} = 0$$

then

$$Z_o = \sqrt{Z_1 Z_2}$$

where

$$Z_1 = R + j\omega L$$

$$\frac{1}{Z_2} = \frac{1}{R_s} + \frac{1}{1/j\omega C}$$

$$\frac{1}{Z_2} = G + j\omega C$$

$$Z_2 = \frac{1}{G} + \frac{1}{j\omega C} = \frac{1}{G + j\omega C}$$

Therefore,

$$Z_o = \sqrt{(R + j\omega L) \frac{1}{G + j\omega C}}$$

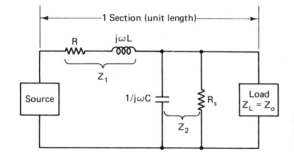

FIGURE 9-7 Equivalent circuit for a single section of transmission line terminated in a load equal to Z_o.

or

$$Z_o = \sqrt{\frac{R + j\omega L}{G + j\omega C}} \tag{9-5}$$

For extremely low frequencies, the resistances dominate and Equation 9-5 simplifies to

$$Z_o = \sqrt{\frac{R}{G}} \tag{9-6}$$

For extremely high frequencies, the inductance and capacitance dominate and Equation 9-5 simplifies to

$$Z_o = \sqrt{\frac{j\omega L}{j\omega C}} = \sqrt{\frac{L}{C}} \tag{9-7}$$

From Equation 9-7 it can be seen that for high frequencies the characteristic impedance of a transmission line approaches a constant, is independent of both frequency and length, and is determined solely by the distributed inductance and capacitance. It can also be seen that the phase angle is $0°$. Therefore, Z_o looks purely resistive and all of the incident energy is absorbed by the line.

From a purely resistive approach, it can easily be seen that the impedance seen looking into a transmission line made up of an infinite number of sections approaches the characteristic impedance. This is shown in Figure 9-8. Again, for simplicity, only

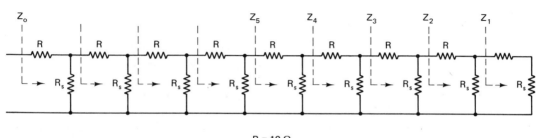

$$R = 10\ \Omega$$
$$R_s = 100\ \Omega$$

FIGURE 9-8 Characteristic impedance of a transmission line of infinite sections or terminated in load equal to Z_o.

the series resistance R and the shunt resistance R_s are considered. The impedance seen looking into the last section of the line is simply the sum of R and R_s. Mathematically, Z_1 is

$$Z_1 = R + R_s = 10 + 100 = 110$$

Adding a second section, Z_2, gives

$$Z_2 = R + \frac{R_s Z_1}{R_s + Z_1} = 10 + \frac{10 \times 110}{10 + 110} = 10 + 52.38 = 62.38$$

and a third section, Z_3, is

$$Z_3 = R + \frac{R_s Z_2}{R_s + Z_2}$$

$$= 10 + \frac{10 \times 62.38}{10 + 62.38} = 10 + 38.42 = 48.32$$

A fourth section, Z_4, is

$$Z_4 = 10 + \frac{10 \times 48.32}{10 + 48.32} = 10 + 32.62 = 42.62$$

    It can be seen that after each additional section, the total impedance seen looking into the line decreases from its previous value. However, each time the magnitude of the decrease is less than the previous value. If the process shown above were continued, the impedance seen looking into the line decreases asymptotically toward 37 Ω, which is the characteristic impedance of the line.

    If the transmission line shown in Figure 9-8 were terminated in a load resistance $Z_L = 37$ Ω, the impedance seen looking into any number of sections would equal 37 Ω, the characteristic impedance. For a single section of line, Z_o is

$$Z_o = Z_1 = R + \frac{R_s \times Z_L}{R_s + Z_L} = 10 + \frac{100 \times 37}{100 + 37} = 10 + \frac{3700}{137} = 37 \ \Omega$$

Adding a second section, Z_2, is

$$Z_o = Z_2 = R + \frac{R_s \times Z_1}{R_s + Z_1} = 10 + \frac{100 \times 37}{100 + 37} = 10 + \frac{3700}{137} = 37 \ \Omega$$

Therefore, if this line were terminated into a load resistance $Z_L = 37$ Ω, $Z_o = 37$ Ω no matter how many sections are included.

    The characteristic impedance of a transmission line can also be determined using Ohm's law. When a source is connected to an infinitely long line and a voltage is applied, a current flows. Even though the load is open, the circuit is complete through the distributed constants of the line. The characteristic impedance is simply the ratio of source voltage (E_o) to line current (I_o). Mathematically, Z_o is

$$Z_o = \frac{E_o}{I_o}$$ (9-8)

The characteristic impedance of a two-wire parallel transmission line with an air dielectric can be determined from its physical dimensions (see Figure 9-9a) and the formula

$$Z_o = 276 \log \frac{D}{r}$$ (9-9)

where

D = distance between the centers of the two conductors
r = radius of the conductor

and $D \gg r$.

EXAMPLE 9-1

Determine the characteristic impedance for an air dielectric two-wire parallel transmission line with a D/r ratio = 12.22.

Solution Substituting into Equation 9-9, we obtain

$$Z_o = 276 \log 12.22 = 300 \ \Omega$$

The characteristic impedance of a concentric coaxial cable can also be determined from its physical dimensions (see Figure 9-9b) and the formula

$$Z_o = \frac{138}{\sqrt{\epsilon_r}} \log \frac{D}{d}$$ (9-10)

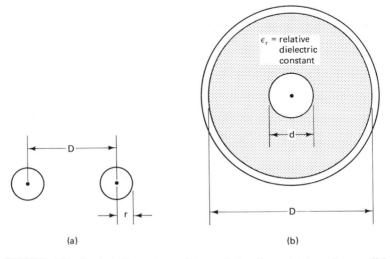

FIGURE 9-9 Physical dimensions of transmission lines: (a) two-wire parallel transmission line; (b) coaxial cable transmission line.

where

D = inside diameter of the outer conductor
d = outside diameter of the inner conductor
ϵ_r = relative dielectric constant of the insulating material

EXAMPLE 9-2

Determine the characteristic impedance for a RG-59A coaxial cable with the following specifications: $L = 0.118 \ \mu H/ft$, $C = 21 \ pF/ft$, $d = 0.25$ in., $D = 0.87$ in., and $\epsilon = 1$.

Solutions   Substituting into Equation 9-8 yields

$$Z_o = \sqrt{\frac{L}{C}} = \sqrt{\frac{0.118 \times 10^{-6} \, H/ft}{21 \times 10^{-12} \, F/ft}} = 75 \ \Omega$$

Substituting into Equation 9-10 gives us

$$Z_o = \frac{138}{\sqrt{1}} \log \frac{0.87 \ in.}{0.25 \ in.} = 75 \ \Omega$$

Transmission lines can be summarized thus far as follows:

1. The input impedance of an infinitely long line at radio frequencies is resistive and equal to Z_o.
2. Electomagnetic waves travel down the line without reflections; such a line is called nonresonant.
3. The ratio of voltage to current at any point along the line is equal to Z_o.
4. The incident voltage and current at any point along the line are in phase.
5. Line losses on a nonresonant line are minimum per unit length.
6. Any transmission line that is terminated in a purely resistive load equal to Z_o acts like an infinite line.
   a. $Z_i = Z_o$.
   b. There are no reflected waves.
   c. V and I are in phase.
   d. There is maximum transfer of power from source to load.

Propagation constant.   *Propagation constant* (sometimes called *propagation coefficient*) is used to express the attenuation (signal loss) and the phase shift per unit length of a transmission line. As a wave propagates down a transmission line, its amplitude decreases with distance traveled. The propagation constant is used to determine the reduction in voltage or current with distance as a TEM wave propagates down a transmission line. For an infinitely long line, all of the incident power is dissipated in the resistance of the wire as the wave propagates down the line. Therefore, with an infinitely long line or a line that looks infinitely long, such as a finite line terminated in a matched load ($Z_o = Z_L$), no energy is returned or reflected back toward the source. Mathematically, the propagation constant is

$$\gamma = \alpha + j\beta \qquad (9\text{-}11\text{a})$$

where

γ = propagation constant
α = attenuation coefficient (nepers per unit length)
β = phase shift coefficient (radians per unit length)

and since the propagation constant is a complex quantity,

$$\gamma = \sqrt{(R + j\omega L)(G + j\omega C)} \qquad (9\text{-}11\text{b})$$

Since a phase shift of 2π rad occurs over a distance of one wavelength, then

$$\beta = \frac{2\pi}{\lambda} \qquad (9\text{-}12)$$

At intermediate and radio frequencies, $\omega L > R$ and $\omega C > G$; thus

$$\alpha = \frac{R}{2Z_o} + \frac{GZ_o}{2} \qquad (9\text{-}13)$$

and

$$\beta = \omega\sqrt{LC} \qquad (9\text{-}14)$$

The current and voltage distribution along a transmission line that is terminated in a load equal to its characteristic impedance (i.e., a matched line) are determined from the formula

$$I = I_s e^{-l\gamma} \qquad (9\text{-}15)$$

$$V = V_s e^{-l\gamma} \qquad (9\text{-}16)$$

where

I_s = current at the source end of the line
V_s = voltage at the source end of the line
γ = propagation constant
l = length of the line at which the current or voltage is determined

For a matched load ($Z_L = Z_o$), the loss in signal voltage or current and phase shift for a given length of cable is equal to γl.

TRANSMISSION-LINE WAVE PROPAGATION

As stated previously, electromagnetic waves travel at the speed of light when propagating through a vacuum, and nearly at the speed of light when propagating through air. However, in metallic transmission lines where the conductor is generally copper and the dielectric materials vary considerably with cable type, an electromagnetic wave travels much more slowly.

Velocity Factor

Velocity factor (sometimes called *velocity constant*) is defined simply as the ratio of the actual velocity of propagation through a given medium to the velocity of propagation through free space. Mathematically, the velocity factor is

$$V_f = \frac{V_p}{c} \qquad (9\text{-}17)$$

where

V_f = velocity factor
V_p = actual velocity of propagation
c = velocity of propagation through free space, $c = 3 \times 10^8$ m/s

and

$$V_f \times c = V_p$$

The velocity at which an electromagnetic wave travels through a transmission line is dependent on the dielectric constant of the insulating material separating the two conductors. The velocity factor is closely approximated with the formula

$$V_f = \frac{1}{\sqrt{\epsilon_r}} \qquad (9\text{-}18)$$

where ϵ_r is the dielectric constant of a given material relative to the dielectric constant of a vacuum (ϵ/ϵ_o).

Dielectric constant is simply the *permittivity* of a material. The relative dielectric constant of air is 1.0006. However, the relative dielectric constant of materials commonly used in transmission lines range from 1.2 to 2.8, giving velocity factors from 0.6 to 0.9. The velocity factors for several common transmission-line configurations are given in Table 9-1, and the relative dielectric constants for several insulating materials are listed in Table 9-2.

The dielectric constant of a material is dependent on the primary constants inductance and capacitance. Inductors store magnetic energy and capacitors store electric

TABLE 9-1   VELOCITY FACTORS

Material	Velocity factor
Air	0.95–0.975
Rubber	0.56–0.65
Polyethylene	0.66
Teflon	0.70
Teflon foam	0.82
Teflon pins	0.81
Teflon spiral	0.81

TABLE 9-2   RELATIVE DIELECTRIC CONSTANTS

Material	Relative dielectric constant
Vacuum	1.0
Air	1.0006
Teflon	2.0
Paper, paraffined	2.5
Rubber	3.0
Mica	5.0
Glass	7.5

energy. It takes a finite amount of time for an inductor or a capacitor to take on or give off energy. Therefore, the velocity at which an electromagnetic wave propagates along a transmission line varies with the inductance and capacitance of the cable. It can be shown that time $T = \sqrt{LC}$. Therefore, inductance, capacitance, and velocity of propagation are mathematically related by the formula

$$\text{velocity} \times \text{time} = \text{distance}$$

Therefore,

$$V_p = \frac{\text{distance}}{\text{time}} = \frac{D}{T} \tag{9-19}$$

Substituting for time yields

$$V_p = \frac{D}{\sqrt{LC}} \tag{9-20}$$

If distance is normalized to 1 m, the velocity of propagation for a lossless line is

$$V_p = \frac{1 \text{ m}}{\sqrt{LC}} = \frac{1}{\sqrt{LC}} \qquad \text{m/s} \tag{9-21}$$

EXAMPLE 9-3

For a given length of RG8A/U coaxial cable with a distributed capacitance $C = 96.6$ pF/m, a distributed inductance $L = 241.56$ nH/m, and a relative dielectric constant $\epsilon_r = 2.3$; determine the velocity of propagation and the velocity factor.

Solution   From Equation 9-21,

$$V_p = \frac{1}{\sqrt{96.6 \times 10^{-12} \times 241.56 \times 10^{-6}}}$$

$$= 2.07 \times 10^8 \text{ m/s}$$

From Equation 9-17,

$$V_f = \frac{2.07 \times 10^8 \, \text{m/s}}{3 \times 10^8 \, \text{m/s}} = 0.69$$

From Equation 9-18,

$$V_p = \frac{1}{\sqrt{2.3}} = 0.66$$

Because wavelength is directly proportional to velocity and the velocity of propagation of a TEM wave varies with dielectric constant, the wavelength of a TEM wave also varies with dielectric constant. Therefore, for transmission media other than free space, Equation 9-3 can be rewritten as

$$\lambda = \frac{V_p}{F} = \frac{cV_f}{F} = \frac{c}{F\sqrt{\epsilon_r}} \tag{9-22}$$

Electrical Length of a Transmission Line

The length of a transmission line relative to the length of the wave propagating down it is an important consideration when analyzing transmission line behavior. At low frequencies (i.e., long wavelengths), the voltage along the line remains relatively constant. However, for high frequencies, several wavelengths of the signal may be present on the line at the same time. Therefore, the voltage along the line may vary appreciably. Consequently, the length of a transmission line is often given in wavelengths rather than in linear dimensions. Transmission-line phenomena apply to long lines. Generally, a transmission line is defined as long if its length exceeds one-sixteenth of a wavelength; otherwise, it is considered short. A given length of transmission line may appear short at one frequency and long at another frequency. For example, a 10-m length of transmission line at 1000 Hz is short (λ = 300,000 m; 10 m is only a small fraction of a wavelength). However, the same line at 6 HGz is long (λ = 5 cm; the line is 200 wavelengths long). It will be apparent later in this chapter, in Chapter 10, and in Appendix A that electrical length is used extensively for transmission line calculations and antenna design.

TRANSMISSION-LINE LOSSES

For analysis purposes, transmission lines are often considered totally lossless. In reality, however, there are several ways in which power is lost in a transmission line. They are conductor loss, radiation loss, dielectric heating loss, coupling loss, and corona.

Conductor Loss

Because current flows through a transmission line and the transmission line has a finite resistance, there is an inherent and unavoidable power loss. This is sometimes called *conductor* or *conductor heating loss* and is simply an I^2R loss. Because resistance is distributed throughout a transmission line, conductor loss is directly proportional to line length. Also, because power dissipation is directly proportional to current, conductor loss is inversely proportional to characteristic impedance. To reduce conductor loss, simply shorten the transmission line or use a larger-diameter wire (keep in mind that changing the wire diameter also changes the characteristic impedance and consequently the current).

Conductor loss is somewhat dependent on frequency. This is because of an action called the *skin effect*. When current flows through an isolated round wire, the magnetic flux associated with it is in the form of concentric circles. This is shown in Figure 9-10. It can be seen that the flux density near the center of the conductor is greater than it is near the surface. Consequently, the lines of flux near the center of the conductor encircle the current and reduce the mobility of the encircled electrons. This is a form of self-induction and causes the inductance near the center of the conductor to be greater than at the surface. Therefore, at radio frequencies, most of the current flows along the surface (outer skin) rather than near the center of the conductor. This is equivalent

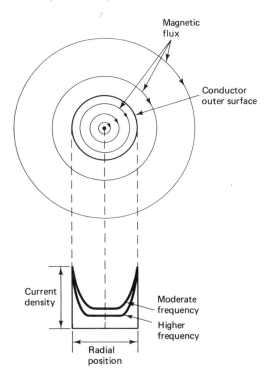

FIGURE 9-10 Isolated round conductor showing magnetic lines of flux, current distributions, and the skin effect.

to reducing the cross-sectional area of the conductor and increasing the opposition to current flow (i.e., resistance). The additional opposition has a 0° phase angle and is therefore a resistance and not a reactance. Therefore, the ac resistance of the conductor is directly proportional to frequency. The ratio of the ac resistance to the dc resistance of a conductor is called the resistance ratio. Above approximately 100 MHz, the center of a conductor can be completely removed and have absolutely no effect on the total conductor loss or EM wave propagation.

Conductor loss in transmission lines varies from as low as a fraction of a decibel per 100 m for rigid air dielectric coaxial cable to as high as 200 dB per 100 m for solid dielectric flexible line.

Radiation Loss

If the separation between conductors in a transmission line is an appreciable fraction of a wavelength, the electrostatic and electromagnetic fields that surround the conductor cause the line to act like an antenna and transfer energy to nearby conductors. The amount of energy radiated depends on the dielectric material, the conductor spacing, and the length of the line. *Radiation losses* are reduced by properly shielding the cable. Therefore, coaxial cables have less radiation loss than do two-wire parallel lines. Radiation loss is also directly proportional to frequency.

Dielectric Heating Loss

A difference of potential between the two conductors of a transmission line causes *dielectric heating*. Heat is a form of energy and must be taken from the energy propagating down the line. For air dielectric lines, the heating loss is negligible. However, for solid lines, dielectric heating loss increases with frequency.

Coupling Loss

Coupling loss occurs whenever a connection is made to or from a transmission line or when two separate pieces of transmission line are connected together. Mechanical connections are discontinuities (places where dissimilar materials meet). Discontinuities tend to heat up, radiate energy, and dissipate power.

Corona

Corona is arcing that occurs between the two conductors of a transmission line when the difference of potential between them exceeds the *breakdown* voltage of the dielectric insulator. Generally, once corona has occurred, the transmission line is destroyed.

INCIDENT AND REFLECTED WAVES

An ordinary transmission line is bidirectional; power can propagate equally well in both directions. Voltage that propagates from the source toward the load is called *incident voltage*, and voltage that propagates from the load toward the source is called *reflected*

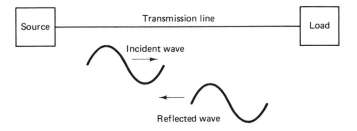

FIGURE 9-11 Source, load, transmission line, and their corresponding incident and reflected waves.

voltage. Similarly, there are incident and reflected currents. Consequently, incident power is power that propagates toward the load, and reflected power is power that propagates toward the source. Incident voltage and current are always in phase. For an infinitely long line, all of the incident power is absorbed by the line and there is no reflected power. Also, if the line is terminated in a purely resistive load equal to the characteristic impedance of the line, the load absorbs all of the incident power (this assumes a lossless line). For a more practical definition, reflected power is the portion of the incident power that was not absorbed by the load. Therefore, the reflected power can never exceed the incident power.

Resonant and Nonresonant Lines

A line with no reflected power is called a *flat* or *nonresonant line*. On a flat line, the voltage and current are constant throughout its length, assuming no losses. When the load is either a short or an open circuit, all of the incident power is reflected back toward the source. If the source were replaced with an open or a short and the line were lossless, energy present on the line would reflect back and forth (oscillate) between the load and source ends similar to the power in a tank circuit. This is called a *resonant line*. In a resonant line, the energy is alternately transferred between the magnetic and electric fields of the distributed inductance and capacitance. Figure 9-11 shows a source, transmission line, and load with their corresponding incident and reflected waves.

Reflection Coefficient

The reflection coefficient (sometimes called the *coefficient of reflection*) is a vector quantity that represents the ratio of incident voltage to reflected voltage or incident current to reflected current. Mathematically, the reflection coefficient is

$$\Gamma = \frac{E_r}{E_i} \qquad \text{or} \qquad \frac{I_r}{I_i} \qquad (9\text{-}23)$$

where

Γ = reflection coefficient
E_i = incident voltage

E_r = reflected voltage
I_i = incident current
I_r = reflected current

From Equation 9-23 it can be seen that the maximum and worst-case value for Γ is 1 ($E_r = E_i$) and the minimum value and ideal condition is when $\Gamma = 0$ ($E_r = 0$).

STANDING WAVES

When $Z_o = Z_L$, all of the incident power is absorbed by the load. This is called a *matched line*. When $Z_o \neq Z_L$, some of the incident power is absorbed by the load and some is returned (reflected) to the source. This is called an *unmatched* or *mismatched line*. With a mismatched line, there are two electromagnetic waves, traveling in opposite directions, present on the line at the same time (these waves are in fact called *traveling waves*). The two traveling waves set up an interference pattern known as a *standing wave*. This is shown in Figure 9-12. As the incident and reflected waves pass each other, a stationary voltage and current are produced on the line. These stationary waves are called standing waves because they appear to remain in a fixed position on the line, varying only in amplitude. The standing wave has minima (nodes) and maxima (antinodes), which are separated by half of a wavelength of the traveling waves.

Standing-Wave Ratio

The *standing-wave ratio* (SWR) is defined as the ratio of the maximum voltage to the minimum voltage or the maximum current to the minimum current of a standing wave on a transmission line. SWR is often called the *voltage standing-wave ratio* (VSWR).

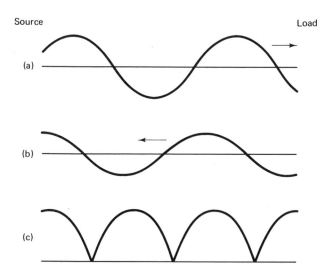

FIGURE 9-12 Developing a standing wave on a transmission line: (a) incident wave; (b) reflected wave; (c) standing wave.

Essentially, SWR is a measure of the mismatch between the load impedance and the characteristic impedance of the transmission line. Mathematically, SWR is

$$\text{SWR} = \frac{V_{\max}}{V_{\min}} \tag{9-24}$$

The voltage maxima ($V_{\max}$) occur when the incident and reflected waves are in phase (i.e., their maximum peaks pass the same point on the line with the same polarity), and the voltage minima ($V_{\min}$) occur when the incident and reflected waves are 180° out of phase. Mathematically, $V_{\max}$ and $V_{\min}$ are

$$V_{\max} = E_i + E_r \tag{9-25}$$
$$V_{\min} = E_i - E_r \tag{9-26}$$

Therefore, Equation 9-24 can be rewritten as

$$\text{SWR} = \frac{V_{\max}}{V_{\min}} = \frac{E_i + E_r}{E_i - E_r} \tag{9-27}$$

From Equation 9-27 it can be seen that when the incident and reflected waves are equal in amplitude (a total mismatch), SWR = infinity. This, of course, is the worst-case condition. Also, from Equation 9-27, it can be seen that when there is no reflected wave ($E_r = 0$), SWR = E_i/E_i or 1. This condition occurs when $Z_o = Z_L$ and is the ideal situation.

The standing-wave ratio can also be written in terms of Γ. Rearranging Equation 9-23 yields

$$\Gamma E_i = E_r$$

Substituting into Equation 9-27 gives us

$$\text{SWR} = \frac{E_i + E_i\Gamma}{E_i - E_i\Gamma}$$

Factoring out E_i yields

$$\text{SWR} = \frac{E_i(1 + \Gamma)}{E_i(1 - \Gamma)} = \frac{1 + \Gamma}{1 - \Gamma} \tag{9-28}$$

Cross-multiplying gives

$$\text{SWR}(1 - \Gamma) = 1 + \Gamma$$
$$\text{SWR} - \text{SWR}\,\Gamma = 1 + \Gamma$$
$$\text{SWR} = 1 + \Gamma + \text{SWR}$$
$$\text{SWR} - 1 = \Gamma(1 + \text{SWR}) \tag{9-29}$$

$$\Gamma = \frac{\text{SWR} - 1}{\text{SWR} + 1} \tag{9-30}$$

EXAMPLE 9-4

For a transmission line with incident voltage $E_i = 5$ V$_p$ and reflected voltage $E_r = 3$ V$_p$, determine:

  (a) The reflection coefficient.
  (b) The SWR.

Solution   (a) Substituting into Equation 9-23 yields

$$\Gamma = \frac{E_r}{E_i} = \frac{3}{5} = 0.6$$

(b) Substituting into Equation 9-27 gives us

$$\text{SWR} = \frac{E_i + E_r}{E_i - E_r} = \frac{5+3}{5-3} = \frac{8}{2} = 4$$

Substituting into Equation 9-30, we obtain

$$\Gamma = \frac{4-1}{4+1} = \frac{3}{5} = 0.6$$

When the load is purely resistive, SWR can also be expressed as a ratio of the characteristic impedance to the load impedance, or vice versa. Mathematically, SWR is

$$\text{SWR} = \frac{Z_o}{Z_L} \quad \text{or} \quad \frac{Z_L}{Z_o} \quad \text{(whichever gives an SWR greater than 1)} \qquad (9\text{-}31)$$

The numerator and denominator for Equation 9-31 are chosen such that the SWR is always a number greater than 1, to avoid confusion and comply with the convention established in Equation 9-27. From Equation 9-31 it can be seen that a load resistance $Z_L = 2Z_o$ gives the same SWR as a load resistance $Z_L = Z_o/2$; the degree of mismatch is the same.

The disadvantages of not having a matched (flat) transmission line can be summarized as follows:

1. One hundred percent of the source incident power does not reach the load.
2. The dielectric separating the two conductors can break down and cause corona as a result of the high-voltage standing-wave ratio.
3. Reflections and rereflections cause more power loss.
4. Reflections cause ghost images.
5. Mismatches cause noise interference.

Although it is highly unlikely that a transmission line will be terminated in a load that is either an open or a short circuit, these conditions are examined because they illustrate the worst possible conditions that could occur and produce standing waves that are representative of less severe conditions.

Standing Waves on an Open Line

When incident waves of voltage and current reach an open termination, none of the power is absorbed; it is all reflected back toward the source. The incident voltage wave is reflected in exactly the same manner as if it were to continue down an infinitely long line. However, the incident current is reflected 180° reversed from how it would have continued if the line were not open. As the incident and reflected waves pass, standing waves are produced on the line. Figure 9-13 shows the voltage and current standing waves on a transmission line that is terminated in an open circuit. It can be seen that the voltage standing wave has a maximum value at the open end and a minimum value one quarter wavelength from the open. The current standing wave has a minimum value at the open end and a maximum value one quarter wavelength from the open. It stands to reason that maximum voltage occurs across an open and there is minimum current.

The characteristics of a transmission line terminated in an open can be summarized as follows:

1. The voltage standing wave is reflected back just as if it were to continue (i.e., no phase reversal).
2. The current standing wave is reflected back 180° from how it would have continued.
3. The sum of the incident and reflected current waveforms is minimum at the open.
4. The sum of the incident and reflected voltage waveforms is maximum at the open.

From Figure 9-13 it can also be seen that the voltage and current standing waves repeat every one-half wavelength. The impedance at the open end $Z = V_{max}/I_{min}$ and is maximum. The impedance one-quarter wavelength from the open $Z = V_{min}/I_{max}$ and is minimum. Therefore, one-quarter wavelength from the open an impedance inversion occurs and additional impedance inversions occur each quarter wavelength.

Figure 9-14 shows the development of a voltage standing wave on a transmission line that is terminated in an open circuit. Figure 9-14 shows an incident wave propagating down a transmission line toward the load. The wave is traveling at approximately the speed of light; however, for illustration purposes, the wave has been frozen at quarter-wavelength intervals. In Figure 9-14a it can be seen that the incident wave has not

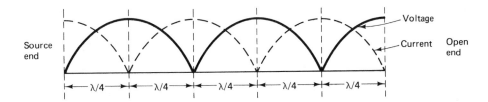

FIGURE 9-13 Voltage and current standing waves on a transmission line that is terminated in an open circuit.

Source

Load

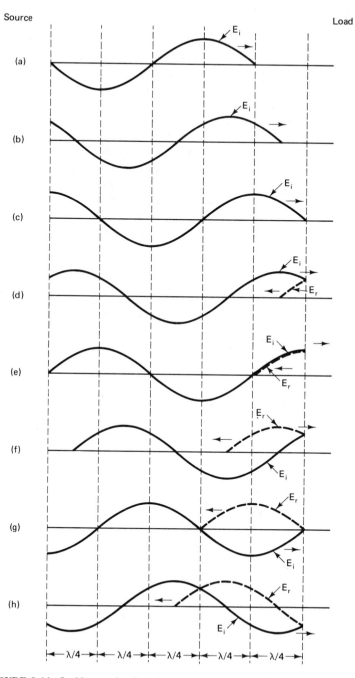

FIGURE 9-14   Incident and reflected waves on a transmission line terminated in an open circuit.

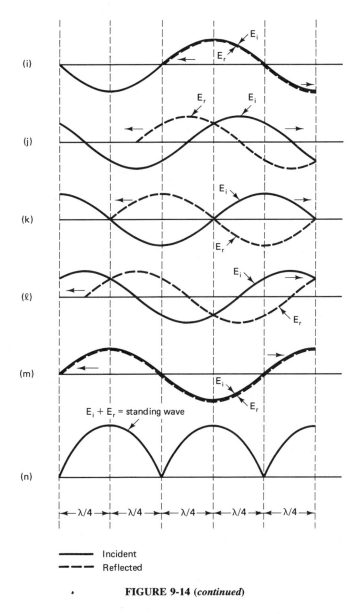

(i)

(j)

(k)

(ℓ)

(m)

$E_i + E_r$ = standing wave

(n)

|← λ/4 →|← λ/4 →|← λ/4 →|← λ/4 →|← λ/4 →|

——— Incident
- - - Reflected

FIGURE 9-14 (*continued*)

reached the open. Figure 9-14b shows the wave one time unit later (for this example, the wave travels one-quarter wavelength per time unit). As you can see, the wave has moved one-quarter wavelength closer to the open. Figure 9-14c shows the wave just as it arrives at the open. Thus far there has been no reflected wave and, consequently, no standing wave. Figure 9-14d shows the incident and reflected waves one time unit after the incident wave has reached the open; the reflected wave is propagating away from the open. Figure 9-14e, f, and g show the incident and reflected waves for the next three time units. In Figure 9-14e it can be seen that the incident and reflected waves are at their maximum positive values at the same time, thus producing a voltage maximum at the open. It can also be seen that one-quarter wavelength from the open the sum of the incident and reflected waves (the standing wave) is always equal to 0 V (a minimum). Figure 9-14h through m show propagation of the incident and reflected waves until the reflected wave reaches the source, and Figure 9-14n shows the resulting standing wave. It can be seen that the standing wave remains stationary (the voltage nodes and antinodes remain at the same points). However, the amplitude of the antinodes vary from maximum positive to zero to maximum negative then repeats. For an open load, all of the incident voltage is reflected ($E_r = E_i$); therefore, $V_{max} = E_i + E_r$ or $2E_i$. A similar illustration can be shown for a current standing wave (however, remember that the current reflects back with a 180° phase inversion).

Standing Waves on a Shorted Line

As with an open line, none of the incident power is absorbed by the load when a transmission line is terminated in a short circuit. However, with a shorted line, the incident voltage and current waves are reflected back in the opposite manner. The voltage wave is reflected 180° reversed from how it would have continued down an infinitely long line, and the current wave is reflected in exactly the same manner as if there were no short.

Figure 9-15 shows the voltage and current standing waves on a transmission line that is terminated in a short circuit. It can be seen that the voltage standing wave has a minimum value at the shorted end and a maximum value one quarter wavelength from the short. The current standing wave has a maximum value at the short and a minimum value one quarter wavelength back. The voltage and current standing waves repeat every quarter wavelength. Therefore, there is an impedance inversion every quarter wavelength. The impedance at the short $Z = V_{min}/I_{max} = $ minimum, and one quarter wavelength back $Z = V_{max}/I_{min} = $ maximum. Again, it stands to reason that a voltage minimum will occur across a short and there is maximum current.

The characteristics of a transmission line terminated in a short can be summarized as follows:

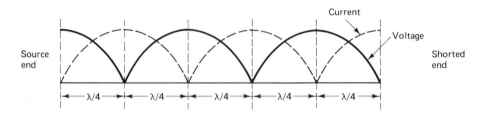

FIGURE 9-15 **Voltage and current standing waves on a transmission line that is terminated in a short circuit.**

1. The voltage standing wave is reflected back 180° reversed from how it would have continued.
2. The current standing wave is reflected back the same as if it had continued.
3. The sum of the incident and reflected current waveforms is maximum at the short.
4. The sum of the incident and reflected voltage waveforms is zero at the short.

For a transmission line terminated in either a short or an open circuit, the reflection coefficient is 1 (the worst case) and the SWR is infinity (also the worst-case condition).

TRANSMISSION-LINE INPUT IMPEDANCE

In the preceding section it was shown that when a transmission line is terminated in either a short or an open circuit, there is an *impedance inversion* every quarter wavelength. For a lossless line, the impedance varies from infinity to zero. However, in a more practical situation where power losses occur, the amplitude of the reflected wave is always less than that of the incident wave. Therefore, the impedance varies from some maximum to some minimum value, or vice versa, depending on whether the line is terminated in a short or an open. The input impedance seen looking into a transmission line that is terminated in a short or an open can be resistive, inductive, or capacitive, depending on the distance from the termination.

Phasor Analysis of Input Impedance—Open Line

Phasor diagrams are generally used to analyze the input impedance of a transmission line because they are relatively simple and give a pictorial representation of the voltage and current phase relationships. Voltage and current phase relations refer to variations in time. Figures 9-13, 9-14, and 9-15 show standing waves of voltage and current plotted versus distance and are therefore not indicative of true phase relationships. The succeeding sections use phasor diagrams to analyze the input impedance of several transmission-line configurations.

Quarter-wavelength transmission line. Figure 9-16a shows the phasor diagram for the voltage and Figure 9-16b shows the phasor diagram for the current on a quarter-wave section of a transmission line terminated in an open circuit. I_i and V_i are the in-phase incident current and voltage waveforms at the input (source) end of the line at a given instant in time, respectively. Any reflected voltage (E_r) present at the input of the line has traveled one-half wavelength (from the source to the open and back) and is, consequently, 180° behind the incident voltage. Therefore, the total voltage (E_t) at the input end is the sum of E_i and E_r. $E_t = E_i + E_r/-180°$ and, assuming a small line loss, $E_t = E_i - E_r$. The reflected current is delayed 90° propagating from the source to the load and another 90° from the load back to the source. Also, the reflected current undergoes an 180° phase reversal at the open. The reflected current has effectively

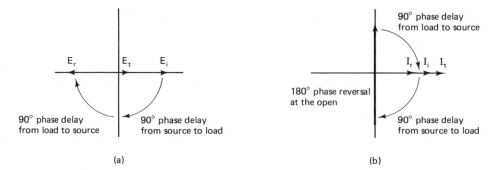

FIGURE 9-16 Voltage and current phase relationships for a quarter-wave line terminated in an open circuit: (a) voltage phase relationships; (b) current phase relationships.

been delayed 360°. Therefore, when the reflected current reaches the source end, it is in phase with the incident current and the total current $I_t = I_i + I_r$. By examining Figures 9-16, it can be seen that E_t and I_t are in phase. Therefore, the input impedance seen looking into a transmission line one quarter wavelength long that is terminated in an open circuit $Z_{in} = E_t \underline{/0°}/I_t \underline{/0°} = Z_{in} \underline{/0°}$. Z_{in} has a 0° phase angle, is resistive, and is minimum. Therefore, a quarter-wavelength transmission line terminated in an open circuit is equivalent to a series *LC* circuit.

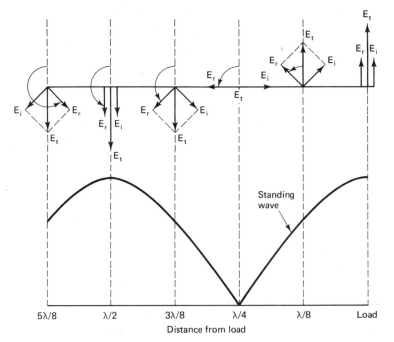

FIGURE 9-17 Vector addition of incident and reflected waves producing a standing wave.

Figure 9-17 shows several voltage phasors for the incident and reflected waves on a transmission line that is terminated in an open circuit and how they produce a voltage standing wave.

Transmission line less than one-quarter wavelength long.   Figure 9-18a shows the voltage phasor diagram and Figure 9-18b the current phasor diagram for a transmission line that is less than one-quarter wavelength long and terminated in an open circuit. Again, the incident current (I_i) and voltage (E_i) are in phase. The reflected voltage wave is delayed 45° traveling from the source to the load (a distance of one eighth wavelength) and another 45° traveling from the load back to the source (an additional one eighth wavelength). Therefore, when the reflected wave reaches the source end, it lags the incident wave by 90°. The total voltage at the source end is the vector sum of the incident and reflected waves. Thus $E_t = \sqrt{E_i^2 + E_r^2} = E_t\underline{/-45°}$. The reflected current wave is delayed 45° traveling from the source to the load and another 45° from the load back to the source (a total distance of one-quarter wavelength). In addition, the reflected current wave has undergone an 180° phase reversal at the open prior to being reflected. The reflected current wave has been delayed a total of 270°. Therefore, the reflected wave effectively leads the incident wave by 90°. The total current at the source end is the vector sum of the present and reflected waves. Thus $I_t = \sqrt{I_i^2 + I_r^2} = I_t\underline{/+45°}$. By examining Figure 9-18, it can be seen that E_t lags I_t by 90°. Therefore, $Z_{in} = E_t\underline{/-45°}/I_t\underline{/+45°} = Z_{in}\underline{/-90°}$. Z_{in} has a $-90°$ phase angle and is therefore capacitive. Any transmission line that is less than one-quarter wavelength and terminated in an open circuit is equivalent to a capacitor. The amount of capacitance depends on the exact electrical length of the line.

Transmission line more than one-quarter wavelength long.   Figure 9-19a shows the voltage phasor diagram and Figure 9-19b shows the current phasor diagram for a transmission line that is more than one-quarter wavelength long and terminated in an open circuit. For this example, a three-eighths wavelength transmission line is used.

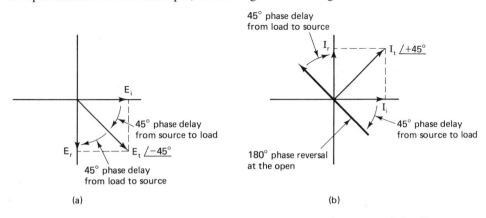

(a)   (b)

FIGURE 9-18 Voltage and current phase relationships for a transmission line less than one-quarter wavelength terminated in an open circuit: (a) voltage phase relationships; (b) current phase relationships.

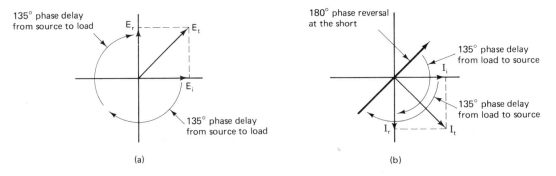

FIGURE 9-19 Voltage and current phase relationships for a transmission more than one-quarter wavelength terminated in an open circuit: (a) voltage phase relationships; (b) current phase relationships.

The reflected voltage is delayed three-quarters wavelength or 270°. Therefore, the reflected voltage effectively leads the incident voltage by 90°. Consequently, the total voltage $E_t = \sqrt{E_i^2 + E_r^2} = E_t\underline{/+45°}$. The reflected current wave has been delayed 270° and undergone an 180° phase reversal. Therefore, the reflected current effectively lags the incident current by 90°. Consequently, the total current $I_t = \sqrt{I_i^2 + I_r^2} = I_t\underline{/-45°}$. Therefore, $Z_{in} = E_t\underline{/+45°}/I_t\underline{/-45°} = Z_{in}\underline{/+90°}$. Z_{in} has a +90° phase angle and is therefore inductive. A transmission line between one-quarter and one-half wavelength that is terminated in an open circuit is equivalent to an inductor. The amount of inductance depends on the exact electrical length of the line.

Open transmission line as a circuit element.   From the preceding discussion and Figures 9-16 through 9-19, it is obvious that an open transmission line can behave like a resistor, an inductor, or a capacitor, depending on its electrical length. Because standing-wave patterns on an open line repeat every half wavelength, the input impedance also repeats. Figure 9-20 shows the variations in input impedance for an open transmission line of various electrical lengths. It can be seen that an open line is resistive and maximum at the open and at each successive half-wavelength interval, and resistive and minimum one-quarter wavelength from the open and at each successive half-wavelength interval. For electrical lengths less than one-quarter wavelength, the input impedance is capacitive and decreases with length. For electrical lengths between one-quarter and one-half wavelength, the input impedance is inductive and increases with length. The capacitance and inductance patterns also repeat every half wavelength.

Phasor Analysis of Input Impedance—Shorted Line

The following explanations use phasor diagrams to analyze shorted transmission lines in the same manner as with open lines. The difference is that with shorted transmission lines, the voltage waveform is reflected back with an 180° phase reversal and the current waveform is reflected back as if there were no short.

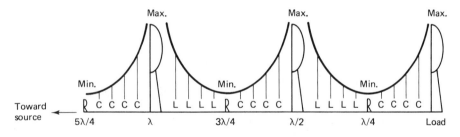

FIGURE 9-20 Input impedance variations for an open-circuited transmission line.

Quarter-wavelength transmission line. The voltage and current phasor diagrams for a quarter-wavelength transmission line terminated in a short circuit are identical to those shown in Figure 9-16, except reversed. The incident and reflected voltages are in phase; therefore, $E_t = E_i + E_r$ and maximum. The incident and reflected currents are 180° out of phase; therefore, $I_t = I_i - I_r$ and minimum. $Z_{in} = E_t \underline{/0°}/I_t \underline{/0°} = Z_{in} \underline{/0°}$ and maximum. Z_{in} has a 0° phase angle, is resistive, and is maximum. Therefore, a quarter-wavelength transmission line terminated in a short circuit is equivalent to a parallel LC circuit.

Transmission line less than one-quarter wavelength long. The voltage and current phasor diagrams for a transmission line less than one-quarter wavelength long and terminated in a short circuit are identical to those shown in Figure 9-18, except reversed. The voltage is reversed 180° at the short and the current is reflected with the same phase as if it had continued. Therefore, the total voltage at the source end of the line lags the current by 90° and the line looks inductive.

Transmission line more than one-quarter wavelength long. The voltage and current phasor diagrams for a transmission line more than one-quarter wavelength long and terminated in a short circuit are identical to those shown in Figure 9-19, except reversed. The total voltage at the source end of the line leads the current by 90° and the line looks capacitive.

Open transmission line as a circuit element. From the preceding discussion it is obvious that a shorted transmission line can behave like a resistor, an inductor, or a capacitor, depending on its electrical length. On a shorted transmission line, standing waves repeat every half wavelength; therefore, the input impedance also repeats. Figure 9-21 shows the variations in input impedance of a shorted transmission line for various electrical lengths. It can be seen that a shorted line is resistive and minimum at the short and at each successive half-wavelength interval, and resistive and maximum one-quarter wavelength from the short and at each successive half-wavelength interval. For electrical lengths less than one-quarter wavelength, the input impedance is inductive and increases with length. For electrical lengths between one-quarter and one-half wavelength, the input impedance is capacitive and decreases with length. The inductance and capacitance patterns also repeat every half wavelength.

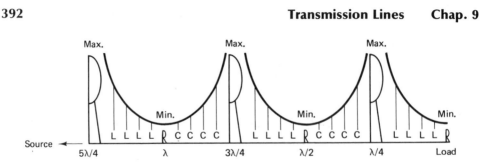

FIGURE 9-21 Input impedance variations for a short-circuited transmission line.

Transmission-line input impedance summary. Figure 9-22 summarizes the transmission-line configurations described in the preceding sections, their input impedance characteristics, and their equivalent *LC* circuits. It can be seen that both shorted and open sections of transmission lines can behave as resistors, inductors, or capacitors, depending on their electrical length.

Transmission-Line Impedance Matching

Power is transferred most efficiently to a load when there are no reflected waves: that is, when the load is purely resistive and equal to Z_o. Whenever the characteristic impedance of a transmission line and its load are not matched (equal), standing waves are present on the line and maximum power is not transferred to the load. As stated previously, standing waves cause power loss, dielectric breakdown, noise, radiation, and *ghost signals*. Therefore, whenever possible a transmission line should be matched to its load. There are two common transmission-line techniques that are used to match a transmission line to a load with an impedance that is not equal to Z_o. They are quarter-wavelength transformer matching and stub matching.

Quarter-wavelength transformer matching. Quarter-wavelength *transformers* are used to match transmission lines to purely resistive loads whose resistance is not equal to the characteristic impedance of the line. Keep in mind that a quarter-wavelength transformer is not actually a transformer, but rather, a quarter-wavelength section of transmission line that acts like a transformer. In a previous section it was shown that the input impedance to a transmission line varies from some maximum value to some minimum value, or vice versa, every quarter wavelength. Therefore, a transmission line one-quarter wavelength long acts like a *step-up* or *step-down transformer*, depending on whether Z_L is greater than or less than Z_o. A quarter-wavelength transformer is not a broadband impedance-matching device; it is a quarter wavelength at only a single frequency. The impedance transformations for a quarter wavelength transmission line are as follows:

1. $R_L = Z_o$: The quarter-wavelength line acts like a transformer with a 1:1 turns ratio.
2. $R_L > Z_o$: The quarter-wavelength line acts like a step-down transformer.
3. $R_L < Z_o$: The quarter-wavelength line acts like a step-up transformer.

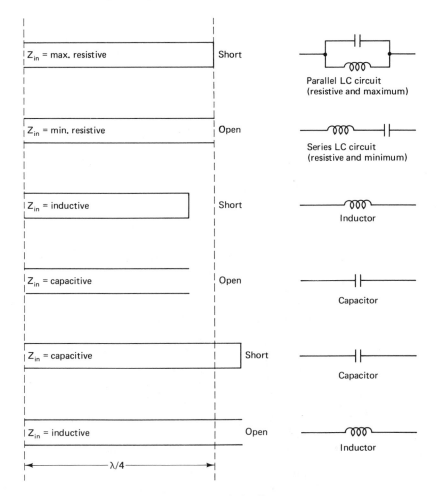

FIGURE 9-22   Transmission-line summary.

Like a transformer, a quarter-wavelength transformer is placed between a transmission line and its load. A quarter-wavelength transformer is simply a length of transmission line one-quarter wavelength long. Figure 9-23 shows how a quarter-wavelength transformer is used to match a transmission line to a purely resistive load. The characteristic impedance of the quarter-wavelength section is determined mathematically from the formula

$$Z_o' = \sqrt{Z_o Z_L}\tag{9-32}$$

where

Z_o' = characteristic impedance of quarter-wavelength transformer
Z_o = characteristic impedance of the transmission line that is being matched
Z_L = load impedance

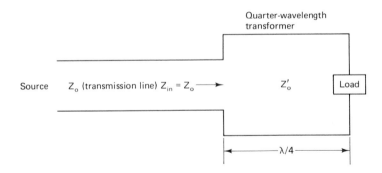

FIGURE 9-23   Quarter-wavelength transformer.

EXAMPLE 9-5

Determine the physical length and characteristic impedance for a quarter-wavelength transformer that is used to match a section of RG-8A/U ($Z_o = 50\ \Omega$) to a 150-Ω resistive load. The frequency of operation is 150 MHz and the velocity factor $V_f = 1$.

Solution   The physical length of the transformer is dependent on the wavelength of the signal. Substituting into Equation 9-2 yields

$$\lambda = \frac{c}{F} = \frac{3 \times 10^8\ \text{m/s}}{150\ \text{MHz}} = 2\ \text{m}$$

$$\frac{\lambda}{4} = \frac{2\ \text{m}}{4} = 0.5\ \text{m}$$

The characteristic impedance of the 0.5-m transformer is determined from Equation 9-32:

$$Z'_o = \sqrt{Z_o Z_L} = \sqrt{(50)(150)} = 86.6\ \Omega$$

Stub matching.   When a load is purely inductive or purely capacitive, it absorbs no energy. The reflection coefficient is 1 and the SWR is infinity. When the load is a complex impedance (which is usually the case), it is necessary to remove the reactive component to match the transmission line to the load. Transmission-line *stubs* are commonly used for this purpose. A transmission-line stub is simply a piece of additional transmission line that is placed across the primary line as close to the load as possible. The susceptance of the stub is used to tune out the susceptance of the load. With *stub matching*, either a shorted or an open stub can be used. However, shorted stubs are preferred because open stubs have a tendency to radiate, especially at the higher frequencies.

Figure 9-24 shows how a shorted stub is used to cancel the susceptance of the load and match the load resistance to the characteristic impedance of the transmission line. In previous sections it was shown how a shorted section of transmission line can look resistive, inductive, or capacitive, depending on its electrical length. A transmission line that is one-half wavelength or shorter can be used to tune out the reactive component of a load.

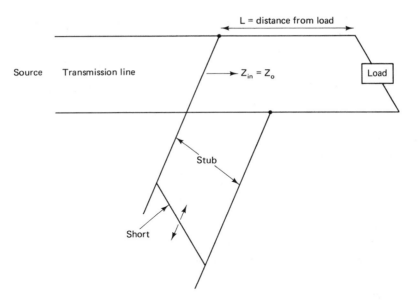

FIGURE 9-24   Shorted stub impedance matching.

The process of matching a load to a transmission line with a shorted stub is as follows.

1. Locate a point as close to the load as possible where the resistive component of the input impedance is equal to the characteristic impedance of the transmission line. $Z_{in} = R + jX$ where $R = Z_o$.

2. Attach the shorted stub to the point on the transmission line identified in step 1.

3. Depending on whether the reactive component of the load is inductive or capacitive, the stub length is adjusted accordingly. $Z_{in} = R + jX_L - jX_C = R$ or $R + jX_C - jX_L = R$.

For a more complete explanation of stub matching using the Smith chart, refer to Appendix A.

QUESTIONS

9-1. Define *transmission line*.

9-2. Describe a transverse electromagnetic wave.

9-3. Define *wave velocity*.

9-4. Define *frequency* and *wavelength* for a transverse electromagnetic wave.

9-5. Describe balanced and unbalanced transmission lines.

9-6. Describe an open-wire transmission line.

9-7. Describe a twin-lead transmission line.

9-8. Describe a twisted-pair transmission line.

9-9. Describe a shielded cable transmission line.

9-10. Describe a concentric transmission line.

9-11. Describe the electrical and physical properties of a transmission line.

9-12. List and describe the four primary constants of a transmission line.

9-13. Define *characteristic impedance* for a transmission line.

9-14. What properties of a transmission line determine its characteristic impedance?

9-15. Define *propagation constant* for a transmission line.

9-16. Define *velocity factor* for a transmission line.

9-17. What properties of a transmission line determine its velocity factor?

9-18. What properties of a transmission line determine its dielectric constant?

9-19. Define *electrical length* for a transmission line.

9-20. List and describe five types of transmission-line losses.

9-21. Describe an incident wave; a reflected wave.

9-22. Describe a resonant transmission line; a nonresonant transmission line.

9-23. Define *reflection coefficient*.

9-24. Describe standing waves; standing-wave ratio.

9-25. Describe the standing waves present on an open transmission line.

9-26. Describe the standing waves present on a shorted transmission line.

9-27. Define *input impedance* for a transmission line.

9-28. Describe the behavior of a transmission line that is terminated in a short circuit that is greater than one-quarter wavelength long; less than one-quarter wavelength.

9-29. Describe the behavior of a transmission line that is terminated in an open circuit that is greater than one-quarter wavelength long; less than one-quarter wavelength long.

9-30. Describe the behavior of an open transmission line as a circuit element.

9-31. Describe the behavior of a shorted transmission line as a circuit element.

9-32. Describe the input impedance characteristics of a quarter-wavelength transmission line.

9-33. Describe the input impedance characteristics of a transmission line that is less than one-quarter wavelength long; greater than one-quarter wavelength long.

9-34. Describe quarter-wavelength transformer matching.

9-35. Describe how stud matching is accomplished.

PROBLEMS

P. 362

9-1. Determine the wavelengths for electromagnetic waves with the following frequencies: 1 kHz, 100 kHz, 1 MHz, and 1 GHz.

9-2. Determine the frequencies for electromagnetic waves with the following wavelengths: 1 cm, 1 m, 10 m, 100 m, and 1000 m.

P. 369-371

9-3. Determine the characteristic impedance for an air-dielectric transmission line with *D/r* ratio of 8.8.

9-4. Determine the characteristic impedance for a concentric transmission line with D/d ratio of 4.

P. 369-371 **9-5.** Determine the characteristic impedance for a coaxial cable with inductance $L = 0.2 \ \mu H/ft$ and capacitance $C = 16 \ pF/ft$.

9-6. For a given length of coaxial cable with distributed capacitance $C = 48.3 \ pF/m$ and distributed inductance $L = 241.56 \ nH/m$, determine the velocity factor and velocity of propagation.

P. 379 **9-7.** Determine the reflection coefficient for a transmission line with incident voltage $E_i = 0.2 \ V$ and reflected voltage $E_r = 0.01 \ V$.

9-8. Determine the standing-wave ratio for the transmission line described in Question 9-7.

P. 381 **9-9.** Determine the SWR for a transmission line with maximum voltage standing-wave amplitude $V_{max} = 6 \ V$ and minimum voltage standing-wave amplitude $V_{min} = 0.5$.

9-10. Determine the SWR for a 50-Ω transmission line that is terminated in a load resistance $Z_L = 75 \ \Omega$.

P. 382 **9-11.** Determine thee SWR for a 75-Ω transmission line that is terminated in a load resistance $Z_L = 50 \ \Omega$.

9-12. Determine the characteristic impedance for a quarter-wavelength transformer that is used to match a section of 75-Ω transmission line to a 100-Ω resistive load.

Chapter 10

WAVE PROPAGATION

INTRODUCTION

In Chapter 9 we explained how a metallic wire is used as a transmission medium to propagate electromagnetic waves from one point to another. However, very often in communications systems it is impractical or impossible to interconnect two pieces of equipment with a physical facility such as a wire; for example, across large spans of water, rugged mountains, or desert terrain, or to and from satellite transponders parked 22,000 mi above the earth. Also, when the transmitters and receivers are mobile, as with two-way radio or mobile telephone, metallic facilities are impossible. Therefore, free space or the earth's atmosphere is often used as a transmission medium. Free-space propagation of electromagnetic waves is often called *radio-frequency* (RF) *propagation* or simply *radio propagation*.

To propagate TEM waves through the earth's atmosphere, it is necessary that energy be radiated from the source; then the energy must be *captured* at the receive end. Radiating and capturing energy are antenna functions and are explained in Chapter 11, and the properties of electromagnetic waves were explained in Chapter 9.

RAYS AND WAVEFRONTS

Electromagnetic waves are invisible. Therefore, they must be analyzed by indirect methods using schematic diagrams. The concept of *rays* and *wavefronts* is an aid to illustrating the effects of electromagnetic wave propagation through free space. A ray is a line

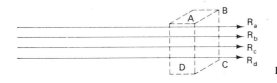

FIGURE 10-1 Plane wave.

drawn along the direction of propagation of an electromagnetic wave. Rays are used to show the relative direction of electromagnetic wave propagation. However, a ray does not necessarily represent the propagation of a single electromagnetic wave. Several rays are shown in Figure 10-1 (R_a, R_b, R_c, etc.). A wavefront shows a surface of constant phase of a wave. A wavefront is formed when points of equal phase on rays propagated from the same source are joined together. Figure 10-1 shows a wavefront with a surface that is perpendicular to the direction of propagation (rectangle *ABCD*). When a surface is plane, its wavefront is perpendicular to the direction of propagation. The closer to the source, the more complicated the wavefront becomes.

Most wavefronts are more complicated than a simple plane wave. Figure 10-2 shows a point source, several rays propagating from it, and the corresponding wavefront. A *point source* is a single location from which rays propagate equally in all directions (i.e., an *isotropic source*). The wavefront generated from a point source is simply a sphere with radius R and its center located at the point of origin of the waves. In free space and a sufficient distance from the source, the rays within a small area of a spherical wavefront are nearly parallel. Therefore, the farther from a source, the closer to plane a wavefront appears.

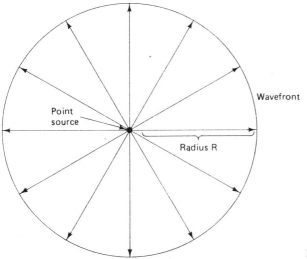

FIGURE 10-2 Wavefront from a point source.

ELECTROMAGNETIC RADIATION

Power Density and Field Intensity

Electromagnetic waves represent the flow of energy in the direction of propagation. The rate at which energy passes through a given surface area in free space is called *power density*. Therefore, power density is energy per unit time per unit of area and is usually given in watts per square meter. *Field intensity* is the intensity of the voltage and magnetic fields of an electromagnetic wave propagating in free space. Voltage intensity is usually given in volts per meter and magnetic intensity in ampere-turns/ meter. Mathematically, the rms power density is

$$\mathcal{P} = \mathcal{E}H \text{ watts per meter squared} \tag{10-1}$$

where

$\mathcal{P}$ = power density (W/m^2)
$\mathcal{E}$ = rms voltage intensity (V/m)
H = rms magnetic intensity (At/m)

Characteristic Impedance of Free Space

The electric and magnetic intensities of an electromagnetic wave in free space are related through the characteristic impedance (resistance) of free space. The characteristic impedance of a lossless transmission medium is equal to the square root of the ratio of its magnetic permeability to its electric permittivity. Mathematically, the characteristic impedance of free space (Z_s) is

$$Z_s = \frac{\mu}{\epsilon} \tag{10-2}$$

where

Z_s = characteristic impedance of free space
μ = magnetic permeability of free space (1.26 × 10^{-6} H/m)
ϵ = electric permittivity of free space (8.85 × 10^{-12} F/m)

Substituting into Equation 10-2, we have

$$Z_s = \sqrt{\frac{1.26 \times 10^{-6}}{8.85 \times 10^{-12}}} = 377 \ \Omega$$

Therefore, using Ohm's law, we obtain

$$\mathcal{P} = \frac{\mathcal{E}^2}{377} = 377H^2 \qquad \text{watts per meter squared} \tag{10-3}$$

and

$$H = \frac{\mathscr{E}}{377} \qquad \text{ampere-turns per meter} \qquad (10\text{-}4)$$

SPHERICAL WAVEFRONT AND THE INVERSE SQUARE LAW

Spherical Wavefront

Figure 10-3 shows a point source that radiates power at a constant rate uniformly in all directions. Such a source is called an *isotropic radiator*. A true isotropic radiator does not exist. However, it is closely approximated by an *omnidirectional antenna*. An isotropic radiator produces a spherical wavefront with radius R. All points distance R from the source lie on the surface of the sphere and have equal power densities. For example, in Figure 10-3 points A and B are equal distance from the source. Therefore, the power density at points A and B are equal. At any instant of time, the total power radiated, P_r watts, is uniformly distributed over the total surface of the sphere (this assumes a lossless transmission medium). Therefore, the power density at any point on the sphere is the total radiated power divided by the total area of the sphere. Mathematically, the power density at any point on the surface of a spherical wavefront is

$$\mathscr{P} = \frac{P_r}{4\pi R^2} \qquad (10\text{-}5)$$

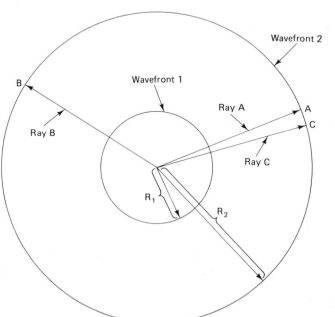

FIGURE 10-3 Spherical wavefront from an isotropic source.

where

P_r = total power radiated

R = radius of the sphere (which is equal to the distance from any point on the surface of the sphere to the source)

$4\pi R^2$ = area of the sphere

and for a distance R_a meters from the source, the power density is

$$\mathcal{P} = \frac{P_r}{4\pi R_a^2}$$

Equating Equations 10-3 and 10-5 gives

$$\frac{P_r}{4\pi R^2} = \frac{\mathcal{E}^2}{377}$$

Therefore,

$$\mathcal{E}^2 = \frac{377 P_r}{4\pi R^2} \quad \text{and} \quad \mathcal{E} = \frac{\sqrt{30 P_r}}{R} \tag{10-6}$$

Inverse Square Law

From Equation 10-3 it can be seen that the farther the wavefront moves from the source, the smaller the power density (R_a and R_c move farther apart). The total power distributed over the surface of the sphere remains the same. However, because the area of the sphere increases in direct proportion to the distance from the source squared (i.e., the radius of the sphere squared), the power density is inversely proportional to the square of the distance from the source. This relationship is called the *inverse square law*. Therefore, the power density at any point on the surface of the outer sphere is

$$\mathcal{P}_2 = \frac{P_r}{4\pi R_2^2}$$

and the power density at any point on the inner sphere is

$$\mathcal{P}_1 = \frac{P_r}{4\pi R_1^2}$$

Therefore,

$$\frac{\mathcal{P}_2}{\mathcal{P}_1} = \frac{P_r/4\pi R_2^2}{P_r/4\pi R_1^2} = \frac{R_1^2}{R_2^2} = \left(\frac{R_1}{R_2}\right)^2 \tag{10-7}$$

From Equation 10-7 it can be seen that as the distance from the source doubles, the power density decreases by a factor of 2^2, or 4. When deriving the inverse square law of radiation (Equation 10-7) it was assumed that the source radiates isotropically, although it is not necessary. However, it is necessary that the velocity of propagation in all directions be uniform. Such a propagation medium is called an *isotropic medium*.

EXAMPLE 10-1

For an isotropic antenna radiating 100 W of power, determine:
    (a) The power density 1000 m from the source.
    (b) The power density 2000 m from the source.

Solution

    (a) Substituting into Equation 10-5 yields

$$\mathscr{P} = \frac{100}{4\pi 1000^2} = 7.96 \ \mu\text{W/m}^2$$

    (b) Again, substituting into Equation 10-5 gives us

$$\mathscr{P} = \frac{100}{4\pi 2000^2} = 1.99 \ \mu\text{W/m}^2$$

or substituting into Equation 10-7, we have

$$\frac{\mathscr{P}_2}{\mathscr{P}_1} = \frac{1000^2}{2000^2} = 7.96 \ \mu\text{W/m}^2$$

$$\mathscr{P}_2 = 7.96 \ \mu\text{W/m}^2 \left(\frac{1000^2}{2000^2}\right) = 1.99 \ \mu\text{W/m}^2$$

WAVE ATTENUATION AND ABSORPTION

Attenuation

The inverse square law for radiation mathematically describes the reduction in power density with distance from the source. As a wavefront moves away from the source, the continuous electromagnetic field that is radiated from that source spreads out. That is, the waves move farther away from each other and, consequently, the number of waves per unit area decreases. None of the radiated power is lost or dissipated because the wavefront is moving away from the source; the wave simply spreads out or disperses over a larger area, decreasing the power density. The reduction in power density with distance is equivalent to a power loss and is commonly called *wave attenuation*. Because the attenuation is due to the spherical spreading of the wave, it is sometimes called the *space attenuation* of the wave. Wave attenuation is generally expressed in terms of the common logarithm of the power density ratio (dB loss). Mathematically, wave attenuation (γ_a) is

$$\gamma_a = 10 \log \frac{\mathscr{P}_1}{\mathscr{P}_2} \tag{10-8}$$

    The reduction in power density due to the inverse-square law presumes free-space propagation (i.e., a vacuum or nearly a vacuum) and is called wave attenuation. The reduction in power density due to nonfree-space propagation is called *absorption*.

Absorption

The earth's atmosphere is not a vacuum. Rather, it is made up of atoms and molecules of various substances, such as gases, liquids, and solids. Some of these materials are capable of absorbing electromagnetic waves. As an electromagnetic wave propagates through the earth's atmosphere, energy is transferred from the wave to the atoms and molecules of the atmosphere. Wave absorption by the atmosphere is analogous to an I^2R power loss. Once absorbed, the energy is lost forever and causes an attenuation in the voltage and magnetic field intensities and a corresponding reduction in power density.

Absorption of radio frequencies in a normal atmosphere is dependent on frequency and relatively insignificant below approximately 10 GHz. Figure 10-4 shows atmospheric absorption in dB/km due to oxygen and water vapor for radio frequencies above 10 GHz. It can be seen that certain frequencies are affected more or less by absorption, creating peaks and valleys in the curves. Wave attenuation due to absorption does not depend on distance from the radiating source, but rather, the total distance that the wave propagates through the atmosphere. In other words, for a *homogeneous medium* (one with uniform properties throughout), the absorption experienced during the first mile of propagation is the same as for the last mile. Also, abnormal atmospheric conditions such as heavy rain or dense fog absorb more energy than a normal atmosphere. Atmospheric absorption (η) for a wave propagating from R_1 to R_2 is $\gamma(R_2 - R_1)$, where γ is the absorption coefficient. Therefore, wave attenuation depends on the ratio R_2/R_1 and wave absorption depends on the distance between R_1 and R_2. In a more practical situation (i.e., an *inhomogeneous medium*), the absorption coefficient varies considerably with location, thus creating a difficult problem for radio systems engineers.

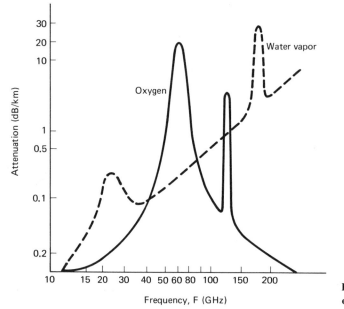

FIGURE 10-4 Atmospheric absorption of electromagnetic waves.

OPTICAL PROPERTIES OF RADIO WAVES

In the earth's atmosphere, ray-wavefront propagation may be altered from free-space behavior by *optical* effects such as *refraction, reflection, diffraction,* and *interference.* Using rather unscientific terminology, refraction can be thought of as *bending,* reflection as *bouncing,* diffraction as *scattering,* and interference as *colliding.* Refraction, reflection, diffraction, and interference are called optical properties because they were first observed in the science of optics, which is the behavior of light waves. Because light waves are high-frequency electromagnetic waves, it stands to reason that optical properties will also apply to radio-wave propagation. Although optical principles can be analyzed completely by application of Maxwell's equations, this is necessarily complex. For most applications, *geometric ray tracing* can be substituted for analysis by Maxwell's equations.

Refraction

Electromagnetic *refraction* is the change in direction of a ray as it passes obliquely from one medium to another with different velocities of propagation. As stated previously, the velocity at which an electromagnetic wave propagates is inversely proportional to the density of the medium in which it is propagating. Therefore, refraction occurs whenever a radio wave passes from one medium into another medium of different density. Figure 10-5 shows refraction of a wavefront at a *plane* boundary between two media with different densities. For this example, medium 1 is less dense than medium 2 (i.e.,

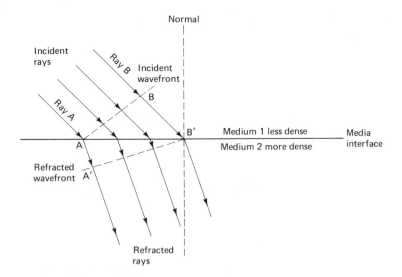

FIGURE 10-5   Refraction at a plane boundary between two media.

$v_1 > v_2$). It can be seen that ray A enters the more dense medium before ray B. Therefore, ray B propagates more slowly than ray A and travels distance $B–B'$ during the same time that ray A travels distance $A–A'$. Therefore, wavefront ($A'B'$) is *tilted* or bent in a downward direction. Since a ray is defined as being perpendicular to the wavefront at all points, the rays in Figure 10-5 have changed direction at the interface of the two media. Whenever a ray passes from a less dense to a more dense medium, it is effectively bent toward the *normal*. (The normal is simply an imaginary line drawn perpendicular to the interface at the point of incidence.) Conversely, whenever a ray passes from a more dense to a less dense medium, it is effectively bent away from the normal. The *angle of incidence* is the angle formed between the incident wave and the normal, and the *angle of refraction* is the angle formed between the refracted wave and the normal.

The amount of bending or refraction that occurs at the interface of two materials of different densities is quite predictable and depends on the *refractive index* (also called the *index of refraction*) of the two materials. The refractive index is simply the ratio of the velocity of propagation of a light ray in free space to the velocity of propagation of a light ray in a given material. Mathematically, the refractive index is

$$n = \frac{c}{v} \qquad (10\text{-}9)$$

where

n = refractive index
c = speed of light in free space
v = speed of light in a given material

The refractive index is also a function of frequency. However, the variation in most applications is insignificant and is therefore omitted from this discussion. How an electromagnetic wave reacts when it meets the interface of two transmissive materials that have different indexes of refraction can be explained with *Snell's law*. Snell's law simply states:

$$n_1 \sin \theta_1 = n_2 \sin \theta_2 \qquad (10\text{-}10)$$

and

$$\frac{\sin \theta_1}{\sin \theta_2} = \frac{n_2}{n_1}$$

where

n_1 = refractive index of material 1
n_2 = refractive index of material 2
θ_1 = angle of incidence
θ_2 = angle of refraction

and since the refractive index of a material is equal to the square root of its dielectric constant,

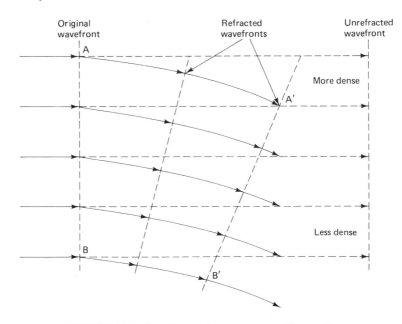

FIGURE 10-6   Wavefront refraction in a gradient medium.

$$\frac{\sin \theta_1}{\sin \theta_2} = \sqrt{\frac{\epsilon_2}{\epsilon_1}} \qquad (10\text{-}11)$$

where

ϵ_1 = dielectric constant of medium 1
ϵ_2 = dielectric constant of medium 2

Refraction also occurs when a wavefront propagates in a medium that has a *density gradient* that is perpendicular to the direction of propagation (i.e., parallel to the wavefront). Figure 10-6 shows wavefront refraction in a transmission medium that has a gradual variation in its refractive index. The medium is more dense near the bottom and less dense at the top. Therefore, rays traveling near the top travel faster than rays near the bottom and consequently, the wavefront tilts downward. The tilting occurs in a gradual fashion as the wave progresses as shown.

Reflection

Reflect means to cast or turn back, and *reflection* is the act of reflecting. Electromagnetic reflection occurs when an incident wave strikes a boundary of two media and some or all of the incident power does not enter the second material. The waves that do not penetrate the second medium are reflected. Figure 10-7 shows electromagnetic wave reflection at a plane boundary between two media. Because all the reflected waves

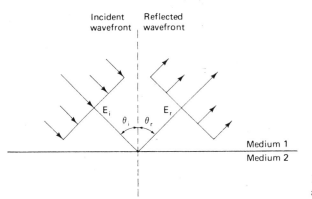

FIGURE 10-7 Electromagnetic reflection at a plane boundary of two media.

remain in medium 1, the velocities of the reflected and incident waves are equal. Consequently, the *angle of reflection* equals the *angle of incidence* ($\theta_i = \theta_r$). However, the reflected voltage field intensity is less than the incident voltage field intensity. The ratio of the reflected to the incident voltage intensities is called the *reflection coefficient*, Γ (sometimes called the *coefficient of reflection*). For a perfect conductor, $\Gamma = 1$. Γ is used to indicate both the relative amplitude of the incident and reflected fields and also the phase shift that occurs at the point of reflection. Mathematically, the reflection coefficient is

$$\Gamma = \frac{E_r e^{j\theta_r}}{E_i e^{j\theta_i}} = \frac{E_r}{E_i} e^{j(\theta_r - \theta_i)} \tag{10-12}$$

where

Γ = reflection coefficient
E_i = incident voltage intensity
E_r = reflected voltage intensity
θ_i = incident phase
θ_r = reflected phase

The ratio of the reflected and incident power densities is Γ. The portion of the total incident power that is not reflected is called the *power transmission coefficient* (T) (or simply the *transmission coefficient*). For a perfect conductor, $T = 0$. The *law of conservation of energy* states that for a perfect reflective surface, the total reflected power must equal the total incident power. Therefore,

$$T + |\Gamma|^2 = 1 \tag{10-13}$$

For imperfect conductors, both $|\Gamma|^2$ and T are functions of the angle of incidence, the electric field polarization, and the dielectric constants of the two materials. If medium 2 is not a perfect conductor, some of the incident waves penetrate it and are absorbed. The absorbed waves set up currents in the resistance of the material and the energy is converted to heat. The fraction of power that penetrates medium 2 is called the *absorption coefficient* (or sometimes the *coefficient of absorption*).

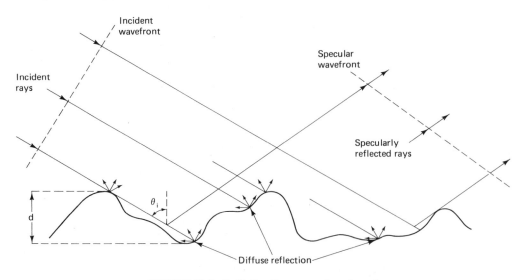

FIGURE 10-8 Reflection from a semirough surface.

When the reflecting surface is not plane (i.e., it is curved) the curvature of the reflected wave is different from that of the incident wave. When the wavefront of the incident wave is curved and the reflective surface is plane, the curvature of the reflected wavefront is the same as that of the incident wavefront.

Reflection also occurs when the reflective surface is *irregular* or *rough*. However, such a surface may destroy the shape of the wavefront. When an incident wavefront strikes an irregular surface, it is randomly scattered in many directions. Such a condition is called *diffuse reflection*, whereas reflection from a perfectly smooth surface is called *specular* (mirror-like) *reflection*. Surfaces that fall between smooth and irregular are called *semirough surfaces*. Semirough surfaces cause a combination of diffuse and specular reflection. A semirough surface will not totally destroy the shape of the reflected wavefront. However, there is a reduction in the total power. The *Rayleigh criterion* states that a semirough surface will reflect as if it were a smooth surface whenever the cosine of the angle of incidence is greater than $\lambda/8d$, where d is the depth of the surface irregularity and λ is the wavelength of the incident wave. Reflection from a semirough surface is shown in Figure 10-8. Mathematically, Rayleigh's criterion is

$$\cos \theta_i > \frac{\lambda}{8d} \qquad (10\text{-}14)$$

Diffraction

Diffraction is defined as the modulation or redistribution of energy within a wavefront when it passes near the edge of an *opaque* object. Diffraction is the phenomenon that allows light or radio waves to propagate (*peek*) around corners. The previous discussions of refraction and reflection assumed that the dimensions of the refracting and reflecting

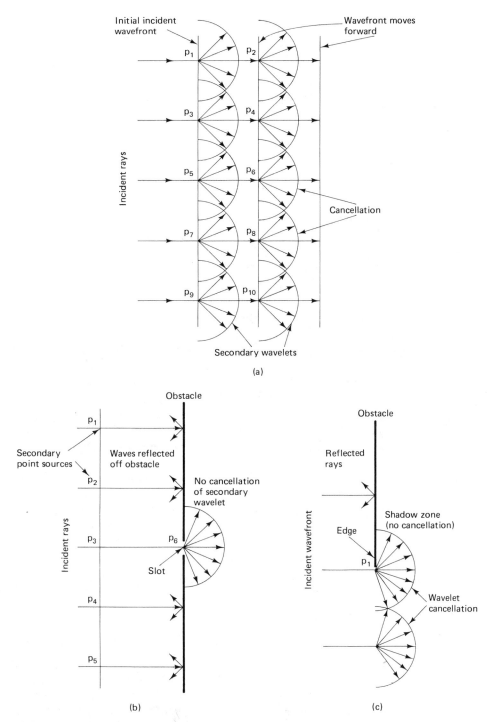

FIGURE 10-9 Electromagnetic wave diffraction: (a) Huygens' principle for a plane wavefront; (b) finite wavefront through a slot; (c) around an edge.

410

surfaces were large in respect to a wavelength of the signal. However, when a wavefront passes near an obstacle or discontinuity with dimensions comparable in size to a wavelength, simple geometric analysis cannot be used to explain the results and *Huygens' principle* (which is deduced from Maxwell's equations) is necessary.

Huygens' principle states that every point on a given spherical wavefront can be considered as a secondary point source of electromagnetic waves from which other secondary waves (wavelets) are radiated outward. Huygens' principle is illustrated in Figure 10-9. Normal wave propagation considering an infinite plane is shown in Figure 10-9a. Each secondary point source (p_1, p_2, etc.) radiates energy outward in all directions. However, the wavefront continues in its original direction rather than spreading out because cancellation of the secondary wavelets occurs in all directions except straight forward. Therefore, the wavefront remains plane.

When a finite plane wavefront is considered, as in Figure 10-9b, cancellation in random directions is incomplete. Consequently, the wavefront spreads out or *scatters*. This scattering effect is called *diffraction*. Figure 10-9c shows diffraction around the edge of an obstacle. It can be seen that wavelet cancellation occurs only partially. Diffraction occurs around the edge of the obstacle, which allows secondary waves to "sneak" around the corner of the obstacle into what is called the *shadow zone*. This phenomenon can be observed when a door is opened into a dark room. Light rays diffract around the edge of the door and illuminate the area behind the door.

Interference

Interfere means to come into opposition, and *interference* is the act of interfering. Radio-wave interference occurs when two or more electromagnetic waves combine in such a way that system performance is degraded. Refraction, reflection, and diffraction are categorized as geometric optics, which means that their behavior is analyzed primarily in terms of rays and wavefronts. Interference, on the other hand, is subject to the principle of *linear superposition* of electromagnetic waves and occurs whenever two or more waves simultaneously occupy the same point in space. The principle of linear superposition states that the total voltage intensity at a given point in space is the sum of the individual wave vectors. Certain types of propagation media have nonlinear properties; however, in an ordinary medium (such as air or the earth's atmosphere), linear superposition holds true.

Figure 10-10 shows the linear addition of two instantaneous voltage vectors whose phase angles differ by angle θ. It can be seen that the total voltage is not simply the sum of the two vectors, but rather, the phasor addition of the two. With free-space propagation, a phase difference may exist simply because the electromagnetic polariza-

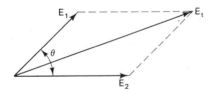

FIGURE 10-10 Linear addition of two vectors with differing phase angles.

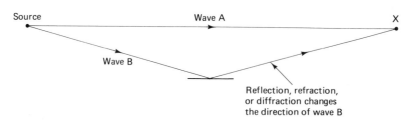

FIGURE 10-11   Electromagnetic wave interference.

tions of two waves differ. Depending on the phase angles of the two vectors, either addition or subtraction can occur. (This implies simply that the result may be more or less than either vector because the two electromagnetic waves can reinforce or cancel.)

Figure 10-11 shows interference between two electromagnetic waves in free space. It can be seen that at point X the two waves occupy the same area of space. However, wave B has traveled a different path than wave A, and therefore their relative phase angles may be different. If the difference in distance traveled is an odd integral multiple of one-half wavelength, reinforcement takes place. If the difference is an even integral multiple of one-half wavelength, total cancellation occurs. More likely the difference in distance falls somewhere between the two and partial cancellation occurs. For frequencies below VHF, the relatively large wavelengths prevent interference from being a significant problem. However, with UHF and above, wave interference can be severe.

PROPAGATION OF WAVES

In radio communications systems, there are several ways in which waves can be propagated, depending on the type of system and the environment. Also, as previously explained, electromagnetic waves travel in straight lines except when the earth and its atmosphere alter their path. There are three ways of propagating electromagnetic waves: ground wave, space wave (which includes both direct and ground-reflected waves), and sky wave propagation.

Figure 10-12 shows the normal modes of propagation between two radio antennas. Each of these modes exists in every radio system; however, some are negligible in certain frequency ranges or over a particular type of terrain. At frequencies below 1.5 MHz, ground waves provide the best coverage. This is because ground losses increase rapidly with frequency. Sky waves are used for high-frequency applications, and space waves are used for very high frequencies and above.

Ground-Wave Propagation

A *ground wave* is an electromagnetic wave that travels along the surface of the earth. Therefore, ground waves are sometimes called *surface waves*. Ground waves must be vertically polarized. This is because the electric field in a horizontally polarized wave would be parallel to the earth's surface, and such waves would be short-circuited by

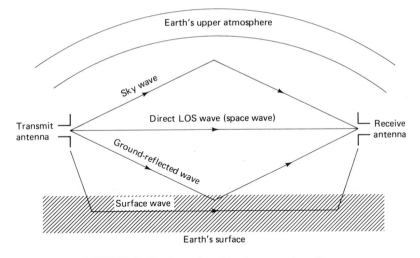

FIGURE 10-12 **Normal modes of wave propagation.**

the conductivity of the ground. With ground waves, the changing electric field induces voltages in the earth's surface, which cause currents to flow that are very similar to those in a transmission line. The earth's surface also has resistance and dielectric losses. Therefore, ground waves are attenuated as they propagate. Ground waves propagate best over a surface that is a good conductor, such as salt water, and poorly over dry desert areas. Ground-wave losses increase rapidly with frequency. Therefore, ground-wave propagation is generally limited to frequencies below 2 MHz.

Figure 10-13 shows ground-wave propagation. The earth's atmosphere has a *gra-*

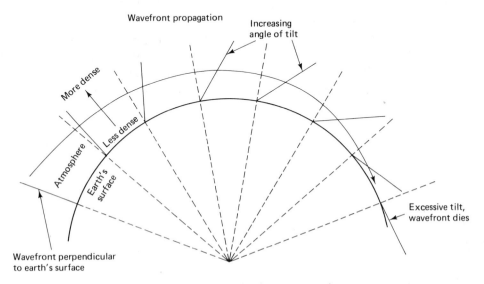

FIGURE 10-13 **Ground-wave propagation.**

dient density (i.e., the density decreases gradually with distance from the earth's surface), which causes the wavefront to tilt progressively forward. Therefore, the ground wave propagates around the earth, remaining close to its surface, and if enough power is transmitted, the wavefront could propagate beyond the horizon or even around the entire earth's circumference. However, care must be taken when selecting the frequency and the terrain over which the ground wave will propagate, to ensure that the wavefront does not tilt excessively and simply turn over, lie flat on the ground, and cease to propagate.

Ground-wave propagation is commonly used for ship-to-ship and ship-to-shore communications, for radio navigation, and for maritime mobile communications. Ground waves are used at frequencies as low as 15 kHz.

The disadvantages of ground-wave propagation are as follows:

1. Ground waves require a relatively high transmission power.
2. Ground waves are limited to low and very low frequencies (LF and VLF), facilitating large antennas (the reason for this is explained in Chapter 11).
3. Ground losses vary considerably with surface material.

The advantages of ground-wave propagation are as follows:

1. Given enough transmit power, ground waves can be used to communicate between any two locations in the world.
2. Ground waves are relatively unaffected by changing atmospheric conditions.

Space-Wave Propagation

Space-wave propagation includes radiated energy that travels in the lower few miles of the earth's atmosphere. Space waves include both direct and ground-reflected waves (see Figure 10-14). *Direct waves* are waves that travel essentially in a straight line between the transmit and receive antennas. Space-wave propagation with direct waves is commonly called *line-of-sight* (LOS) *transmission*. Therefore, space-wave propagation

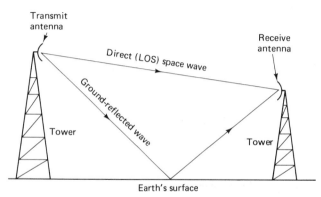

FIGURE 10-14 Space-wave propagation.

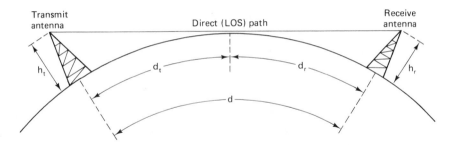

FIGURE 10-15 Space waves and radio horizon.

is limited by the curvature of the earth. Ground-reflected waves are those waves that are reflected by the earth's surface as they propagate between the transmit and receive antennas.

Figure 10-14 shows space-wave propagation between two antennas. It can be seen that the field intensity at the receive antenna depends on the distance between the two antennas (attenuation and absorption) and whether the direct and ground-reflected waves are in phase (interference).

The curvature of the earth presents a horizon to space wave propagation commonly called the *radio horizon*. Due to atmospheric refraction, the radio horizon extends beyond the *optical horizon* for the common *standard atmosphere*. The radio horizon is approximately four-thirds that of the optical horizon. Refraction is caused by the troposphere, due to changes in its density, temperature, water vapor content, and relative conductivity. The radio horizon can be lengthened simply by elevating the transmit or receive antennas (or both) above the earth's surface with towers or by placing them on top of mountains or high buildings.

Figure 10-15 shows the effect of antenna height on the radio horizon. The line-of-sight radio horizon for a single antenna is given as

$$d = \sqrt{2h} \qquad (10\text{-}15)$$

where

$\quad\quad d$ = distance to radio horizon (mi)
$\quad\quad h$ = antenna height above sea level (ft)

Therefore, for a transmit and receive antenna, the distance between the two antennas is

$$d = d_t + d_r$$

or

$$d = \sqrt{2h_t} + \sqrt{2h_r} \qquad (10\text{-}16)$$

where

$\quad\quad d$ = total distance (mi)
$\quad\quad d_t$ = radio horizon for transmit antenna (mi)

d_r = radio horizon for receive antenna (mi)
h_t = transmit antenna height (ft)
h_r = receive antenna height (ft)

or

$$d = 4\sqrt{h_t} + 4\sqrt{h_r} \qquad (10\text{-}17)$$

where d_t and d_r are distance in kilometers and h_t and h_r are height in meters.

From Equations 10-16 and 10-17 it can be seen that the space-wave propagation distance can be extended simply by increasing either the transmit or receive antenna height, or both.

Because the conditions in the earth's lower atmosphere are subject to change, the degree of refraction can vary with time. A special condition called *duct propagation* occurs when the density of the lower atmosphere is such that electromagnetic waves are trapped between it and the earth's surface. The layers of the atmosphere act like a duct and an electromagnetic wave can propagate for great distances around the curvature of the earth within this duct. Duct propagation is shown in Figure 10-16.

Sky-Wave Propagation

Electromagnetic waves that are directed above the horizon level are called *sky waves*. Typically, sky waves are radiated in a direction that produces a relatively large angle with reference to the earth. Sky waves are radiated toward the sky, where they are either reflected or refracted back to earth by the ionosphere. The ionosphere is the region of space located approximately 50 to 400 km (30 to 250 mi) above the earth's surface. The ionosphere is the upper portion of the earth's atmosphere. Therefore, it absorbs large quantities of the sun's radiant energy, which ionizes the air molecules, creating free electrons. When a radio wave passes through the ionosphere, the electric field of the wave exerts a force on the free electrons, causing them to vibrate. The vibrating electrons decrease current, which is equivalent to reducing the dielectric constant. Reducing the dielectric constant increases the velocity of propagation and causes electromagnetic waves to bend away from the regions of high electron density toward regions of low electron density (i.e., increasing refraction). As the wave moves farther from earth, ionization increases. However, there are fewer air molecules to ionize.

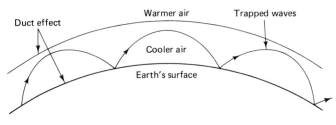

FIGURE 10-16   Duct propagation.

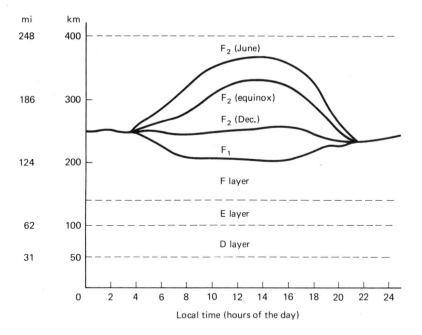

FIGURE 10-17   Ionospheric layers.

Therefore, in the upper atmosphere, there is a higher percentage of ionized molecules than in the lower atmosphere. The higher the ion density, the more refraction. Also, due to the ionosphere's nonuniform composition and its temperature and density variations, it is *stratified*. Essentially, there are three layers that comprise the ionosphere (the D, E, and F layers), which are shown in Figure 10-17. It can be seen that all three layers of the ionosphere vary in location and in *ionization density* with the time of day. They also fluctuate in a cyclic pattern throughout the year, and according to the 11-year *sunspot cycle*. The ionosphere is most dense during times of maximum sunlight (i.e., during the daylight hours and in the summer).

D layer. The *D layer* is the lowest layer of the ionosphere and is located between 30 and 60 mi (50 to 100 km) from the earth's surface. Because it is the layer farthest from the sun, there is very little ionization in this layer. Therefore, the D layer has very little effect on the direction of propagation of radio waves. However, the ions in the D layer can absorb appreciable amounts of electromagnetic energy. The amount of ionization in the D layer depends on the altitude of the sun above the horizon. Therefore, it disappears at night. The D layer reflects VLF and LF waves and absorbs MF and HF waves. (See Table 1-1 for VLF, LF, MF, and HF frequency regions.)

E layer. The *E layer* is located between 60 and 85 mi (100 to 140 km) above the earth's surface. The E layer is sometimes called the *Kennelly–Heaviside layer* after the two scientists who discovered it. The E layer has its maximum density at approximately 70 km at noon, when the sun is at its highest point. Like the D layer, the E layer almost totally disappears at night. The E layer aids MF surface-wave propagation and reflects HF waves somewhat during the daytime. The upper portion of the E layer is sometimes considered separately and called the sporadic E layer because it seems to come and go rather unpredictably. The sporadic E layer is caused by *solar flares* and *sunspot activity*. The sporadic E layer is a thin layer with a very high ionization density. When it appears, there generally is an unexpected improvement in long-distance transmission.

F layer. The *F layer* is actually made up of two layers: the F_1 and F_2 layers. During the daytime, the F_1 layer is located between 85 and 155 mi (140 to 250 km) above the earth's surface, and the F_2 layer is located 85 to 185 mi (140 to 300 km) above the earth's surface during the winter and 155 to 220 mi (250 to 350 km) in the summer. During the night, the F_1 layer combines with the F_2 layer to form a single layer. The F_1 layer absorbs and attenuates some HF waves, although most of the waves pass through to the F_2 layer, where they are refracted back to earth.

PROPAGATION TERMS AND DEFINITIONS

Critical Frequency and Critical Angle

Frequencies above the UHF range are virtually unaffected by the ionosphere because of their extremely short wavelengths. At these frequencies, the distance between ions are appreciably large, and consequently, the electromagnetic waves pass through them with little noticeable effect. Therefore, it stands to reason that there must be an upper frequency limit for sky-wave propagation. *Critical frequency* (F_c) is defined as the highest

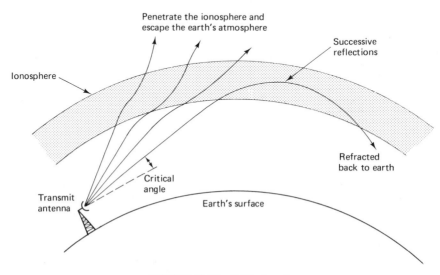

FIGURE 10-18   Critical angle.

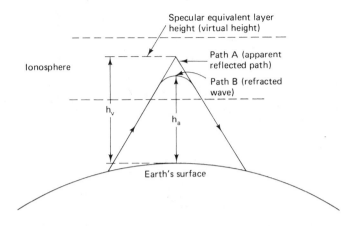

Specular equivalent layer
height (virtual height)

Ionosphere

Path A (apparent
reflected path)

Path B (refracted
wave)

h_v

h_a

Earth's surface

FIGURE 10-19 Virtual and actual height.

frequency that can be propagated directly upward and still be returned to earth by the ionosphere. The critical frequency depends on the ionization density and therefore varies with the time of day and the season. If the vertical angle of radiation is decreased, frequencies at or above the critical frequency can still be refracted back to the earth's surface because they will travel a longer distance in the ionosphere and thus have a longer time to be refracted. Therefore, critical frequency is used only as a point of reference for comparison purposes. However, every frequency has a maximum vertical angle at which it can be propagated and still be refracted back by the ionosphere. This angle is called the *critical angle*. The critical angle θ_c is shown in Figure 10-18.

Virtual Height

Virtual height is the height above the earth's surface from which a refracted wave appears to have been reflected. Figure 10–19 shows a wave that has been radiated from the earth's surface toward the ionosphere. The radiated wave is refracted back to earth and follows path *B*. The actual maximum height that the wave reached is height h_a. However, path *A* shows the projected path that a reflected wave could have taken and still been returned to earth at the same location. The maximum height that this hypothetical reflected wave would have reached is the virtual height (h_v).

Maximum Usable Frequency

The *maximum usable frequency* (MUF) is the highest frequency that can be used for sky-wave propagation between two specific points on the earth's surface. It stands to reason, then, that there are as many values possible for MUF as there are points on earth and frequencies—an infinite number. MUF, like the critical frequency, is a limiting frequency for sky-wave propagation. However, the maximum usable frequency is for a specific angle of incidence (the angle between the incident wave and the normal). Mathematically, MUF is

$$\text{MUF} = \frac{\text{critical frequency}}{\cos \theta} \qquad (10\text{-}18\text{a})$$

$$= \text{critical frequency} \times \sec \theta \qquad (10\text{-}18\text{b})$$

where θ is the angle of incidence.

Equations 10-18 are called the *secant law*. The secant law assumes a flat earth and a flat reflecting layer, which of course, can never exist. Therefore, MUF is used only for making preliminary calculations.

Skip Distance

The *skip distance* (t_s) is the minimum distance from a transmit antenna that a sky wave of given frequency (which must be greater than the critical frequency) will be returned to earth. Figure 10-20a shows several waves with different angles of incidence

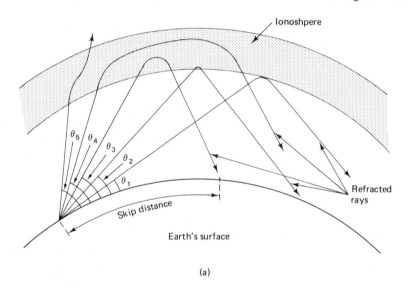

(a)

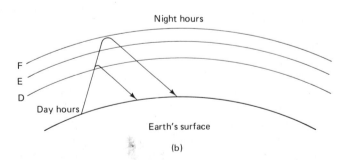

(b)

FIGURE 10-20   Skip distance: (a) skip distance; (b) daytime versus nighttime propagation.

being radiated from the same point on earth. It can be seen that the point where the wave is returned to earth moves closer to the transmitter as the angle of incidence (θ) is increased. Eventually, however, the angle of incidence is sufficiently high that the wave penetrates through the ionosphere and totally escapes the earth's atmosphere.

Figure 10-20b shows the effect on the skip distance of the disappearance of the D and E layers during nighttime. Effectively, the *ceiling* formed by the ionosphere is raised, allowing sky waves to travel higher before being refracted back to earth. This effect explains how far-away radio stations are sometimes heard during the night that cannot be heard during daylight hours. Figure 10-20c shows the effects of *multipath* and *multiple skip* sky-wave propagation.

QUESTIONS

10-1. Describe an electromagnetic ray; a wavefront.

10-2. Describe power density; voltage intensity.

10-3. Describe a spherical wavefront.

10-4. Explain the inverse square law.

10-5. Describe wave attenuation.

10-6. Describe wave absorption.

10-7. Describe refraction. Explain Snell's law for refraction.

10-8. Describe reflection. Explain Snell's law for reflection.

10-9. Describe diffraction. Explain Huygens' principle.

10-10. Describe the composition of a good reflector.

10-11. Describe the atmospheric conditions that cause electromagnetic refraction.

10-12. Define electromagnetic wave interference.

10-13. Describe ground-wave propagation. List its advantages and disadvantages.

10-14. Describe space-wave propagation.

10-15. Explain why the radio horizon is at a greater distance than the optical horizon.

10-16. Describe the various layers of the ionosphere.

10-17. Describe sky-wave propagation.

10-18. Explain why atmospheric conditions vary with time of day, month of year, and so on.

10-19. Define *critical frequency*; *critical angle*.

10-20. Describe virtual height.

10-21. Define *maximum usable frequency*.

10-22. Define *skip distance* and give the reasons that it varies.

PROBLEMS

10-1. Determine the power density for a radiated power of 1000 W at distance of 20 km from an isotropic antenna.

$$\wp = \frac{P_r}{4\pi R^2}$$

1000 w 30 km

10-2. Determine the power density for Problem 10-1 for a point that is 30 km from the antenna.

3^2

10-3. Describe the effects on power density if the distance from a transmit antenna is tripled.

10-4. Determine the radio horizon for a transmit antenna that is 100 ft high and a receiving antenna that is 50 ft high; 100 m and 50 m.

P. 420

10-5. Determine the maximum usable frequency for a critical frequency of 10 MHz and an angle of incidence of 45°. $MUF = \dfrac{Critical\ Frequency}{cos\ \theta}$

10-6. Determine the voltage intensity for the same point in Problem 10-1.

P. 402

10-7. Determine the voltage intensity for the same point in Problem 10-2. $\varepsilon = \dfrac{\sqrt{30 P_r}}{R}$

10-8. For a radiated power $P_r = 10$ kW, determine the voltage intensity at a distance 20 km from the source.

4^2

10-9. Determine the change in power density when the distance from the source increases by a factor of 4.

10-10. If the distance from the source is reduced to one-half its value, what affect does this have on the power density?

P. 403

10-11. The power density at a point from a source is 0.001 μW and the power density at another point is 0.00001 μW; determine the attenuation in decibels. $10\ Log\ \dfrac{P_1}{P_2}$

10-12. For a dielectric ratio $\sqrt{\epsilon_2/\epsilon_1} = 0.8$ and an angle of incidence $\theta_i = 26°$, determine the angle of refraction, θ_r.

$d = \sqrt{2h}$ *P. 415*

10-13. Determine the distance to the radio horizon for an antenna located 40 ft above sea level.

10-14. Determine the distance to the radio horizon for an antenna that is 40 ft above the top of a 4000-ft mountain peak.

P. 415

10-15. Determine the maximum distance between identical antennas equally distant above sea level for Problem 10-13.

$$d_t = \sqrt{2h_1} + \sqrt{2h_2}$$

OPEN CIRCUIT $\frac{1}{6}$ wavelength

(V lags I by 90°)

Capacitive

I_r I_t

30°

30° I_i

60°

00°

E_r E_T

$\frac{1}{6}$ wavelength (60°)

Short Circuit

I lags V 90° IND.

E_r E_T

30°

60° I_i

60°

I_r I_t

Chapter 11

ANTENNAS

INTRODUCTION

In essence, an *antenna* is a metallic conductor system capable of transmitting and receiving electromagnetic waves. An antenna is used to interface a transmitter to free space or free space to a receiver. An antenna couples energy from the output of a transmitter to the earth's atmosphere or from the earth's atmosphere to the input of a receiver. An antenna is a *passive reciprocal device*: *passive* in that it cannot actually amplify a signal, at least not in the true sense of the word (however, as you will see later in this chapter, an antenna can have *gain*); and *reciprocal* in that the transmit and receive characteristics of an antenna are identical, except where *feed* currents to the antenna element are tapered to modify the transmit pattern.

Basic Antenna Operation

Basic antenna operation is best understood by looking at the voltage standing-wave patterns on a transmission line, which are shown in Figure 11-1a. The transmission line is terminated in an open circuit, which represents an abrupt discontinuity to the incident voltage wave in the form of a phase reversal. The phase reversal causes some of the incident voltage to be radiated, not reflected back toward the source. The radiated energy propagates away from the antenna in the form of transverse electromagnetic waves. The *radiation efficiency* of an open transmission line is extremely low. Radiation efficiency is the ratio of radiated to reflected energy. To radiate more energy, simply spread the conductors farther apart. Such an antenna is called a *dipole* (meaning two poles) and is shown in Figure 11-1b.

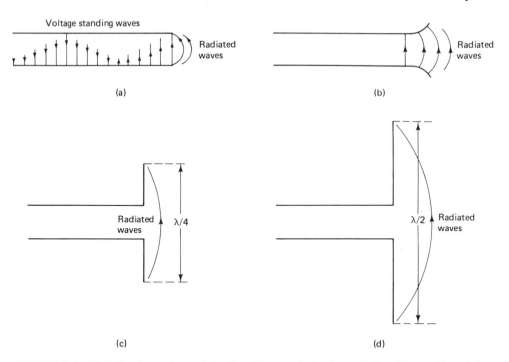

FIGURE 11-1   Radiation from a transmission line: (a) transmission-line radiation; (b) spreading conductors; (c) Marconi antenna; (d) Hertz antenna.

In Figure 11-1c the conductors are spread out in a straight line to a total length of one-quarter wavelength. Such an antenna is called a basic *quarter-wave dipole* or a *Marconi antenna*. A half-wave dipole is called a *Hertz antenna* and is shown in Figure 11-1d.

TERMS AND DEFINITIONS

Radiation Pattern

A *radiation pattern* is a *polar* diagram or graph representing field strengths or power densities at various points in space relative to an antenna. If the radiation pattern is plotted in terms of electric field strength ($\mathscr{E} = V/m$) or power density ($\mathscr{P} = W/m^2$), it is called an *absolute* radiation pattern. If it plots field strength or power density in respect to a reference point, it is called a *relative* radiation pattern. Figure 11-2a shows an absolute radiation pattern for an unspecified antenna. The pattern is plotted on *polar* coordinate paper with the heavy solid line representing points of equal power density ($10 \ \mu W/m^2$). The circular gradients indicate distance in 2-km steps. It can be seen that maximum radiation is in a direction 90° from the reference. The power density 10 km from the antenna in a 90° direction is $10 \ \mu W/m^2$. In a 45° direction, the point of equal

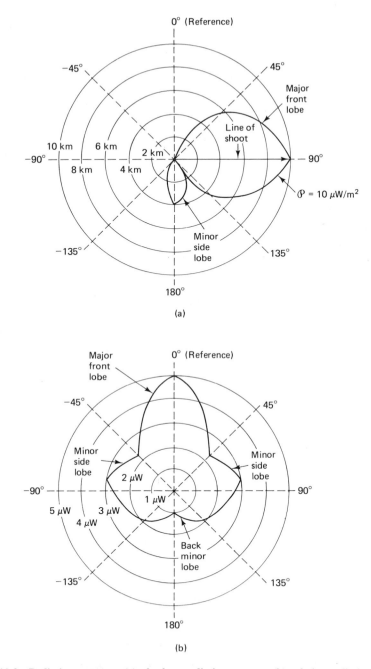

FIGURE 11-2 Radiation patterns: **(a)** absolute radiation pattern; **(b)** relative radiation pattern; **(c)** relative radiation pattern in decibels; **(d)** relative radiation pattern for an omnidirectional antenna.

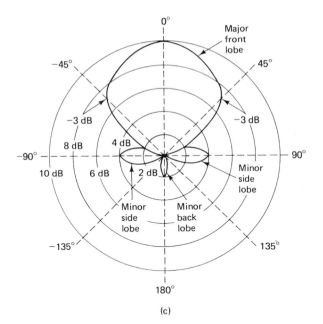

(c)

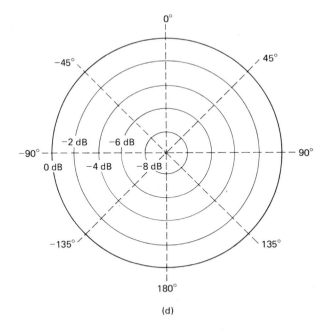

(d)

FIGURE 11-2 (*continued*)

power density is 5 km from the antenna; at 180°, only 4 km; and in a $-90°$ direction, there is essentially no radiation.

In Figure 11-2a the primary beam is in a 90° direction and is called the *major lobe*. There can be more than one major lobe. There is also a *secondary* beam or *minor* lobe in a $-180°$ direction. Normally, minor lobes represent undesired radiation or reception. Because the major lobe propagates and receives the most energy, that lobe is called the *front* lobe (i.e., the front of the antenna). Lobes adjacent to the front lobe are called *side* lobes (the 180° minor lobe is a side lobe), and lobes in a direction exactly opposite the front lobe are called *back* lobes (there is no back lobe shown on this pattern). The ratio of the front lobe to the back lobe is simply called the *front-to-back ratio*, and the ratio of the front lobe to a side lobe is called the *front-to-side ratio*. The line bisecting the major lobe or pointing from the center of the antenna in the direction of maximum radiation is called the *line of shoot*.

Figure 11-2b shows a relative radiation pattern for an unspecified antenna. The heavy solid line represents points of equal distance from the antenna (10 km), and the circular gradients indicate power density in 1 $\mu W/m^2$ divisions. It can be seen that maximum radiation (5 $\mu W/m^2$) is in the direction of the reference (0°) and the antenna radiates the least power (1 $\mu W/m^2$) in a direction 180° from the reference. Consequently, the front-to-back ratio is $5:1 = 5$. Generally, relative field strength and power density are plotted in decibels (dB), where dB $= 20 \log (\epsilon/\epsilon_{max})$ or $10 \log (\mathscr{P}/\mathscr{P}_{max})$. Figure 11-2c shows a relative radiation pattern for power density in decibels. In a direction $\pm 45°$ from the reference, the power density is -3 dB (half power) relative to the power density in the direction of maximum radiation (0°). Figure 11-2d shows a relative radiation pattern for power density for an omnidirectional antenna. As stated previously, an omnidirectional antenna radiates energy equally in all directions; therefore, the radiation pattern is simply a circle (actually, a sphere). Also, with an omnidirectional antenna, there are no front, back, or side lobes because radiation is equal in all directions.

The radiation patterns shown in Figure 11-2 are two-dimensional. However, radiation from an actual antenna is three-dimensional. Therefore, radiation patterns are taken in both the horizontal (from the top) and the vertical (from the side) planes. For the omnidirectional antenna shown in Figure 11-2d, the radiation patterns in the horizontal and vertical planes are circular and equal because the actual radiation pattern for an isotropic radiator is a sphere.

Near and Far Fields

The radiation field that is close to an antenna is not the same as the radiation field that is at a great distance. The term *near field* refers to the field pattern that is close to the antenna, and the term *far field* refers to the field pattern that is at great distance. During one half of a cycle, power is radiated from an antenna where some of the power is stored temporarily in the near field. During the second half of the cycle, power in the near field is returned to the antenna. This action is similar to the way in which an inductor stores and releases energy. Therefore, the near field is sometimes called the *induction field*. Power that reaches the far field continues to radiate outward and is never returned to the antenna. Therefore, the far field is sometimes called the *radiation field*. Radiated power is usually the more important of the two; therefore, antenna radiation patterns are generally given for the far field. The near field is defined as the area within a distance D^2/λ from the antenna where λ is the wavelength and D the antenna diameter in the same units.

Radiation Resistance and Antenna Efficiency

All of the power supplied to an antenna is not radiated. Some of it is converted to heat and dissipated. *Radiation resistance* is somewhat "unreal" in that it cannot be measured directly. Radiation resistance is an ac antenna resistance and is equal to the ratio of the power radiated by the antenna to the square of the current at its feed point. Mathematically, radiation resistance is

$$R_r = \frac{P}{i^2} \qquad (11\text{-}1)$$

where

R_r = radiation resistance
P = rms power radiated by the antenna
i = rms antenna current at the feedpoint

Radiation resistance is the resistance which, if it replaced the antenna, would dissipate exactly the same amount of power that the antenna radiates.

Antenna efficiency is the ratio of the power radiated by an antenna to the sum of the power radiated and the power dissipated or the ratio of the power radiated by the antenna to the total input power. Mathematically, antenna efficiency is

$$\eta = \frac{P_r}{P_r + P_d} \times 100 \qquad (11\text{-}2)$$

where

η = antenna efficiency (%)
P_r = power radiated by antenna
P_d = power dissipated in antenna

Figure 11-3 shows a simplified electrical equivalent circuit for an antenna. Some of the input power is dissipated in the dc resistances (ground resistance, corona, imperfect dielectrics, eddy currents, etc.) and the remainder is radiated. The total antenna power is the sum of the dissipated and radiated powers. Therefore, in terms of resistance and current, antenna efficiency is

$$\eta = \frac{i^2 R_r}{i^2(R_r + R_{dc})} = \frac{R_r}{R_r + R_{dc}} \qquad (11\text{-}3)$$

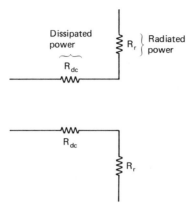

FIGURE 11-3 Simplified equivalent circuit of an antenna.

where

η = antenna efficiency
i = antenna current
R_r = radiation resistance
R_{dc} = dc antenna resistance

Directive Gain and Power Gain

The terms *directive gain* and *power gain* are often misunderstood and, consequently, misused. *Directive gain* is the ratio of the power density radiated in a particular direction to the power density radiated to the same point by a reference antenna, assuming both antennas are radiating the same amount of power. The relative power density radiation pattern for an antenna is actually a directive gain pattern if the power density reference is taken from a standard reference antenna, which is generally an isotropic antenna. The maximum directive gain is called *directivity*. Mathematically, directive gain is

$$\mathcal{D} = \frac{\mathcal{P}}{\mathcal{P}_{ref}} \tag{11-4}$$

where

$\mathcal{D}$ = directive gain
$\mathcal{P}$ = power density at some point with a given antenna
$\mathcal{P}_{ref}$ = power density at the same point with a reference antenna

Power gain is the same as directive gain except that the total power fed to the antenna is used (i.e., antenna efficiency is taken into account). It is assumed that the given antenna and the reference antenna have the same input power and that the reference antenna is lossless (i.e., $\eta = 100\%$). Mathematically, power gain (A_p) is

$$A_p = \mathcal{D}\eta \tag{11-5}$$

If an antenna is lossless, it radiates 100% of the input power and the power gain is equal to the directive gain. The power gain for an antenna is also given in decibels relative to some reference antenna. Therefore, power gain is

$$A_p = 10 \log \frac{\mathscr{P}\eta}{\mathscr{P}_{\text{ref}}} \qquad (11\text{-}6)$$

For an isotropic reference the directivity of a half-wave dipole is approximately 1.5 (1.76 dB). It is usual to state the power gain in decibels if referred to a λ/2 dipole. However, if referenced to an isotropic radiator, the decibel figure is stated as dBi, or dB/isotropic radiator and is 1.76 dB greater than if a half-wave dipole were used for the reference. It is important to note that the power radiated from an antenna can never exceed the input power. Therefore, the antenna does not actually amplify the input power. An antenna simply concentrates its radiated power in a particular direction. Therefore, points that are located in areas where the radiated power is concentrated realize an apparent gain relative to the power density at the same points had an isotropic antenna been used. If gain is realized in one direction, a corresponding reduction in power density (a loss) must be realized in another direction. The direction in which an antenna is "pointing" is always the direction of maximum radiation. Because an antenna is a reciprocal device, its radiation pattern is also its reception pattern. For maximum *captured* power, a receive antenna must be pointing in the direction from which reception is desired. Therefore, receive antennas have directivity and power gain just like transmit antennas.

Effective Isotropic Radiated Power

Effective isotropic radiated power (EIRP) is defined as an equivalent transmit power and is expressed mathematically as

$$\text{EIRP} = P_r A_t \qquad \text{watts} \qquad (11\text{-}7a)$$

where

P_r = total radiated power
A_t = transmit antenna directive gain

or

$$\text{EIRP (dBm)} = 10 \log \frac{P_r}{0.001} + 10 \log A_t \qquad (11\text{-}7b)$$

Equation 11-7a can be rewritten using antenna input power and power gain as

$$\text{EIRP} = P_{\text{in}} A_p \qquad (11\text{-}7c)$$

EIRP or simply ERP (effective radiated power) is the equivalent power that an isotropic antenna would have to radiate to achieve the same power density in the chosen direction at a given point as another antenna. For instance, if a given antenna has a power gain of 10, the power density is 10 times greater than it would have been had the antenna been an isotropic radiator. An isotropic antenna would have to radiate 10 times as much power in the desired direction to achieve the same power density. Therefore, the given antenna "effectively" radiates 10 times as much power as an isotropic antenna with the same input power and efficiency.

To determine the power density at a given point, Equation 10-5 is expanded to include the transmit antenna gain and rewritten as

$$\mathscr{P} = \frac{P_r A_t}{4\pi R^2} \qquad (11\text{-}8a)$$

EXAMPLE 11-1

If a transmit antenna has a directive gain $A_t = 10$ and radiated power $P_r = 100$ W, determine:
(a) The EIRP.
(b) The power density at a point 10 km away.
(c) The power density had an isotropic antenna been used with the same input power and efficiency.

Solution   (a) Substituting into Equation 11-7a yields

$$\text{EIRP} = P_r A_t = 100 \text{ W} \times 10 = 1000 \text{ W}$$

(b) Substituting into Equation 11-8a gives us

$$\mathscr{P} = \frac{P_r A_t}{4\pi R^2} = \frac{\text{EIRP}}{4\pi R^2} = \frac{1000 \text{ W}}{4\pi(10,000 \text{ m})^2} = 0.796 \ \mu\text{W/m}^2$$

(c) Substituting into Equation 10-5, we obtain

$$\mathscr{P} = \frac{P_r}{4\pi R^2} = \frac{100}{4\pi(10,000 \text{ m})^2} = 0.0796 \ \mu\text{W/m}^2$$

It can be seen that the power density at a point 10,000 m from the transmit antenna is 10 times greater with the first antenna than with the isotropic radiator. To achieve the same power density, the isotrope would have to radiate 1000 W. Therefore, the first antenna effectively radiates 1000 W.

As stated previously, antennas are reciprocal devices; an antenna has the same power gain and directivity when it is used to receive electromagnetic waves as it has for transmitting electromagnetic waves. Consequently, the power received or captured by an antenna is the product of the power density in the space immediately surrounding the antenna and the antenna directive gain. Therefore, Equation 11-8a can be expanded to

$$\text{captured power} = c = \frac{P_r A_t A_r}{4\pi R^2} \qquad (11\text{-}8b)$$

In Example 11-1, if an antenna that was identical to the transmit antenna were used to receive the signal, the captured power would be

$$c = \mathscr{P}A_r$$
$$= (0.0796 \ \mu\text{W/m}^2)(10) = 0.796 \ \mu\text{W}$$

The captured power is not all useful; some of it is dissipated in the receive antenna. The actual useful received power is the product of the received power density, the receive antenna's direct gain, and the receive antenna's efficiency or the receive power density times the receive antenna's power gain.

Antenna Polarization

The *polarization* of an antenna refers simply to the orientation of the electric field radiated from it. An antenna may be *linearly* (generally, either horizontally or vertically, assuming that the antenna elements lie in a horizontal or vertical plane), *elliptically*, or *circularly polarized*. If an antenna radiates a vertically polarized electromagnetic wave, the antenna is defined as vertically polarized; if an antenna radiates a horizontally polarized electromagnetic wave, the antenna is said to be horizontally polarized; if the radiated electric field rotates in an elliptical pattern, it is elliptically polarized; and if the electric field rotates in a circular pattern, it is circularly polarized.

Antenna Beamwidth

Antenna *beamwidth* is simply the angular separation between the two half-power (-3 dB) points on the major lobe of an antenna's plane radiation pattern, usually taken in one of the "principal" planes. The beamwidth for the antenna whose radiation pattern is shown in Figure 11-4 is the angle formed between points A, X, and B (angle θ). Points A and B are the half-power points (the power density at these points is one-half of what it is an equal distance from the antenna in the direction of maximum radiation). Antenna beamwidth is sometimes called -3-dB beamwidth or half-power beamwidth.

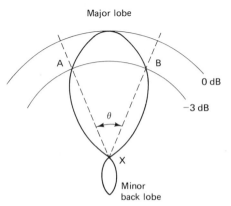

FIGURE 11-4   Antenna beamwidth.

Antenna Bandwidth

Antenna *bandwidth* is vaguely defined as the frequency range over which antenna operation is "satisfactory." This is normally taken between the half-power points, but it sometimes refers to variations in the antenna's input impedance.

Antenna Input Impedance

Radiation from an antenna is a direct result of the flow of RF current. The current flows to the antenna through a transmission line, which is connected to a small gap between the conductors that make up the antenna. The point on the antenna where the transmission line is connected is called the antenna input terminal or simply the *feedpoint*. The feedpoint presents an ac load to the transmission line called the *antenna input impedance*. If the transmitter's output impedance and the antenna's input impedance are equal to the characteristic impedance of the transmission line, there will be no standing waves on the line, and maximum power is transferred to the antenna and radiated.

Antenna input impedance is simply the ratio of the antenna's input voltage to input current. Mathematically, input impedance is

$$Z_{\text{in}} = \frac{E_i}{I_i} \qquad (11\text{-}9)$$

where

$$Z_{\text{in}} = \text{antenna input impedance}$$
$$E_i = \text{antenna input voltage}$$
$$I_i = \text{antenna input current}$$

Antenna input impedance is generally complex. However, if the feedpoint is at a current maximum and there is no reactive component, the input impedance is equal to the sum of the radiation resistance and the dc resistance.

BASIC ANTENNAS

Elementary Doublet

The simplest type of antenna is the *elementary doublet*. The elementary doublet is an electrically short dipole and is often referred to simply as a *short dipole*. "Electrically short" means short compared to one-half wavelength (generally, any dipole that is less than one-tenth wavelength long is considered electrically short). In reality, an elementary doublet cannot be achieved. However, the concept of a short dipole is useful in understanding more practical antennas.

An elementary doublet is a short dipole that has uniform current throughout its length. However, the current is assumed to vary sinusoidally in time and at any instant is

$$i = I \sin (2\pi ft + \theta)$$

where

i = instantaneous current
I = peak amplitude of the RF current
f = frequency (Hz)
t = instantaneous time
θ = phase angle

With the aid of Maxwell's equations, it can be shown that the far (radiation) field is

$$\mathscr{E} = \frac{60\pi Il \sin \theta}{\lambda R} \qquad (11\text{-}10)$$

where

$\mathscr{E}$ = electric field intensity (V rms/m)
I = dipole current (A rms)
l = length of the dipole (m)
R = distance from the dipole (m)
λ = wavelength (m)
θ = phase angle

Plotting Equation 11-10 gives the relative electric field intensity pattern for an elementary dipole, and is shown in Figure 11-5. It can be seen that radiation is maximum at right angles to the dipole and falls off to zero at the ends.

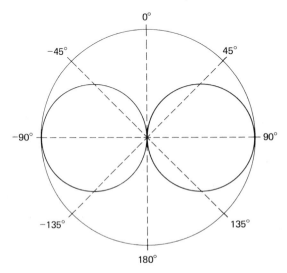

FIGURE 11-5 Relative radiation pattern for an elementary doublet in a plane perpendicular to the dipole axis.

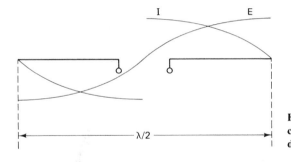

FIGURE 11-6 Idealized voltage and current distributions along a half-wave dipole.

The relative power density pattern can be derived from Equation 11-10 by substituting $\mathscr{P} = \mathscr{E}^2/120\pi$. Mathematically, we have

$$\mathscr{P} = \frac{30\pi I^2 l^2 \sin^2 \theta}{\lambda^2 R^2} \qquad (11\text{-}11)$$

Half-wave Dipole

The linear half-wave dipole is one of the most widely used antennas at frequencies above 2 MHz. At frequencies below 2 MHz, the physical length of a half-wavelength antenna is prohibitive. The half-wave dipole is generally referred to as a *Hertz antenna*.

A Hertz antenna is a *resonant* antenna. That is, it is a multiple of quarter-wavelengths long and open circuited at the far end. Standing waves of voltage and current exist along a resonant antenna. Figure 11-6 shows the idealized voltage and current distributions along a half-wave dipole for each half-cycle of the input signal. Each pole of the antenna looks like an open quarter-wavelength section of transmission line. Therefore, there is a voltage maximum and current minimum at the ends and a voltage minimum and current maximum in the middle. Consequently, assuming that the feedpoint is in the center of the antenna, the input impedance is E_{min}/I_{max} and a minimum value. The impedance at the ends of the antenna is E_{max}/I_{min} and a maximum value. Figure 11-7 shows the impedance curve for a center-fed half-wave dipole. The impedance varies from a maximum value at the ends of approximately 2500 Ω to a minimum value at the feedpoint of approximately 73 Ω (of which between 68 and 70 Ω is the radiation resistance).

A wire radiator such as a half-wave dipole can be thought of as an infinite number

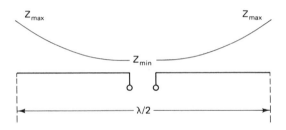

FIGURE 11-7 Impedance curve for a center fed half-wave dipole.

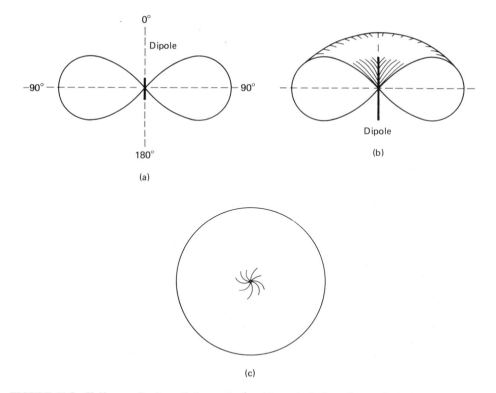

FIGURE 11-8 **Half-wave dipole radiation patterns: (a) vertical view of a vertically mounted dipole; (b) cross-sectional view; (c) horizontal view.**

of elementary doublets placed end to end. Therefore, the radiation pattern can be obtained by integrating Equation 11-10 over the length of the antenna. The free-space radiation pattern for a half-wave dipole depends on whether the antenna is placed horizontally or vertically in respect to the earth's surface. Figure 11-8a shows the vertical (from the side) radiation pattern for a vertically mounted half-wave dipole. Note that there are two major lobes radiating in opposite directions which are at right angles to the antenna. Also note that the lobes are not circles. Circular lobes are obtained only for the ideal case when the current is constant throughout the antenna's length and this is, of course, unachievable in a practical antenna. Figure 11-8b shows the cross-sectional view. Note that the radiation pattern has a figure-eight pattern and resembles the shape of a doughnut. Maximum radiation is in a plane parallel to the earth's surface. The higher the angle of elevation (θ), the less the radiation, and for $\theta = 90°$ there is no radiation. Figure 11-8c shows the horizontal (from the top) radiation pattern for a vertically mounted half-wave dipole. The pattern is circular because radiation is uniform in all directions perpendicular to the antenna.

Ground effects on a half-wave dipole.   The radiation patterns shown in Figure 11-8 are for free-space conditions. In the earth's atmosphere, wave propagation is affected

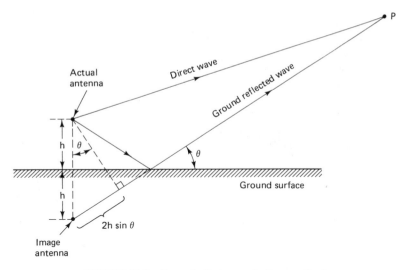

FIGURE 11-9   Ground effects on a half-wave dipole.

by antenna orientation, atmospheric absorption, and ground effects such as reflection. The effect of ground reflection on the radiation pattern for an ungrounded half-wave dipole is shown in Figure 11-9. The antenna is mounted an appreciable number of wavelengths (height h) above the surface of the earth. The field strength at any given point in space is the sum of the direct and ground reflected waves. The ground reflected wave appears to be radiating from an image antenna distance h below the earth's surface. This apparent antenna is a mirror image of the actual antenna. The ground reflected wave is inverted 180° and travels a distance $2h \sin \theta$ farther than the direct wave to reach the same point in space (point P). The resulting radiation pattern is a summation of the radiations from the actual antenna and the mirror antenna. Note that this is the classical ray-tracing technique used in optics.

Figure 11-10 shows the vertical radiation patterns for a horizontally mounted half-wave dipole one-quarter and one-half wavelength above the ground. For an antenna mounted one-quarter wavelength above the ground, the lower lobe is completely gone and the field strength directly upward is doubled. Figure 11-10a shows the vertical distribution in a plane parallel (and through) the antenna, and Figure 11-10b shows the vertical distribution in a plane at right angles to the antenna. Figure 11-10c shows the vertical radiation pattern for a horizontal dipole one-half wavelength above the ground. The figure shows that the pattern is now broken into two lobes and the direction of maximum radiation (end view) is now at 30° to the horizontal instead of directly upward. There is no component along the ground for horizontal polarization because of the phase shift of the reflected component. Ground-reflected waves have similar effects on all antennas. The best way to eliminate or reduce the effect of ground reflected waves is to mount the antenna far enough above the earth's surface to obtain free-space conditions. However, in many applications, this is impossible. Ground reflections are sometimes desirable to get the desired elevation angle for the major lobe maximum response.

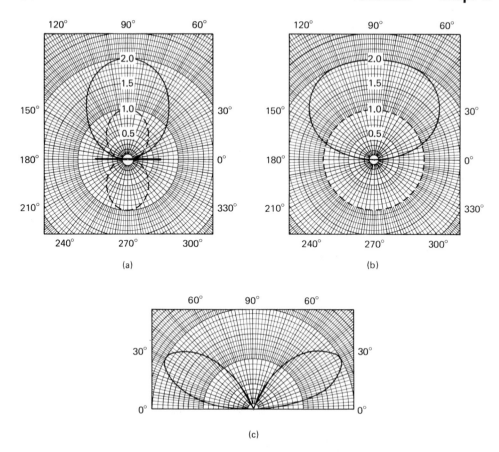

FIGURE 11-10   Vertical radiation pattern for a half-wave dipole.

The height of an ungrounded antenna above the earth's surface also affects the antenna's radiation resistance. This is due to the reflected waves cutting through or intercepting the antenna and altering its current. Depending on the phase of the ground reflected wave, the antenna current can increase or decrease, causing a corresponding increase or decrease in the input impedance.

Grounded Antenna

A *monopole* (single pole) antenna one-quarter wavelength long mounted vertically with the lower end either connected directly to ground or grounded through the antenna coupling network is called a *Marconi antenna*. The characteristics of a Marconi antenna are similar to those of the Hertz antenna because of the ground reflected waves. Figure 11-11 shows the voltage and current standing waves for a quarter-wave grounded antenna. It can be seen that if the Marconi antenna is mounted directly on the earth's surface, the actual antenna and its *image* combine and produce exactly the same standing-wave

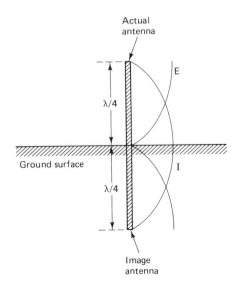

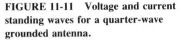

FIGURE 11-11 Voltage and current standing waves for a quarter-wave grounded antenna.

patterns as those of the half-wave ungrounded (Hertz) antenna. Current maxima occur at the grounded ends, which causes high current flow through ground. To reduce power losses, the ground should be a good conductor, such as rich, loamy soil. If the ground is a poor conductor, such as sandy or rocky terrain, an artificial *ground plane* system made of heavy copper wires spread out radially below the antenna may be required. Another way of artificially improving the conductivity of the ground area below the antenna is with a *counterpoise*. A counterpoise is a wire structure placed below the antenna and erected above the ground. The counterpoise should be insulated from earth ground. A counterpoise is a form of capacitive ground system; capacitance is formed between the counterpoise and the earth's surface.

Figure 11-12 shows the radiation pattern for a quarter-wave grounded (Marconi) antenna. It can be seen that the lower half of each lobe is canceled by the ground reflected waves. This is generally of no consequence because radiation in the horizontal direction is increased, thus increasing radiation along the earth's surface (ground waves)

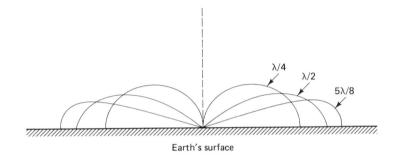

FIGURE 11-12 Grounded antenna radiation patterns.

and improving area coverage. It can also be seen that increasing the antenna length improves horizontal radiation at the expense of sky-wave propagation. This is also shown in Figure 11-12. Optimum horizontal radiation occurs for an antenna that is approximately five-eighths wavelength long. For a one-wavelength antenna, there is no ground-wave propagation.

A Marconi antenna has the obvious advantage over a Hertz antenna of being only half as long. The disadvantage of a Marconi antenna is that it must be located close to the ground.

Antenna Loading

Thus far we have considered antenna length in terms of wavelengths rather than physical dimensions. By the way, how long is a quarter-wavelength antenna? For a transmit frequency of 1 GHz, one-quarter wavelength is 0.075 m (2.95 in.). However, for a transmit frequency of 1 MHz, one-quarter wavelength is 75 m, and at 100 kHz, one-quarter wavelength is 750 m. It is obvious that the physical dimensions for low-frequency antennas are not practical, especially for mobile radio applications. However, it is possible to increase the electrical length of an antenna by a technique called *loading*. When an antenna is loaded, its physical length remains unchanged although its effective electrical length is increased. There are several techniques used for loading antennas.

Loading coils.   Figure 11-13a shows how a coil (inductor) added in series with a dipole antenna effectively increases the antenna's electrical length. Such a coil is appropriately called a *loading coil*. The loading coil effectively cancels out the capacitance component of the antenna input impedance. Thus the antenna looks like a resonant circuit, is resistive, and can now absorb 100% of the incident power. Figure 11-13b shows the current standing-wave patterns on an antenna with a loading coil. The loading coil is generally placed at the bottom of the antenna, allowing the antenna to be easily

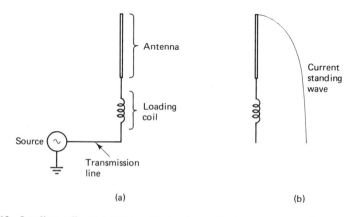

(a)   (b)

FIGURE 11-13   Loading coil: (a) antenna with loading coil; (b) current standing wave with loading coil.

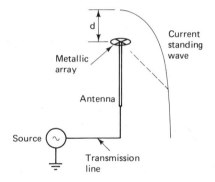

FIGURE 11-14 Antenna top loading.

tuned to resonance. A loading coil effectively increases the radiation resistance of the antenna by approximately 5 Ω. Note also that the current standing wave has a maximum value at the coil, increasing power losses, creating a situation of possible corona, and effectively reducing the radiation efficiency of the antenna.

Top loading.   Loading coils have several shortcomings that can be avoided by using a technique called antenna *top loading*. With top loading, a metallic array that resembles a spoked wheel is placed on top of the antenna. The wheel increases the shunt capacitance to ground, reducing the overall antenna capacitance. Antenna top loading is shown in Figure 11-14. Notice that the current standing-wave pattern is pulled up along the antenna as though the antenna length had been increased distance *d*, placing the current maximum at the base. Top loading results in a considerable increase in the radiation resistance and radiation efficiency. It also reduces the voltage of the standing wave at the antenna base. Unfortunately, top loading is awkward for mobile applications.

The current loop of the standing wave can be raised even further (improving the radiation efficiency even more) if a *flat top* is added to the antenna. If a vertical antenna is folded over on top to form an L or T, as shown in Figure 11-15, the current loop will occur nearer the top of the radiator. If the flat-top and vertical portions are each one-quarter wavelength long, the current maximum will occur at the top of the vertical radiator.

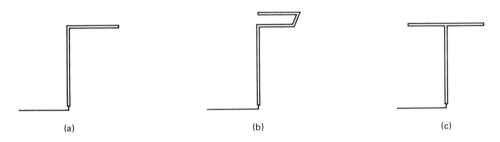

(a)   (b)   (c)

FIGURE 11-15   Flat top antenna loading.

ANTENNA ARRAYS

An antenna *array* is formed when two or more antenna elements are combined to form a single antenna. An antenna element is an individual radiator such as a half- or quarter-wave dipole. The elements are physically placed in such a way that their radiation fields interact with each other, producing a total radiation pattern that is the vector sum of the individual fields. The purpose of an array is to increase the directivity of an antenna and concentrate the radiated power within a smaller geographic area.

In essence, there are two types of antenna elements: *driven* and *parasitic* (non-driven). Driven elements are elements that are directly connected to the transmission line and receive power from or are driven by the source. Parasitic elements are not connected to the transmission line; they receive energy only through mutual induction with a driven element. A parasitic element that is longer than the driven element from which it receives energy is called a *reflector*. A reflector effectively reduces the signal strength in its direction and increases it in the opposite direction. Therefore, it acts like a concave mirror. This action occurs because the wave passing through the parasitic element induces a voltage that is reversed 180° in respect to the wave that induced it. The induced voltage produces an in-phase current and the element radiates (it actually reradiates the energy it just received). The reradiated energy sets up a field that cancels in one direction and reinforces in the other. A parasitic element that is shorter than its associated driven element is called a *director*. A director increases field strength in its direction and reduces it in the opposite direction. Therefore, it acts like a convergent convex lens. This is shown in Figure 11-16.

Radiation directivity can be increased in either the horizontal or the vertical plane, depending on the placement of the elements and whether they are driven or not, and if not, whether they are reflectors or directors.

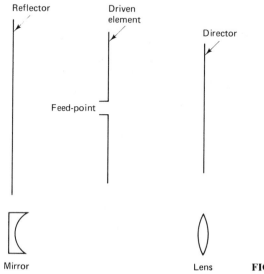

FIGURE 11-16   Antenna array.

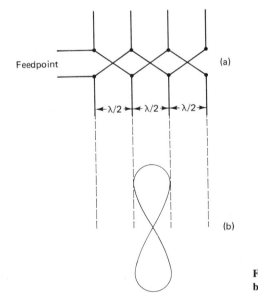

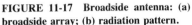

FIGURE 11-17 Broadside antenna: (a) broadside array; (b) radiation pattern.

Broadside Array

A *broadside array* is one of the simplest types of antenna arrays. It is made by simply placing several resonant dipoles of equal size (both length and diameter) in parallel with each other and in a straight line (i.e., collinear). All elements are fed in phase from the same source. As the name implies, a broadside array radiates at right angles to the plane of the array, and radiates very little in the direction of the plane. Figure 11-17a shows a broadside array that is comprised of four driven half-wave elements separated by one-half wavelength. Therefore, the signal that is radiated from element 2 has traveled one-half wavelength farther than the signal radiated from element 1 (i.e., they are radiated 180° out of phase). Crisscrossing the transmission line produces an additional 180° phase shift. Therefore, the currents in all of the elements are in phase and the radiated signals are in phase and additive in a plane at right angles to the plane of the array. Although the horizontal radiation pattern for each element by itself is omnidirectional, when combined their fields produce a highly directive bidirectional radiation pattern (Figure 11-17b). Directivity can be increased even further by increasing the length of the array (i.e., adding more elements).

End-Fire Array

An *end-fire array* is essentially the same element configuration as the broadside array except that the transmission line is not crisscrossed between elements. As a result, the fields are additive in line with the plane of the array. Figure 11-18 shows an end-fire array and its resulting radiation pattern.

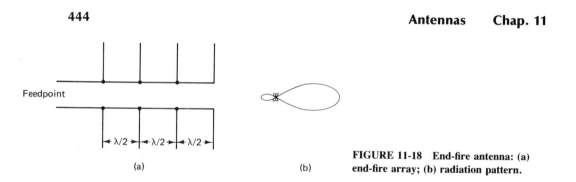

(a)   (b)

FIGURE 11-18   End-fire antenna: (a)
end-fire array; (b) radiation pattern.

Nonresonant Array—The Rhombic Antenna

The *Rhombic antenna* is a nonresonant antenna that is capable of operating satisfactorily over a relatively wide bandwidth, making it ideally suited for HF transmission (range 3 to 30 MHz). The Rhombic antenna is made up of four nonresonant elements each several wavelengths long. The entire array is terminated in a resistor if unidirectional operation is desired. The most widely used arrangement for the Rhombic antenna resembles a transmission line that has been pinched out in the middle and is shown in Figure 11-19. The antenna is mounted horizontally and placed one-half wavelength or more above the ground. The exact height depends on the precise radiation pattern desired. Each set of elements acts like a transmission line terminated in its characteristic impedance; thus waves are radiated only in the forward direction. The terminating resistor absorbs approximately one-third of the total antenna input power. Therefore, a Rhombic antenna has a maximum efficiency of 67%. Gains of over 40 (16 dB) have been achieved with Rhombic antennas.

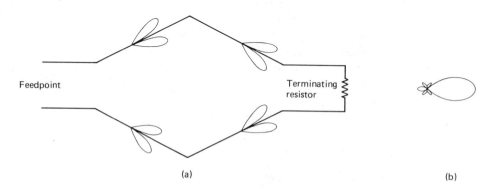

(a)   (b)

FIGURE 11-19   Rhombic antenna: (a) Rhombic array; (b) radiation pattern.

SPECIAL-PURPOSE ANTENNAS

Folded Dipole

A two-wire *folded dipole* and its associated voltage standing-wave pattern are shown in Figure 11-20a. The folded dipole is essentially a single antenna made up of two elements. One element is fed directly, while the other is conductively coupled at the ends. Each element is one-half wavelength long. However, because current can flow around corners, there is a full wavelength of current on the antenna. Therefore, for the same input power, the input current will be one-half that of the basic half-wave dipole and the input impedance is four times higher ($4 \times 72 = 288$). The input impedance of a folded dipole is equal to the half-wave impedance (72 Ω) times the number of folded wires squared. For example, if there are three dipoles, as shown in Figure 11-20b, the input impedance is $3^3 \times 72 = 648$ Ω. Another advantage of a folded dipole over a basic half-wave dipole is wider bandwidth. The bandwidth can be increased even further by making the dipole elements larger in diameter (such an antenna is appropriately called a *fat dipole*). However, fat dipoles have slightly different current distributions and input impedance characteristics than thin ones.

Yagi–Uda antenna. A widely used antenna that commonly uses a folded dipole as the driven element is the *Yagi–Uda antenna*, named after two Japanese scientists who invented and described its operation. (The Yagi–Uda is generally called simply Yagi.) A Yagi antenna is a linear array consisting of a dipole and two or more parasitic elements: one reflector and one or more directors. A simple three-element Yagi is shown in Figure 11-21a. The driven element is a half-wavelength folded dipole. (This element is referred to as the driven element because it is connected to the transmission line.

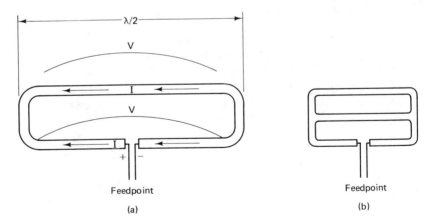

FIGURE 11-20 (a) Folded dipole; (b) three-element folded dipole.

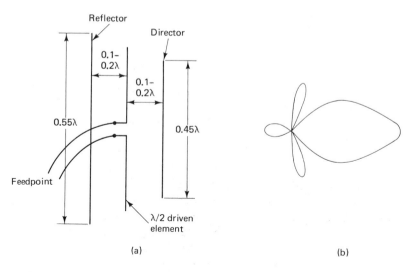

FIGURE 11-21   Yagi–Uda antenna: (a) three-element Yagi; (b) radiation pattern.

However, it is generally used for receiving only.) The reflector is a straight aluminum rod approximately 5% longer than the dipole, and the director is cut approximately 5% shorter than the driven element. The spacing between elements is generally between 0.1 and 0.2 wavelength. Figure 11-12b shows the radiation pattern for a Yagi antenna. The typical directivity for a Yagi is between 7 and 9 dB. The bandwidth of the Yagi can be increased by using more than one folded dipole, each cut to a slightly different length. Therefore, the Yagi antenna is commonly used for VHF television reception because of its wide bandwidth (the VHF TV band extends from 54 to 216 MHz).

Log-Periodic Antenna

A class of frequency-independent antennas called *log-periodics* evolved from the initial work of V. H. Rumsey, J. D. Dyson, R. H. DuHamel, and D. E. Isbell at the University of Illinois in 1957. The primary advantages of log-periodic antennas is the independence of their radiation resistance and radiation pattern to frequency. Log-periodic antennas have bandwidth ratios of 10:1 or greater. The bandwidth ratio is the ratio of the highest to the lowest frequency over which an antenna will satisfactorily operate. The bandwidth ratio is often used rather than simply stating the percentage of the bandwidth to the center frequency. Log-periodics are not simply a type of antenna but rather a class of antenna, because there are many different types, some that are quite unusual. Log-periodic antennas can be unidirectional or bidirectional and have a low-to-moderate directive gain. High gains may also be achieved by using them as an element in a more complicated array.

The physical structure of a log-periodic antenna is repetitive, which results in repetitive behavior in its electrical characteristics. In other words, the design of a log-periodic antenna consists of a basic geometric pattern that repeats, except with a different

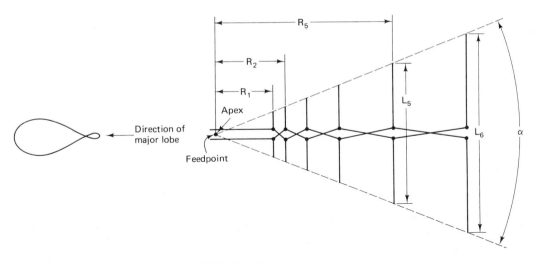

FIGURE 11-22   Log-periodic antenna.

size pattern. A basic log-periodic dipole array is probably the closest that a log-periodic comes to a conventional antenna and is shown in Figure 11-22. It consists of several dipoles of different length and spacing that are fed from a single source at the small end. The transmission line is crisscrossed between the feed points of adjacent pairs of dipoles. The radiation pattern for a basic log-periodic antenna has maximum radiation outward from the small end. The lengths of the dipoles and their spacing are related in such a way that adjacent elements have a constant ratio to each other. Dipole lengths and spacings are related by the formula

$$\frac{R_2}{R_1} = \frac{R_3}{R_2} = \frac{R_4}{R_3} = \frac{1}{\tau} = \frac{L_2}{L_1} = \frac{L_3}{L_2} = \frac{L_4}{L_3} \tag{11-12}$$

or

$$\tau = \frac{R_{n+1}}{R_n} = \frac{L_{n+1}}{L_n}$$

where

R = dipole spacing
L = dipole length
τ = design ratio (number < 1)

The ends of the dipoles lie along a straight line, and the angle where they meet is designated α. For a typical design, $\tau = 0.7$ and $\alpha = 30°$. With the preceding structural stipulations, the antenna input impedance varies repetitively when plotted as a function of frequency, and when plotted against the log of the frequency, varies periodically (hence the name "log-periodic"). A typical plot of the input impedance is shown in Figure 11-23. Although the input impedance varies periodically, the variations are not

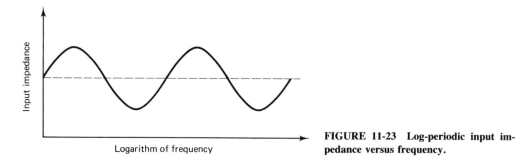

FIGURE 11-23 Log-periodic input impedance versus frequency.

necessarily sinusoidal. Also, the radiation pattern, directivity, power gain, and beamwidth undergo a similar variation with frequency.

The magnitude of a log-frequency period is dependent on the design ratio and, if two successive maxima occur at frequencies F_1 and F_2, they are related by the formula

$$\log F_2 - \log F_1 = \log \frac{F_2}{F_1} = \log \frac{1}{\tau} \qquad (11\text{-}13)$$

Therefore, the measured properties of a log-periodic antenna at frequency F will have identical properties at frequency τF, $\tau^2 F$, $\tau^3 F$, and so on. Log-periodic antennas, like Rhombic antennas, are used mainly for HF and VHF communications. However, log-periodic antennas do not have a terminating resistor and are therefore more efficient. Very often TV antennas advertised as so-called "high-gain" or "high-performance" antennas are log-periodic antennas.

Loop Antenna

The most fundamental *loop antenna* is simply a single-turn coil of wire that is sufficiently shorter than one wavelength and carries RF current. Such a loop is shown in Figure 11-24. If the radius (r) is small compared to a wavelength, current is essentially in phase throughout the loop. A loop can be thought of as many elemental dipoles connected together. Dipoles are straight; therefore, the loop is actually a polygon rather than circular.

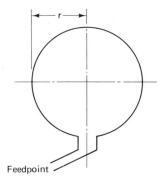

FIGURE 11-24 Loop antenna.

However, a circle can be approximated if the dipoles are assumed to be sufficiently short. The loop is surrounded by a magnetic field which is at right angles to the wire and the directional pattern is independent of its exact shape. Generally, loops are circular; however, any shape will work. The radiation pattern for a loop antenna is essentially the same as that of a short horizontal dipole.

The radiation resistance for a small loop is

$$R_r = \frac{31,200A^2}{\lambda 4}$$

(11-14)

where A is the area of the loop. For very low-frequency applications, loops are often made with more than one turn of wire. The radiation resistance of a multiturn loop is simply the radiation resistance for a single turn loop times the number of turns squared. The polarization of a loop antenna, like an elementel dipole, is linear. However, a vertical loop is vertically polarized and a horizontal loop is horizontally polarized.

Small vertically polarized loops are very often used as direction-finding antennas. The direction of the received signal can be found by orienting the loop until a null or zero value is found. This is the direction of the received signal. Loops have an advantage over most other types of antennas in direction finding in that loops are generally much smaller and therefore more easily adapted to mobile communications applications.

QUESTIONS

11-1. Define *antenna*.

11-2. Describe basic antenna operation using standing waves.

11-3. Describe a relative radiation pattern; an absolute radiation pattern.

11-4. Define *front-to-back ratio*.

11-5. Describe an omnidirectional antenna.

11-6. Define *near field*; *far field*.

11-7. Define *radiation resistance*; *antenna efficiency*.

11-8. Define and contrast *directive gain* and *power gain*.

11-9. What is the directivity for an isotropic antenna?

11-10. Define *effective isotropic radiated power*.

11-11. Define *antenna polarization*.

11-12. Define *antenna beamwidth*.

11-13. Define *antenna bandwidth*.

11-14. Define *antenna input impedance*. What factors contribute to an antenna's input impedance?

11-15. Describe the operation of an elementary doublet.

11-16. Describe the operation of a half-wave dipole.

11-17. Describe the effects of ground on a half-wave dipole.

11-18. Describe the operation of a grounded antenna.

11-19. What is meant by *antenna loading*?

11-20. Describe an antenna loading coil.

11-21. Describe antenna top loading.

11-22. Describe an antenna array.

11-23. What is meant by *driven element*; *parasitic element*?

11-24. Describe the radiation pattern for a broadside array; an end-fire array.

11-25. Define *nonresonant antenna*.

11-26. Describe the operation of the Rhombic antenna.

11-27. Describe a folded dipole antenna.

11-28. Describe a Yagi–Uda antenna.

11-29. Describe a log-periodic antenna.

11-30. Describe the operation of a loop antenna.

PROBLEMS

11-1. For an antenna with input power $P_i = 100$ W, rms current $i = 2$ A, and dc resistance $R_{dc} = 2$ Ω; determine:
   (a) The antenna's radiation resistance.
   (b) The antenna's efficiency.
   (c) The power radiated from the antenna, P_r.

11-2. Determine the directivity in decibels for an antenna that produces power density $\mathcal{P} = 2$ μW_m^2 at a point when a reference antenna produces 0.5 μW_m^2 at the same point.

11-3. Determine the power gain in decibels for an antenna with directive gain $\mathcal{D} = 40$ and efficiency $\eta = 65\%$.

11-4. Determine the effective isotropic radiated power for an antenna with power gain $A_p = 43$ dB and radiated power $P_r = 200$ W.

11-5. Determine the effective isotropic radiated power for an antenna with directivity $\mathcal{D} = 33$ dB, efficiency $\eta = 82\%$, and input power $P_i = 100$ W.

11-6. Determine the power density at a point 20 km from an antenna that is radiating 1000 W and has power gain $A_p = 23$ dB.

11-7. Determine the power density at a point 30 km from an antenna that has input power $P_{in} = 40$ W, efficiency $\eta = 75\%$, and directivity $\mathcal{D} = 16$ dB.

11-8. Determine the power captured by a receiving antenna for the following parameters:

   Power radiated $P_r = 50$ W
   Transmit antenna directive gain $A_t = 30$ dB
   Distance between transmit and receive antennas $d = 20$ km
   Receive antenna directive gain $A_r = 26$ dB

11-9. Determine the directivity (in decibels) for an antenna that produces a power density at a point that is 40 times greater than the power density at the same point when the reference antenna is used.

11-10. Determine the effective radiated power for an antenna with directivity $\mathcal{D} = 400$, efficiency $\eta = 0.60$, and input power $P_{in} = 50$ W.

11-11. Determine the efficiency for an antenna with radiation resistance $R_r = 18.8 \, \Omega$, dc resistance $R_{dc} = 0.4 \, \Omega$, and directive gain $\mathcal{D} = 200$.

11-12. Determine the power gain A_p for Problem 11-11.

11-13. Determine the efficiency for an antenna with radiated power $P_r = 44$ W, dissipated power $P_d = 0.8$ W, and directive gain $\mathcal{D} = 400$.

11-14. Determine power gain A_p for Problem 11-13.

Chapter 12

BASIC TELEVISION
PRINCIPLES

INTRODUCTION

The word *television* comes from the Greek word *tele* (meaning distant) and the Latin word *vision* (meaning sight). Therefore, television simply means to see from a distance. In its simplest form, television is the process of converting *images* (either stationary or in motion) to electrical signals, then transmitting those signals to a distant receiver, where they are converted back to images that can be perceived with the human eye. Thus television is a system in which images are transmitted from a central location, then received at distant receivers, where they are reproduced in their original form.

HISTORY OF TELEVISION

The idea of transmitting images or pictures was first experimented with in the 1880s when Paul Nipkow, a German scientist, conducted experiments using revolving *disks* placed between a powerful light source and the subject. A spiral row of holes was punched in the disk, which permitted light to scan the subject from top to bottom. After one complete revolution of the disk, the entire subject had been scanned. Light reflected from the subject was directed to a light-sensitive cell, producing current that was proportional in intensity to the reflected light. The fluctuating current operated a neon lamp, which gave off light in exact proportion to that reflected from the subject. A second disk exactly like the one in the transmitter was used in the receiver and the two disks revolved in exact synchronization. The second disk was placed between the neon lamp and the eye of the observer, who thus saw a reproduction of the subject.

The images reproduced with Nipkow's contraption were barely recognizable, although his scanning and synchronization principles are still used today.

In 1925, C. Francis Jenkins in the United States, and John L. Baird in England, using scanning disks connected to vacuum-tube amplifiers and photoelectric cells were able to reproduce images that were recognizable, although still of poor quality. Scientists worked for several years trying to develop effective mechanical scanning disks that with improved mirrors and lens and a more intense light source would improve the quality of the reproduced image. However, in 1933, Radio Corporation of America (RCA) announced a television system, developed by Vladimir K. Zworykin, that used

TABLE 12-1   FCC CHANNEL AND FREQUENCY ASSIGNMENTS

Channel number	Frequency band (MHz)	Channel number	Frequency band (MHz)	Channel number	Frequency band (MHz)
1[a]	44–50	29	560–566	57	728–734
2	54–60	30	566–572	58	734–740
3	60–66	31	572–578	59	740–746
4	66–72	32	578–584	60	746–752
5	76–82	33	584–590	61	752–758
6	82-88	34	590–596	62	758–764
7	174–180	35	596–602	63	764–770
8	180–186	36	602–608	64	770–776
9	186–192	37	608–614	65	776–782
10	192–198	38	614–620	66	782–788
11	198–204	39	620–626	67	788–794
12	204–210	40	626–632	68	794–800
13	210–216	41	632–638	69	800–806
14	470–476	42	638–644	70	806–812
15	476–482	43	644–650	71	812–818
16	482–488	44	650–656	72	818–824
17	488–494	45	656–662	73[a]	824–830
18	494–500	46	662–668	74[a]	830–836
19	500–506	47	668–674	75[a]	836–842
20	506–512	48	674–680	76[a]	842–848
21	512–518	49	680–686	77[a]	848–854
22	518–524	50	686–692	78[a]	854–860
23	524–530	51	692–698	79[a]	860–866
24	530–536	52	698–704	80[a]	866–872
25	536–542	53	704–710	81[a]	872–878
26	542–548	54	710–716	82[a]	878–884
27	548–554	55	716–722	83[a]	884–890
28	554–560	56	722–728		

[a] No longer assigned to television broadcasting.

an electronic scanning technique. Zworykin's system required no mechanical moving parts and is essentially the system used today.

In 1941, commercial broadcasting of *monochrome* (black and white) television signals began in the United States. In 1945, the FCC assigned 13 VHF television channels: 6 low-band channels, 1 to 6 (44 to 88 MHz), and 7 high-band channels, 7 to 13 (174 to 216 MHz). However, in 1948 it was found that channel 1 (44 to 50 MHz) caused interference problems; consequently, this channel was reassigned to mobile radio services. In 1952, UHF channels 14 to 83 (470 to 890 MHz) were assigned by the FCC to provide even more television stations. In 1974, the FCC reassigned to cellular telephone frequency bands at 825 to 845 MHz and 870 to 890 MHz, thus eliminating UHF channels 73 to 83 (however, existing licenses are renewable). Table 12-1 shows a complete list of the FCC channel and frequency assignments used in the United States. In 1947, R. B. Dome of General Electric Corporation proposed the method of *intercarrier* sound transmission for television broadcasting that is used today. In 1949, experiments began with color transmission, and in 1953, the FCC adopted the *National Television Systems Committee* (NTSC) system for color television broadcasting, which is also still used today.

MONOCHROME TELEVISION TRANSMISSION

Television Transmitter

Monochrome television broadcasting involves the transmission of two separate signals: an *aural* (sound) signal and a *video* (picture) signal. Every television transmitter broadcasts two totally separate signals for the picture and sound information. Aural transmission uses frequency modulation and video transmission uses amplitude modulation. Figure 12-1 shows a simplified block diagram for a monochrome television transmitter. It shows two totally separate transmitters (an FM transmitter for the sound information and an AM transmitter for the picture information) whose outputs are combined in a *diplexer bridge* and fed to a single antenna. A diplexer bridge is a network that is used to combine the outputs from two transmitters operating at different frequencies that use the same antenna system. The video information is limited to frequencies below 4 MHz and can originate from either a *camera* (for live transmissions), a video tape or cassette recorder, or a video disk recorder. The video switcher is used to select the desired video information source for broadcasting. The audio information is limited to frequencies below 15 kHz and can originate from either a microphone (again, only for live transmissions), from sound tracks on tape or disk recorders, or from a separate audio cassette or disk recorder. The audio mixer/switcher is used to select the appropriate audio source for broadcasting. Figure 12-1 also shows *horizontal* and *vertical* synchronizing signals which are combined with the picture information prior to modulation. These signals are used in the receivers to synchronize the horizontal and vertical *scanning rates* (synchronization is discussed in detail later in the chapter).

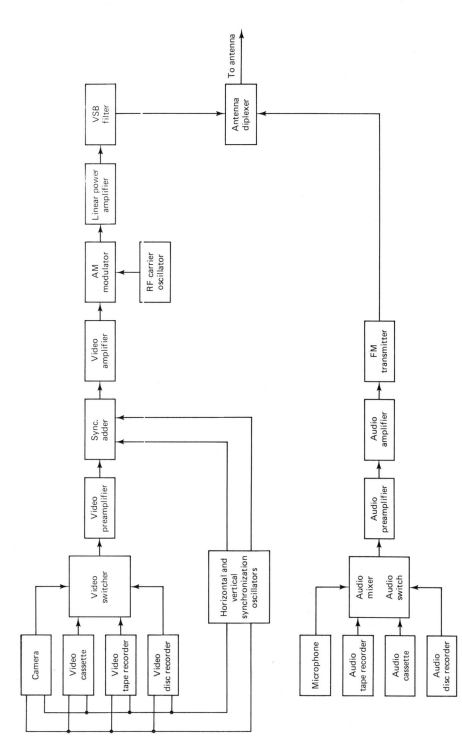

FIGURE 12-1 Simplified block diagram for a monochrome television transmitter.

Television Broadcast Standards

Figure 12-2 shows the frequency spectrum for a standard television broadcast channel. Its total bandwidth is 6 MHz. The picture carrier is spaced 1.25 MHz above the lower limit for the channel, and the sound carrier is spaced 0.25 MHz below the upper limit. Therefore, the picture and sound carriers are always 4.5 MHz apart. The color *subcarrier* is located 3.58 MHz above the picture carrier. Commercial television broadcasting uses AM vestigial sideband transmission for the picture information. The lower sideband is 0.75 MHz wide and the upper sideband 4 MHz. Therefore, the low video frequencies (rough outline of the image) are emphasized relative to the high video frequencies (fine detail of the image). The FM sound carrier has a bandwidth of approximately 50 kHz (±25 kHz deviation for 100% modulation). Both amplitude and phase modulation are used to encode the color information onto the 3.58-MHz color subcarrier. The bandwidth and composition of the color spectrum are discussed later in the chapter.

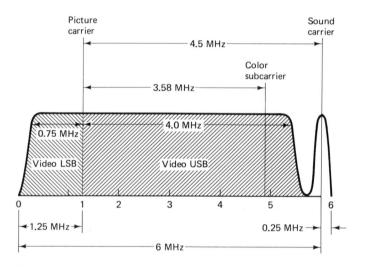

FIGURE 12-2   Standard television broadcast channel.

THE COMPOSITE VIDEO SIGNAL

Composite means made up of disparate or different parts. The composite video signal includes three separate parts. These parts are (1) the luminance signal, (2) the synchronization pulses, and (3) the blanking pulses. These three signals are combined in such a way to form the composite or total video signal.

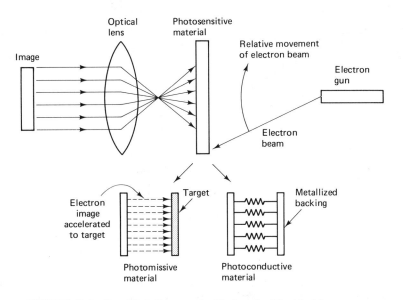

FIGURE 12-3   Simplified diagram of a black-and-white television camera.

The Luminance Signal

The *luminance signal* is the picture information or video signal. This signal originates in the camera and varies in amplitude proportional to the intensity (brightness) of the image. With *negative* transmission, the lower amplitudes correspond to the whitest parts of the image and the higher amplitudes correspond to the darkest. With *positive* transmission, the lower amplitudes correspond to the darkest parts and the higher amplitudes to the whitest. Negative transmission is the FCC standard for modulation of the final picture carrier. However, at intermediate points in both the transmitter and receiver, both negative or positive transmission signals occur and can be observed with a standard oscilloscope.

Figure 12-3 shows a simplified diagram of a black-and-white television camera tube. Light is reflected from an image through an optical lens onto the surface of a photosensitive material. The surface can be made from either a photomissive or photoconductive material and is divided into smaller discrete segments called *picture elements*. A *photomissive* material emits photoelectrons proportional to the intensity of the light striking its surface. A *photoconductive* material has a resistance that is inversely proportional to the intensity of the light striking it. When excited by an electron beam, a picture element outputs a signal that is proportional to the intensity of the light striking it. Therefore, if elements are individually scanned (excited) in sequence, the amplitude of the output signal will vary in accordance with the intensity of the image being scanned.

Figure 12-4 shows the amplitude changes in the luminance signal for a single

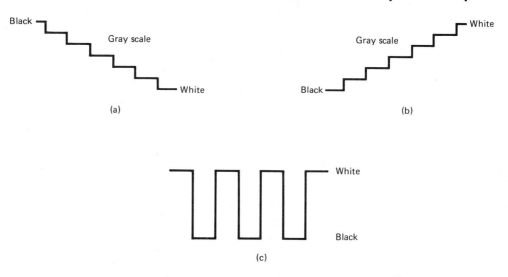

FIGURE 12-4   Luminance signal for a black-and-white camera: (a) gray scale, negative transmission; (b) gray scale, positive transmission; (c) checkerboard pattern, positive transmission.

horizontal scan at the output of a black-and-white camera for a image with varying light intensities. Figure 12-4a shows the amplitude changes for a negative transmission signal as the image changes from black, to a gray scale, to pure white. Figure 12-4b shows the same signal except for positive transmission. A positive transmission luminance signal for an alternating black/white checkerboard pattern is shown in Figure 12-4c. It can be seen that the amplitude of the luminance signal simply alternates from minimum to maximum with the brightness of the image.

Scanning

To produce the complete image, the entire surface of the camera tube must be *scanned*. Figure 12-5 shows a simple scanning sequence. The scanning is done in essentially the same manner in which a page from a book is read. This is called *sequential* horizontal scanning. The entire image is scanned in a sequential series of horizontal lines, one under the other. When the electron beam strikes the back of a picture element, a signal is produced whose amplitude is proportional to the light intensity striking the front of the element. Figure 12-5a shows the scanning beam beginning the active portion of the scan from the upper left corner and moving diagonally to the far right (line *A–B*). This is called the *active* portion of the scan line because this is the time in which the image is converted to electrical signals. Once the beam has reached the far-right side of the photosensitive surface, it immediately returns or retraces to the left side (point *C*). The return time is called horizontal *retrace* or *flyback time*. When the electron scanning beam has reached the bottom-right portion of the photosensitive surface (point *Z*), the beam is returned to top left (point *A*) and the sequence repeats. The return time is called the *vertical retrace time*. While the beam is retracing from the left to the right

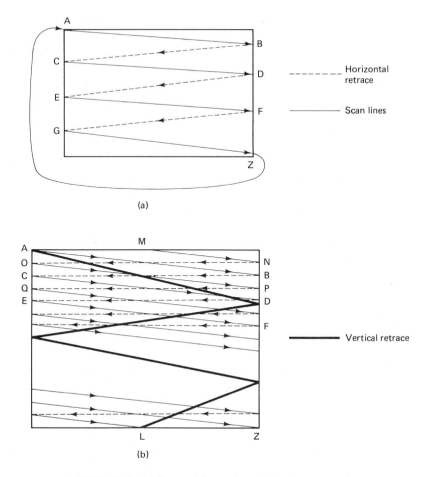

(a)

(b)

FIGURE 12-5 (a) Sequential scanning; (b) interlaced scanning.

side of the image and from the bottom to the top, it is *shut off* or *blanked*. Consequently, there is no video signal produced during either the horizontal or the vertical retrace times. The active and blanked portions of a single horizontal scan constitute one complete horizontal scan line. The number of horizontal scan lines depends on the detail desired and several other factors that are discussed later in the chapter.

In the United States, a total of 525 horizontal scan lines constitutes one *picture frame*, which is divided into two *fields* of 262.5 horizontal lines each. This scanning technique is called *interlaced scanning* and is shown in Figure 12-5b. Horizontal scanning produces the left-to-right movement of the electron beam and vertical scanning produces the downward movement. The vertical scanning rate is 30 Hz. Therefore, 30 frames per second are produced. Because the human eye can barely perceive a 30-Hz flicker, the frame is divided into two fields. 262.5 horizontal scan lines beginning at the top left (point *A*) and ending at the bottom middle (point *L*) constitute one picture field (the odd field). The second field (the even field) comprises the remaining 262.5 horizontal scan lines interlaced between the scan lines of the first field. The second field begins at the top middle (point M) and ends at the bottom right (point Z). Between fields, the electron beam retraces from the bottom of the picture back to the top in a zigzag pattern. This is called the *vertical retrace time*. Each field is vertically scanned at a 60-Hz rate. Therefore, although the entire picture changes every $\frac{1}{30}$ s, only half of the picture changes every $\frac{1}{60}$ s. This scanning technique allows 525 lines to be scanned at a 30-Hz rate without producing a noticeable flicker in the picture. To scan 525 horizontal scan lines in $\frac{1}{30}$ s, a 15,750-Hz scanning frequency is required ($30 \times 525 = 15,750$).

Horizontal lines are called *raster lines*, and 525 horizontal lines constitute a *raster*. The raster is the luminance that you see when there is no picture (i.e., when you are tuned to an unassigned channel). A raster simply means that there is horizontal and vertical scanning and brightness, but not necessarily a picture or an image on the screen.

Scanning waveforms.   The scanning beam for the camera and the receiver CRT must move both horizontally and vertically at *uniform* rates. This is called *linear scanning*. Linear scanning is necessary to ensure that picture elements are not "squashed" together or "bunched" to one side or to the top or bottom of the screen. *Magnetic deflection* is generally used to move the electron beam. Magnetic deflection produces essentially the same results as *electrostatic deflection*, which is commonly used in the CRT circuitry of an oscilloscope. Electrostatic deflection cannot be used with large CRTs such as those found in most television sets. With magnetic deflection, a linear rise in current through the deflection coils produces a linear change in magnetic flux. The force from the magnetic flux pulls the scanning beam from the left to right and from the top to bottom of the screen in a continuous, uniform motion. Figure 12-6a shows an ideal horizontal scanning current waveform. The positive slope of the sawtooth moves the beam from left to right in a smooth, constant motion. The negative slope of the waveform produces a magnetic field with the opposite polarity; thus it pulls the beam back to the left side. This is the retrace time. The rate at which the beam moves is proportional to the slope. During retrace it is desirable that the beam move as rapidly as possible. Therefore, the slope during retrace is much steeper than during the active portion of the scan line. In Figure 12-6a it can be seen that the most negative current corresponds to the left side of the screen and the most positive current to the right side. When zero current is flowing, the beam is in the center of the screen. Each scan line takes the time of one cycle of the sawtooth wave. Therefore, the frequency of the sawtooth is equal to the horizontal scan rate, 15,750 Hz, and the time for each scan line is 63.5 μs.

Figure 12-6b shows the vertical scanning waveform. Like the horizontal waveform, it is a sawtooth wave which ensures a uniform movement of the beam in the vertical direction. The bottom of the waveform corresponds to the top of the screen, the top of the waveform corresponds to the bottom of the screen, and zero current again corresponds to the center of the screen. Each vertical cycle corresponds to one complete vertical scan plus the vertical retrace time. Therefore, the vertical sawtooth frequency is 60 Hz.

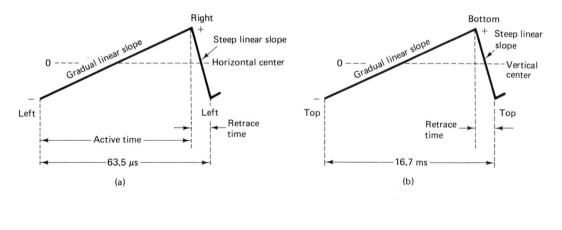

(a)

(b)

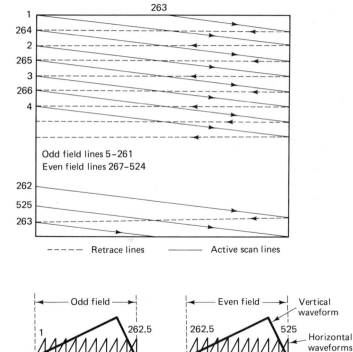

(c)

FIGURE 12-6 (a) Horizontal scanning waveform; (b) vertical scanning waveform; (c) interlaced scanning waveform.

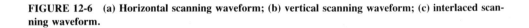

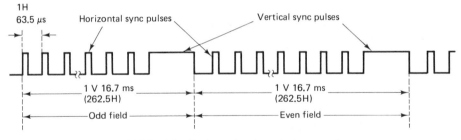

FIGURE 12-7 Horizontal and vertical sync pulses.

Figure 12-6c shows the interlaced scanning patterns for the odd and even fields of a standard U.S. television broadcast system. The vertical scanning waveforms are also shown superimposed over the horizontal scanning waveforms.

Synchronizing pulses. To reproduce the original image, the horizontal and vertical scanning rates at the receiver must be equal to those at the camera. Also, the scan lines at the camera and the receiver must begin and end in exact time synchronization. Therefore, a horizontal synchronizing pulse of 15,750 Hz and a vertical synchronizing pulse of 60 Hz are added to the luminance signal at the transmitter. The synchronizing (sync) pulses are stripped off at the receiver and used to synchronize its scanning circuits.

Figure 12-7 shows the horizontal and vertical sync pulses for both the odd and even fields. A new field is scanned once every $\frac{1}{60}$ s. Therefore, the time between vertical sync pulses is $1V$, where $V = \frac{1}{60}$ s or 16.7 ms. The time between horizontal sync pulses is $1H$, where $H = 63.5$ μs. $1V$ is sufficient time to scan 262.5 horizontal lines (1/15,750 = 63.5 μs and 16.7 ms/63.5 μs = 262.5). Each horizontal sync pulse produces one sawtooth horizontal scanning waveform, and each vertical sync pulse produces one sawtooth vertical scanning waveform.

Blanking Pulses

Blanking pulses are video signals that are added to the luminance and synchronizing pulses with the proper amplitude to ensure that the receiver is blacked out during the vertical and horizontal retrace times. The image is not scanned by the camera during retrace, and therefore there is no luminance information transmitted for those times. Blanking pulses are essentially video signals with amplitudes that do not produce any luminance (brightness) on the CRT. Horizontal and vertical sync pulses occur during their respective blanking times.

The Composite Signal

The *composite video signal* includes luminance (brightness) signals, horizontal and vertical sync pulses, and blanking pulses. Figure 12-8 shows the composite video signal for a single horizontal scan line, $1H$ (63.5 μs). (Notice that a positive transmission signal is shown.) The figure shows that the active portion of the scan line occurs during the positive slope of the scanning waveform and horizontal retrace occurs during the

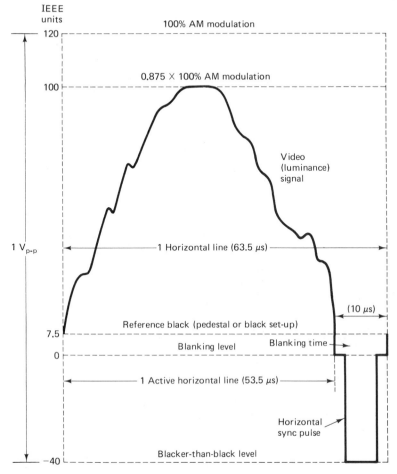

FIGURE 12-8 Composite video signal.

negative slope (i.e., the blanking time). The brightness range for standard television broadcasting is 160 IEEE units peak to peak. 160 IEEE units is generally normalized to 1 V_{p-p}. The exact value of 1 IEEE unit is unimportant; however, the relative value of a video signal in IEEE units determines its brightness. For example, maximum brightness (pure white) is 120 IEEE units, and no brightness is produced for signals below the reference black level (7.5 IEEE units). The *reference* black level is also called the *pedestal* or *black setup* level. The blanking level is 0 IEEE units, which is below the black level or, in other words, *blacker than black*. Sync pulses are negative-going pulses that occupy 25% of the total IEEE range. A sync pulse has a maximum level of 0 IEEE units and a minimum level of −40 IEEE units. Therefore, the entire sync pulse is below black and thus produces no brightness. The brightness range occupies 75% of the total IEEE scale and extends from 0 to 120 IEEE units, with 120 units corresponding to 100% AM modulation of the RF carrier. However, to ensure that

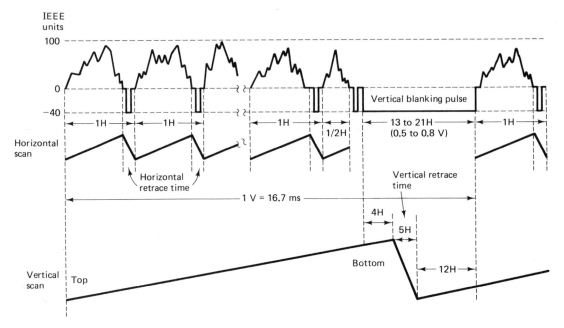

FIGURE 12-9   Composite video for the even field.

overmodulation does not occur, the FCC has established the maximum brightness (pure white) level to be 87.5% or 100 IEEE units (0.875 × 160 = 140 units, −40 + 140 = 100 units).

Figure 12-9 shows the composite video signal for the even field, which equals 1V or 16.7 ms and is sufficient time for 262.5 horizontal scan lines (262.5H). However, the vertical blanking pulse width is between 0.05 and 0.08V or 833 to 1333 μs. Therefore, the vertical blanking pulse occupies the time of 13 to 21 horizontal scan lines, which leaves 241.5 to 249.5 active horizontal scan lines. The figure also shows that most of the active scan lines occur during the positive slope of the vertical scanning waveform, and the vertical blanking pulse occurs during the negative slope of the waveform (i.e., the retrace time).

Horizontal blanking time.   Figure 12-10 shows the blanking time for a single horizontal scan line. The total blanking time is approximately 0.16H or 9.5 to 11.5 μs. Therefore, the active (visible) time for a horizontal line is approximately 0.84H or 52 to 54 μs. Figure 12-10 shows that the sync pulse does not occupy the entire blanking time. The width of the actual sync pulse is approximately 0.08H or 4.25 to 5.25 μs. The time between the beginning of the blanking time and the leading edge of the sync pulse is called the *front porch* and is approximately 0.02H with a minimum time of 1.27 μs. The time between the trailing edge of the sync pulse and the end of the blanking time is called the *back porch* and is approximately 0.06H with a minimum time of 3.81 μs.

Period	Time
Horizontal line	1H 63.5 μs
Horizontal blanking	0.16H 9.5–11.5 μs
Sync pulse	0.08H 4.75 ± 0.5 μs
Front porch	0.02H 1.27 μs minimum
Back porch	0.06H 3.81 μs minimum

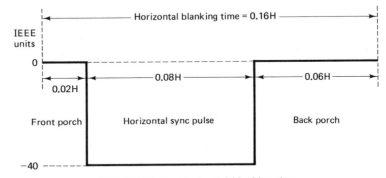

FIGURE 12-10   Horizontal blanking time.

Vertical blanking time.   Figure 12-11 shows the first 10*H* of a vertical blanking pulse for a negative transmission waveform. The figure shows that the entire blanking pulse is below the level for reference black (i.e., below 7.5 IEEE units). Each vertical blanking pulse begins with six *equalizing* pulses, a vertical sync pulse, and six more equalizing pulses. The equalizing pulses ensure a smooth, synchronized transition be-

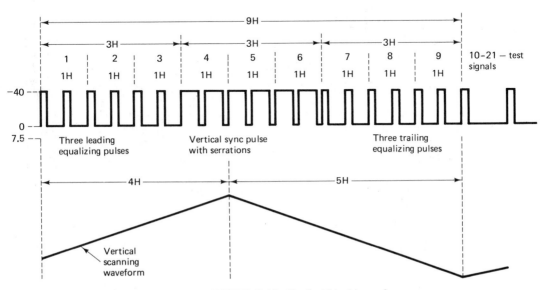

FIGURE 12-11   Vertical blanking pulse.

tween the odd and even fields. Equalizing pulses are explained in more detail in a later section of the chapter. The equalizing pulse rate is 31.5 kHz, which is twice the horizontal scanning rate. Therefore, each equalizing pulse takes $\frac{1}{2}H$ and the 12 pulses occupy a total time of 6H. The actual vertical sync pulse occupies the time of 3H. The *serrations* in the vertical sync pulse ensure that the receiver maintains horizontal synchronization during the vertical retrace time. Vertical serrations are explained in more detail in a later section. A total time of nine horizontal scan lines (9H) is required to transmit the equalizing and vertical sync pulses. From the figure it can be seen that the first 4H occur at the end of a vertical scan (i.e., at the bottom of the CRT). The following 5H occur during the retrace time, and all 9H occur during the vertical blanking time and are therefore not visible. The exact vertical blanking time is determined by the transmitting station; however, it is generally 21H. Horizontal lines 10 through 21 of each field are often used to send studio test signals and automatic color and brightness signals.

RF Transmission of the Composite Video

The transmitted AM picture carrier is shown in Figure 12-12 for negative polarity modulation, which is the FCC standard. As stated previously, negative transmission of the RF carrier simply means that changes toward white in the picture decrease the amplitude of the AM picture carrier. The advantage of negative transmission is that noise pulses

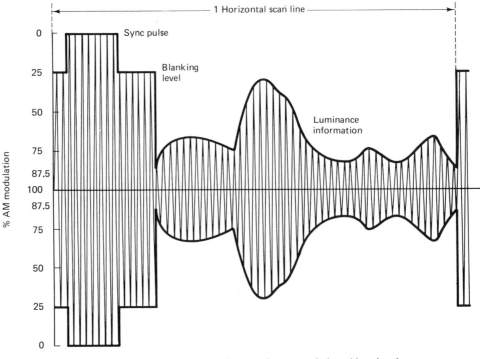

FIGURE 12-12   RF envelope for negative transmission video signal.

in the RF signal increase the carrier toward black, which makes the noise less annoying to the viewer than changes toward white. Also, with negative transmission, brighter images occur more often than darker ones; thus negative transmission uses less power than positive transmission. Notice that the AM envelope shown in Figure 12-12 has the shape of the composite video signal, and the luminance, blanking, and sync signals can easily be identified. Also, note that during the tips of the horizontal sync pulse there is no AM modulation, and the luminance signal never exceeds 87.5% AM modulation.

MONOCHROME TELEVISION RECEPTION

Figure 12-13 shows a block diagram for a monochrome television receiver. The receiver can be separated into five primary sections: RF, IF, video, horizontal and vertical deflection, and sound.

RF Section

A block diagram of an RF section is shown in Figure 12-14. The RF section includes the UHF and VHF antennas, the antenna coupling circuits, the preselectors, an RF amplifier, and a mixer/converter. A Yagi–Uda antenna is used for the VHF channels, and a simple loop antenna is used for the UHF channels.

The purposes of the RF or front-end section are channel selection (i.e., tuning), image frequency rejection, to isolate the local oscillator from the antenna (i.e., preventing the local oscillator signal from radiating), to convert RF signals to IF signals, amplification, and antenna coupling. VHF signals are captured by the antenna, coupled to the receiver input, bandlimited by the preselector, then amplified by the RF amplifier and fed to the mixer/converter. The mixer/converter beats the local oscillator frequency with the RF to produce the difference frequency, which is the IF. Channel selection is accomplished by changing the bandpass characteristics of the preselector and RF amplifier by switching capacitors or inductors in their tuned circuits and, at the same time, changing the local oscillator frequency. The preselector and local oscillator tuning circuits are ganged together. Commercial television receivers use high-side injection (the local oscillator is tuned to the IF above the desired RF channel frequency).

UHF signals are captured by the loop antenna, then immediately mixed down to IF. The UHF mixer is generally a simple diode mixer. When receiving UHF signals, the RF amplifier is simply an additional IF amplifier.

IF Section

The IF section of a television receiver provides most of the receiver's selectivity and gain. The block diagram for a three-stage IF amplifier is shown in Figure 12-15a. The IF section is generally several cascaded high-gain tuned amplifiers. In modern receivers, the IF section processes both the picture and sound IF signals. Such receivers are called

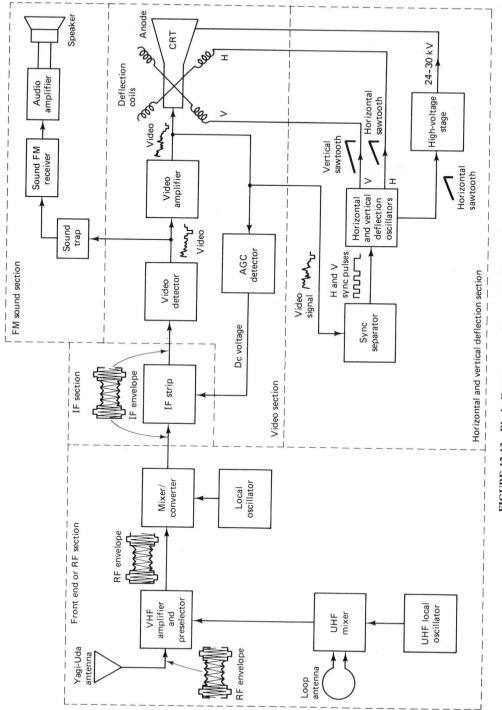

FIGURE 12-13 Block diagram monochrome television receiver.

468

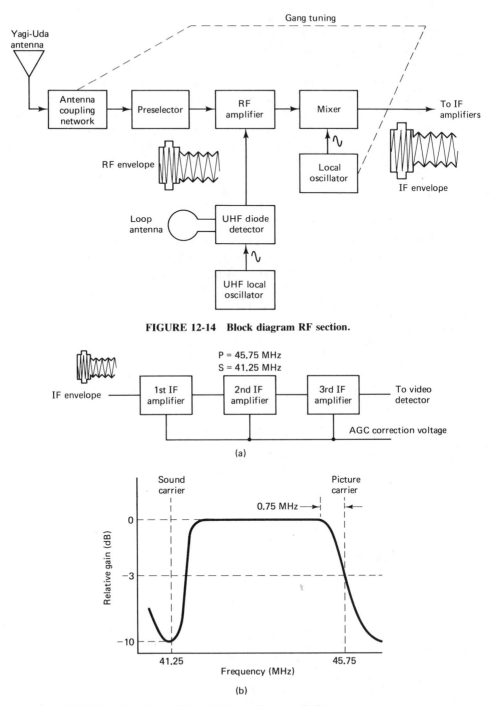

FIGURE 12-14 Block diagram RF section.

P = 45.75 MHz
S = 41.25 MHz

FIGURE 12-15 IF amplifiers: (a) block diagram; (b) frequency response curve.

intercarrier receivers. The standard IFs used in commercial television receivers are 45.75 MHz for the picture and 41.25 MHz for the sound. The IF carriers are separated by 4.5 MHz just as the RF carriers are. IF amplifiers use tuned bandpass filters that bandlimit the signal and prevent adjacent channel interference. A typical IF response curve is shown in Figure 12-15b. Special narrowband bandstop filters called *wavetraps* are used to trap or block the adjacent channel picture and sound carrier frequencies (39.75 MHz and 47.25 MHz, respectively). Wavetraps are also used to attenuate the sound and picture carriers of the selected channel and limit the IF passband to approximately 3 MHz. 3 MHz is used rather than 4 MHz to minimize interference from the color signal, which is explained later in this chapter.

The Video Section

The video section includes a video detector and a series of video amplifiers. A simplified block diagram for a video section is shown in Figure 12-16a. The detector down-converts the picture IF signals to video frequencies and the first sound IF to a second sound IF. The second sound IF is fed to the FM receiver, where the aural information is removed and fed to the audio amplifiers. The video detector is generally a single-diode peak detector. The IF input signal provides the ac voltage necessary to drive the diode into

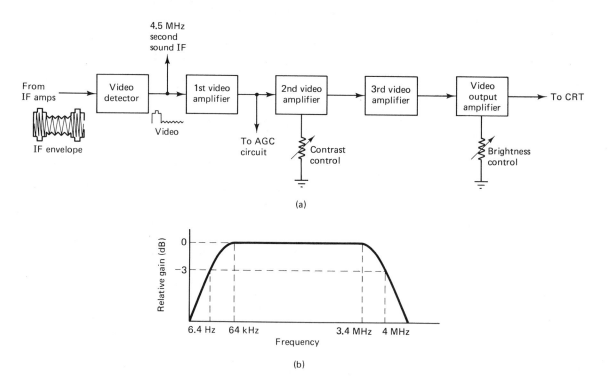

FIGURE 12-16 Video amplifiers: (a) block diagram; (b) frequency response curve.

conduction as a half-wave peak rectifier. The output from the video detector is the composite video signal which is fed to the video amplifiers. The video amplifiers provide the gain necessary for the luminance signal to drive the CRT. Video amplifiers are generally direct-coupled to provide dc restoration of the picture brightness. The contrast and brightness controls are located in the video section, and the AGC takeoff point is generally at the output of the first video amplifier. The brightness control simply allows the viewer to vary the dc bias voltage of the video signal. The contrast control adjusts the gain of the video amplifiers. The picture and sound IF signals mix in the diode detector, which is a nonlinear device, and produce a difference signal of 4.5 MHz, which is the second sound IF. A typical frequency response curve for a video amplifier section is shown in Figure 12-16b.

The Horizontal and Vertical Deflection Circuits

A simplified block diagram showing the vertical and horizontal deflection circuits is shown in Figure 12-17. The deflection section includes a sync separator, horizontal and vertical deflection oscillators, and a high-voltage stage. The horizontal and vertical synchronizing pulses are removed from the composite video signal by the sync separator circuit. The horizontal and vertical sync pulses are then separated with filters and fed to their respective deflection circuits. The deflection circuits convert the sync pulses to sawtooth scanning signals and provide the dc high voltage required for the anode of the CRT.

Sync separator. Figure 12-18a shows the schematic diagram for a single-transistor sync separator, which is a simple clipper circuit. Q_1 is a class C amplifier and the R_1C_1 coupling circuit provides signal bias. The positive portion of the composite video signal (i.e., the sync pulses) forward biases Q_1, causing base current to flow, which charges C_1 to the polarity shown. Between sync pulses, C_1 discharges slightly through R_1. The long R_1C_1 time constant keeps C_1 charged to approximately 90% of the peak positive value. Therefore, once C_1 has charged, the luminance signal drives Q_1 further into cutoff. Thus Q_1 conducts only during the more positive sync pulses. Consequently, the sync pulses are the only portion of the composite video signal that appears at the collector of Q_1. The base–emitter circuit of Q_1 is effectively a diode rectifier. The rectifier operation is shown in Figure 12-18b. Once removed, the horizontal and vertical sync pulses are separated with filters. A high-pass filter (differentiator) detects the 15,750-Hz horizontal sync pulses and a low-pass filter (integrator) detects the 60-Hz vertical sync pulses.

Vertical deflection oscillator. The output from the integrator is a 60-Hz waveform, which is fed to the vertical deflection oscillator. The deflection oscillator produces a 60-Hz linear sawtooth deflection voltage, which produces the vertical scan on the CRT. Figure 12-19 shows a schematic diagram for a transistorized *blocking oscillator*, which is a circuit often used to produce the sawtooth scanning waveform. A blocking oscillator is simply a *triggered oscillator* that produces a sawtooth output waveform

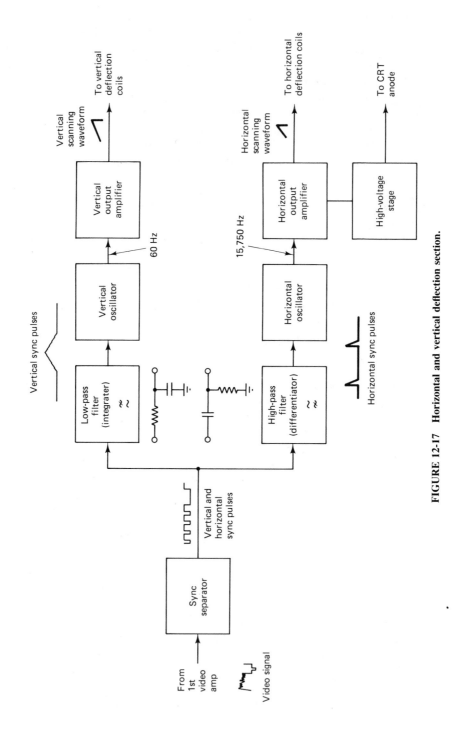

FIGURE 12-17 Horizontal and vertical deflection section.

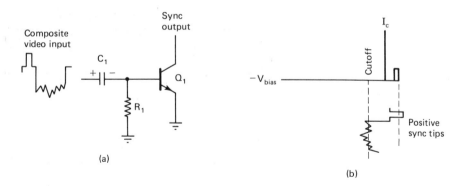

FIGURE 12-18   Sync separator circuit: (a) schematic diagram; (b) bias operation.

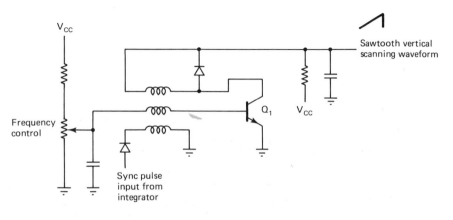

FIGURE 12-19   Vertical blocking oscillator.

that is synchronized to the incoming vertical sync pulse rate. The frequency control sets the *threshold* or *trigger level* for the oscillator. However, the frequency of oscillation is determined by the recovered vertical sync pulses. The output from the vertical oscillator is fed to a vertical output amplifier, which produces the high-voltage sawtooth waveform required to drive the vertical deflection coils.

The integrator (low-pass filter) passes only the vertical sync pulses and produces a 60-Hz trigger pulse for the blocking oscillator. Figure 12-20a shows the operation of the integrator without equalizing pulses. It can be seen that the sync pulses from the odd and even fields begin charging the capacitor from different initial levels; thus the two trigger pulses reach the threshold level for the oscillator at different times (Δt). This causes the vertical oscillator to change frequency and lose synchronization between fields, which is noticeable to the viewer as a slight vertical roll. To prevent loss of synchronization, 12 equalizing pulses are transmitted during the vertical blanking interval, six immediately before and six immediately after the vertical sync pulse. Figure 12-20b shows the waveform produced across the capacitor when the equalizing pulses are

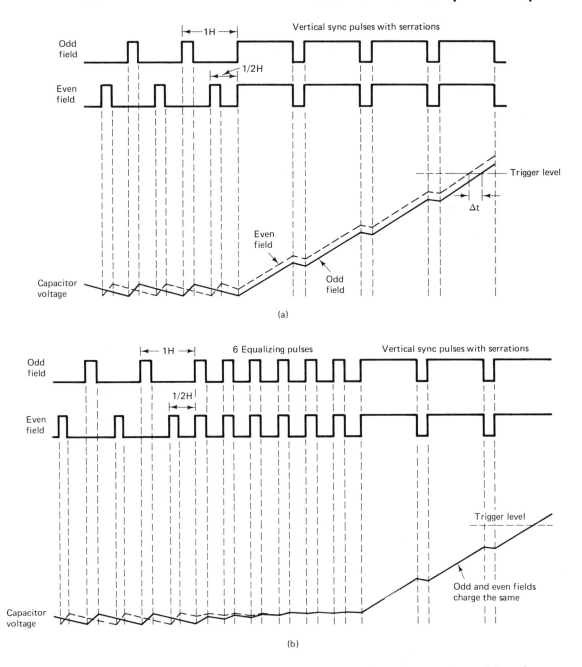

FIGURE 12-20 **Integrator operation: (a) without equalizing pulses; (b) with equalizing pulses.**

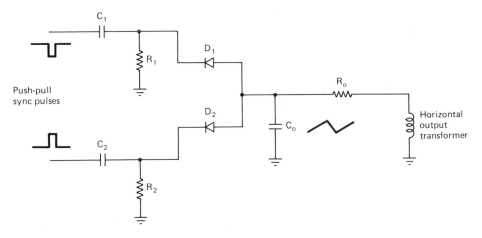

FIGURE 12-21 Push-pull horizontal AFC circuit.

included. The sync pulses from each field begin charging the capacitor at precisely the same time with exactly the same initial voltage; thus the capacitor voltage reaches the threshold level at the same time for each field, preventing the vertical sync oscillator from losing synchronization during the transition between the odd and even fields.

Horizontal deflection oscillator. Deflection oscillators, such as the one shown in Figure 12-20, are highly susceptible to noise. Noise pulses can be mistaken for synchronizing pulses and trigger the oscillator at the wrong time, thus changing the horizontal scanning rate. To improve noise immunity, automatic frequency control (AFC) circuits are often used for the horizontal deflection oscillator in television receivers. Figure 12-21 shows the schematic diagram for a push-pull sync discriminator commonly used for horizontal AFC. A phase splitter generates two 180° out-of-phase sync pulses which are required for push-pull operation. The dual-diode sync discriminator produces a horizontal sawtooth waveform across output capacitor C_o. Consequently, the sawtooth frequency is synchronized to the recovered horizontal sync pulses. The output from the AFC circuit is fed to the horizontal deflection circuit, where it provides horizontal scanning for the CRT. The AFC output is also fed to the receiver high-voltage section, where the anode voltage for the CRT is produced.

COLOR TELEVISION TRANSMISSION AND RECEPTION

Color Television Transmitter

In essence, a color television transmitter is identical to the black-and-white transmitter shown in Figure 12-1, except that a color camera is used to produce the video signal. With color broadcasting, all of the colors are produced by mixing different amounts of the three *primary colors*: red, blue, and green. A color camera is actually three cameras

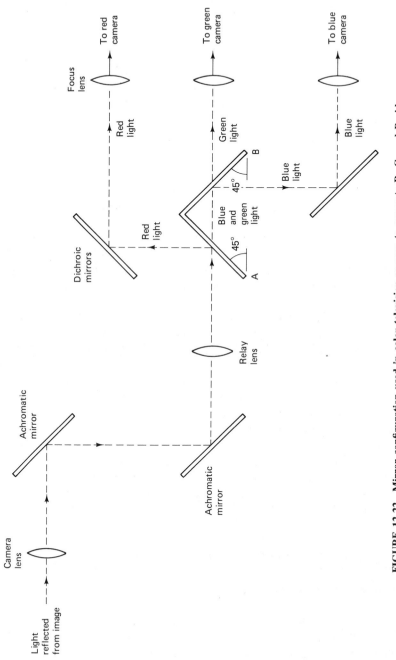

FIGURE 12-22 Mirror configuration used in color television camera to separate R, G, and B video signals.

476

in one, each with separate video output signals. When an image is scanned, separate camera tubes are used for each of the primary colors. The red camera produces the R video signal, the green camera produces the G video signal, and the blue camera produces the B video signal. The R, G, and B video signals are combined in an encoder to produce the composite color signal, which when combined with the luminance signal, amplitude modulates the RF carrier.

Color Camera

Figure 12-22 shows a configuration of mirrors that can be used to split an image into the three primary colors. The *chromatic* mirrors reflect light of all colors. The *dichroic* mirrors are coated to reflect light of only one frequency (color) and allow all other frequencies (colors) to pass through. Light reflected from the image passes through a single camera lens, is reflected by the two achromatic mirrors, and passes through the relay lens. Dichroic mirrors *A* and *B* are mounted on opposing 45° angles. Mirror *A* reflects red light, while blue and green light pass straight through to mirror *B*. Mirror *B* reflects blue light and allows green light to pass through. Consequently, the image is separated into red, green, and blue light frequencies. Once separated, the three color frequency signals modulate their respective camera tubes and produce the R, G, and B video signals.

Color Encoding

Figure 12-23 shows a simplified block diagram for a color television transmitter. The R, G, and B video signals are combined in specific proportions in the *color matrix* to produce the brightness (luminance) or *Y* video signal and the *I* and *Q* chrominance (color) video signals. The luminance signal corresponds to a monochrome video signal. The *I* and *Q* color signals amplitude modulate a 3.58-MHz color subcarrier to produce the total color signal, *C*. The *I* signal modulates the subcarrier directly in the *I* balanced modulator, while the *Q* signal modulates a quadrature (90° out of phase) subcarrier in the *Q* balanced modulator. The *I* and *Q* modulated signals are linearly combined to produce a quadrature amplitude modulation (QAM) signal, *C*, which is a combination of both phase and amplitude modulation. The *C* signal is combined with the *Y* signal to produce the total composite video signal (*T*).

The luminance signal. The *Y* or *luminance signal* is formed by combining 30% of the R video signal, 59% of the G video signal, and 11% of the B video signal. Mathematically, *Y* is expressed as

$$Y = 0.30R + 0.59G + 0.11B \qquad (12\text{-}1)$$

The percentages shown in Equation 12-1 correspond to the relative brightness of the three primary colors. Consequently, a scene reproduced in black and white by the

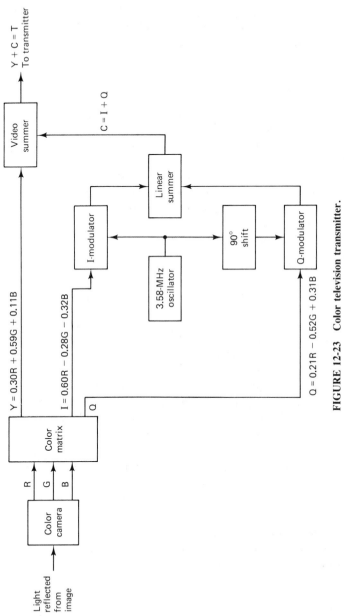

FIGURE 12-23 Color television transmitter.

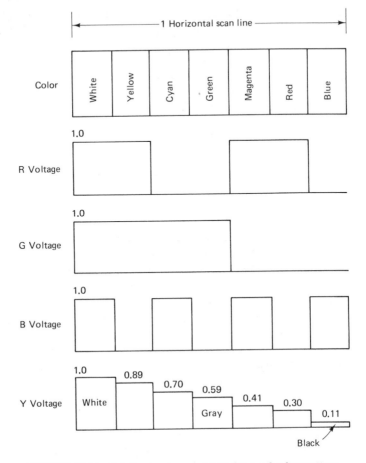

FIGURE 12-24 Relative luminance values for a color bar pattern.

Y signal has exactly the same brightness as the original image. Figure 12-24 shows how the *Y* signal voltage is formed from several values of R, G, and B.

The *Y* signal has a maximum relative amplitude of unity or 1, which is 100% white. For maximum values of R, G, and B (1*V* each), the value for brightness is determined from Equation 12-1 as follows:

$$Y = 0.30(1) + 0.59(1) + 0.11(1) = 1.00$$

The voltage values for *Y* shown in Figure 12-24 are the relative luminance values for each of the colors. If only the *Y* signal is used to reproduce the pattern in a receiver, it would appear on the CRT as seven monochrome bars shaded from white on the left to gray in the middle and black at the right. The *Y* signal is transmitted with a bandwidth of 0 to 4 MHz. However, most receivers bandlimit the *Y* signal to 3.2 MHz to minimize interference with the 3.58-MHz color signal. The *I* signal is transmitted with a bandwidth of 1.5 MHz, while the Q signal is transmitted with a bandwidth of 0.5 MHz. However, most receivers limit both the I and Q signals to 0.5-MHz bandwidth.

The chrominance signal.   The *chrominance* or C *signal* is a combination of the I and Q color signals. The I or in-phase color signal is produced by combining 60% of the R video signal, 28% of the inverted G video signal, and 32% of the inverted B video signal. Mathematically, I is expressed as

$$I = 0.60R - 0.28G - 0.32B \qquad (12\text{-}2)$$

The Q or in quadrature color signal is produced by combining 21% of the R video signal, 52% of the inverted G video signal, and 31% of the B video signal. Mathematically, Q is expressed as

$$Q = 0.21R - 0.52G + 0.31B \qquad (12\text{-}3)$$

The I and Q signals are combined to produce the C signal, and since the I and Q signals are in quadrature, the C signal is the phasor sum of the two (i.e., $C = \sqrt{I^2 + Q^2}$). Therefore, the amplitude and phase of the C signal is dependent on the amplitudes of the I and Q signals, which are in turn proportional to the R, G, and B video signals. Figure 12-25 shows the *color wheel* for television broadcasting. The R-Y and B-Y signals are used in most color television receivers for demodulating the R, G, and B video signals and are explained later in the chapter. In the receiver, the C signal reproduces colors proportionate to the amplitudes of the I and Q signals. The hue (color tone) is determined by the phase of the C signal, and the depth or saturation is proportional to

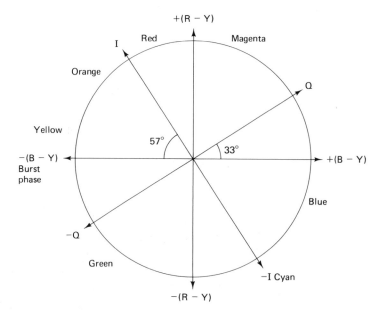

FIGURE 12-25   Standard television broadcasting color wheel.

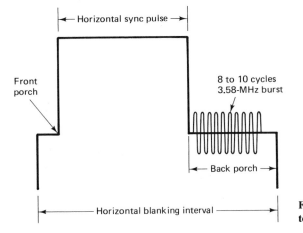

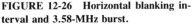

FIGURE 12-26 Horizontal blanking interval and 3.58-MHz burst.

the magnitude of the *C* signal. The outside of the circle corresponds to a relative value of 1.0.

The color burst. The phase of the 3.58-MHz color subcarrier is the reference phase for color demodulation. Therefore, the color subcarrier must be transmitted together with the composite video so that a receiver can reconstruct the subcarrier with the proper frequency and reference phase and thus determine the phase (color) of the received signal. Eight to ten cycles of the 3.58-MHz subcarrier are inserted on the back porch of each horizontal blanking pulse. This is referred to as the *color burst*. In the receiver, the burst is removed and used to synchronize a local 3.58-MHz color oscillator. The color burst is shown in Figure 12-26. Figure 12-27 shows the composite RF frequency spectrum for color television broadcasting.

Scanning frequencies for color transmission. The frequency of the color subcarrier is determined by harmonic relations among the color subcarrier and the horizontal and vertical scanning rates. The exact value for the color subcarrier is 3.579545 MHz.

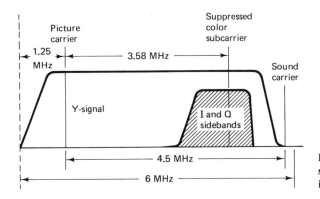

FIGURE 12-27 Composite RF frequency spectrum for color television broadcasting.

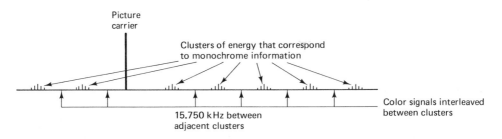

Picture
carrier

Clusters of energy that correspond
to monochrome information

15.750 kHz between
adjacent clusters

Color signals interleaved
between clusters

FIGURE 12-28   Frequency interleaving of color and luminance signals.

The sound subcarrier (4.5 MHz) is the 286th harmonic of the horizontal line frequency. Therefore, the horizontal line rate (F_H) for color transmission is not exactly 15.750 kHz. Mathematically, F_H is

$$F_H = \frac{4.5 \text{ MHz}}{286} = 15,734.26 \text{ Hz}$$

The exact value of the vertical scan rate (F_V) is

$$F_V = \frac{15,734.26}{262.5} = 59.94 \text{ Hz}$$

The color subcarrier frequency (C) is chosen as the 455th harmonic of one-half of the horizontal scan rate. Therefore,

$$C = \frac{15,734.26}{2} \times 455 = 3.579545 \text{ MHz}$$

Frequency interlacing.   The Y portion of the video signal produces clusters of energy at 15.73426-kHz intervals throughout the 4-MHz video bandwidth. By producing color signals around a 3.579545-MHz color subcarrier, the color energy is *clustered* within the void intervals between the black-and-white information. This is called frequency *interlacing* or sometimes frequency *interleaving* and is a form of *multiplexing* (i.e., the color and black-and-white information is frequency-division multiplexed into the total video spectrum). Figure 12-28 shows the spectrum for frequency interlacing.

Color Television Receivers

A color television receiver is essentially the same as a black-and-white receiver except for the picture tube and the addition of the color decoding circuits. Figure 12-29 shows the simplified block diagram for the color circuits in a color television receiver.

The composite video signal is fed to the *chroma* bandpass amplifier, which is tuned to the 3.58-MHz subcarrier and has a bandpass of 0.5 MHz. Therefore, only the C signal is amplified and passed on to the B-Y and R-Y demodulators. The 3.58-MHz color burst is separated from the horizontal blanking pulse by keying on the burst separator only during the horizontal flyback time. A synchronous 3.58-MHz color subcarrier is

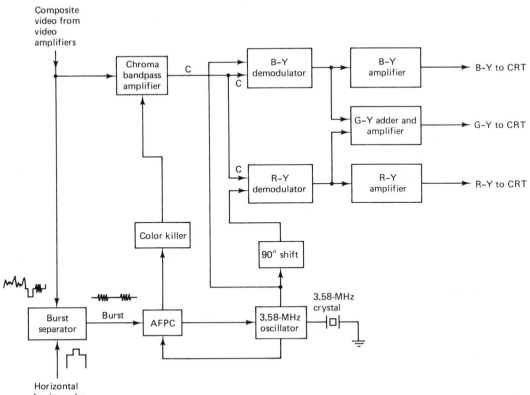

FIGURE 12-29   Color demodulator circuits.

reproduced in the color AFC circuit, which consists of a 3.58-MHz color oscillator and a color AFPC (automatic frequency and phase control) circuit. The *color killer* shuts off the chroma amplifier during monochrome reception (no colors are better than wrong colors). The C signal is demodulated in the B-*Y* and R-*Y* demodulators by mixing it with the phase coherent 3.58-MHz subcarrier. The B-*Y* and R-*Y* signals produce the R and B video signal by combining them with the *Y* signal in the following manner:

$$B - Y + Y = B$$
$$R - Y + Y = R$$

The G video signal is produced by combining the B-*Y* and R-*Y* signals with the proper proportions.

QUESTIONS

12-1. Briefly describe the meaning of the word *television*.

12-2. Describe a diplexer bridge.

12-3. What components make up the composite video signal?

12-4. Describe negative transmission; positive transmission. Which is the FCC standard for RF transmission?

12-5. Name and briefly describe two types of photosensitive materials used in television cameras.

12-6. Describe sequential horizontal scanning.

12-7. What is meant by the *active portion* of a horizontal line?

12-8. What is meant by *retrace time*?

12-9. Describe interlaced scanning.

12-10. Define *raster lines*; *raster*.

12-11. Why should the horizontal and vertical scanning waveforms be linear?

12-12. Why are horizontal and vertical synchronizing pulses included with the composite video signal?

12-13. Describe a blanking pulse.

12-14. Describe the IEEE scale used for television broadcasting.

12-15. What is meant by *reference black*; *black setup*; *pedestal*; *blacker than black*?

12-16. Describe the horizontal blanking time.

12-17. Describe the vertical blanking time.

12-18. Why are equalizing pulses transmitted during the vertical blanking interval?

12-19. Why are there serrations in the vertical sync pulses?

12-20. Draw the block diagram for a monochrome television receiver and describe its basic operation and the primary purpose of each section.

12-21. Describe the operation of a vertical blocking oscillator.

12-22. Describe the operation of a horizontal AFC circuit.

12-23. What are the three primary colors for television broadcasting?

12-24. Describe the basic operation of a color television camera.

12-25. Describe the Y or luminance signal.

12-26. Describe the I, Q, and C signals.

12-27. What is the color burst? How is it transmitted? What is its purpose?

12-28. Explain why the scanning frequencies used with color television are slightly different than those used for black-and-white transmission.

12-29. Describe frequency interlacing.

12-30. Draw the block diagram of the color decoding circuits in a color television receiver and briefly describe the decoding operation.

PROBLEMS

12-1. Draw the RF frequency spectrum for channel 8. Include the picture, sound, and the color frequencies and their respective bandwidths.

12-2. How long is $0.8H$? $0.8V$?

12-3. How many horizontal scan lines occur during $0.12V$?

12-4. If the back porch of the horizontal blanking pulse is $0.06H$, what is the maximum number of cycles of the color burst signal that can be transmitted?

12-5. If the vertical blanking interval is $20H$ long, how long is this?

12-6. What is the color subcarrier frequency for channel 55; channel 66?

12-7. What is the sound subcarrier frequency for channel 55; Channel 66?

12-8. Determine the value for Y for the following R, G, and B signals: $R = 0.8V$, $G = 0.6V$, and $G = 0.2V$.

12-9. Determine I and Q for the R, G, and B values given in Problem 12-8.

12-10. Determine the C signal for Problem 12-9.

Appendix A

THE SMITH CHART

INTRODUCTION

Mathematical solutions for transmission line impedances are laborious. Therefore, it is common practice to use charts to graphically solve transmission line impedance problems. Equation A-1 is the formula for determining the impedance at a given point on a transmission line.

$$Z = Z_O \left[\frac{Z_L + jZ_O \tan\beta\, S}{Z_O + jZ_L \tan\beta\, S} \right]$$ (A-1)

where

Z = line impedance at a given point
Z_L = load impedance
Z_O = line characteristic impedance
βS = distance from the load to the point where the impedance value is to be calculated

There are several charts available on which the properties of transmission lines are graphically presented. However, the most useful graphical representations are those that give the impedance relations that exist along a lossless transmission line for varying load conditions. The *Smith chart* is the most widely used transmission-line calculator of this type. The Smith chart is a special kind of impedance coordinate system that portrays the relationship of impedance at any point along a uniform transmission line to the impedance at any other point on the line.

The Smith chart was developed by Philip H. Smith at Bell Telephone Laboratories

486

IMPEDANCE OR ADMITTANCE COORDINATES

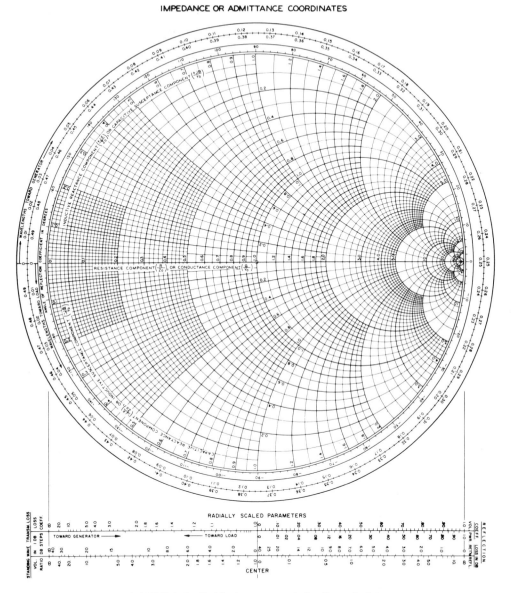

FIGURE A-1 Smith chart, transmission line calculator.

and was originally described in an article entitled "Transmission Line Calculator"(*Electronics*, January, 1939). A Smith chart is shown in Figure A-1. This chart is based on two sets of *orthogonal* circles. One set represents the ratio of the resistive component of the line impedance (R) to the characteristic impedance of the line (Z_O), which for a lossless line is also purely resistive. The second set of circles represents the ratio of the reactive component of the line impedance ($\pm jX$) to the characteristic impedance of the line (Z_O). Parameters plotted on the Smith chart include:

1. Impedance (or admittance) at any point along a transmission line
   a. Reflection coefficient magnitude (Γ)
   b. Reflection coefficient angle in degrees
2. Length of transmission line between any two points in wavelengths
3. Attenuation between any two points
   a. Standing-wave loss coefficient
   b. Reflection loss
4. Voltage or current standing-wave ratio
   a. Standing-wave ratio
   b. Limits of voltage and currrent due to standing waves

SMITH CHART DERIVATION

The characteristic impedance of a transmission line, Z_O, is made up of both *real* and *imaginary* components of either sign (i.e., $Z = \pm R \pm jX$). Figure A-2(a) shows three typical circuit elements, and Figure A-2(b) shows their impedances graphed on a *rectangular* coordinate plane. All values of Z that correspond to passive networks must be plotted on or to the right of the imaginary axis of the Z plane (this is because a negative real component implies that the network is capable of supplying energy). In order to

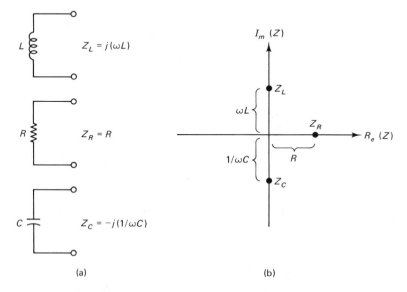

(a)   (b)

FIGURE A-2   **(a) Typical circuit elements; (b) impedances graphed on rectangular coordinate plane. (Note: ω is the angular frequency at which Z is measured.)**

display the impedance of all possible passive networks on a rectangular plot, the plot must extend to infinity in three directions ($+R$, $+jX$, and $-jX$). The Smith chart overcomes this limitation by plotting the complex *reflection coefficient*,

$$\Gamma = \frac{z - 1}{z + 1} \qquad \text{(A-2)}$$

where

z = the impedance normalized to the characteristic impedance (i.e., $z = Z/Z_O$)

Equation A-2 shows that for all passive impedance values, z, the magnitude of Γ is between 0 and 1. Also, since $|\Gamma| \leq 1$, the entire right side of the z plane can be mapped onto a circular area on the Γ plane. The resulting circle has a radius $r = 1$ and a center at $\Gamma = 0$, which corresponds to $z = 1$ or $Z = Z_O$.

Lines of constant $R_e(z)$.   Figure A-3(a) shows the rectangular plot of four lines of constant resistance $R_e(z) = 0$, 0.5, 1, and 2. For example, any impedance with a real part $R_e = 1$ will lie on the $R = 1$ line. Impedances with a positive reactive component (X_L) will fall above the real axis, while impedances with a negative reactive component (X_C) will fall below the real axis. Figure A-3b shows the same four values of R mapped onto the Γ plane. $R_e(z)$ are now circles of $Re(\Gamma)$. However, inductive impedances are still transferred to the area above the horizontal axis, and capacitive impedances are transferred to the area below the horizontal axis. The primary difference between the two graphs is that with the circular plot, the lines no longer extend to infinity. The infinity points all meet on the plane at a distance of 1, to the right of the origin. This implies, of course, that for $z = \infty$ (whether real, inductive, or capacitive), $\Gamma = 1$.

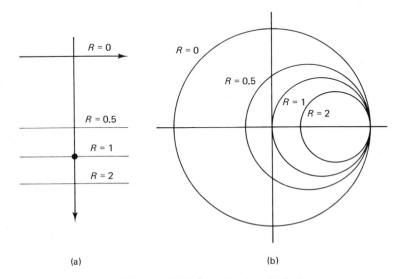

(a)   (b)

FIGURE A-3   (a) Rectangular plot; (b) Γ plane.

Lines of constant $I_m(z)$.   Figure A-4(a) shows the rectangular plot of three lines of constant inductance ($+jX = 0.5$, 1, and 2), three lines of constant capacitance ($-jX = -0.5$, -1, and -2), and a line of zero reactance ($jX = 0$). Figure A-4(b) shows the same seven values of jX plotted onto the plane. It can be seen that all values of infinite magnitude again meet at $\Gamma = 1$. The entire rectangular z plane curls to the right and its three axes (which previously extended infinitely) meet at the intersection of the $\Gamma = 1$ circle and the horizontal axis.

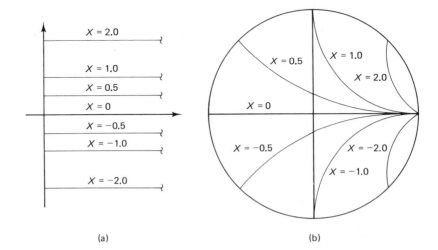

(a)   (b)

FIGURE A-4   (a) Rectangular plot; (b) Γ plane.

Impedance inversion (admittance).   *Admittance* (Y) is the mathematical inverse of Z (i.e., $Y = 1/Z$). Y, or for that matter any complex number, can be found graphically using the Smith chart by simply plotting z on the complex Γ plane, then rotating this point 180° about $\Gamma = 0$. By rotating every point on the chart by 180°, a second set of coordinates (the y coordinates) can be developed that are an inverted mirror image of the original chart. See Figure A-5(a). Occasionally, the admittance coordinates are superimposed on the same chart as the impedance coordinates. See Figure A-5(b). Using the combination chart, both impedance and admittance values can be read directly by using the proper set of coordinates.

Complex conjugate.   The *complex conjugate* can easily be determined using the Smith chart by simply reversing the sign of the angle of Γ. On the Smith chart, Γ is usually written in polar form and angles become more negative (phase lagging) when

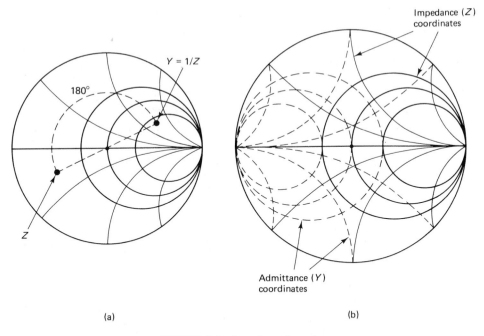

FIGURE A-5 Impedance inversion.

rotated in a clockwise direction around the chart. Hence, $0°$ is on the right end of the real axis and $\pm 180°$ is on the left end. For example, let $\Gamma = 0.5\underline{/+150°}$. The complex conjugate, Γ', is $0.5\underline{/-150°}$. In Figure A-6, it is shown that Γ' is found by mirroring Γ about the real axis.

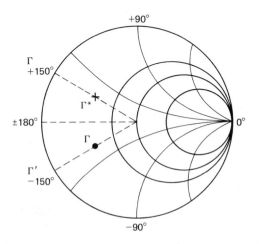

FIGURE A-6 Complex conjugate.

Plotting Impedance, Admittance, and SWR
on the Smith Chart

Any impedance Z can be plotted on the Smith chart by simply *normalizing* the impedance value to the characteristic impedance (i.e., $z = Z/Z_O$) and plotting the real and imaginary

IMPEDANCE OR ADMITTANCE COORDINATES

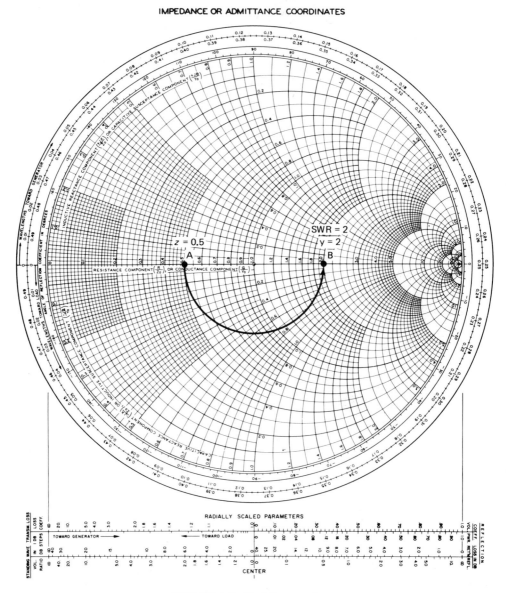

FIGURE A-7   Resistive impedance.

parts. For example, for a characteristic impedance $Z_O = 50$ ohms and an impedance $Z = 25$ ohms resistive, the normalized impedance z is determined as follows:

$$z = \frac{Z}{Z_O} = \frac{25}{50} = 0.5$$

Because z is purely resistive, its plot must fall directly on the horizontal axis ($\pm jX = 0$). $Z = 25$ is plotted on Figure A-7 at point A (i.e., $z = 0.5$). Rotating $180°$ around the chart gives a normalized admittance value $y = 2$ (where $y = Y/Y_O$). y is plotted on Figure A-7 at point B.

As previously stated, a very important characteristic of the Smith chart is that any lossless line can be represented by a circle having its origin at $1 \pm j0$ (i.e., the center of the chart) and radius equal to the distance between the origin and the impedance plot. Therefore, the *standing wave ratio*, SWR, corresponding to any particular circle is equal to the value of Z/Z_O at which the circle crosses the horizontal axis on the right side of the chart. Therefore, for this example, SWR $= 0.5$ (i.e., $Z/Z_O = 25/50 = 0.5$. It also should be noted that any impedance or admittance point can be rotated $180°$ by simply drawing a straight line from the point through the center of the chart to where the line intersects the circle on the opposite side.

For a characteristic impedance $Z_O = 50$ and an inductive load $Z = +j25$, the normalized impedance z is determined as follows:

$$z = \frac{Z}{Z_O} = \frac{+jX}{Z_O} = \frac{+j25}{50} = +j0.5$$

Because z is purely inductive, its plot must fall on the $R = 0$ axis which is the outer circle on the chart. $z = +j0.5$ is plotted on Figure A-8 at point A and its admittance $y = -j2$ is graphically found by simply rotating $180°$ around the chart (point B). SWR for this example must lie on the far right end of the horizontal axis which is plotted at point C and corresponds to SWR ∞, which is inevitable for a purely reactive load. SWR is plotted at point C.

For a complex impedance $Z = 25 + j25$, z is determined as follows:

$$z = \frac{25 + j25}{50} = 0.5 + j0.5$$

$$z = 0.707 \underline{/45°}$$

therefore

$$Z = 0.707 \underline{/45°} \times 50 = 35.35 \underline{/45°}$$

and

$$Y = \frac{1}{35.35 \underline{/45°}} = 0.02829 \underline{/-45°}$$

thus

$$y = \frac{Y}{Y_O} = \frac{0.02829}{0.02} = 1.414$$

and

$$y = 1 - j1$$

IMPEDANCE OR ADMITTANCE COORDINATES

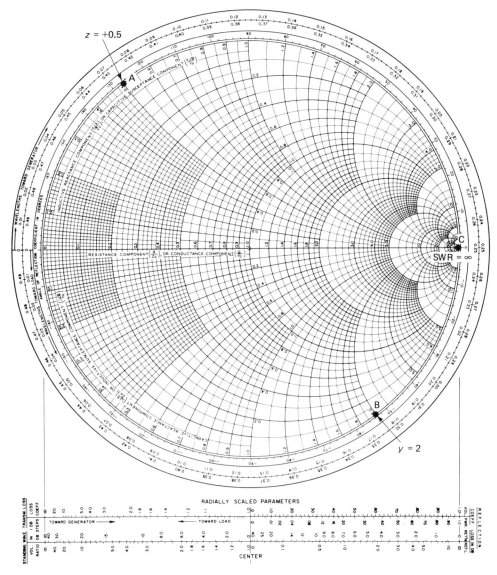

FIGURE A-8   Inductive load.

z is plotted on the Smith chart by locating the point where the $R = 0.5$ arc intersects the $X = 0.5$ arc on the top half of the chart. $z = 0.5 + j0.5$ is plotted on Figure A-9 at point A, and y is plotted at point B ($1 - j1$). From the chart, SWR is approximately 2.6 (point C).

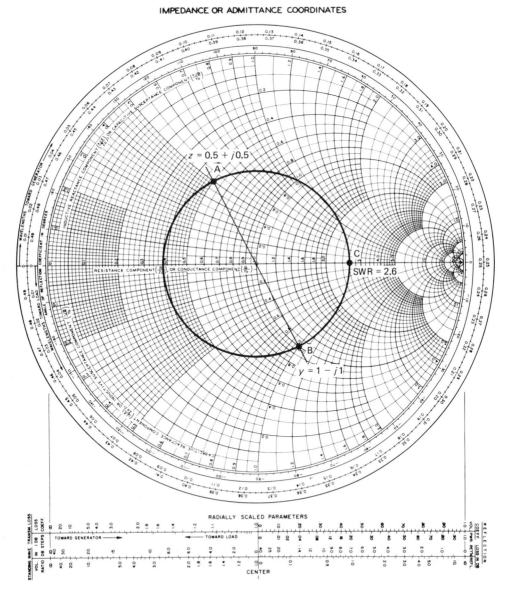

FIGURE A-9 Complex impedance.

Input Impedance and the Smith Chart

The Smith chart can be used to determine the input impedance of a transmission line at any distance from the load. The two outermost scales on the Smith chart indicate distance in *wavelengths* (see Figure A-1). The outside scale gives distance from the load toward the generator and increases in a clockwise direction, and the second scale

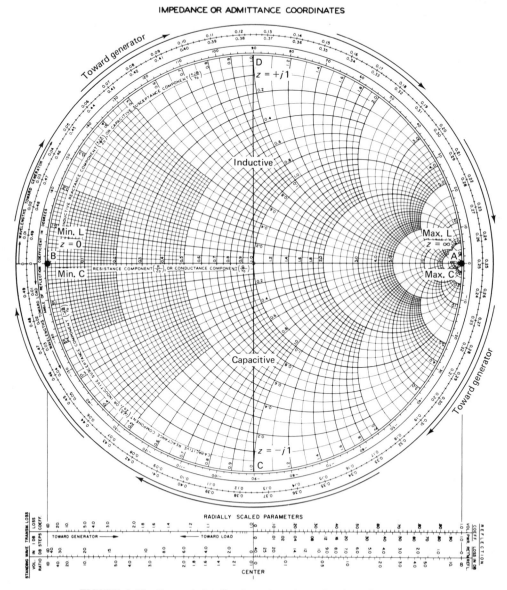

FIGURE A-10   Transmission line input impedance for shorted and open line.

gives distance from the source toward the load and increases in a counterclockwise direction. However, neither scale necessarily indicates the position of either the source or the load. One complete revolution (360°) represents a distance of one-half wavelength (0.5 λ), half of a revolution (180°) represents a distance of one-quarter wavelength (0.25 λ), etc.

A transmission line that is terminated in an open circuit has an impedance at the open end which is purely resistive and equal to infinity (Chapter 9). On the Smith chart, this point is plotted on the right end of the $X = 0$ line (point A on Figure A-10). As you move toward the source (generator) the input impedance is found by rotating around the chart in a clockwise direction. It can be seen that input impedance immediately becomes capacitive and maximum. As you rotate farther around the circle (move toward the generator), the capacitance decreases to a normalized value of unity (i.e., $z = -j1$) at a distance of one-eighth wavelength from the load (point C on Figure A-10) and a minimum value just short of one-quarter wavelength. At a distance of one-quarter wavelength, the input impedance is purely resistive and equal to 0 Ω (point B on Figure A-10). As stated in Chapter 9, there is an impedance inversion every one-quarter wavelength on a transmission line. Moving just past one-quarter wavelength, the impedance becomes inductive and minimum, then the inductance increases to a normalized value of unity (i.e., $z = +j1$) at a distance of three-eighths wavelength from the load (point D on Figure A-10) and a maximum value just short of one-half wavelength. At a distance of one-half wavelength, the input impedance is again purely resistive and equal to infinity (return to point A on Figure A-10). The results of the preceding analysis are identical to those achieved with phasor analysis in Chapter 9 and plotted in Figure 9-20.

A similar analysis can be done with a transmission line that is terminated in a short circuit, although the opposite impedance variations are achieved as with an open load. At the load, the input impedance is purely resistive and equal to 0. Therefore, the load is located at point B on Figure A-10 and point A represents a distance one-quarter wavelength from the load. Point D is a distance of one-eighth wavelength from the load and point C a distance of three-eighths wavelength. The results of such an analysis are identical to those achieved with phasors in Chapter 9 and plotted in Figure 9-21.

For a transmission line terminated in a purely resistive load not equal to Z_o, Smith chart analysis is very similar to the process described in the preceding section. For example, for a load impedance $Z_L = 37.5$ Ω resistive and a transmission line characteristic impedance $Z_O = 75$ Ω, the input impedance at various distances from the load is determined as follows:

1. The normalized load impedance z is

$$z = \frac{Z_L}{Z_O} = \frac{37.5}{75} = 0.5$$

2. $z = 0.5$ is plotted on the Smith chart (point A on Figure A-11). A circle is drawn that passes through point A with its center located at the intersection of the $R = 1$ circle and the $x = 0$ arc.

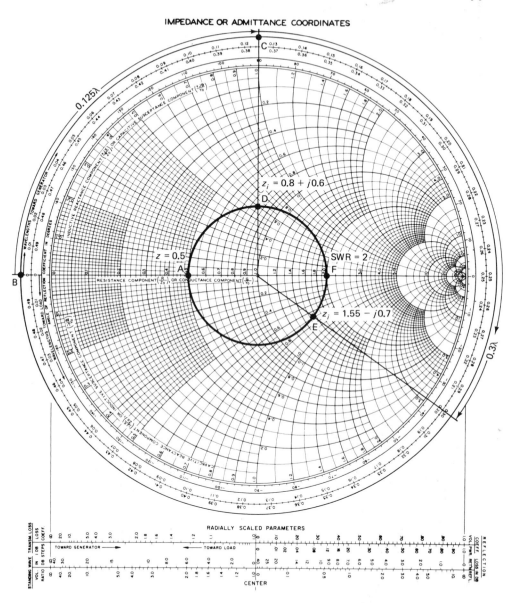

IMPEDANCE OR ADMITTANCE COORDINATES

$z_i = 0.8 + j0.6$

$z = 0.5$

SWR = 2

$z_i = 1.55 - j0.7$

RADIALLY SCALED PARAMETERS

TOWARD GENERATOR ⟶   ⟵ TOWARD LOAD

CENTER

FIGURE A-11   Input impedance calculations.

3. SWR is read directly from the intersection of the $z = 0.5$ circle and the $X = 0$ line on the right side (point F), SWR $= 2$. The impedance circle can be used to describe all impedances along the transmission line. Therefore, the input impedance (Z_i) at a distance of 0.125 λ from the load is determined by extending the z circle to the outside of the chart, moving point A to a similar position on the outside scale (point B on Figure A-11), and moving around the scale in a clockwise direction a distance of 0.125 λ.

4. Rotate from point B a distance equal to the length of the transmission line (point C on Figure A-11). Transfer this point to a similar position on the $z = 0.5$ circle (point D on Figure A-11). The normalized input impedance is located at point D $(0.8 + j0.6)$. The actual input impedance is found by multiplying the normalized impedance by the characteristic impedance of the line. Therefore, the input impedance Z_i is

$$Z_i = (0.8 + j0.6)75 = 60 + j45$$

Input impedances for other distances from the load are determined in the same way. Simply rotate in a clockwise direction from the initial point a distance equal to the length of the transmission line. At a distance of 0.3 λ from the load, the normalized input impedance is found at point E ($z = 1.55 - j0.7$ and $Z_i = 116.25 - j52.5$). For distances greater than 0.5 λ, simply continue rotating around the circle with each complete rotation accounting for 0.5 λ. A length of 1.125 λ is found by rotating around the circle two complete revolutions and an additional 0.125 λ.

EXAMPLE A-1

Determine the input impedance and SWR for a transmission line 1.25 λ long with a characteristic impedance $Z_O = 50 \; \Omega$ and a load impedance $Z_L = 30 + j40 \; \Omega$.

Solution The normalized load impedance z is

$$z = \frac{30 + j40}{50} = 0.6 + j0.8$$

z is plotted on Figure A-12 at point A and the impedance circle is drawn. SWR is read off the Smith chart from point B.

$$SWR = 3.4$$

The input impedance 1.3 λ from the load is determined by rotating from point C 1.25 in a clockwise direction. Two complete revolutions account for 1 λ. Therefore, the additional 0.25 λ is simply added to point C.

$$0.12 \; \lambda + 0.25 \; \lambda = 0.37 \; \lambda \qquad \text{point D}$$

Point D is moved to a similar position on the $z = 0.6 + j0.8$ circle (point E) and the input impedance is read directly from the chart.

$$z_i = 0.55 - j0.9$$
$$Z_i = z_i Z_O = 50(0.55 - j0.9) = 27.5 - j45$$

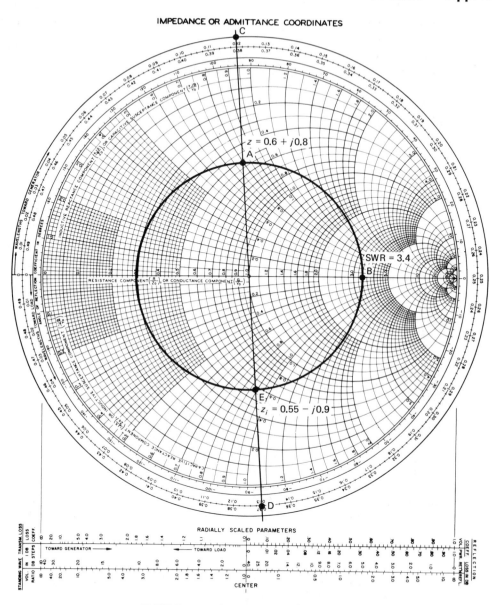

FIGURE A-12   Smith chart for Example A-1.

Quarter-wave Transformer Matching with the Smith Chart

As described in Chapter 9, a length of transmission line acts like a transformer (i.e., there is an impedance inversion every one-quarter wavelength). Therefore, a transmission line with the proper length located the correct distance from the load can be used to match a load to the impedance of the transmission line. The procedure for matching a load to a transmission line with a quarter-wave transformer using the Smith chart is outlined in the following steps.

1. A load $Z_L = 75 + j50$ Ω can be matched to a 50-Ω source with a quarter-wave transformer. The normalized load impedance z is

$$z = \frac{75 + j50}{50} = 1.5 + j1$$

2. $z = 1.5 + j1$ is plotted on the Smith chart (point A, Figure A-13) and the impedance circle is drawn.

3. Extend point A to the outermost scale (point B). The characteristic impedance of an ideal transmission line is purely resistive. Therefore, if a quarter-wave transformer is located at a distance from the load where the input impedance is purely resistive, the transformer can match the transmission line to the load. There are two points on the impedance circle where the input impedance is purely resistive: either end of the horizontal $X = 0$ line (points C and D on Figure A-13). Therefore, the distance from the load to a point where the input impedance is purely resistive is determined by simply calculating the distance in wavelengths from point B on Figure A-13 to either point C or D, whichever is the shortest. The distance from Point B to point C is

point C	0.250 λ
−point D	−0.192 λ
distance	0.058 λ

If a quarter-wave transformer is placed 0.058 λ from the load, the input impedance is read directly from Figure A-13, $z_i = 2.4$ (point E).

4. Note that 2.4 is also the SWR of the mismatched line and is read directly from the chart.

5. The actual input impedance $Z_i = 50(2.4) = 120$ Ω. The characteristic impedance of the quarter-wave transformer is determined from Equation 9-32.

$$Z_O' = \sqrt{Z_O Z_1} = \sqrt{50 \times 120} = 77.5 \; \Omega$$

Thus, if a quarter-wave length of a 77.5-Ω transmission line is placed 0.058 λ from the load, the line is matched. It should be noted that a quarter-wave transformer does not totally eliminate standing waves on the transmission line. It simply eliminates them from the transformer back to the source. Standing waves are still present on the line between the transformer and the load.

IMPEDANCE OR ADMITTANCE COORDINATES

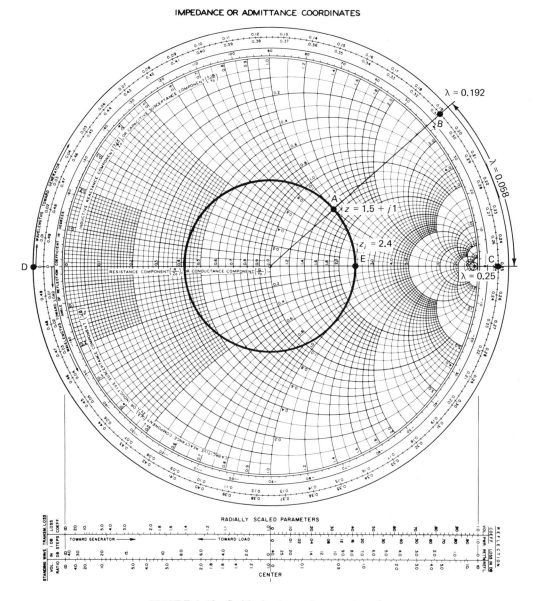

FIGURE A-13   Smith chart, quarter-wave transformer.

EXAMPLE A-2

Determine the SWR, characteristic impedance of a quarter-wave transformer, and the distance the transformer must be placed from the load to match a 75-Ω transmission line to a load $Z_L = 25 - j50$.

Solution The normalized load impedance z is

$$z = \frac{25 - j50}{75} = 0.33 - j0.66$$

z is plotted at point A on Figure A-14 and the corresponding impedance circle is drawn. The SWR is read directly from point F

$$SWR = 4.6$$

The closest point on the Smith chart where z_i is purely resistive is point D. Therefore, the distance that the quarter-wave transformer must be placed from the load is

point D	0.5 λ
$-$point B	0.4 λ
distance	0.1 λ

The normalized input impedance is found by moving point D to a similar point on the z circle (point E), $z_i = 2.2$. The actual input impedance is

$$Z_i = 2.2(75) = 165\ \Omega$$

The characteristic impedance of the quarter-wavelength transformer is again found from Equation 9-32.

$$Z_O' = \sqrt{75 \times 165} = 111.24$$

Stub Matching with the Smith Chart

As described in Chapter 9, shorted and open stubs can be used to cancel the reactive portion of a complex load impedance and thus match the load to the transmission line. Shorted stubs are preferred because open stubs have a greater tendency to radiate.

Matching a complex load, $Z_L = 50 - j100$, to a 75-Ω transmission line using a shorted stub is accomplished quite simply with the aid of a Smith chart. The procedure is outlined in the following steps:

1. The normalized load impedance z is

$$z = \frac{50 - j100}{75} = 0.67 - j1.33$$

2. $z = 0.67 - j1.33$ is plotted on the Smith chart shown in Figure A-15 at point A and the impedance circle is drawn in. Because stubs are shunted across the load (i.e., placed in parallel with the load), admittances are used rather than impedances to simplify the calculations, and the circles and arcs on the Smith chart are now used for conductance and susceptance.

IMPEDANCE OR ADMITTANCE COORDINATES

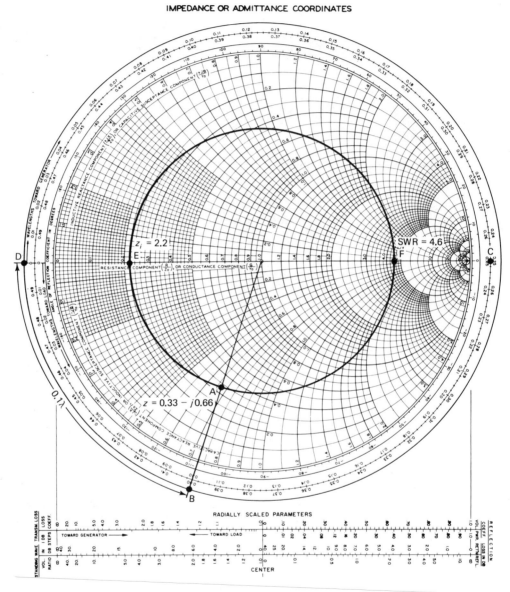

FIGURE A-14 Smith chart for Example A-2.

3. The normalized admittance y is determined from the Smith chart by simply rotating the impedance plot, z, 180°. This is done on the Smith chart by simply drawing a line from point A through the center of the chart to the opposite side of the circle (point B).

4. Rotate the admittance point clockwise to a point on the impedance circle where

IMPEDANCE OR ADMITTANCE COORDINATES

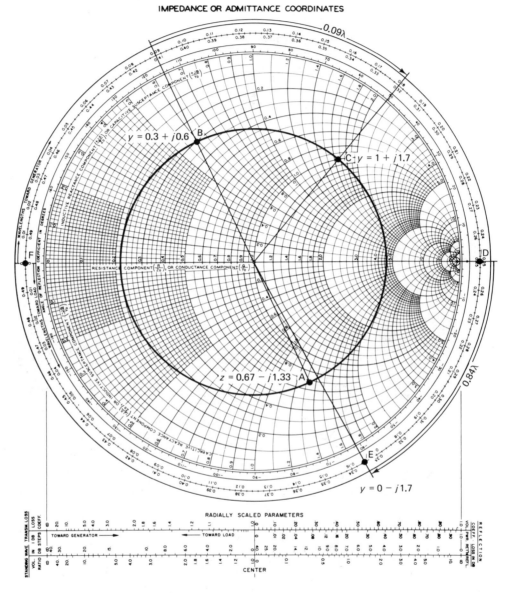

FIGURE A-15   Stub matching, Smith chart.

it intersects the $R = 1$ circle (point C). The real component of the input impedance at this point is equal to the characteristic impedance Z_O, $Z_{in} = R \pm jX$ where $R = Z_O$). At point C, the admittance $y = 1 + j1.7$.

5. The distance from point B to point C is how far from the load the stub must be placed. For this example, the distance is $0.18 \lambda - 0.09 \lambda = 0.09 \lambda$. The stub must have an impedance with a zero resistive component and a susceptance that has the opposite polarity (i.e., $y = 0 - j1.7$).

6. To find the length of the stub with an admittance $y = 0 - j1.7$, move around the outside circle of the Smith chart (the circle where $R = 0$) beginning at point D until an admittance $y = 1.7$ is found (point E). You begin at point D because a shorted

IMPEDANCE OR ADMITTANCE COORDINATES

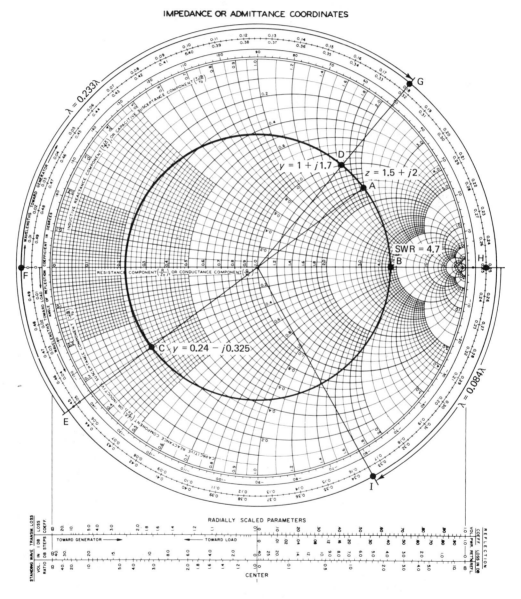

RADIALLY SCALED PARAMETERS

FIGURE A-16   Smith chart for Example A-3.

stub has minimum resistance ($R = 0$) and, consequently, a susceptance B = infinity. Point D is such a point. (If an open stub were used, you would begin your rotation at the opposite side of the $X = 0$ line − point F).

7. The distance from point D to point E is the length of the stub. For this example, the length of the stub is $0.334\ \lambda - 0.25\ \lambda = 0.084\ \lambda$.

EXAMPLE A-3

For a transmission line with a characteristic impedance $Z_O = 300\ \Omega$ and a load with a complex impedance $Z_L = 450 + j600$; determine SWR, the distance a shorted stub must be placed from the load to match the load to the line, and the length of the stub.

Solution The normalized load impedance z is

$$z = \frac{450 + j600}{300} = 1.5 + j2$$

$z = 1.5 + j2$ is plotted on Figure A-16 at point A and the corresponding impedance circle is drawn.

SWR is read directly from the chart at point B.

$$SWR = 4.7$$

Rotate point A 180° around the impedance circle to determine the normalized admittance.

$$y = 0.24 - j0.325 \text{ (point C)}$$

To determine the distance from the load to the stub, rotate clockwise around the outer scale beginning from point C until the circle intersects the $R = 1$ circle (point D).

$$y = 1 + j1.7$$

The distance from point C to point D is the sum of the distances from point E to point F and point F to point G.

$$
\begin{aligned}
\text{E to F} &= 0.5\ \lambda - 0.447\ \lambda &= 0.053\ \lambda \\
+\text{F to G} &= \underline{0.18\ \lambda - 0\ \lambda} &= \underline{0.18\ \ \lambda} \\
&\text{total distance} &= 0.233\ \lambda
\end{aligned}
$$

To determine the length of the shorted stub, calculate the distance from the $y = \infty$ point (point H) to the $y = 0 - j1.7$ point (point I).

$$\text{stub length} = 0.334\ \lambda - 0.25\ \lambda = 0.084\ \lambda$$

SOLUTIONS TO ODD-NUMBERED PROBLEMS

CHAPTER 1

1-1. $V_1 = 10.19V_p$, $V_2 = 0V_p$, $V_3 = 3.4V_p$, $V_4 = 0V$, $V_5 = 2.04V_p$

1-3. %2nd order = 50%, %3rd order = 25%, %THD = 55.9%

1-5. 2nd order intermodulation cross products = 1883, 1890, 1892, 1899, 1901, and 1908 Hz

1-7. (a) $N = 9.3 \times 10^{-15}$ w, $N = -110.3$ dBm
    (b) 2.5 dB decrease
    (c) 3 dB increase

1-9. Duty cycle = 20%

1-11. %2nd IMD = 0.77%

CHAPTER 2

2-1. (a) decrease of 1600 Hz
    (b) decrease of 3200 Hz
    (c) increase of 3200 Hz

2-3. $F_o = 159.2$ kHz

2-5. $C_d = 0.003$ μF

2-7. Nonlinear amplifier with two input frequencies: 1 and 5 kHz. The cross product frequencies are 4 and 6 kHz.

CHAPTER 3

3-1. $m = 0.25$, $M = 25\%$

3-3. (a) $V_c = 25V_p$ (b) $E_m = 15V_p$ (c) $m = 0.6$, (d) $M = 60\%$

3-5. $V_c = 16V_p$, $V_{usb} = V_{lsb} = 3.2V_p$

3-7. (a) $V_{usb} = V_{lsb} = 4V_p$ (b) $V_c = 12V_p$ (c) $E_m = 8V_p$
(d) $m = 0.67$ (e) $M = 67\%$

3-9. (a) $P_{sbt} = 20W_p$ (b) $1020W_p$

3-11. $A_o = 33$, $A_{max} = 51$, $A_{min} = 15$

3-13. (a) $A_{max} = 112$, $A_{min} = 48$
(b) $V_{max} = 0.244V_p$, $V_{min} = 0.096V_p$

3-15. (a) 190 to 210 kHz
(b) 193 and 207 kHz
(c) Maximum bandwidth B $= 20$ kHz
(d)

CHAPTER 4

4-1. SF $= 12$, %Selectivity $= 1200\%$

4-3. $T_e = 900°K$

4-5. IF carrier $= 600$ kHz, $IF_{usf} = 604$ kHz, $IF_{lsb} = 596$ kHz

4-7. (a) Image frequency $= 1810$ kHz
(b) IFRR $= 122$

4-9. (a) B $= 20$ kHz
(b) B $= 14276$ Hz

CHAPTER 5

5-1. (a) $F = 185$ kHz
(b) $\Delta F = 90$ kHz
(c) $200{-}60/6 = 120/6 = 22.5$ kHz/V

5-3. (a) $K_v = 12000$ Hz/rad
(b) $\Delta F = 10$ kHz
(c) $V_{out} = 0.67$ V
(d) $V_d = 0.167$ V
(e) $\theta_e = 0.833$ rad
(f) $\Delta F_{max} = 18.84$ kHz

5-5. $\Delta F = 2$ kHz

5-7. $\Delta F_{max} = 31.4$ kHz

5-9. $F = 1.57$ MHz or 1.23 MHz

5-11. Channel 14

5-13. $F_c = 10$ kHz

CHAPTER 6

6-1. (a) $F_{out} = 396$ to 404 kHz

(b) 397.2 kHz and 402.8 kHz

6-3. (a)

96 104 kHz   100 104 kHz   3.896 3.9   4.1 4.104 MHz

Balanced modulator 1   BPF 1   Balanced modulator 2

4.1 4.104 MHz   25.896 25.9   34.1 34.104 MHz

BPF 2   Balanced modulator 3

34.1 34.104 MHz

BPF 3

(b) BPF1 $= 101.5$ kHz, BPF2 $= 4.1015$ MHz, BPF3 $= 34.1015$ MHz

6-5. (a)   (b) 28.0022 MHz

28 28.003 MHz

6-7. (a)   (b) 497 kHz

496 500 kHz

6-9. (a)

196 204 kHz   196 200 kHz   196 204 kHz   200 204 kHz

Balanced modulator A   BPF A   Balanced modulator B   BPF B

196 204 kHz   196 200 204 kHz

Hybrid   Linear summer

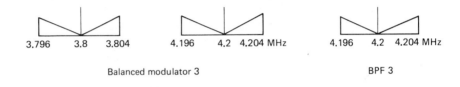

Balanced modulator 3

BPF 3

Balanced modulator 4

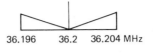

BPF 4

(b) BPFA = 197.5 kHz
BPFB = 203 kHz
BPF3 = 4.1975, 4.2, and 4.203 MHz
BPF4 = 36.1975, 36.2, and 36.203 MHz

6-11. IF = 10.602 MHz, BF0 = 10.6 MHz

6-13. F_{out} = 202 and 203 kHz
PEP = 5.76w
P_{ave} = 2.88w

CHAPTER 7

7-1. 0.5 kHz/V

7-3. (a) ΔF = 40 kHz (b) CS = 80 kHz (c) m = 20

7-5. %mod = 80%

7-7. (a) 4 sets

(b) and (c)

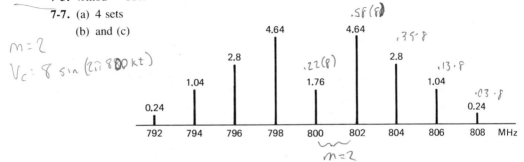

7-9. $\Delta F = 50$ kHz

7-11. DR $= 2$, B $= 100$ kHz

7-13.

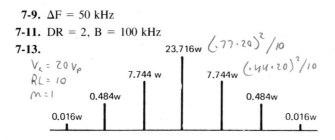

7-15. 0.0377 r

7-17. (a) Combining network out m $= 0.0036$ r
   Power amplifier out m $= 7.2$ r

   (b) Combining network out $\Delta F = 7.2$ Hz
   Power amplifier out $\Delta F = 14.4$ kHz

   (c) $F_c = 90$ MHz

CHAPTER 8

8-1. 22 dB

8-3. $F_{lo} = 103.45$ MHz, IF $= 114.15$ MHz

8-5. 20.3335 to 20.4665 MHz

8-7. $V_{out} = 0.8$V

8-9. 13.3 dB

CHAPTER 9

9-1. 300,000 m, 3,000 m, 300 m, 0.3 m

9-3. $Z_o = 261$ Ω

9-5. $Z_o = 111.8$ Ω

9-7. $\Gamma = 0.05$

9-9. SWR $= 12$

9-11. SWR $= 1.5$

CHAPTER 10

10-1. $\mathcal{P} = 0.2$ μW/m^2

10-3. The power density decreases by a factor of 27 (3^3).

10-5. MUF $= 14.14$ MHz

10-7. $\mathcal{E} = 0.0058$ V/m

10-9. The power density decreases by a factor of 16 (4^2).

10-11. $\mathcal{D} = 20$ dB

10-13. d $= 8.94$ miles

10-15. d $= 17.89$ miles

CHAPTER 11

11-1. (a) $R_r = 25$ ohms
(b) $\eta = 92.6\%$
(c) $P_r = 92.6$ W

11-3. $A_p = 38.13$ dB

11-5. EIRP $= 164,000$ W

11-7. 0.106 μW/m

11-9. $\mathscr{D} = 16$ dB

11-11. $\eta = 97.9\%$

11-13. $\eta = 98.2\%$

CHAPTER 12

12-1.

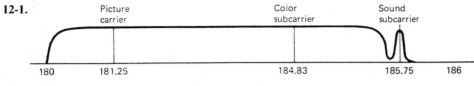

Picture carrier   Color subcarrier   Sound subcarrier

180   181.25   184.83   185.75   186

12-3. 31.5 horizontal scan lines

12-5. 1.27 ms

12-7. Channel 55 $= 721.75$ MHz, Channel 66 $= 787.75$ MHz

12-9. I $= 0.248$ V, Q $= -0.082$ V

INDEX

Absorption, 403, 415
Absorption coefficient, 408
Achromatic mirrors, 477
Acoustical energy, 290
Acquisition time, 181
Adjacent channel interference, 302
Advanced Mobile Phone Service (AMPS), 356
AFC loop, 306
AGC detector, 259
AM detector, 126
AM detector circuits, 159
AM double-sideband full carrier (AM DSBFC), 82
AM envelope, 83, 87, 92, 96
American Radio Telephone Service (ARTS), 356
AM independent-sideband (ISB), 220
Amplitude distortion, 18, 129
Amplitude limiters, 335
Amplitude modulation, 2, 81
Amplitude modulation reception, 126
AM receivers, 130
AM single-sideband full carrier (AM SSBFC), 217
AM single-sideband reduced carrier (AM SSBRC), 220
AM single-sideband reinserted carrier (AM SSBRC), 220
AM single-sideband suppressed carrier (AM SSBSC), 219
AM vestigial-sideband (VSB), 220
Angle of incidence, 406, 408, 421
Angle of reflection, 408
Angle of refraction, 406
Angle modulation, 264, 265
Angular displacement, 265
Antennas, 423
  arrays, 442
  bandwidth, 433

beamwidth, 432
efficiency, 428
elements, 442
feedpoint, 433
gain, 429
input impedance, 433
loading, 440
polarization, 432
Armstrong indirect FM transmitter, 310
Asynchronous receivers, 130
Atmospheric noise, 21
Attenuation, 403, 415
Attenuation factor, 127
Aural signal, 454
Automatic call completion, 355
Automatic desensing, 165
Automatic frequency control (AFC), 304, 306, 475
Automatic gain control (AGC), 38
Automatic gain control circuits, 163, 165, 168
Automatic volume control (AVC), 164, 165

Backporch, 464
Balanced bridge modulator, 230
Balanced diode mixer, 152
Balanced lattice modulator, 224
Balanced modulator, 152, 224, 233
Balanced ring modulator, 224
Balanced transmission line, 363
Balums, 363
Bandlimiting, 17, 18
Bandpass limiter/amplifier (BPL), 338
Bandwidth, 4
Bandwidth improvement, 128
Bandwidth reduction, 157, 158
Barkhausen criterion, 36

Baseband, 2
Base frequency, 198
Beat frequency, 180, 202, 204
Beat frequency oscillator, 254
Bessel function, 274
Bessel function table, 275
Black body noise, 22
Blacker than black, 463
Black setup, 463
Blanking pulses, 456, 462
Blocking oscillator, 471
Boltzmann's constant, 23
Broadside array, 442
Brownian movement, 22
Brownian noise, 22
Buffer amplifier, 113
Bypass capacitor, 115

Camera, 454
Captured power, 430–32
Capture effect, 335
Capture range, 181, 182, 185
Carrier, 2, 3, 81, 90
Carrier leak, 226
Carrier null, 276
Carrier rest frequency, 265
Carrier shift, 120
Carrier supplies, 35
Carrier swing, 269
Carson's rule, 278–80
Cascaded amplifier, 126
Cascoded amplifier, 148
Cascoded limiter, 339
CCIR, 3
Cells, 356
Cellular radio, 303, 355
Ceramic filter, 248
Channel interleaving, 345

Characteristic impedance:
free space, 400
transmission line, 368, 369
Chromatic mirrors, 477
Chrominance signal, 477, 480
Circularly polarized, 423
Citizen's band, 81, 198
Clipping, 318
Clock circuits, 35
Closed loop feedback control system, 172
Closed loop gain, 183, 185
Coaxial line, 362, 366
Coefficient of absorption, 408
Coefficient of coupling, 152, 154, 155
Coefficient of modulation, 86, 90
Coefficient of reflection, 379
Coefficient of refraction, 408
Coherent receiver, 130
Coherent SSB BFO receiver, 256
Coincidence detector, 326
Collector modulation, 104
Collinear, 443
Color burst, 481
Color killer, 483
Color matrix, 480
Color subcarrier, 456
Color television receiver, 482
Color television transmission, 475
Color wheel, 480
Colpitts oscillator, 41, 43, 44
Commercial broadcast band (AM), 30
Commercial broadcast band (FM), 302
Common carrier, 354
Common mode rejection ratio, 231
Complex waves, 7
Compliance, 48
Composite information signal, 2
Composite video, 456, 462
Concentric line, 366
Conductor loss, 376, 378
Conservation of energy, 408
Contact bounce, 350
Conversion gain, 149
Conversion loss, 149
Corona, 376, 378
Correlated noise, 18
Cosmic noise, 21
Counter electromotive force (CEMF), 39
Counterpoise, 439
Coupling loss, 376, 378
Critical angle, 418
Critical coupling, 155
Critical frequency, 418
Crosby direct FM transmitter, 304
Cross products, 19, 20, 76–78
Crystals:
cuts, 45, 46
equivalent circuit, 48
filter, 247
lattice, 45, 46
oscillator, 49, 116
oscillator module, 52–54
oven, 47
wafer, 47

Damped oscillation, 41
Deemphasis, 289, 319
Deep space noise, 21
Deflection circuits, 471
Degenerative feedback, 164
Delayed AGC, 165
Demodulation, 1, 2, 81
Demodulator, 81

Desensing, 165
Desensitized, 167
Detector distortion, 162
Deviation ratio, 280
Deviation sensitivity, 267
Diagonal clipping, 163
Dichroic mirrors, 477
Dielectric heating loss, 376, 378
Differential amplifier, 231
Diffraction, 405, 409
Diffuse reflection, 409
Diplexer bridge, 454
Direct FM, 290
Direct FM transmitter, 304
Direct frequency synthesizer, 196
Directive gain, 429
Directivity, 429
Director, 442
Direct waves, 414
Distributed parameters, 367
Double conversion AM receiver, 168
Double limiting, 339
Double tuned circuits, 152
Down converted, 135
Down converter, 160
Downward modulation, 120
Driven element, 442
Duct propagation, 416
Duty cycle, 12, 13, 16

Effective isotropic radiated power (EIRP), 430
Effective radiated power (ERP), 430
E-field, 361
Elasticity, 48
Electrical noise, 18
Electromagnetic radiation, 400
Electromagnetic wave, 361
Electronic communications, 1
Electronic switching centers, 356
Elementary dipole, 434
Elementary doublet, 433
Elliptically polarized, 432
Emitter modulation, 98
End-fire array, 443
Envelope detection, 160
Envelope detection receiver, 257
Epoch, 31
Equalizing pulses, 465, 473
Equipartition law, 22
Equivalent noise temperature, 130
Esaki, diode, 55
Even function, 8
Excess noise, 25
Exhalted carrier, 220
Extended source, 341
External noise, 21
Extraterrestrial noise, 21

Far field, 427
Fat dipole, 445
FCC, 3, 45, 272
Federal Communications Commission, 3
Feedback oscillator, 36
Feedthrough capacitor, 148
FET push-pull balanced modulator, 227
Fidelity, 129
Field, 459
Field intensity, 400
First carrier null, 276
First detector, 134

Flicker noise, 25
Flux linkage, 154
Flyback time, 458
Flywheel effect, 41, 108
FM (see also Frequency modulation), 264
FM capture effect, 336
FM demodulators, 319
FM double conversion receivers, 320
FM mobile radio communications, 303
FM-PLL demodulators, 326
FM quieting, 336
FM receivers, 319
FM stereo, 340
FM thresholding, 335, 336
FM transmitters, 304
direct, 304
indirect, 310
PLL, 309
Folded dipole, 445
Forward AGC, 165
Foster–Seeley discriminator, 322
Fourier series, 7, 11, 13
Frame, 459
Frame time, 31
Free running multivibrator, 173
Free running oscillator, 35, 36
Frequency acquisition, 180
Frequency conversion, 135
Frequency correction network, 313
Frequency deviation, 265, 269
Frequency deviator, 272
Frequency discriminator, 306, 319
Frequency distortion, 129
Frequency division multiplexing, 30, 204, 342, 482
Frequency domain, 6, 7, 75
Frequency generation, 35
Frequency interlacing, 482
Frequency lock, 179
Frequency modulation (see also FM), 2, 264
Frequency modulator, 272
Frequency multipliers, 70, 98, 172
Frequency reuse, 356
Frequency spectrum, 6, 14, 16
Frequency stability, 45
Frequency synthesizer, 70, 187, 196
Frequency translation, 2
Frequency up-converter, 117
Front end, 126
Front porch, 464
Front-to-back ratio, 427
Front-to-side ratio, 427
Fundamental frequency, 8, 75

Galactic noise, 22
Gang tuned, 135, 151
Geometric ray tracing, 405
Ghost signals, 392
Gradient density, 414
Grounded antenna, 438
Ground plane, 439
Ground reflected wave, 414, 437
Ground waves, 414

Half duplex, 348
Half wave dipole, 424, 435
Half wave symmetry, 8
Handoff, 356
Harmonic, 8, 70, 75, 76

Harmonic distortion, 18, 20, 21, 75, 77, 146, 148
Harmonic frequency, 18
Harmonic generator, 198
Hartley oscillator, 41, 42, 44
Hartley's law, 4
Hertz antenna, 424, 435, 440
Heterodyne, 133
Heterodyning, 134
Hexagonal cells, 356
H-field, 361
High beat injection, 135
High Capacity Mobile Phone Service, 355
High fidelity, 73
High index FM, 278
High level modulator, 103, 104
High level transmitter, 115
High power AM DSBFC transistor modulator, 109
High side injection, 135, 141
Hold-in range, 183–85
Homogeneous medium, 404
Honeycomb, 356
Horizontal blanking time, 464
Horizontal deflection oscillator, 475
Huygen's principle, 411
Hybrid coil, 232

IEEE units, 463
IF amplifier circuits, 152
IF section, 126, 135
IF strip, 135
Image frequency, 133, 141–44
Image frequency rejection, 142, 144, 146
Image frequency rejection ratio (IFRR), 142, 146
Improved Mobile Telephone System (IMTS), 355
Incident waves, 362, 378
Independent sideband (ISB), 220
Independent sideband transmitter, 252
Index of refraction, 406
Indirect FM, 295
Indirect frequency synthesizer, 201
Inductance, 154
Induction field, 427
Inductive coupling, 152
Industrial noise, 22
Information capacity, 4
Information signal, 2
Inhomogeneous medium, 404
Insertion loss, 130
Instantaneous frequency, 265, 267
Instantaneous frequency deviation, 265, 267
Instantaneous phase, 265, 266
Instantaneous phase deviation, 265, 266, 270
Intelligence signal, 3
Intercapacitive feedback, 109
Intercarrier receiver, 470
Intercarrier sound, 454
Interference, 405, 411
Interlaced scanning, 459, 482
Intermediate frequency (IF), 126, 146
Intermodulation distortion, 18–21, 76, 77, 146, 148
Internal noise, 22
International Radio Consultative Committee (CCIR), 3
International Telecommunications Conventions, 3
Inverse square law, 401, 402

Ionization density, 417
ISB transmission, 221
Isotropic medium, 402
Isotropic radiator, 401
Isotropic source, 399

Johnson noise, 22

Kennelly–Heaviside layer, 417

Large scale integration oscillators, 55
Large scale integration programmable timer, 20
Large scale integration stereo demodulator, 348
Lattice modulator, 224
LC oscillator, 39
Limiter, 319, 335
Limiter circuits, 338
Limiting, 285, 318
Linear amplification, 72
Linear integrated circuit balanced modulator, 231
Linear integrated circuit FM generator, 295
Linearly polarized, 432
Linear mixing, 71, 73
Linear scanning, 460
Linear summation, 73
Linear summer, 232
Linear summing, 70
Line of shoot, 427
Line of sight (LOS) transmission, 414
LM1496, 231–44
Loading coils, 440
Lobes, 427
Local oscillator, 133–36, 139, 140
Local oscillator tracking, 136
Lock range, 184, 185
Log periodic antenna, 446
Longitudinal currents, 363
Longitudinal wave, 361
Loop antenna, 448
Loop filter, 187
Loop gain, 182
Loop operation, 179
Loose coupling, 155
Low beat injection, 135
Lower sideband, 83
Lower side frequency, 83, 90
Low frequency noise, 25
Low index FM, 278, 303, 304
Low level modulator, 103
Low level transmitter, 113
Low power AM DSBFC transistor modulator, 98
Low side injection, 135, 139
Luminance signal, 456, 457, 477
Lumped parameters, 367

Magnetic flux, 152
Major lobes, 427
Man–man noise, 22
Marconi antenna, 424, 438–40
Mass-to-friction ratio, 48
Maximum usable frequency, 419
Mechanical energy, 290
Mechanical filter, 249

Medium power AM DSBFC transistor modulator, 104
Metallic circuit currents, 363
Microwave radio communications, 304
Minor lobes, 427
Mirror symmetry, 8
Mixer/converter circuits, 148
Mixer/converter section, 126, 134
Mixing, 70
Mobile television service, 353
Modulate, 2, 81
Modulated signal, 2, 3, 81–83
Modulated wave, 2, 3, 82
Modulation, 1, 2, 81
Modulation coefficient (AM), 163
Modulation index, 270, 271
Monochrome, 454
Monochrome reception, 467
Monochrome transmission, 454
Monophonic, 340
Monopole antenna, 438
Multifrequency synthesizer, 204
Multipath propagation, 421
Multiple crystal frequency synthesizer, 196, 198
Multiple skip propagation, 421
Multiplexing, 29
Multistage tuning, 132
Mutual inductance, 154, 155

Narrowband FM, 303
National Television Systems Committee (NTSC), 454
Natural frequency, 173, 183
Near field, 427
Negative carrier shift, 120
Negative feedback, 164
Negative resistance, 55
Negative resistance oscillator, 54
Negative temperature coefficient, 47
Negative transmission (TV), 457
Neutralization, 148
Neutralization capacitor, 148
Neutralize, 109
Neutralizing capacitor, 109
Noise, 4
Noise factor, 26
Noise figure, 26–29, 128, 146
Noise temperature, 130
Noise triangle, 284
Noise voltage, 24, 146
Noncoherent receivers, 130
Nonlinear amplification, 21, 75
Nonlinear distortion, 18, 146, 148
Nonlinear mixing, 74, 83, 196
Nonresonant arrays, 444
Nonresonant line, 379
Nonsinusoidal periodic waves, 7
Nonuniform gain, 129

Odd function, 8
Odd symmetry, 8
Off ground neutralization, 148
Omnidirectional antenna, 401
Open loop gain, 36, 181, 184, 185, 201
Optical horizon, 415
Optimum coupling, 156
Oscillator, 35, 36, 173
Oscillator action, 39

Overtone mode, 47
Overtone crystal oscillator, 47

Padder capacitor, 139
Parallel two-wire line, 362, 364
Parasitic element, 442
Peak detector, 159–61
Peak envelope power (PEP), 222, 259
Peak envelope volt (PEV), 222, 259
Pedestal, 463
Percent modulation (AM), 86, 87
Percent modulation (FM), 272
Periodic waveform, 7
Permanence, 45
Phase comparator, 172, 174, 179, 181, 183, 184, 186
Phase deviation, 265
Phase distortion, 129
Phase error, 178, 180, 182
Phase-locked loop, 172
Phase-locked loop frequency synthesizer, 201
Phase modulation, 2, 264
Phase modulation, 272
Phase shift discriminator, 322
Phase shifter, 251
Phase shift keying (PSK), 224
Photoconductive, 457
Photomissive, 457
Picture elements, 457
Picture frame, 459
Pierce crystal oscillator, 115
Pierce oscillator, 49
  discrete, 50
  IC, 50, 51
Piezoelectric effect, 45
Pilot, 204, 344, 346
Pilot carrier, 220
Pilot carrier receiver, 257
Pilot supplies, 35
PLL-FM demodulator, 326
PM (*see also* Phase modulation), 264
Point source, 340, 399
Point symmetry, 8
Polarization, 432
Positive carrier shift, 120
Positive feedback, 36
Positive temperature coefficient, 47
Positive transmission (TV), 457
Post AGC, 165
Postdetection S/N, 335
Power density, 400
Power gain, 429
Power transmission coefficient, 408
Preamplifier, 113
Precipitation noise, 25
Predistorter, 313
Preemphasis, 289
Prescaled frequency synthesizer, 203
Prescaling, 203
Preselector, 126, 133, 136
Primary colors, 475
Primary constants, 364, 367, 374
Product detectors, 152
Product modulators, 152, 227
Propagation coefficient, 372
Propagation constant, 372
Pull-in range, 182, 185
Pull-in time, 181
Pulse code modulation (PCM), 31
Push-to-talk:
  electronic, 350
  manual, 348

Quadrature amplitude modulation (QAM), 224, 477
Quadrature FM demodulator, 326
Quarter-wave dipole, 424
Quarter-wavelength transformer matching, 392
Quarter-wavelength transmission line, 387
Quieting, 335

Radiation directivity, 442
Radiation efficiency, 423
Radiation field, 428
Radiation loss, 376, 378
Radiation pattern, 424, 427, 443, 447
Radiation resistance, 428
Radio common carrier, 354
Radio communications, 1
Radio frequency (RF), 3, 81
Radio frequency choke (RFC), 115, 148
Radio horizon, 415
Radio propagation, 398
Random noise, 22
Random white noise, 22
Raster, 460
Raster lines, 460
Ratio detector, 324
Rayleigh criterion, 409
Rays, 398
Reactance modulator, 293
Receiver threshold, 129
Rectangular pulses, 11
Rectangular waveform, 11, 14, 16
Rectifier distortion, 163
Reference black, 463, 465
Reflected waves, 362, 378
Reflection, 405, 407, 442
Reflection coefficient, 379, 408
Refraction, 405
Refractive index, 406
Regenerated pilot, 257
Regenerative feedback, 30
Reinserted carrier, 220
Resilience, 48
Resistance noise, 22, 25
Resolution, 197, 202
Resonant antenna, 435
Resonant lines, 379
Rest frequency, 265
Retrace time, 458
RF, 3
RF amplifier, 126
RF amplifier circuits, 145
RF oscillator, 113
RF propagation, 398
RF section, 133
Rhombic antenna, 444
Ribbon cable, 364
Ring modulator, 224
RLC half-bridge, 49, 51

Satellite radio communications, 304
Scanning frequencies (color television), 481
Scanning rates (TV), 454, 458
Secant law, 420
Secondary constants, 367
Second-order intermodulation distortion, 20
Selectivity, 126
Self-excited mixer, 149
Self inductance, 154

Sensitivity, 128
Sequential scanning, 458
Serrations, 466
Shadow zone, 411
Shape detector, 160, 161
Shape factor, 127
Shielded cable pair, 365
Short dipole, 433
Shot noise, 25
Sideband, 83, 90
Sideband inversion, 136
Side frequency, 83, 90, 274
Signal analysis, 5
Signal-to-noise ratio, 25–28
  postdetection, 335
  predetection, 335
Signal waveform, 6
Simple AGC, 163
Single-crystal frequency synthesizer, 197
Single sideband filters, 246
Single sideband full carrier (SSBFC), 217
Single sideband receivers, 254
  BFO, 254
  coherent, 256
  envelope detection, 255
  multichannel pilot carrier, 257
Single sideband reduced carrier (SSBRC), 220
Single sideband reinserted carrier (SSBRC), 220
Single sideband suppressed carrier (SSBSC), 219
Single sideband systems, 217
Single sideband transmitters, 232
  filter method, 233
  phase shift method, 249
  the ''third'' method, 249
Single tuned transformers, 154
Sinusoidal signals, 5
$(\sin x)/x$ function, 13
Skin effect, 132
Skip distance, 420
Sky waves, 412, 414
Slope detectors, 319
  balanced, 321, 322
  single-ended, 319, 322
Smith chart, 486
Solar flare-ups, 418
Solar noise, 21
Source information, 2
Space attenuation, 403
Space quadrature, 362
Space waves, 412, 414
Spectral components, 7
Specular reflection, 409
Spherical wavefront, 401
Square law devices, 129
Squelch circuits, 167
Standard atmosphere, 415
Standing waves, 380
Standing wave ratio (SWR), 380
Static phase error, 182
Stereo, 340
Stub matching, 394
Subcarrier, 303, 341
Subsidary Communications Authorization (SCA), 341
Sun spot activity, 21, 418
Superheterodyne AM receiver, 133–35
Surface wave, 412
Surge impedance, 368
Switched crystal frequency synthesizer, 198
Synchronization (TV), 454, 456
Synchronizing pulses, 462

Synchronous receiver, 130
Sync separator, 471
Synthesizer, 187
System noise, 3

Tank circuit, 39
Television, 452
Television broadcasting, 303
Television broadcast standards, 456
Temperature coefficient, 45, 47, 48
Thermal noise, 22, 24
Third-order intercept distortion, 129
Three point compensation, 140
Three point tracking, 139
Tight coupling, 155
Time division multiplexing, 31
Time domain, 6, 18
Top loading, 441
Tracking, 130
Tracking error, 139
Tracking filter, 186
Tracking range, 183
Transducer, 290
Transformer coupling, 152
Transit time noise, 25
Transmission coefficient, 408
Transmission frequencies, 3
Transmission line, 360
Transmission line equivalent circuit, 367
Transmission line input impedance, 387
Transmission line losses, 376
Transverse electromagnetic waves (TEM), 360
Trapezoidal patterns, 117–20
Traveling waves, 380
Trimmer capacitor, 139
Triple limiting, 339
Tuned circuit, 107

Tuned primary–untuned secondary, 155
Tuned radio frequency receiver, 131, 132
Tunnel diode, 54, 55
Tweeters, 340
Twin lead, 364
Twisted pair cable, 364
Two modulus prescaler, 203
Two-way FM mobile communications, 303
Two-way FM radio communications, 348

Unbalanced line, 363
Uncorrelated noise, 18, 21
Untuned primary–tuned secondary, 154
Up-converted, 135
Up-converter, 160
Upper cutoff frequency, 18
Upper sideband, 83
Upper side frequency, 83, 90
Upward modulation, 120

Vacuum tube AM DSBFC modulator, 109
Varactor diode, 53, 54
Varactor diode FM modulator, 292
Varicap, 53
Velocity constant, 374
Velocity factor, 374
Velocity of propagation, 375
Vertical deflection oscillator, 471
Vertical blanking time, 465
Vertical retrace time, 460
Vestigial sideband, 222
Video signal, 454
Virtual height, 419
Voice operated transmitter (VOX), 352

Voltage controlled oscillator (VCO), 172, 173
Voltage standing wave ratio (VSWR), 380

Wave attenuation, 403
Wavefronts, 398
Wavelength, 362
Wavelets, 411
Wave propagation, 398
Wavetraps, 470
Wave velocity, 362
White noise, 22
White noise source, 24
Wideband phase shifter, 251
Wien bridge, 38
Wien bridge oscillator, 37–39
Wipe off, 306
Woofers, 340

XR-1310 Stereo demodulator, 348–50
XR-2206 Monolithic function generator, 295–301
XR-2207 Voltage controlled oscillator, 56–69
XR-2211 Integrated circuit PLL, 187–96
XR-2212 Phase-locked loop, 326, 328–34
XR-2240 Programmable timer, 205–13

Yagi–Uda antenna, 445
Y-signal (*see also* Luminance signal), 479

Zero beat, 180
Zero coefficient crystals, 48